Addison Wesley Longman Limited
Edinburgh Gate
Harlow
Essex   CM20 2JE
United Kingdom
and Associated Companies throughout the world

Published in the United States of America
by Addison Wesley Longman, New York

First published 1991
Reprinted 1992
Second Edition 1997

ISBN 0 582 27722 1

British Library Cataloguing-in-Publication Data

A catalogue record for this book is available from the British Library

Library of Congress Cataloging-in-Publication Data

A catalogue entry for this title is available from the Library of Congress

Set by 32 in 9/11pt Times
Produced by Longman Singapore Publishers (Pte) Ltd
Printed in Singapore

Dedication

For JCM and MM

No day passes without remembrance

Contents

Preface   ix
Acknowledgements   xi

Chapter 1   Nature, culture and environmental change   1

             1.1   Introduction   1
             1.2   The development of ideas about environmental change   1
             1.3   Modern concepts: environmental systems and Gaia   4
             1.4   Agents and processes of environmental change   6
             1.5   People/environment relationships   8
             1.6   Conclusion   11
             Further reading   11

Chapter 2   Quaternary geology and climatic change   12
             2.1   Introduction   12
             2.2   Quaternary subdivisions based on the terrestrial
                   record   12
             2.3   The record of climatic change from the oceans   16
             2.4   The record of climatic change from ice cores   21
             2.5   Tree rings, historical and meteorological records   24
             2.6   Causes of climate change   28
             2.7   Environmental change in high latitudes   30
             2.8   Environmental change in middle latitudes   32
             2.9   Environmental change in low latitudes   36
             2.10 Sea-level changes   38
             2.11 Conclusion   40
             Further reading   41

Chapter 3   Environmental change in the late- and post-glacial periods   42

             3.1   Introduction   42
             3.2   The interglacial cycle   42
             3.3   Climatic change during the late-glacial period   43
             3.4   Regional expression of changes during the
                   late-glacial period   53
             3.5   Faunal changes during the late-glacial period   60
             3.6   Climatic change during the early Holocene   63
             3.7   Regional expression of changes during the
                   early Holocene   67
             3.8   Climatic change during the later Holocene   71

3.9 Regional expression of changes during the
later Holocene   73
3.10 Conclusion   77
Further reading   77

Chapter 4   Prehistoric communities as agents of environmental
change   78

4.1 Introduction   78
4.2 The evolution of modern humans   78
4.3 The relationship between environment and
Palaeolithic groups   89
4.4 The relationship between environment and
Mesolithic groups   97
4.5 Domestication of plants and animals:
the beginnings of agriculture   102
     4.5.1 Centres of plant domestication:
            the Near East   107
     4.5.2 Centres of the Far East   107
     4.5.3 The sub-Saharan centre   107
     4.5.4 Centres of the Americas   107
     4.5.5 The domestication of animals   107
4.6 The Neolithic period   112
4.7 The Bronze Age   118
4.8 The Iron Age   122
4.9 Conclusion   128
Further reading   128

Chapter 5   Environmental change in the historic period   129

5.1 Introduction   129
5.2 The impact of the Greeks and Romans   129
5.3 The Middle Ages (ca. 400–1400)   134
5.4 The period 1400–1750   139
5.5 Immediate consequences of industrialisation,
1750–1914   143
5.6 Rural changes after 1750   145
5.7 Changes in Africa following European settlement   147
5.8 Changes in the Americas following European
settlement   151
5.9 Changes in Australia and New Zealand following
European settlement   157
5.10 Conclusion   160
Further reading   161

Chapter 6   Environmental change due to post-1700 industrialisation 162

6.1 Introduction   162
6.2 Changes where mineral extraction occurs   162

	6.3	Changes distant from the source of mineral extraction	167
	6.4	Reclamation of mine-damaged land	168
	6.5	Changes due to fossil-fuel use: global warming	171
		6.5.1 Greenhouse gases: sources and sinks	171
		6.5.2 The potential impact of global warming	177
	6.6	Changes due to fossil-fuel use: acidification	186
	6.7	Ozone and lead problems	193
	6.8	Changes caused by disposal of waste materials	199
	6.9	Conclusion	204
		Further reading	204
Chapter 7		The environmental impact of agriculture in the developed world	205
	7.1	Introduction	205
	7.2	Landscape change: loss of habitats and biodiversity	205
	7.3	Soil degradation, erosion and conservation	209
	7.4	Desertification	217
	7.5	Water quality: cultural eutrophication	220
	7.6	Soil and water quality: salinisation	225
	7.7	The impact of crop-protection agents	231
	7.8	Conclusion	238
		Further reading	238
Chapter 8		The environmental impact of agriculture in the developing world	239
	8.1	Introduction	239
	8.2	Landscape change: loss of natural habitats and biodiversity	239
	8.3	Soil degradation, erosion and conservation	245
	8.4	Desertification	251
	8.5	Soils: salinisation, alkalinisation and waterlogging	255
	8.6	The impact of agriculture on water quality	261
	8.7	Conclusion	264
		Further reading	265
Chapter 9		Other agents of change: forestry, recreation and tourism, biotechnology	266
	9.1	Introduction	266
	9.2	Forestry and afforestation in the developed world	266
	9.3	Forestry and afforestation in the developing world	275
	9.4	The impact of recreational activities	280
	9.5	The impact of tourism	284
	9.6	The environmental implications of biotechnology	289
		9.6.1 Agriculture	289
		9.6.2 Other applications	296

9.7 Conclusion   302
Further reading   303

Chapter 10 Conclusion and prospect   304
10.1 Introduction   304
10.2 Environmental factors: a series of perspectives   306
      10.2.1 The Quaternary period   307
      10.2.2 The impact of industrialisation   308
      10.2.3 The impact of agriculture   308
      10.2.4 The implications of social and political factors   309
10.3 Environmental factors: their future impacts   312
      10.3.1 The future of the climate   313
      10.3.2 The future effects of industrialisation   315
      10.3.3 The future effects of agriculture   315
      10.3.4 The future effects of social and political factors   318
10.4 Envoi   321
Further reading   322

References   323
Index   375

Preface

The world has changed a great deal since the first edition of *Global Environmental Change* was published in 1991. The physical processes of environmental change have in many cases accelerated, and attitudes to environmental change have altered. Changes in attitude have been prompted to a large extent by the Earth Summit, the United Nations Conference on Environment and Development which took place in Rio de Janeiro in 1992, and by the continued media coverage of environmental issues. These developments are encouraging and emphasise the continued importance of environmental change at all scales and its contribution to global transformation.

It was with fear and trepidation that I set out to write GEC2. I was fearful because of the success of GEC1 and thought perhaps that I should quit while ahead. The trepidation sprang from the realisation that GEC2 needed to be a new book and not just a new edition because of the huge volume of literature produced over the past few years. Even in the eighteen months since I began GEC2 I have had to rewrite several sections before publication. As well as revising the structure and content, I have borne in mind the many comments I received from individuals via Longman's questionnaire that accompanied inspection copies of GEC1. I appreciate the opinions and sympathise with those who expressed reservations about the limited coverage of certain topics. This is particularly so in relation to the policy and management dimensions of global environmental change. Where appropriate I have referred to these issues; it was never my intention to make them a major focus of either GEC1 or GEC2 but rather to concentrate on the processes and effects of environmental change. There are also set word limits within which I am obliged to work and to do full justice to all the topics covered, I really would need to write an encyclopaedia!

Antoinette M. Mannion

Note Dating: K years = 000s years (1×10^3)

Acknowledgements

Several people have contributed to the production of this book. In particular I should like to acknowledge the contribution of Sonia Luffrum who, with the assistance of Michelle Norris, has patiently typed yet another lengthy manuscript for me. The tables and diagrams have been drawn using Apple Mac hardware and software by Heather Browning. This book, like its precursor, has benefited from the advice of Dr Erlet Cater on development issues and tourism. As ever, I am indebted to Dr Michael D. Turnbull, who has again acted as unofficial reviewer and proof-reader as well as official indexer.

We are grateful to the following for permission to reproduce copyright material:

The Academic Press and the authors for Figure 4.5 from Figure 14.1 in Lanpo and Weiwen (1985); Blackie and Son Ltd for Table 6.1 from a table in *Land Use and Landscape Planning*, ed. Lovejoy (1973) published by Leonard Hill (Blunden 1985); Elsevier Science Publishers and the author, Professor Shackleton for Figure 2.8 from part of fig. 2 in Shackleton *et al* (1983); Addison Wesly Longman for Figure 7.1 from a figure in Morgan (1986); Macmillan Magazines Ltd and the authors for Figure 2.1 from part of figure 4 in Brassell *et al* (1986), Figure 2.8 from fig. 1c in Chappell and Shackleton (1986), Figure 3.3a from fig. 2 in Atkinson *et al* (1987); Pergamon Press and the author, Professor Bowen for Tables 2.2 and 2.4 from part of Chart 1 in Bowen *et al* (1986a & b, and Sibrava 1986) Pergammon Press plc; Pergamon Press and the author, Professor Walker for Figure 3.3b from fig. 7 in Walker *et al* (1993); John Wiley & Sons Ltd for Figure 3.1 from fig. 1.2 in Birks (1986), and Figure 3.3c from fig. 4 in Lowe *et al* (1995) and Table 3.3 from parts of figures 2 and 3 in Walker *et al* (1994); Opulus Press and the author, B. Huntley, for Figure 3.7 from a figure in Huntley (1990).

Whilst every effort has been made to trace the owners of copyright material, in a few cases this has proved impossible and we take this opportunity to offer our apologies to any copyright holders whose rights we may have unwittingly infringed.

In Nature there are neither rewards nor
Punishments – there are consequences

Robert Green Ingersoll (1833–1899)
From: Lectures and Essays, 'Some Reasons Why'

CHAPTER 1

Nature, culture and environmental change

1.1 Introduction

Environmental change is a continual process that has been in operation since the Earth first came into existence some 5×10^9 years ago. From the time of this genesis, dynamic systems of energy and material transfer have operated at local, regional and global scales to effect gradual, and sometimes catastrophic transformations of the atmosphere, hydrosphere, lithosphere and biosphere. For most of Earth history the agents of change have been the elements of wind, ice, water, tectonic forces, micro-organisms, plants and animals; all of them have operated reciprocally within dynamic ecosystems that characterise the biosphere. Environmental change and the evolution of organisms occurred as stimuli were generated either indigenously through mutual dependence or exogenously due to the Earth's orbital geometry and/or changes in the Sun's luminosity.

Some 5×10^6 years ago, however, a new agent of environmental change emerged. This was the first hominid species (see Fig. 4.2) from which *Homo sapiens sapiens*, i.e. modern humans, eventually evolved. Initially this species was an integral component of the biosphere like any other animal but the capacity of humans to develop technology with which to turn Earth surface components into resources has rendered the species a particularly powerful agent of environmental change. As knowledge has progressed and as science and technology have developed, society has achieved both an improved insight into the processes and consequences of environmental change and the means of altering the environment drastically. This can occur directly through activities such as habitat loss and urban expansion. There is also much inadvertent environmental change through agencies such as pollutants from agriculture and industry.

Since *Homo sapiens sapiens* evolved, and quite when this occurred is disputed, the relationship between nature, i.e. the biota, and environment has changed. This is particularly so for the past 10 K years, since when humans have attempted to dominate ecosystem processes. The relationship has become tripartite rather than bipartite. Nature and environment have been joined by culture, i.e. the human element. In addition, the manifestation on the ground of this tripartite relationship has changed spatially and temporally. The changing relationship between environment, nature and culture is the theme of this book.

1.2 The development of ideas about environmental change

Human beings must always have been concerned with their environment since it was their immediate surroundings that provided the resources necessary for survival. As heterotrophs, humans need to acquire energy, which, as integral components of the biosphere, they do by selectively consuming plants and animals. They also exploit substances derived from the Earth, collectively known as mineral resources, and in the later part of their history they have turned to the fossil plant communities of past eons, i.e. coal, oil and natural gas, to obtain fuel energy.

For preliterate times the only direct testimony to the relationship between people and their environment lies in prehistoric cave paintings. Examples include the recently discovered Chauvet cave in the Ardèche of France, where the oldest rock paintings in the world are found and which are dated at ca. 32 K years, and the famous caves of Lascaux in the Dordogne, the paintings in which are dated to ca. 17 K years ago. These works of art reflect the lives of hunter-gatherers who lived during the time of the last major ice advance and depict the prey and predators of the period, e.g. aurochs, deer, horses and wolves.

With the rise of the ancient Mediterranean civilisations of Greece and Rome came the first written accounts of places, trade routes, crops, etc. Herodotus (ca. 485–425 BC), a Greek scholar often described as the father of history, recorded a variety of environmental features such as the regime of the River Nile, and later another Greek scholar, Aristotle (384–322 BC) advanced the idea of a spherical instead of a flat Earth. He also introduced mathematical concepts for the measurement of global features such as latitude and longitude. This tradition was continued by Eratosthenes (276–194 BC), who produced a massive eight-volume work entitled *Geography* containing details of map projections and methods for the calculation of the Earth's dimensions. Similarly Strabo (64 BC to AD 20), a Roman scholar, produced a seventeen-volume work entitled *Geographica*, which examined much of the then known world and involved a recognition of dualism in geography: that of people (human geography) and place (physical geography).

Ideas changed little during the Middle Ages when European scholars returned to the concept of a flat Earth to conform with ecclesiastical teaching. Once again the civilisations of the Mediterranean, principally the Arab world, augmented the information of the Greeks, while in China geographical knowledge was well advanced, although inaccessible to European scholars until Marco Polo's (1255–1325) expeditions. The Renaissance, however, brought a scientific revival to Europe, including renewed interest in Ptolemy's *Geography* which provided explorers such as Columbus with a basic mathematical approach to location. Travellers provided a potpourri of descriptive information on people and places which, in addition to the publication of Mercator's map projections in 1569, led to the production of the first globes and new maps.

The bipartite nature of geography, first intimated in Strabo's work, involving human and physical divisions, was formalised by a Dutch scholar Varenius (1622–1650), who originated the ideas of

regional or 'special' geography and systematic or 'general' geography. The latter, he advocated, could be studied using natural laws and was thus a more exact science. This approach was continued by Immanuel Kant (1724–1804) who argued strongly for a scientific base to the study of geographical or environmental phenomena, which he considered to be just as essential as the exact sciences. This stance was continued by von Humboldt (1769–1859), who developed an inductive approach to explaining natural phenomena. A renowned explorer, von Humboldt published a five-volume work entitled *Kosmos* (1845–1862) in which he attempted not only to describe natural phenomena, such as rocks, plants and animals, but to explain their occurrence and to undertake comparative studies.

However, the deductive and mechanistic philosophy earlier advocated by Newton (1642–1727) was continued in the work of Charles Darwin (1809–1882). In 1859 he published his classic work *On the Origin of Species* in which he charted the development of life, and advanced theories on evolution. For the student of environmental change this is a most significant publication since it suggests a relationship between environment and organisms and, moreover, charts a developmental sequence. If organisms changed due to environmental parameters, then environment itself must have changed. Although many of these scholars held views that were inherently deterministic, i.e. a belief in the overall control of environment, especially climate, on human activity the term 'environmental determinism' originated with the work of Darwin that highlighted the operation of the laws of nature in relation to organisms. Social scientists and philosophers such as Herbert Spencer (1820–1903) began to apply Darwin's ideas on the environmental control of biota to humanity. Spencer suggested there were many similarities between organisms and societies and that to be successful only the 'fittest' in a free-enterprise system would survive. Indeed it was Spencer (1864) who first coined the phrase 'survival of the fittest'. Many notable geographers and natural scientists were profoundly influenced by these ideas, including William Morris Davis (see below). Most importantly these ideas invoked changing landscapes and societies. So it is perhaps no coincidence that many notable scientists began to advance theories relating to environmental change at much the same time and, perhaps more pertinently, though not explicitly, introduced the idea of dynamism into environmental studies.

By the end of the eighteenth century the diluvial theory, the proposal that the Biblical flood was a major agent in shaping the face of the Earth, was being questioned. Scientists such as James Hutton and John Playfair were among the first to advance the theory of glaciation. Hutton's observations of erratic boulders in the Jura Mountains, noted earlier by a Swiss minister, Bernard Friederich Kuhn, led him to invoke glacier ice as the agent of transportation (Hutton, 1795). Nevertheless, it was not until the 1820s that the glacial theory became more widely proposed and, although Agassiz presented the theory to the Swiss Society of Natural Sciences in 1837, numerous earlier workers had already published evidence for glacial processes. These included Jean-Pierre Perraudin, a Swiss mountaineer, Jean de Charpentier, a naturalist, and Ignace Venetz, a highway engineer, all of whom proposed that Swiss glaciers had extended beyond their present positions. Despite this wider acclaim the glacial theory was still rejected by many in favour of Charles Lyell's (1833) explanation for erratics, drift, etc., as being the products of floating icebergs. This idea was also given credence by reports of boulder-containing icebergs from contemporary explorers like Darwin himself.

However, William Buckland, who was appointed professor of geology at Oxford in 1820, was the first to acknowledge that neither the diluvial nor iceberg drift theories could provide satisfactory explanations for all the evidence and eventually, in 1840, he and Charles Lyell accepted Agassiz's views. Despite residual resistance in the scientific community, the glacial theory became widely accepted by the mid-1860s. By this time further developments were in hand. Evidence for changing sea levels was compiled by Jamieson (1865) based on evidence from Scotland, North America and Scandinavia; von Richtofen (1882) advanced a windborne origin for loess deposits; Gilbert (1890) presented evidence to show that the Great Salt Lake of Utah is only a remnant of a much larger lake, and in Britain Archibald Geikie (1863) suggested the idea of multiple glaciation. Multiple glaciation involved several cold stages or glacials, separated by warm stages or interglacials and was further substantiated by James Geikie (1874). Moreover, Penck and Brückner (1909) interpreted Alpine terrace sequences in terms of oscillating warm and cold stages.

Inevitably, as evidence on environmental change accrued, attention also focused on the underlying cause of climatic change. Adhémar, a French mathematician, was the first to involve the astronomical theory in studies of the ice ages. In 1842 he proposed that changes in the orbit of the Earth round the Sun may be responsible for climatic change of such great magnitude. A similar approach was advanced by a Scottish geologist, James Croll, who in 1864 suggested that changes in the Earth's orbital eccentricity might cause ice ages. This theory he explained in full in his book *Climate and Time* (Croll, 1875). This theory was accepted by both Archibald and James Geikie, giving it much credence in the geological community. Owing to the inability of geologists to substantiate Croll's predictions in the absence of reliable dating techniques, his theories fell into disuse until their revival in the 1920–1940 period by Milutin Milankovitch, a Yugoslavian astronomer (Section 2.6). Although numerous other theories for causes of climatic change have been proposed, Milankovitch's ideas have become widely accepted since the 1950s, when evidence from deep-sea core stratigraphy was first obtained.

The late eighteenth and early nineteenth centuries also witnessed the establishment of new methods, based principally on biological remains, for examining the nature of environmental change, and thus establishing the general field of palaeoecology. Plant macrofossils were among the first biological indicators to be used to interpret Quaternary palaeoenvironments. Blytt (1876) and Sernander (1908), both Scandinavian geologists, used macrofossils in peat deposits to explain the forest history of northwest Europe and associated climatic changes. Similarly, von Post (1916) developed pollen analysis as a means of examining environmental change. The two are complementary techniques and have played a major role in examining vegetation change during interglacial stages and the post-glacial period, as will be shown in Chapter 2. Numerous other fossil groups of plants and animals have also been used as indicators of past environments in the last 50 years; see the review in Berglund (1986).

Apart from the inception of the glacial theory and its development during the nineteenth century, there were other notable developments in the early twentieth century, which also significantly altered scientific thought. For example, Davis in 1909 advanced a theory which he called 'the geographical cycle'. This encapsulates an idealised landscape originating with mountain uplift and culminating in

Table 1.1 Types of environmental systems and their characteristics

Type of system	Characteristics	Example
Morphologic	Observable physical properties such as height, slope, angle, sediment type	A desert dune system
Cascading	Has a flux or cascade of energy and/or matter	The solar energy cascade
Process-response	Combines morphologic and cascading systems: morphological variables may control and be controlled by the flux of energy and/or matter	Erosion of a river bed as increased energy is made available through enhanced discharge
Control	Combines a process-response system with the controlling presence of an intelligence (usually humans)	Urban development or deforestation causing changes in local hydrological cycles
Ecosystem	Involves interactions between biotic factors (including humans) and abiotic factors such as landforms, soils, etc.	A small pond at the local scale to the global biosphere/lithosphere at the large scale

lowland plains. Although this is no longer an accepted theory, it is historically significant insofar as it invokes the idea of continual processes and the idea of continual change. This theoretical treatment of landscape development has a counterpart in ecological studies. In 1916 Clements wrote: 'As an organism the climax formation arises, grows, matures and dies. Its response to the habitat is shown in processes or functions and in structures which are the record as well as the results of these functions'. Thus Clements expressed the nature of vegetation communities as ever-changing entities, and while there is much debate about the acceptability of his ideas (Mannion, 1986), he was responsible for injecting the idea of dynamism into ecological systems. These scholars were influenced by environmental determinism (see above), a philosophy that many other natural and social scientists rejected and replaced with an alternative known as possibilism, a term coined by the French historian Lucien Le Febvre. This philosophy recognised the constraints imposed by the physical environment on human activity but espoused the view that there are many possibilities for human manipulation of the environment.

Ideas relating to environmental change have changed substantially in the last 200 years. The acceptance of the glacial theory not only led to the development of Quaternary geology but also initiated related studies into climatic and ecological change. Similarly, the recognition that change is often gradual and continuous has prompted investigation of environmental processes and the development of techniques to determine the nature, direction and rate of environmental change.

1.3 Modern concepts: environmental systems and Gaia

Since the 1930s the concept of general systems has been imbued in environmental studies. In 1935 Tansley first introduced the ecosystem concept into ecology:

> The more fundamental conception is ... the whole system including not only the organism complex, but also the whole complex of physical factors forming what we call the environment .... We cannot separate [the organisms] from their special environment in which they form one physical system .... It is the system so formed which provides the basic units of nature on the face of the earth .... These ecosystems as we may call them, are of the most various kinds and sizes.

Ecosystems thus represent one particular type of system in which living organisms play a fundamental role. In addition, the ecosystem concept promotes a holistic approach, linking both biotic and abiotic components which interact to create an indentifiable whole.

Geographers have also adopted this approach, and as Chorley and Kennedy (1971) point out, it has many advantages for the study of physical geography. On the one hand, it provides a means of subdividing a complex entity, the environment, into identifiable parts while maintaining a holistic approach by emphasising interrelationships. On the other hand, it lends itself to quantification which may ultimately be used for prediction and so may provide a significant contribution to environmental management programmes. Different types of systems have been recognised and they are described in Table 1.1. Almost all environmental systems are functionally open systems characterised by import and export of both energy and matter, although occasionally some may be closed systems, allowing only the exchange of energy.

All natural systems are in a state of dynamic equilibrium: a balance is achieved between inputs, outputs, elements and processes. To maintain this state all systems have controls known as negative feedback loops wherein output influences input. Negative feedback has a stabilising influence, and in ecosystems it is often effected via population changes. For example, the depletion of a particular food source by an expanding animal population may cause an increase in mortality and thus force a return to the status quo. In some cases, catastrophic events may be part of the negative feedback loops in ecosystems. These have been called eustresses by Rapport *et al.* (1985). An example of such a stress is the dependence of many boreal forest species on periodic fire to release seeds from cones and thus ensure reproductive success. Conversely, positive feedback will produce a new system. For example, the removal of vegetation from a catchment may create changes in the local hydrological cycle and lead to accelerated soil erosion. The remaining soil system may become so degraded that it is incapable of supporting a vegetation cover similar to the original. Positive feedback is a major component of environmental change.

Most environmental systems, however, have both resilience and resistance to change and it may be that positive feedback occurs over such a long time period, especially in relation to the human life span, that change is gradual and almost imperceptible. This occurs because environmental systems are complex and response to positive feedback is often slow, an effect called lag time. In most environmental systems a lagged response is usually more common than immediate change. For example, the elimination of grazing or firing in moorland ecosystems may ultimately cause woodland to become established, but it will take some 20–50 years. Similarly, it is only in the last three decades that the ravages of 'acid rain' on Scandinavian lakes have been recognised, although the actual process of acidification began several decades ago with increased use of fossil fuels (Chapter 6). It may be some considerable time before a threshold between one ecosystem and its replacement is crossed.

Some lithological changes in the geological record may well represent the operation of positive feedback over very long periods. Conversely, catastrophic changes such as volcanic eruption may occur to create completely new landscapes. The creation of the island of Surtsey off the coast of Iceland between 1963 and 1966 (Fridriksson, 1987) and the alarmingly rapid demise of tropical forests in Amazonia due to human agencies, are both examples in which external factors, not just positive feedback, have overwhelmed internal ecosystem mechanisms. Thresholds were crossed rapidly in these cases and a lagged response was replaced by an immediate response.

The Gaia hypothesis represents another type of systems approach to examining the global environmental condition. It was first formulated in the early 1970s by James E. Lovelock, an independent scientist based in Cornwall, UK (Lovelock, 1972; Margulis and Lovelock, 1974). From its inception Gaia has been highly praised and heavily criticised for reasons discussed in Mannion and Bowlby (1992) and refuted in Lovelock (1991). This hypothesis is inherently deterministic in a way that contains elements of the old environmental determinism discussed above but which adds a new dimension: the coupling of the environment and its biota in mutual evolution. The key to this mutualism, a form of holism, is the chemistry of the atmosphere, which in turn is intimately linked with many of the Earth's biogeochemical cycles. This has been called neodeterminism (Mannion, 1994) since climate plays a pivotal role in the Earth–biota relationship but the relationship itself can bring about environmental change, i.e. change (including climatic change) can be generated within the system.

There is a great deal of research that highlights the importance of biota in effecting fluxes between biogeochemical pools (Schneider and Boston, 1991). Research on global climate change (Section 6.3) has demonstrated the importance of the global carbon cycle in climate regulation, notably through the accumulation and/or reduction of the heat-trapping gases carbon dioxide and methane. Inevitably, components of the biota have been involved in these exchanges, often in combination with physical and chemical processes. There are also numerous examples from the Earth's geological history which reflect the close relationships between life, its evolution and changing atmospheric composition. Westbroek (1991), for example, describes the role of life as a geological force. A notable example between life, environment and evolution concerns the evolution of green plants. Their ability to photosynthesise not only led to the storage of carbon in biomass, thus taking it out of active circulation, but also led to a considerable increase in the amount of atmospheric oxygen. This eventually allowed mammals to evolve, including humans. Humans, through their release of sequestrated carbon from coal, oil and natural gas, are also changing atmospheric composition. Lovelock's view of Gaia as a robust system that will continue to maintain life does not necessarily mean that life will continue to exist in its present form. The new atmospheric composition that arises from human (biotic) activity may eventually cease to support humans and/or those other components of the biota on which human communities depend. As in the case of the systems discussed earlier, there is likely to be a threshold and a time lag before any changes become noticeable, making it difficult to detect them.

Gaia may be controversial, and it may not offer any quick solutions to the many problems that beset the Earth, problems with which this book is concerned, but it does furnish new perspectives. These perspectives should not be ignored or their challenges relegated. One factor that Gaia makes abundantly clear is the significance of perturbations to biogeochemical cycles in environmental change. This is possibly where Gaian science can be translated into policy decisions.

1.4   Agents and processes of environmental change

Positive feedback, whether it is gradual or catastrophic, is necessary for environmental change to occur. Since all environmental components are part of feedback loops, each may be involved in exerting stress and may be recipients of the resulting strain. Either individually or in a group, the role of environmental components in initiating stress and ultimate change will vary in magnitude both spatially and temporally. The outcome will depend on the internal resilience of the given environmental system and, if thresholds are passed, there will be changes in the energy and matter networks within and between systems. Changes in the energy networks involve changes in the solar energy input (i.e. climatic change) and its subsequent transmission via food webs in ecosystems; changes

in the matter networks will involve changes in biogeochemical cycles.

During the past 2 to 3 M years, the major agents of environmental change have been climate and humans, both of which directly affect the processes operative in environmental systems. This is brought about by influences on energy transfers, the hydrological cycle, sediment transport systems, soil processes and ecosystem function. Changes in solar energy receipt are the key to climatic change since they directly influence the hydrological cycle, wind systems and the amount of energy available for ecosystem function, on a global basis. These in turn affect the global cycles of the nutrients, especially the carbon cycle. Human agencies of environmental change have only become significant in the past 15 K years or so, but their impact has greatly increased as technology has developed. Environmental change has occurred as humans by their ingenuity, i.e. technology, have transposed components of the Earth's surface into resources. Their impact has been direct as natural ecosystems have been cleared to provide land for agriculture and urbanisation, and as land-based resources, such as minerals, have been extracted. Human communities have also brought about indirect and often inadvertent environmental change as a result of resource use and agricultural and technological innovations. Much of this is in the form of pollution. These themes will be developed in Chapters 6 to 9.

The human impact on environment is chiefly due to the need to manipulate energy. Table 1.2 charts a classification of ecosystem types based on energy characteristics devised by Odum (1975, 1993). Here a distinction is drawn between solar-powered ecosystems, human-subsidised solar-powered ecosystems and fuel-powered urban-industrial systems. Only green plants can manufacture food energy, a process operating in all ecosystems, which people can manipulate by transposing ecosystems into agroecosystems and channelling energy flow into specific plant or animal harvest for human consumption. This can be achieved relatively simply by hunter-gathering activities or in a more sophisticated way by enhancing nutrient availability, and by introducing plant and animal breeding and mechanisation. This latter has become increasingly necessary in order to sustain population growth and the urban industrial systems, and involves a considerable addition of fossil-fuel energy. Moreover, the continued exploitation of fossil fuels for industrial purposes and the development of new

Table 1.2 Classification of ecosystems based on energy characteristics and the development of human communities

	Annual energy flow (kcal m^{-2})	Estimated average (kcal m^{-2})	Stages in human community development	Energy available (kcal per day per capita)
1. Unsubsidised natural solar-powered ecosystems, e.g. open oceans, upland forests. Anthropogenic factor: hunter-gathering, shifting cultivation	1×10^3 to 1×10^4	2×10^3	Hunter-gatherers	3×10^3
2. Naturally subsidised solar-powered ecosystems, e.g. tidal estuary, coral reef, some rain forests. In these ecosystems natural processes augment solar energy input. Tides and waves, for example, cause an import of organic matter and/or recycling of nutrients, whereas energy from the Sun is used in the production of organic matter. Anthropogenic factor: fishing, hunter-gathering	1×10^4 to 4×10^4	2×10^4	Early agriculturalists	3×10^3
3. Human-subsidised solar-powered ecosystems, e.g. agriculture, aquaculture, silviculture. These are food- and fibre-producing ecosystems which are subsidised by fuel or energy provided by human communities, e.g. mechanised farming, use of pesticides, fertilisers	1×10^4 to 4×10^4	2×10^4	Agriculturalists	2×10^4 to 2.6×10^4
4. Fuel-powered urban-industrial systems, e.g. cities, suburbs, industrial estates. These are wealth-generating systems (as well as pollution-generating) in which the Sun has been replaced as the chief energy source by fuel. These ecosystems (including socioeconomic systems) are totally dependent upon types 1, 2 and 3 for life support, including fuel and food provision and waste disposal	1×10^5 to 3×10^6	2×10^6	Industrialists	7.7×10^4
			Post-industrialists (includes nuclear power and alternative energy sources)	23×10^4

Source: energy characteristics based on Odum (1993); estimated energy characteristics at different stages of human development based on Simmons (1995a)

energy sources such as nuclear power have themselves a considerable environmental impact (Chapter 6). The harnessing of energy sources has also facilitated mechanisation, which in turn has increased the leisure time for a large proportion of the world's population, especially in the developed world. As a result, leisure, sporting and tourist activities have intensified considerably and have thus become new agents of environmental change.

1.5 People/environment relationships

Precisely when modern humans (*Homo sapiens sapiens*) evolved is the cause of much debate (Section 4.2). It is estimated to have been ca. 400 K years ago though the first hominids, the ancestors of modern humans, emerged about 5×10^6 years ago. Although both of these events are important time horizons in environmental and cultural history they represent only a tiny proportion of the Earth's history of ca. 5×10^9 years. Moreover, because there is a comparative paucity of evidence (Section 4.2) for the interaction of hominids and early modern humans with their environment, little can be said about the people/environment relationship in its early period. Some evidence suggests changing technology and changing strategies for food procurement during the early part of human history, and this reflects changing people/environment relationships. However, there is no obvious starting-point for an examination of people/environment relationships in a temporal context. As ever, and because of the holistic complexion of the people/environment relationship, any development, e.g. the establishment of agriculture (Section 4.3), is usually a product of past practice. There have rarely been true revolutions in the people/environment relationship, so it is more appropriate to consider it as a developmental continuum. This, however, makes it particularly abstruse to find an appropriate starting-point for discussion.

The energy relationships between people and environment are as good a starting-point as any. In the broadest terms, the temporally changing disposition of energy relationships is given in Table 1.2. As heterotrophic organisms, humans are dependent on autotrophs, i.e. green plants that photosynthesise, for their food-energy source, even though they may consume herbivorous and carnivorous animals which comprise intermediaries

in the food chain or web but which are themselves entirely dependent on green plants. Thus, at a fundamental level, energy in all its forms as biomass is a vital consideration in people/environment relationships. Moreover, the initial harnessing of extant biomass energy through agriculture and the later annexation of biomass from ancient ecosystems, i.e. fossil fuels, has been and continues to be major though not exclusive means whereby people have increased the possibilities (see possibilism in Section 1.2) for their development as individuals and within communities. In consequence, the increased range of possibilities has diminished the intensity of environmental determinism (Section 1.2).

It is also significant that the harnessing of energy has freed a proportion of any given population to develop and pursue other facets of community life. The turning to account of biomass energy, either ancient or modern, to produce food and fuel underpins all human endeavour. Considering that there is a close relationship between environment, notably climate, and food production and that fossil fuels are the product of past environments then environmental determinism is alive and well and should not be ignored in studies of environmental change. However, the development within society that has occurred as energy has been appropriated has created changes in the way people perceive and relate to their environment. Indeed it is not an easy matter to define the term 'environment' as Harvey (1993) has discussed. It can be an all-embracing term meaning the physical, chemical and biological entity of the entire Earth's surface. Conversely, it is used to describe the condition of an indoor office or the artificially created milieu of a greenhouse or holiday camp. Undoubtedly, the term now means all things to all people and sometimes has little to do with space or place and all to do with people. This social environment, for example, may be the same in a New York office or nightclub as it is in similar establishments in London or Paris. It is axiomatic that the meanings of the term vary spatially now, depending on the level of development. The definition of environment provided by a stockbroker in the City of London will be quite different to that of a shifting cultivator in Brazil's Amazon forest. It is also axiomatic that peoples' perception of, and their interaction with, environment has changed on a temporal basis as has their representation of it in art, literature etc. (Simmons, 1993a).

Recognising this broad phylogeny does not, however, make it any less difficult to identify an adequate model for the changing nature of the people/environment relationship or indeed to construct a new one. There is already an abundant literature on the subject and some of the major approaches have been covered briefly in Sections 1.2 and 1.3. In brief there are those which emanate from scholars who place the existential necessities of life, food, water, etc., at the core. Others consider the people/environment relationship from the cultural or social perspective; people and the way they interact via the flows of ideas, commodities, etc., are the focus of such efforts, sometimes to the virtual exclusion of any consideration of Earth, air, water, etc. (To a physical geographer this is anathema and to a neodeterminist it is heresy!) Many of these latter approaches have been discussed by Walmsley and Lewis (1993). Moreover, Lash and Urry (1994) consider the people/environment relationship in the changed global political economy of the 1990s when flows of people and commodities dominate, operating on a reflexive basis. Their consideration of 'nature' or the physical environment focuses on its subjugation for human benefit, a philosophy particularly associated with the era of industrialisation and mass production (though it has a much earlier origin; see Section 1.2) and known as modernity. They do, however, consider that the veritable explosion in flows creating globalisation, including information flows, have generated a changed attitude to the global physical environment, notably a bringing together of nature and culture in a post-modern approach. This is akin to the mutualistic, dare it be said neodeterministic, view espoused in Section 1.3. Perhaps the views of the social scientists and the environmental scientists are not so different after all! Interestingly, this fledgling new awareness is also considered by Oelschlaeger (1991) in his consideration of the meaning of 'wilderness', in Roszak's (1993) examination of 'ecopsychology' and Simmons' (1993b) discussion on environmental history. Much of this new awareness is due to one of the most important 'flows' that has accelerated in the last few decades, the growth of tourism. Coupled with information technology and exchange, this has altered peoples' attitude not only to their immediate space but to the global environment. This has been achieved actively via 'flows' of people to tourist destinations and passively through media reporting, especially television.

Allied to these developments is the rise of environmental issues in local, national and global politics and the high profile that such issues achieve in the media. There have been many high-profile events of global environmental significance in the last decade. The first of these was the World Commission on Environment and Development (WCED) in 1987. This is documented in the so-called Brundtland Report which formalised that now clichéd notion of sustainable development. WCED states, 'sustainable development is development which meets the needs of the present without compromising the ability of future generations to meet their own needs' (World Commission on Environment and Development, 1987). As O'Riordan (1991) has discussed, sustainable development, despite its many meanings (e.g. Eden, 1994), has become a central theme in what he describes as the new environmentalism. This unequivocally links ecology and economics (e.g. Pearce, 1993), an implicit relationship within any scheme for examining the people/environment relationship but which needs to be developed into an explicit, even pivotal, role in its post-modern context (see above). The strength of this concept is its holism, requiring the reconciliation of economic development, the quality of life and the state of the environment within various political frameworks that begin with individual responsibility and proceed to collective local levels, etc. In the same year as the WCED the first stage of an international agreement relating to an environmental issue was also signed. This was the Montreal Protocol (United Nations Environment Programme, 1987) which was designated to curb the emissions of substances, notably chlorofluorocarbons, capable of depleting stratospheric ozone (Chapter 6). Yet another internationally significant event was the United Nations Conference on Environment and Development (UNCED), the so-called Earth Summit, held in Rio de Janeiro in 1992 (United Nations Conference on Environment and Development, 1992) to discuss the state of the global environment and ways in which it could be improved. Although it may be argued that gatherings such as this pay only lip service to environmental issues, and often address symptoms rather than underpinning causes, they do at least recognise the growing and global significance of the environment and its relevance to society's welfare and development.

The increased awareness of environmental issues

in the past few decades has also prompted a resurgence of interest in the philosophy of nature and the environment (e.g. Attfield, 1994). An additional but related factor, especially in the context of ethics, is the growing concern with animal rights. Midgely (1992, 1994), for example, refutes the Kantian philosophy (Section 1.2) of nature, including animals, as a means of obtaining society's ends and advocates a holistic attitude involving interdependence. The rapid rates of extinction (both plant and animal) that are currently occurring (Sections 7.2 and 8.2) suggest that the exploitation of biota is causing a loss of opportunities for future development. Consequently, society would do well to heed the views of Midgely and others that the best use of nature is through less confrontational and more accommodating policies. Other issues now debated by philosophers include value in nature and the valuing of nature, especially in an aesthetic sense (e.g. Rolston, 1994), the use of resources (e.g. Sterling, 1992) and environmental justice (e.g. Jamieson, 1994). Perhaps the efforts of these and other philosophers will help in the formulation of widely acceptable environmental ethics, which in turn will assist in the construction of a people/environmental paradigm that is socially just and accords equal status to society and the environment. Certainly, and as is reflected throughout this chapter, no satisfactory framework is currently available.

So far the issue of why a generally acceptable framework for examining the people/environment relationship is desirable has not been adequately examined. One reason for such a framework is to determine the most appropriate ways in which society can manipulate its environment in an enduring fashion. Associated with this is the encouragement of adjustments to engender best practice within varying temporal and spatial scales. Such aspirations are synonymous with sustainable development (see above) and demand an element of prediction. Best practices evolve through experiences on the one hand and some means of forecasting on the other hand. However, the environment itself is so complex that modelling and prediction, as illustrated by attempts to project the possible impact of global warming (Chapter 6), are fraught with difficulties. Add the human dimension and the complexity increases. Indeed the natural and cultural environment are individually and collectively characterised by complexity and chaos both of which reduce the chances of accurate

prediction, as does the operation of human ingenuity. It is now widely recognised that instead of being ordered and linearly law-abiding, Nature (and the environment) is characterised by states of disequilibria. This reflects the operation of chaos, (Gleick, 1987), whereby nearly identical initial conditions within a given system can give rise to very variable responses. Some researchers (e.g. Treumann, 1991) consider that the resulting poor level of prediction militates against sensible planning or sustainable development (e.g. Simmons, 1995a).

This might militate against the case for environmental ethics (see above) or the notion of sustainable development which attempts to provide for the well-being of future generations (e.g. Slaughter, 1994), but it does not mean that such goals should be forgotten or ignored. It might be tempting to take the laissez-faire approach of Beckerman (1995), who posits that there is too much emphasis on sustainable development in the sense that it appears often to conflict with economic growth. He thus adopts a somewhat extreme definition of sustainable development and its related concept of the 'precautionary principle'. The 'precautionary principle' is a proactive attitude toward environment requiring what Beckerman describes as 'drastic action' to curtail environmental degradation that may never actually occur. He has some influential supporters, notably Maddox (1995), who points out the now deridable prophecies of environmentalists in the 1970s who believed that some types of resources would run out, e.g. fossil fuels and certain minerals; but even Maddox recognises the failure of Beckerman to take seriously the threat of global warming. Perhaps the diminution of fossil-fuel resources would have been a good thing; possibly Maddox is a closet environmental determinist and the failure of the doom merchant scenarios of the 1970s is an excellent example of the operation of the unpredictable factor of human ingenuity.

All of these ideas, principles and possibilities provide considerable 'food for thought'. At their core lies the notion of individual and group environmental responsibility, whatever form it may take. Proactivity and reactivity are both inherent practices within ethical, responsible people/ environment relationships. However, none of the possibilities discussed above for achieving such a people/environmental paradigm are ideal; it is probably Utopian to anticipate an ideal expression, with a predictive capacity and widespread application in varying spatial and temporal contexts.

Although it is stimulating to appraise the various possibilities it is also vital and pragmatic to ponder the following facts that have recently been revealed from a survey of human disturbance of world ecosystems (Hannah *et al.*, 1994). Using three categories i.e. undisturbed, partially disturbed and human dominated, it appears that 64.9 per cent of Europe, the most disturbed of the continents, is human dominated and only 15.6 per cent remains undisturbed. South America and Australasia, the least disturbed continental land areas, respectively contain 15.1 and 12.0 per cent of land that is human dominated whereas 62.5 and 62.3 per cent are respectively classified as undisturbed. Overall, 52 per cent of the Earth's land area remains undisturbed, though the habitable land area is ca. 66 per cent disturbed, either partially or substantially. The character of the undisturbed land is a product of natural environmental history, whereas the character of the disturbed land is a product of both natural and cultural environmental history. Separating the two is often difficult, especially in relation to the environmental changes of all but the past 200 years or so. Nevertheless, the data quoted above reflect the role of society as a major agent of environmental change. The disturbance by humans of such a large proportion of the Earth's surface reflects the people/environment relationship, its changes over time and all the social and cultural factors, etc., that lead to land transformation.

1.6   Conclusion

There is no doubt that the processes operative at the Earth's surface (including the atmosphere) are dynamic. For a large part of Earth history these processes, notably energy and material exchanges, have been self-regulating, often through the medium of life, which both changes and is changed by these relationships. With the advent of modern humans and their capacity to harness food, and later to harness fuel, a new and particularly efficacious agent of environmental change came into existence.

Through the species' ingenuity, enabling the transposition of components of the Earth's surface into resources, i.e through technology, the capacity of society to alter its environment, often in response to social and economic stimuli etc., has increased substantially. Moreover, society's view on how it relates to its environment has changed markedly. Recognising dynamism as characteristic of Earth surface processes, rather than Earth as a static constant environment, society is now beginning to reappraise its attitude to nature and the environment. Much of this reconsideration has been prompted by concerns about global environmental change, caused by global warming, and the media publicity it has received along with many other environmental issues. The people/environment relationship is complex and difficult, if not impossible, to predict for even the next few decades. Nevertheless, this relationship concerns every individual and it is now becoming apparent that all individuals are linked in some way just as local or patch environmental changes eventually add up to global environmental change. These issues are the subject-matter of this book.

Further reading

Livingstone, D.N. (1992) *The Geographical Tradition.* Blackwell, Oxford.

Lovelock, J. (1995) *The Ages of Gaia: a biography of our living earth*, 2nd edn. Oxford University Press, Oxford.

Mannion, A.M. and Bowlby, S.R. (eds) (1992) *Environmental Issues in the 1990s* J. Wiley, Chichester.

Martin, G.J. and James, P.E. (eds) (1993) *All Possible Worlds: a history of geographical ideas*, 3rd ed. J. Wiley, Chichester.

Meyer, W.B. and Turner, B.L. II (eds) (1994) *Global Land-Use and Land-cover Change.* Cambridge University Press, Cambridge.

Schneider, S.H. and Boston, P.J. (eds) (1991) *Scientists on Gaia.* MIT Press, Cambridge MA.

Turner, B.L. II , Clark, W.C., Kates, R.W., Richards, J.F., Mathews, J.T. and Meyer, W.B. (eds) (1990) *The Earth as Transformed by Human Action.* Cambridge University Press, Cambridge.

CHAPTER 2

Quaternary geology and climatic change

2.1 Introduction

Many of the eighteenth-century geologists who inaugurated the glacial theory (Section 1.2) recognised the significance of climatic change in shaping the Earth's surface during the past 3×10^6 years. The superficial deposits with their stratigraphic variations that these and later workers recognised have been related to periods of different climatic regimes, particularly cold stages (glacials and stadials) and warm stages (interglacials and interstadials). Although it is now accepted that in most parts of the world the terrestrial record is fragmentary, rendering it liable to misinterpretation, it has played a major role in unravelling the record of environmental change.

That the terrestrial record is often incomplete is reinforced by evidence from ocean-sediment stratigraphy, which allows the examination of uninterrupted Quaternary sequences and which is providing a basis for global correlations. Ocean-sediment stratigraphy has contributed substantially to studies of environmental change and has led to the reinstatement of an astronomical cause (Section 2.6) for climatic change. The record of environmental change in polar and continental ice cores has also made a major contribution to the elucidation of environmental change, especially for the past 160 K years. Such data complement those from ocean sediments and constitute another link in the global jigsaw of environmental change. Further evidence, usually of a local nature, derives from historical records and meteorological records.

All of these lines of evidence, and others too numerous to mention, have been used to examine the regional expression of environmental change.

Although the volume of evidence available from the southern hemisphere is increasing there is still a bias towards the northern hemisphere, where the terrestrial record in particular is more extensive. And there is a growing body of evidence from low latitudes which, although not directly glaciated as were the high latitudes, experienced considerable shifts in climate. For example, the Quaternary period witnessed expansions and contractions of the world's deserts and there were shifts in the vegetation belts of the tropics and subtropics. There were also major changes in the global hydrological cycle as sea levels rose and fell in symphony with the growth and decline of the polar ice caps.

2.2 Quaternary subdivisions based on the terrestrial record

Before examining the problems involved in delimiting the various stages of the Quaternary, the stratigraphic position of its commencement, the Pliocene–Pleistocene (Tertiary–Quaternary) boundary must be considered. This has been, and is still, the subject of much debate. Age estimates, based on floral and faunal elements that indicate a sharp change from warm to cold conditions, range from ca. 1.6 to 2.5×10^6 years. In 1985 the International Commission on Stratigraphy (Bassett, 1985; Cowie and Bassett, 1989) formally adopted a stratigraphic horizon where claystone overlies a sapropel bed (sapropel = black mud rich in organic calcium carbonate) in a geological section at Vrica in Calabria, Italy, as the official marker immediately below the initial appearance of *Cytheropteron testudo*, a thermophobic (cold-loving) foraminiferan. This horizon is dated to ca. 1.8×10^6

Table 2.1 The Quaternary period and its subdivisions

Date (K years)	Period (or system)	Epoch (or series)	Other subdivisions	Palaeomagnetic polarity epochs	Events
0					
10	Q	Holocene			
	U	P			
	A	L	Upper		
	T	E	Pleistocene		
	E	I			
125	R	S			
	N	T	Middle		
	A	O	Pleistocene		
750	R	C		Brunhes	
	Y	E		———	Jaramillo
		N	Lower	Matuyama	
		E	Pleistocene		
1800					Olduvai
2010	PLIOCENE				Réunion

years. However, this demarcation has been questioned by Jenkins (1987), who disputes the identification of *C. testudo*, believing it to be an unknown species related to *C. wellmani*, which is not a cold-water indicator. Moreover, estimates for the onset of glaciation have been placed as far back as 3.5×10^6 years, based on the presence of ice-rafted debris in ocean cores (e.g. Opdyke *et al.*, 1966), glacial deposits (e.g. Kvasov and Blazhchishin, 1978) and potassium–argon dated tillites (e.g. Clapperton, 1979). If a date of 1.8×10^6 years is accepted as the base of the Quaternary, these data place the onset of glaciation in the Pliocene. There is also general agreement from widespread evidence in the North Atlantic region for the onset of glaciation at ca. 2.45×10^6 years BP. Shackleton *et al.* (1984), for example, suggest that ice-sheet initiation began at ca. 2.4×10^6 years and this is supported by numerous other studies such as that of Loubere (1988). His analysis of foraminiferal assemblages and oxygen isotope ratios from Deep Sea Drilling Project Site 584 in the northeastern Atlantic indicate that the advent of glaciation was the result of progressively deteriorating climatic cycles prior to 2.5×10^6 years BP (Section 2.6). Similarly, Zagwijn (1985) has dated an early glaciation in the Netherlands (the Praetiglian) at ca. 2.3×10^6 years. Based on a commencement date of ca. 1.8×10^6 years, the major, and now most widely accepted, subdivisions of the Quaternary period are given in Table 2.1.

There are also numerous problems associated with the delimitation and correlation of Quaternary deposits intra- and intercontinentally. This is due to the fragmentary nature of the record and the inadequacy of dating techniques. Since the 1950s the trend has been towards the development of regional sequences, each with their own terminology and some intercontinental correlations have been made which can be validated through dating. It is now clear that the Weichselian, Devensian, Würm, Wisconsin and Valdai stadials, which represent the last glaciation in northwest Europe, Britain, the Alpine Foreland, North America and European Russia respectively, were contemporaneous. Similarly, the succeeding warm stage, known as the Holocene in Europe and North America, is contemporaneous with the Flandrian in Britain.

The difficulties of correlating deposits are also accentuated with older material: as research proceeds there is often need for re-evaluation. Problems of this kind are well illustrated by two examples. Firstly, Zagwijn (1975) has compared the Quaternary sequence of the Netherlands with that of East Anglia in Britain, and has demonstrated that the East Anglian sequence has two hiatuses in the early and middle Quaternary, representing more than 1×10^6 years. Secondly, Green *et al.* (1984) have presented evidence from Marsworth in Buckinghamshire, UK, to show there is an interglacial deposit, of an age hitherto unrecorded in Britain, which underlies Coombe rock and deposits of Ipswichian age. A date of 140 to 170 K years BP has been obtained for this 'new interglacial' (Table 2.2), which places it in what has traditionally been the penultimate glaciation in Britain, the Wolstonian (Mitchell *et al.*, 1973). This is now considered to be a complex period in terms of environmental change, with a cold period, evidenced

Table 2.2 Correlations between various Quaternary sequences in the northern hemisphere

NORTH EUROPEAN STAGES	UK	NETHERLANDS	NORTH GERMANY	POLAND	USSR	ALPINE REGION (AUSTRIA & GERMANY)	NORTH AMERICA	MARINE STAGES	DATES (10³ YEARS)
IG HOLOCENE	FLANDRIAN / LATE DEVENSIAN			MAIN STADIAL (VISTULIAN)	LATE VALDAI G.	MAX WURM G.	HOLOCENE / LATE WISCONSIN	1 / 2	13 / 32–35
G1 WEICHSEL	MIDDLE DEVENSIAN	WEICHSELIAN	WEICHSEL GLACIATION	PREGRUNDZIAD STADIAL		FIRST WURM G.	MIDDLE WISCONSIN / EARLY WISCONSIN	3 / 4	64–65 / 75–79
	EARLY DEVENSIAN		LOW TERRACE	KASZUW STADIAL / KASZULY STADIAL	LOESS FLUVIAL DEPOSITS	STILLFIED A.	EOWISCONSIN	5 a, b, c, d	
IG1 EEM	IPSWICHIAN	EEMIAN	EEMIAN	EEMIAN	MIKULINO	MONDSEE & SOMBERG	SANGAMOAN	5 e	122
G2 SAALE (WARTHE)	RIDGEACRE? (WOLSTONIAN COMPLEX)	(SAALIAN COMPLEX)	SAALE 3 G. / RUGEN 1 G. / SAALE 2 G.	WARTA GLACIATION (CENTRAL POLISH GLACIATION)	MOSCOW	LATE RISS / LATE RISS / RISS? / LATE RISS TERRACE	LATE ILLINOAN	6	128–132
IG2 BANTEGA/ HOOGEVEN	STANTON HARCOURT	HOOGEVEN	TREENE WENNING-STEDTER & KITTMITZER PALAEOSOLS	LUBLIN I.G. / POLICHNA I.S.	ODINTSOVU I.S. (DNEIPER STAGE)	PARABRAUN-EARTH	ILLINOAN	7	195–198
G3 SAALE (DRENTHE)		DRENTHE GLACIATION	DRENTHE G. / MAIN TERRACE	ODRA G. / PODWINEK I.S. / PRE-MAXIMUM STADIAL	DNEIPER GLACIATION	ANTEPENULTIMATE G / EARLY RISS / MINDEL HIGH TERRACE	EARLY ILLINOAN	8	251–262
IG3 DOMNITZ (WACKEN)	HOXNIAN		FREYBURGER BODEN	MAZOVIAN	ROMNY	REDDISH PARABRAUN-EARTH		9	297–302
G4 FUHNE (MEHLECK)			ELDERITZERT TERRACE ERERKNER ORGANIC SEDIMENTS	PRE-MAXIMUM STADIAL? / WILGA G.?	ORCHIK STAGE (PRONYA GLACIATION)	GL4 TERRACE (PRE-RISS TERRACE)		10	338–347
IG4 HOLSTEIN (MULDSBERG)	SWANSCOMBE	HOLSTEIN	HOLSTEIN	FERDYNANDOW	LICHVIN	REDDISH PARAEBRAUN-EARTH		11	367–352
									440–428

G5	ESTER II	ANGLIAN	ELSTER	ELSTER 2 G. FLUVIAL GRAVELS	SAN G. KOCK STADIAL? (SOUTH POLISH GLACIATION)	OKA GLACIATION (OKA GLACIATION)	GL5 GLACIATION OF ALPS LATE MINDEL GUNZ MINDEL GL5 TERRACE	PRE-ILLINOIAN	12	472–480
1G5	CROMERIAN IV	CROMERIAN	CROMERIAN IV	VOIGSTEDT	LUSZANA I.G.?	I.G.?	RIESENBADEN PALAEOSOL		13	502–?
G6	GLACIAL C		GLACIAL C	ELSTER 1 G. FLUVIAL GRAVELS	NALECZORIG SERINKI STADIAL?		LOWER INTERTERRACE GRAVEL EARLY MINDEL DONAU	–	14	542–562
1G6	CROMERIAN III	WAVERLEY WOOD	CROMERIAN III	VOIGSTEDT	PODLASIE 1 G.		RIESENBODEN PALAEOSOL		15	592–630
G7	GLACIAL B	KESGRAVE GROUP?	GLACIAL B	FLUVIAL GRAVELS	PRE-CROMERIAN GLACIATIONS		MIDDLE INTERTERRACE GRAVEL		16	627–687
1G7	CROMERIAN II		CROMERIAN II				RIIESENBODEN PALAEOSOL		17	647–718
G8	GLACIAL A		GLACIAL A				UPPER TERRACE GRAVEL GL8 GLACIATION		18	790
1G8	CROMERIAN I		CROMERIAN I				RIESENBODEN PALAEOSOL		19	
G9	DORST		GLACIAL	HELME FLUVIAL GRAVELS			EARLY GUNZ?		20	
1G9	LEERDAM		LEERDAM	UPPER MUSHELTONE			INTERGLACIAL		21	

Source: based on Bowen *et al.* (1986a), Bowen (1994) and Šibrava (1986) by the Coombe rock of Marsworth, separating the Ipswichian from the 'new' interglacial. There is some dispute over the dating of Marsworth, and plant macrofossils (Field, 1993) do not allow any significant conclusions to be drawn in relation to its interglacial or interstadial status. However, De Rouffignac *et al.* (1995) have recently reported yet another site, at Upper Strensham in Worcestershire, with temperate plant remains (oak trunks) of probable interglacial status. Amino acid age estimates suggest a correlation with oxygen isotope stage 7, in line with Marsworth and another site at Stanton Harcourt (Bowen *et al.*, 1989), northwest of Oxford, where an excavation of mammoth skeletons is now under way.

A major project of the International Geological Correlation Programme (Šibrava *et al.*, 1986) has focused on the correlation of Quaternary stages across the northern hemisphere. Some of their results are summarised in Table 2.2, which also shows the regional terminologies currently in use, approximate dates and the relationship between the terrestrial and the marine stages. The marine stages will be discussed in more detail in Section 2.3. In Europe, for example, this work presents evidence for nine glacial stages and nine interglacials during the past ca. 750 K years. This generally broad correlation must imply a hemispheric, or more likely global, control of environmental change with climatic change being the most obvious mechanism. This will be further discussed in Section 2.6.

In addition to the delimitation of glacial and interglacial stages on the basis of glacial deposits, interglacial organic beds, etc., there are other continental sediment sequences which can assist in the elucidation of Quaternary environmental change. These include loess deposits, calcite sequences and lacustrine sediments. Lacustrine sediments generally represent the Holocene, or parts of the Holocene, in the Arctic and temperate zones but in some regions e.g. Japan, the Mediterranean zone and parts of the tropics there are lake beds which represent the entire Quaternary period and even the later part of the Tertiary period. These will be considered in Sections 2.8 and 2.9. Groundwater saturated with calcite may also leave a record of environmental change as the calcite is deposited. The oxygen which forms part of the calcium carbonate contains an oxygen isotope signature just as it does in foraminifera in ocean sediments (Section 2.3) and in ice cores (Section 2.4). The isotope record can be used as a proxy temperature

record (explained in Section 2.3). Moreover, dating of calcite deposits may be possible using uranium series. Winograd *et al.* (1992) produced an oxygen isotope profile from a calcite unit at Devils Hole, Nevada, which contains a 500 K year chronology determined by uranium series. After an initial suggestion that the chronology of glacial–interglacial stages was markedly different from that represented in a wide range of ocean-sediment cores (Section 2.3), adjustments indicate a close agreement between the two records.

Much attention has also focused on loess deposits, especially those of the Loess Plateau of north-central China. Loess is a wind-blown deposit consisting mainly of silt. In China the deposit is extensive and reaches depths of up to ca. 280 m; it originated from dust blown during arid glacial periods from the Tibetan plateau to the west, whereas during interglacial periods weathering occurred. Consequently, in some parts of the Loess Plateau, variations in loess stratigraphy represent the past 2.5×10^6 years of environmental change (Ding *et al.*, 1992). Various dating techniques and palaeomagnetic measurements have been applied to a wide range of sites (e.g. Zhu *et al.*, 1994; Chen *et al.*, 1995) between which correlation is now possible. Ding *et al.* (1994) have established, on the basis of grain size data, a well-defined timescale for the loess units at Baoji for the past ca. 2.5×10^6 years (Figure 2.1). Their data reflect strengthened monsoonal flow during glacial periods in comparison with interglacial periods when cooling of mid-latitude air masses by reduced ice sheets was minimal. It is now possible to correlate loess sequences with the oxygen-isotope stratigraphy of the ocean sediments as shown in Figure 2.1. In addition to similar links which can be established between ocean sediments, lacustrine sequences (Sections 2.8 and 2.9), calcite deposits (see above) and ice cores (Section 2.4), reflects the advances that have been made in the past two decades towards establishing a framework for global environmental change.

2.3 The record of climatic change from the oceans

Since the 1960s palaeoenvironmental studies have been revolutionised by evidence from ocean sediment cores. The ocean basins receive some 6 to 11×10^9 tonnes (t) of sediment annually, the terrigenous

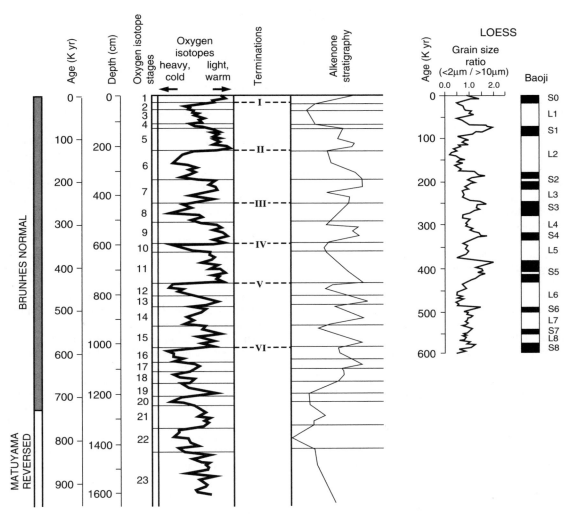

Fig. 2.1 Relationships between oxygen isotope stages (based on Shackleton and Opdyke, 1973), terminations (Broecker and Van Donk, 1970), alkenone stratigraphy (Brassell *et al.*, 1986) and Chinese loess stratigraphy (Ding *et al.*, 1994).

component of which is derived from adjacent continental land masses. As this material settles, biogenic material, consisting of the remains of marine organisms and pollen grains derived from continental vegetation, is incorporated into the mineral matrix. These generally uninterrupted sequences of sediment have been widely investigated.

Much of the research has focused on calcareous foraminiferan tests, on the shells of protozoan organisms that are important components of the marine zooplankton, and in particular on the oxygen isotope ($^{18}O/^{16}O$) composition of the calcium

carbonate. Although the interpretation of oxygen isotope ratios is complicated, the premise is that during cold periods ocean waters become enriched in ^{18}O (the heavier isotope) and this enrichment is reflected in the ^{18}O content of the foraminiferan tests. Conversely, in warmer periods the ^{18}O content is relatively depleted. The mechanism for the observed fractionation is that the lighter $H_2^{16}O$ water evaporates in preference to the heavier $H_2^{18}O$ form, and that this inherent thermodynamic tendency leads to a marked isotopic imbalance in atmospheric water vapour at lower ambient temperatures. During cold periods the relatively $H_2^{16}O$-laden air deposits the water in glaciers and ice sheets, whence it returns to the oceans during warm periods.

Oxygen isotope analyses have been undertaken on many ocean-sediment cores from various parts of the world and most reveal similar changes, implying that the ^{18}O signal is a reflection of changes in continental ice volume. As these variations are generally synchronous, oxygen isotope stages (Table 2.2) provide a means of correlating sediments from different regions and a means of devising a stratigraphic framework for the Quaternary, to which terrestrial deposits can be related. An independently dated timescale has been derived for the various oxygen isotope stages based on palaeomagnetic measurements and radiometric dating techniques. Figure 2.1 illustrates the range of oxygen-isotope marine stages and appropriate dates. Of particular interest is the nature of the transition from glacial to interglacial periods, which in all oceanic cores appears to be extremely rapid. Such changes have been described by Broecker and van Donk (1970) as terminations; the most recent is termination I, which occurs at the end of the last glacial stage (Weichselian, Devensian, Würm, Valdai, Wisconsin). Age estimates for terminations are also given in Figure 2.1. So-called Heinrich events (Heinrich, 1988) have been identified in many North Atlantic cores (see Fig. 2.2) These are layers rich in ice-rafted debris that are considered to represent increased discharge from icebergs and to correspond with periods of cooling in surface water. Dokken and Hald (1996) have recorded six periods of sea-ice disintegration in polar North Atlantic cores, e.g., in isotope stages 4, 3 and 2, and they attribute this to the influx of North Atlantic surface water into the polar North Atlantic. In addition, Rasmussen et al. (1996) have noted a correlation between Heinrich events and changes in the oxygen isotope record in two cores from the Faeroe-Shetland Channel. The occurrence of these layers at important horizons marking climatic change has led Broecker (1994) to suggest that Heinrich events may be sufficiently significant to trigger climatic change of a global nature. Further work by Fronval et al. (1995) lends weight to this suggestion as has reported evidence for synchronous Fennoscandinavian and Laurentide (from North America) iceberg discharges. Paillard and Labeyrie (1994) have suggested that the discharge of icebergs into the North Atlantic initially curtails thermohaline

circulation and thereby causes cooling, i.e. the Heinrich events. As the icebergs melt and their impact diminishes, thermohaline circulation resumes leading to abrupt warming.

Bond et al. (1993) have indicated the positions of Heinrich events in the stratigraphy of a Greenland (GRIP) ice core (Section 2.4). As Figure 2.2 illustrates, there is evidence for alternate warming and cooling; indeed the Younger Dryas cold period (Section 3.3) may be equivalent to a Heinrich event. Table 2.2 and Figure 2.1 reflect the relatively short duration of interglacial periods, i.e. approximately 20 K years, whereas glacial stages are ca. 100 K years long. Moreover, these records show that glacial stages were not characterised simply by one long uninterrupted ice advance and there is corroborating terrestrial evidence for relatively short-lived warmer periods, the interstadials, during these cold stages. In Britain, for example, Ehlers et al. (1991) give evidence for three such stages (named after specific localities): the Chelfordian interstadial of oxygen isotope stage 5c and the Brimpton interstadial of oxygen isotope stage 5a. The other is the Upton Warren interstadial dated at 42 to 43 K years. (They are not marked in Table 2.2 in order to simplify the main correlations.) There are also numerous examples of interstadial deposits in Europe (Šibrava et al., 1986). However, the polar ice cores (Section 2.4) indicate there may have been as many as 22 interstadials in the Arctic region and 9 in the Antarctic region during the last ice advance (Bender et al., 1994). These data reflect the complexity of environmental change and the non-uniformity of cold stages wherein two or more glacial expansions may have taken place. Such environmental changes are also reflected in the oxygen isotope record of marine-sediment cores which show minor as well as major peaks and troughs of global ice volume.

Other components of ocean-sediment cores which furnish palaeoenvironmental data include diatoms, chemical composition, terrigenous material and pollen grains. For example, attention has focused on saturated long-chain alkenones (lipid components of cell membranes) as palaeoclimatic indicators, especially for the reconstruction of sea-surface temperatures (Brassell et al., 1986; Eglinton et al., 1992). Alkenones, with a large number of carbon atoms (C_{37} to C_{39}) have been identified from the marine coccolithophore *Emilani huxleyi* and it has been established that the degree of unsaturation (i.e. where there are multiple rather than single bonds

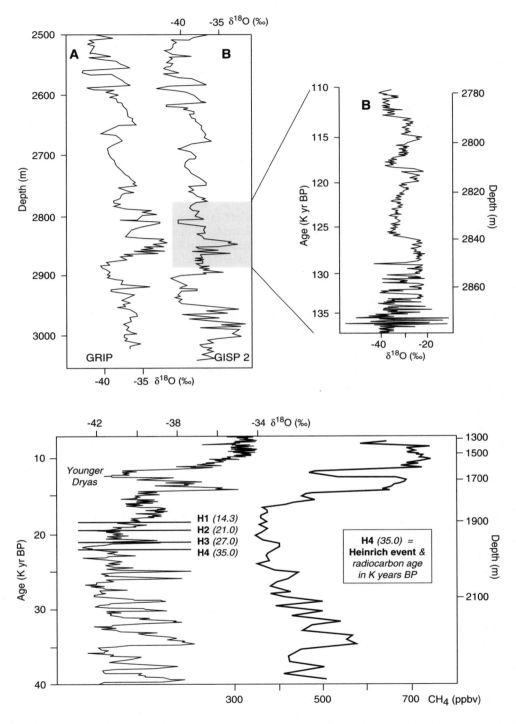

Fig. 2.2 Palaeoenvironmental data from the GRIP and GISP2 (Greenland) ice cores (adapted from Johnsen et al., 1995; Bond et al., 1993).

between adjacent carbon atoms) is temperature dependent. Under laboratory conditions the molecules are more unsaturated at higher temperatures than they are at lower temperatures. In addition, Brassell *et al.* (1986) report that variations in alkenone saturation occur in different climatic regions of the oceans, again reflecting temperature dependence. Variations in the alkenone unsaturation index (U^K_{37}) were also determined for an ocean-sediment core extracted from the eastern equatorial Atlantic and compared with the oxygen isotope ratios of foraminiferan tests. These downcore variations exhibit similar trends and in glacial stages, with higher ^{18}O values, the U^K_{37} index is low. The converse occurs in interglacial periods, as shown in Figure 2.1. Brassell *et al.* suggest that changes in alkenone structure occur as a response of aquatic organisms to environmental stress. By altering the molecular composition of their lipid (fat) bilayers, of which alkenones are an important component, the organisms can maintain the fluidity of their membranes. Alkenone-containing organisms are more directly influenced by sea-surface temperatures than are ^{18}O values. Alkenone analysis may therefore be a more accurate indicator of these temperatures. Emeis *et al.* (1995) have reconstructed sea-surface temperatures for a 500 K year period represented in a sediment core from the Arabian Sea. During the interglacial periods temperatures were as high as 27 °C, and ca. 22 to 24 °C during glacial periods. Schneider *et al.* (1995) have demonstrated a similar sea-surface temperature difference between glacial and interglacial periods from an equatorial South Atlantic core.

Foraminifera tend to dominate cores from warm-water regions whereas diatoms are prevalent in cold-water regions, yet only recently have diatoms been the subject of investigation in relation to their value as a proxy record of climate and changing oceanic productivity. Shemesh *et al.* (1993, 1995) report on the value of oxygen isotope signatures from the diatoms of a Southern Ocean core to generate oxygen isotope stratigraphies and sea-surface temperature reconstructions. This will facilitate, along with alkenone stratigraphy, a means of palaeoenvironmental reconstruction hitherto only possible from cores containing carbonates (e.g. foraminiferan tests). Singer and Shemesh (1995) also report that the isotopic composition of organic carbon within siliceous diatom frustules (the diatom shell) can be used to reconstruct patterns of palaeoproductivity.

Their results suggest that, in the Antarctic oceanic zone represented by their cores, there was no increase in productivity during any of the last five glacial stages. The implication of these results is that the oceans did not deplete, via primary productivity, the atmosphere of carbon dioxide during glacial periods, though this is a controversial issue. Just as controversial is the role of ocean circulation and nutrient availability in global climatic change (Boyle, 1990). Cadmium in the oceans has been used as a proxy indicator of phosphorus availability because it is directly linked with the amount of phosphorus, an important nutrient. For example, Boyle reports that higher nutrient concentrations occurred in mid-depth (3000 m) waters during glacial times than in interglacial times in the North Atlantic Ocean but not in the South Atlantic. Barium has also been used as a palaeoceanic indicator. Both cadmium and barium concentrations reflect various components of ocean circulation such as the role of North Atlantic Deep Water (NADW).

The terrigenous component of ocean-sediment cores also provides palaeoenvironmental data. Reference has already been made to ice-rafted debris (Section 2.2) as indicative of the onset of glaciation. Nam *et al.* (1995) have reported several pulses of coarse-grained terrigenous material from a core located on the east Greenland continental margin during the past 225 K years. This was higher during the last glacial maximum than during previous glacial maxima. Grain-size analysis of the mineral fraction of modern sediments from the abyssal North Pacific have allowed Rea and Hovan (1995) to distinguish between aeolian (wind-derived) material and hemipelagic material in North Pacific and South Atlantic cores. Piper *et al.* (1994) have identified sediment sources for continental margin material deposited during the past 1×10^6 years off the coast of southeastern Canada on the basis of sand and clay mineralogy. They have also been able to relate the type and rate of sedimentation to variations in sea-surface temperatures inferred from foraminifera and dinoflagellates, and they deduce that extensive ice sheets accrued in oxygen isotope stages 10, 8 and 6 with less severe glaciation in stages 4 and 2 (see Table 2.2 for stages).

The pollen content of ocean-sediment cores can also be used to facilitate environmental reconstruction and correlation between marine and terrestrial deposits. As stated in Section 2.2, much of the early part of the Quaternary period is not represented in Britain but some evidence for it derives from borehole data from the North Sea. Ekman and Scourse (1993), for example, have described three pollen zones below the Bruhnes–Matuyama boundary (Table 2.1) which may include two interglacial stages. Unlike the later interglacials of the Quaternary period which are characterised by oak (*Quercus*), elm (*Ulmus*), alder (*Alnus*), etc., these early interglacials are characterised by hickory (*Carya*), hop hornbean type (*Ostrya type*), wing-nut (*Pterocarya*) and *Eucommia*. Moreover, pollen analysis of an ocean-sediment core from a site east of New Zealand has allowed Heusser and van de Geer (1994) to reconstruct the vegetation and climatic history of New Zealand's South Island for the past ca. 350 K years. Four glacial–interglacial cycles have been identified that reflect changes between glacial-stage herb-dominated pollen assemblages and conifer plus broadleaf interglacial assemblages. This and earlier work by Nelson *et al.* (1985) suggests that these middle to late Quaternary environmental changes were generally synchronous with those of the northern hemisphere.

2.4 The record of climatic change from ice cores

A growing volume of data from ice cores is contributing to studies of environmental change and facilitating correlations between polar, continental and ocean-sediment records. The polar ice sheets and those of high tropical mountains are nourished by precipitation from the atmosphere, the composition of which is thus recorded as successive layers of ice accumulate. Such records provide information on environmental change over the past ca. 200 K years and base line data from pre- and post-industrial levels for the biogeochemical cycling of metals such as lead. This information facilitates the determination of the effect that industrial society exerts on atmospheric lead concentrations, as discussed in Section 6.7.

The majority of ice cores have been obtained from the polar ice caps, the main exceptions being those from Peru (Thompson *et al.*, 1995) and the Tibetan plateau (Thompson *et al.*, 1989). However, the ice cores providing the longest temporal record of environmental change are the Vostok core from Antarctica (e.g. Jouzel *et al.*, 1990) and the GRIP and GISP2 cores from Greenland. However, the latter are controversial because there is a

disagreement between the records near the bases of the cores, representing the Eemian (last) interglacial and the latter part of the penultimate glacial period (Dansgaard *et al.*, 1993; Grootes *et al.*, 1993), as shown in Figure 2.2. The controversy rests on the rapidly changing oxygen isotope stratigraphy of the GRIP core's Eemian record, which indicates an unstable climate and which is not paralleled in the Eemian record of GISP2. Not only are there rapid changes in GRIP oxygen isotope stratigraphy but there are substantial oscillations in electrical conductivity (Taylor *et al.*, 1993). Although this suggests the record is reliable, Fuchs and Leuenberger (1996) believe otherwise on the basis of oxygen isotope transitions that are inconsistent with ice-sheet formation and decay. However, if the GISP2 record is reliable, as is generally believed (e.g. Larsen *et al.*, 1995), then reconciliation of the two records is not possible. Moreover, the instability recorded in GRIP is not reflected in the Eemian section of the Vostok core which indicates a warm and stable period (Johnsen *et al.*, 1992). The absence of convincing evidence from other climate indicators from the northern regions for instability (e.g. McManus *et al.*, 1994) indicates that instability may have been confined to the Arctic ice cap. Further evidence is awaited to clarify the nature of the Eemian record, though there is some evidence for instability from several long northwest European pollen diagrams (Field *et al.*, 1994) when compared with the Holocene. Clarification is essential before relationships can be drawn between the Eemian and the Holocene in order to determine the extent of anthropogenically driven climate change in the Holocene.

As Figure 2.3 shows, a range of parameters have been measured in the Vostok core, some of which have also been measured in the Greenland ice cores. Measurements of oxygen isotope ratios provide a stratigraphic and dating control as well as reflecting past temperature changes. In effect they are a mirror image of the oxygen isotope ratios from ocean sediments (explained in Section 2.3) and the data given in Figure 2.3 reflect the observed $^{18}O/^{16}O$ ratios relative to the isotopic composition of a water standard (standard mean ocean water). On the basis of oxygen isotope variations, the record from the Vostok core has been subdivided into several stages. Stage A corresponds to the Holocene, stage G represents the last interglacial period and stage H is the terminal part of the penultimate glacial stage (marine stage 6; see Table 2.2). Stages B to F

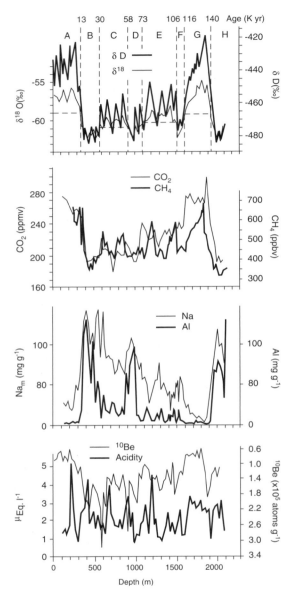

Fig. 2.3 Palaeoenvironmental data from the Vostok ice core, Antarctica (adapted from Lorius *et al.*, 1989; Chappellaz *et al.*, 1990; Raisbeck *et al.*, 1987; Legrand *et al.*, 1988b).

represent the last glacial period in which stages C and E are well-defined interstadials. It is estimated that the change from stages H to G involved a 10 °C temperature difference, an 8 °C difference between stages A and B and ca. 2 °C and 4 °C respectively between the interstadial stages C and E and the

glacial stadials B, D and F (Lorius *et al.* 1985). Similar order of magnitude variations are reflected in the oxygen isotope records from the Greenland cores (Bender *et al.*, 1994), though the latter reflect a higher incidence of interstadials (Section 2.3). In addition, the Vostok deuterium profile (Fig. 2.3) parallels that of the oxygen isotopes (Jouzel *et al.*, 1987). This is because deuterium, which is a heavy isotope of hydrogen, is similarly affected by temperature variations. Both the deuterium and oxygen isotope records indicate that the last interglacial was ca. 2 °C warmer than the present Holocene. The ^{10}Be profile, considered to be a proxy record for precipitation (Yiou *et al.*, 1985), indicates that the amount of precipitation received during the last ice age was only about half that received during the last or present interglacial.

Other parameters analysed from the Vostok core include the carbon dioxide (Barnola *et al.*, 1987) and methane (Raynaud *et al.*, 1988) contents of air bubbles encapsulated in the ice, which reflect the concentrations of these gases in the atmosphere at the time the bubbles formed. Figure 2.3 illustrates how concentrations of these gases have varied considerably, especially between glacial and interglacial periods. The transition between the last glacial period and the Holocene, for example, is characterised by a rise in carbon dioxide concentrations from ca. 190 ppm to ca. 270 ppm; a slightly steeper increase occurred at the transition from the penultimate glaciation to the last interglacial. In both instances, methane concentrations double as they do in the GRIP core (Chappellaz *et al.*, 1993), data from which are given in Figure 2.3. Moreover, in this case the GRIP and GISP2 cores show parallel changes (Brook *et al.*, 1996). The implication of this relationship is that the carbon cycle and global climatic change are linked. The nature of this link is, however, the subject of much debate, as is the role of these gases in terms of whether they are forcing or reinforcing factors. For example, it is now accepted that Milankovitch cycles (Section 2.6) are responsible for global cooling as the Earth enters an ice age but what is controversial is the source of, and the mechanisms whereby, carbon (as carbon dioxide and methane) was fluxed to the atmosphere where it enhanced the greenhouse effect to amplify global warming. Sowers and Bender (1995) have discussed this issue and show that the start of the increase in both gases began 2 to 3 K years before warming in Greenland. Thus the flux of gases to the atmosphere contributed to deglaciation and global warming. The mechanisms Sowers and Bender examine to account for the rise in these gases concern a decrease in sea-surface water alkalinity which controls calcium carbonate formation in deep waters, and the so-called coral reef hypothesis. This also concerns calcium carbonate, notably its precipitation in shallow water which would have been reduced as sea levels rose. Both of these possibilities have their drawbacks and although they are feasible, it is unlikely that they are entirely responsible for changing atmospheric concentrations of carbon dioxide. Along with data from other ice and ocean-sediment cores, this has led Charles et al. (1996) to conclude that climatic fluctuations in the southern hemisphere preceded those of the northern hemisphere by ca. 1.5 K years.

Speculation on this aspect of glacial–interglacial cycles has been rife; apart from the crucial issue of what causes or contributes to climatic change, the ice core data have now raised the issue of mechanisms within the global carbon cycle. In the spirit of Gaia (Section 1.3), the Earth's biota is involved in this regulation of the atmosphere. In this context, Charlson et al. (1987) have suggested that oceanic phytoplankton populations can contribute to the regulation of climate because they produce dimethyl-sulphide (DMS). If phytoplankton productivity increases, as might occur during glacial periods (see below), DMS productivity will increase with a concomitant increase in cloud cover. This would cause an increase in the reradiation of solar energy into the upper atmosphere and so contribute to the reduction of temperature and the maintenance of low temperatures at the Earth's surface. Thus, enhanced marine productivity could be an important mechanism in the depletion of the atmospheric carbon pool. There remains the question of what happens at the glacial–interglacial transition to release this carbon from its repository in ocean sediments, in which it becomes incarcerated as phytoplankton die. However, a further controversy has been generated by the inferred role of phytoplankton in climatic change. This hinges on nutrient availability in ice-age oceans because nutrient availability is a major limiting factor to phytoplankton growth in modern oceans. The controversy was originated in the late 1980s by Martin (reviewed in Martin, 1990) who suggested that the abundance of dust blown from continental regions during the ice ages increased the amount of iron (and probably other nutrients) in ocean waters. This would have stimulated primary production, depleting atmospheric carbon dioxide. It is not yet possible to

confirm Martin's views because there is conflicting evidence from experiments recently conducted to test the effect of iron enrichment on phytoplankton populations. Martin et al. (1994) report positive results whereas Watson et al. (1994) report negative results. Moreover, there is conflicting evidence from ocean-sediment cores for increased productivity during glacial stages (Section 2.3). For example, Lyle (1988) presents evidence for increased organic carbon accumulation during glacial episodes in equatorial ocean sediment cores whereas Kumar et al. (1993) and Singer and Shemesh (1995) show that productivity in several Southern Ocean core regions was lower than at present. However, there is evidence for high DMS production during glacial periods as concentrations of methane sulphonic acid (MSA), an oxidation product of DMS, have been determined in several ice cores, e.g. Dome C and D10 cores from Antarctica (Saigne and Legrand, 1987), in which concentrations are between two and five times higher than today. Increases of non-sea salt sulphate of 20–46 per cent in the Vostok core are also associated by Legrand et al. (1988a) with biotically produced DMS. The changes in the atmospheric concentration of methane are equally difficult to explain. There is probably some relationship with the extent of, and gaseous emissions from, global wetlands (Lorius and Oeschger, 1994), which may also influence carbon dioxide concentrations. Indeed Franzen (1994) has suggested that the decline and rise of peat accumulation is a major control on glacial–interglacial cycles.

Figure 2.3 illustrates a number of other parameters which have been determined for the Vostok core (De Angelis et al., 1987; Legrand et al., 1988b). They include ions such as sodium, to identify marine inputs such as sodium chloride and sodium sulphate, and aluminium, a surrogate for continent-derived dust as well as indicators of nitric, sulphuric and hydrochloric acids. Over the 160 K years represented by this core, indicators of both marine and terrestrial inputs achieved high concentrations during periods of cold climate, especially between 110 and 15 K years ago (the last glacial period) and they were considerably reduced during the last and present (Holocene) interglacials. These increases reflect arid conditions on nearby continents and exposed continental shelf areas that were subject to aeolian deflation. High concentrations of marine salts may reflect enhanced wind circulation from the ocean to the ice sheet. However, the concentrations of acids show no such variation. They relate to periods of volcanic activity and, since there is no obvious relationship with changes in climate, it would appear that volcanism and climate are not related in the long term, though in the short term, i.e. decadal scales or less, there is abundant evidence to show that volcanic eruptions do indeed lead to localised but short-lived, cooling.

2.5 Tree rings, historical and meteorological records

There are many sources of information on climatic change which are temporally and spatially fragmentary. Tree rings and historical records are particularly valuable because they provide evidence for the historic period before the collection of meteorological data. Meteorological records are important because they represent actual records, and often the conditions under which the data were collected are also documented. This allows correction to make them compatible with modern records. There is a vast literature on each of these categories and the following examples illustrate the contribution each can make to studies of environmental change.

The determination of past climates from tree-ring sequences, i.e. dendrochronology and dendroecology, rests on the variations that occur between annual rings. Parameters that can be measured include ring width, the density of latewood, isotopic analyses of the cellulose, e.g. carbon and oxygen isotopes, and metal ions. Thus, tree rings can provide a dating control as well as palaeoclimatological or palaeoenvironmental data; see Schweingruber (1988) for details of techniques, etc. There have been many applications throughout the world in a wide range of palaeoenvironmental contexts. Among the longest tree-ring chronologies available are those from northwest Europe (Becker, 1993) and the southwest United States (Fritts, 1976). However, most records used for reconstructing climate in detail tend to relate to the past 1000 years or so. Hughes *et al.* (1994) have reconstructed rainfall in north-central China since AD 1600 from tree-ring density and width of *Pinus armandii*. Their results are given in Figure 2.4, which also gives a regional wetness/dryness index derived from documentary evidence and shows a high degree of agreement between the two records. In particular, two major droughts occurred: the first was in the 1680s and the second was in the 1920s. Jones *et al.* (1995) have used tree-ring data from 97 sites in North

America and Europe to produce a chronology of the occurrence of cool summers since 1600. Their results are summarised in Figure 2.4, which reflects five extreme low-density years. Of these, four are known to follow major volcanic eruptions and the fifth is likely to have succeeded an eruption. The index for 1601 shows a larger deviation from normal than for any other of the cool years and was probably caused by the eruption of Huaynaputina in Peru. These data reflect the significance of volcanic eruptions in determining average annual temperatures in the years following the eruption.

In the southern hemisphere, Villalba (1994) has reconstructed climatic fluctuations in the mid-latitudes of South America based on tree-rings and glaciological records. Periods of above average wetness and dryness are given in Figure 2.4 along with periods of above average and below average temperature. The warming trend is possibly confined to the northern hemisphere. Norton and Palmer (1992) have also discussed dendroclimatic evidence from Australasia for the period AD 1500 to the present. They show, for example, that since 1750 there have been fluctuating summer temperatures with each warm or cool period lasting about 10 years. However, a study has recently been published by Briffa *et al.* (1995), who have presented a 1000 year temperature record from Siberia. The reconstructed mean summer temperature of the period 1901–1990 is higher than for any other period since AD 914. This is positive evidence for global warming, though Lara and Villalba's (1993) analysis of *Fitzroya cupressoides* (alerce) tree rings from southern Chile does not show evidence for a warming trend in the last few decades.

In relation to historical records, there are three categories of information: observations on weather events such as the frequency of spring frosts, the recording of events such as droughts or floods and records based on the phenological rhythms of certain plants e.g. the dates of first flowering, especially of cherry blossom in Japan. Using a wide range of sources, Currie (1995) has analysed 81 long and 202 short dryness/wetness indices compiled from more than 2200 local annals and other writings by China's Central Meteorological Institute for the period 1470 to 1979. The dry and very dry indices are given in Figure 2.5, which shows how the years 1633 to 1643 experienced severe droughts; these were perhaps responsible for the fall of the Ming Dynasty in 1644. Note that Figure 2.4a also shows a period of drought during this period. In addition,

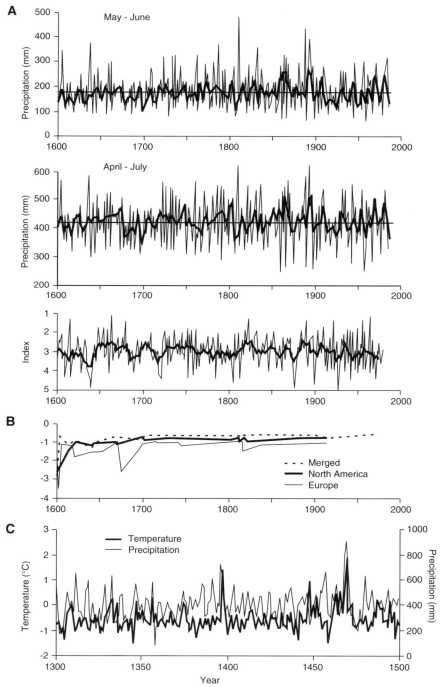

Fig. 2.4 (A) Tree-ring evidence for precipitation patterns in China since AD 1600 (adapted from Hughes *et al.*, 1994). (B) A chronology of cool summers derived from tree-ring records in North America and Europe (data from Jones *et al.*, 1995). (C) Summer temperatures and precipitation patterns in the mid-latitudes of South America during the past 1000 years reconstructed from tree rings (based on Villalba, 1994).

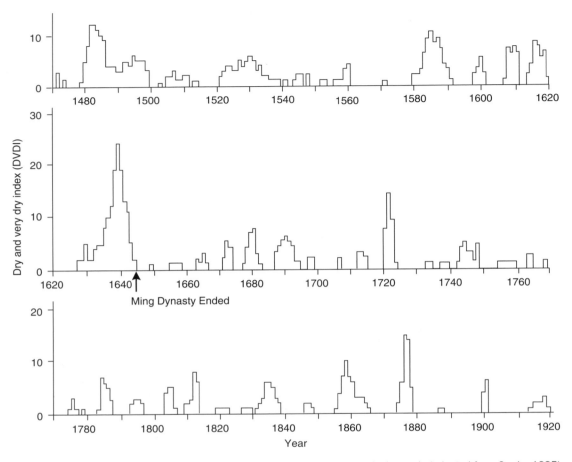

Fig. 2.5 Dry and very dry indices (DVDI) of climate in China derived from historical records (adapted from Currie, 1995).

Currie suggests that the 18.6 year lunisolar cycle and the 10–11 year solar cycle (changes in sunspot numbers) are strongly represented in these records, as in other records such as US crop yields, dates of wine harvests, etc. (Currie *et al.*, 1993).

Pfister (1992) has used documentary evidence, such as weather diaries, weather reports compiled by clergy, dates for the formation and melting of snow cover, phenology, etc., to reconstruct monthly temperature and precipitation in central Europe for the period 1525–1979. He has drawn a variety of conclusions from this study which utilised a total of 33 000 records. For the period 1525–1900 the winter and spring months tended to be colder and drier than those for 1900–1979. The climate before 1900 was also variable, especially around 1600, with pronounced extremes. In recent years the

temperatures in all seasons have been generally higher than the mean for 1901–1960, against which all the temperature and precipitation values are compared. Pfister states, 'Compared to the long-term average of the past half millennium this season [winter] has been 1.3 °C warmer and 25% wetter since 1965. The recent succession of three warm winters (1987–88 to 1989–90) with almost no snow-cover in the lowlands is unique in the last seven hundred years.' Taken together, the tree-ring study of Briffa *et al.* (1995) and Pfister's (1992) data do indeed indicate a warming trend in the northern hemisphere. Pfister's (1992) study also included phenological data based on the flowering of both cherry and vine, the beginning of the rye harvest and the vine harvest. For example, the earlier vines flowered, the earlier the vine harvests, as occurred in

1636–1638 and in seven other years during the period 1525–1979. Pfister also reports that in 1990 sweet cherries began flowering around 20 March. This, he suggests, is equivalent to the earliest springs ever known.

Records of streamflow may also be used to detect climatic change and, on the basis of variability, to manage water resources. Pupacko (1993) has examined data for an eastward-draining and a westward-draining stream in the Sierra Nevada of the USA for the period 1939–1990. There was an increase in the variability of winter streamflow in the mid-1960s when mean monthly values between December and March increased in comparison with the period 1939–1964. There has also been an overall increase in streamflow during the winter and early spring which, Pupacko suggests, could be due to a small increase in temperature causing the snowpack to melt early in the season. There is also evidence to indicate that the timing of snowmelt on the western slopes is particularly responsive to changes in temperature because the snow lie occurs at lower altitudes than on the eastern side. Basins on the eastern side of the mountains thus enjoy a more reliable water supply during periods experiencing a warming trend. However, a similar study based on streamflow statistics of 30 unregulated Australian rivers reveals an equivocal evidence for a warming trend (Chiew and McMahon, 1993). However, since the interannual variability of Australian streams is double that of streams in the northern hemisphere, a large change is required in the Australian variability in order to pinpoint a significant change.

The most reliable records of climatic change are, however, derived from instrumental records. Unfortunately, such systematic records do not generally extend very far back in time, and vary in extent on a regional basis. An important factor in studies of global warming is the question of natural variability of both temperature and precipitation. Sequences of recorded data can contribute to quantifying this variability so that signals of global warming can be readily identified. Plaut *et al.* (1995), using a 335 year central England temperature record have identified two climatic oscillations with interannual frequencies, i.e. 7–8 years, and interdecadal, i.e. 15–25 years. The interannual frequency is likely to be related to wind-driven circulation in the North Atlantic, whereas the interdecadal frequency reflects thermohaline circulation. Forecasts based on the data derived

from the recorded measurements indicate that temperatures in central England will rise for 1995–96 and will decrease as the year 2000 approaches; they will then rise into the middle of the following decade before a decrease by 2010. Plaut *et al.* state that 'The predicted oscillations exceed by far any local greenhouse warming effect expected by the year 2010.' This reflects the difficulties of both detecting and predicting the impact of global warming. Rowell *et al.* (1995) have also attempted to determine variability, but in precipitation rather than temperature, for tropical north Africa using instrumental data for 1906–1992. Apart from the issue of global warming, the Sahel, one of the three regions investigated in this study, is particularly susceptible to drought so improvements to precipitation prediction are vital. The results show that the variability of seasonal rainfall fluctuates on timescales of 5 years or less. The underlying cause of these fluctuations is sea-surface temperatures (SSTs) with minor contributions from land-surface feedbacks including soil moisture and internal atmospheric conditions.

The importance of SSTs as a measure of global temperature change between glacial and interglacial periods is considered in Section 2.3 and it is clear, from studies such as those mentioned above, that they play a major role in influencing continental climates. Recent work by Parrilla *et al.* (1994) and Jones (1995) indicates a warming trend in both SSTs since the 1950s. For example, Jones has examined data (collected since 1957) from 20 stations, 17 of which exhibit an increase in mean temperatures with an average increase of 0.57 °C for the 37 year period, all of which occurred before the early 1970s. The warming has been spatially greatest in the region of the Antarctic peninsula. Parrilla *et al.* (1994) collected temperature data along a transatlantic hydrographic section in 1992 and compared their data with two similar sets of data collected in 1957 and 1981. Their analyses show that waters between 800 and 2500 m depth have persistently warmed over the period of records. The most pronounced warming occurred at 1100 m depth and at an extrapolated rate of 1 °C per century. Such warming trends may well have begun earlier if instrumental records from European nations are to be believed. For example, Wheeler (1995) has discussed some of the oldest instrumental records from Spain which come from Cadiz and San Fernando. For the period 1789–1816 there is evidence for 2 °C cooling for the years 1805–1816, possibly due to volcanic eruptions producing dust veils. Since 1816, however, a warming trend is indicated by the data. The possibility should not be overlooked that this and other land-based records of temperature reflect a warming trend due more to the urban heat island effect than to 'real' warming.

2.6 Causes of climate change

Fluctuations and changes in climate occur both spatially and temporally, the causes of which are a source of much speculation and controversy. What is unequivocal is that the past 2 to 3×10^6 years (and more) have been characterised more by change than by constancy. It is equally apparent that climatic change, whether it is a response to natural or cultural stimuli, is complex. It is not yet understood which factors, either singly or in combination, create positive feedback, nor is it understood how they interact. Moreover, because the only indices of climatic, and indeed environmental change, for the past 2 to 3×10^6 years are proxy records the task of identifying underlying causes is formidable.

Numerous theories concerning climatic change were advanced during the last century. Several have emphasised changes in the quantity and quality of solar radiation, especially in relation to sunspot cycles. Section 2.5 mentioned Currie's (1995) identification of the 18.6 year lunisolar cycle and the 11 year solar cycles in Chinese dryness/wetness indices, for example. Sunspot maxima and minima have been associated with floods, droughts, poor harvests, etc. Labitzke and van Loon (1988) provided evidence that the 11 year sunspot cycle is linked to variations in surface weather via the quasi-biennial oscillation (QBO). The QBO is an oscillation of the zonal wind component in the stratosphere above the equatorial region with a periodicity of ca. 27 months (Maunder, 1994). Using data for the past 36 years, Labitzke and van Loon have shown there are positive correlations between warmer winters during the Sun's more active periods and between colder winters when the Sun is least active and when the QBO is in a westerly direction. The reverse of these trends occurs when the QBO is in its easterly phase. The implications of these trends in relation to surface weather involves their effects on pressure systems. For example, the higher temperatures of the westerly QBO phase, when the

Sun is most active, help to create higher pressures over North America and lower pressures over the adjacent oceans. Since pressure differentials control air movements, which in turn control surface temperatures, the resulting northerly airflows over the Atlantic seaboard of the United States bring colder and more frosty weather than usual. This trend is confirmed by van Loon and Labitzke's examination of winter temperatures for Charlston. However, it has been suggested that solar forcing may not be important in creating such weather patterns (James and James, 1989). Consequently, sunspot activity is still not generally considered respectable as a major cause of climatic change. Nevertheless, it is unlikely that changes in solar irradiance have been completely without impact. The ^{10}Be record in polar ice cores, for example, is considered to be a proxy for solar output and, as Figure 2.2 illustrates, there are clear relationships between changes in ^{10}Be concentration and other temperature indicators. In addition, Beer *et al.* (1990) have linked ^{10}Be deposition with the 11 year sunspot in the Dye 3 ice core from Greenland. Beer *et al.* (1990) state that increased levels of ^{10}Be occur when solar activity declines; and because the intensity of the solar wind is reduced there is an increase in the generation of cosmogenic isotopes such as ^{10}Be and ^{14}C. Not only is there a relationship, albeit complicated by the effects of precipitation, between the ^{10}Be in the Vostok ice core and temperature change, but there is also a possible relationship between ^{14}C concentrations and fluctuations in glaciers (Wigley and Kelly, 1990). The nature of this relationship and the way it varies have yet to be determined; for now, changes in solar irradiance, alias sunspots cycles, remain as enigmatic as ever.

The quality and quantity of solar radiation reaching the Earth's surface are affected by variables besides those mentioned above. In particular, volcanic eruptions emit dust and sulphate aerosols into the atmosphere; the dust scatters and partially reflects incoming solar radiation whereas the aerosols act as cloud-condensation nuclei. Both cause reduced temperatures for short-lived periods unless the volcanic eruptions are very large. Jones' (1995) work on tree-ring data to produce a chronology of cool summers since AD 1600, most of which were due to volcanic eruptions, has been referred to in Section 2.5. The longest record of volcanic activity so far reported is that of Zielinski *et al.* (1994, 1996) from the GISP2 Greenland ice core (Section 2.4), in which a number of major eruptions left a signature in the ice as acidity. In particular, they note that several substantial eruptions in the seventeenth century may have contributed to cooling during the 'Little Ice Age'. Zielinski *et al.* (1995) have also compared the ice-core evidence, i.e glass shards, for the eruption of Eldgjà in Iceland, dated to AD 938 ± 4, with proxy records of climatic change from the region. They were unable to find evidence in, for example, the tree-ring record for cooler climates in the years following the eruption. This is in contrast to proxy evidence for reduced temperatures following the eruption of Laki in 1783 for which surface-temperature records and the ice-core record are in agreement. Zielinski *et al.* suggest that either Eldgjà's eruption occurred over a longer period of time than for Laki, so its impact was reduced, or that Laki's eruption occurred during a period of climatic vulnerability as the tree-ring record shows that climate was experiencing a cooling trend prior to Laki's eruption. There are other mechanisms whereby volcanic eruptions may influence climate. For example, the enhanced volumes of ejecta, notably dust, to the atmosphere may have repercussions in the oceans. Some of the dust will settle into the water body, providing nutrients such as iron and other cations, which may stimulate primary productivity in marine phytoplankton (Section 2.4 discusses the iron fertilisation hypothesis). Their uptake of carbon dioxide could reduce its concentration in the atmosphere and contribute to global cooling by diminishing the greenhouse effect.

All of the factors outlined above probably contribute to climatic change via internal adjustments within the climatic system. Nevertheless, it is unlikely that any of them are major forcing factors which create sufficient positive feedback to effect change at the magnitude of a glacial–interglacial swing. It is now widely accepted that astronomical forcing, the Milankovitch theory of Section 1.2, is the most important primary cause of Quaternary glacial–interglacial cycles and probably those of earlier geological periods. The details of the theory have been examined by many workers, including most recently Schwarzacher (1993) and House (1995). It was, however, brought to the attention of the scientific community by Hays *et al.* (1976), who recognised the three main cycles of Milankovitch in the oxygen isotope records of ocean-sediment cores (Section 2.3). These three components and their periodicities are illustrated in Figure 2.6. It is the change in the orbital

A Orbital eccentricity

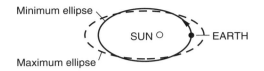

Periodicity of ca. 100 K years

B Axial tilt

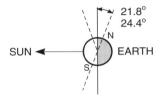

Periodicity of ca. 42 K years

C Precession of the equinoxes due to the wobble of earth's axis

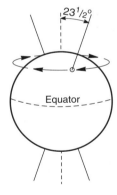

Periodicity of ca. 21 K years

Fig. 2.6 Astronomical forcing factors involved in the Milankovitch theory of climatic change.

eccentricity that is thought to drive the glacial–interglacial cycle. The periodicities of axial tilt and the precession of the equinoxes influence the pattern of stadials (phases of maximum ice advance) and interstadials (phases of relatively short-lived warming). These cycles influence the amount of solar radiation received at the Earth's surface, especially in the high latitudes of the northern hemisphere. Surprisingly, however, orbital eccentricity has the least effect on insolation. Other factors must therefore contribute to cooling either as forcing or reinforcing factors. Many of the possibilities have already been discussed in Sections 2.3 to 2.5 and above. Obvious candidates for consideration are the greenhouse gases, notably carbon dioxide and methane; Figure 2.2 shows how changes in their atmospheric concentration parallel global cooling and warming. There may also be a relationship between ocean circulation, atmospheric concentrations of greenhouse gases and global temperature change (Mannion, 1992a). The production and dampening of North Atlantic Deep Water (NADW) in particular is considered to be a mechanism whereby temperature change over the Arctic ice cap is translated into global change (e.g. Charles and Fairbanks, 1992; Keigwin *et al.*, 1994).

A synthesis of North Atlantic ocean-core evidence by Ruddiman and Raymo (1988) has confirmed the operation of Milankovitch cycles and has shown that the past 3.5×10^6 years were affected by three distinct climatic regimes. During the earliest period between 2.47 and 3.5×10^6 years BP, when ice sheets in the northern hemisphere were small or non-existent, there were only small-scale quasi-periodic oscillations. Between 2.47 and 0.734×10^6 years BP the rhythmic climatic changes were controlled mainly by the 41 K year cycle of orbital obliquity, and during the most recent period (0.734×10^6 year to date) larger-amplitude climatic changes resulted from 100 K year cycles controlled by orbital eccentricity. These changes firstly produced the initiation of glaciation in the northern hemisphere at ca. 2.47×10^6 years BP and secondly produced an intensification of glaciation at 0.734×10^6 year BP. What is not clear, however, is why changes in these regimes should have occurred. In particular, what caused the onset of glaciation? In a series of papers, Raymo and associates (e.g. Raymo *et al.*, 1988; Raymo and Ruddiman, 1992) have advanced the idea of tectonic uplift coupled with increased weathering. They cite the uplift of the Tibetan and Colorado plateaux in the Pliocene. This could have altered the planetary wave structure so that the jet stream directed cold air from high to mid latitudes in Europe and America. This would have cooled the northern hemisphere landmasses and so enhanced their sensitivity to changes in insolation caused by

astronomical forcing. However, an additional mechanism is necessary to explain the removal of carbon dioxide from the atmosphere, another forcing factor. Raymo and associates suggest this is allied with chemical weathering rates; as chemical weathering increases, carbon dioxide is consumed from the atmosphere in a series of reactions which can be summarised as follows:

$$CaSiO_3 + CO_2 \underset{\text{chemical weathering}}{\rightleftharpoons} CaCO_3 + SiO_2$$

Carbon dioxide dissolved in groundwater attacks silicates in rocks to produce bicarbonate. This enters the oceans in drainage water and is incorporated into the calcium carbonate shells of marine organisms (Ruddiman and Kutzbach, 1991). Other possibilities (Raymo, 1994) include the closure of the Panamanian Isthmus, which would have influenced oceanic circulation as would the subsidence of the Bering Strait. The issue of what started the ice ages thus remains enigmatic.

2.7 Environmental change in high latitudes

Areas of high latitude are those which have been most directly affected by the advance and retreat of glaciers and ice caps. Indeed, the Arctic and Antarctic zones are currently experiencing glaciation, and it is from these areas that much can be learnt about glacial processes. However, in latitudes adjacent to the polar ice masses, the terrestrial record of glaciations that occurred before the last major ice advance is frequently blurred or non-existent because the erosive power of ice is such that it will destroy earlier stratigraphic evidence.

Donner's (1995) recent synthesis of the Quaternary history of Scandinavia indicates there is clear evidence for three major cold stages: the Elsterian, Saalian and Weichselian (Table 2.2). The Weichselian is the most recent but least extensive of the three. Although there is some evidence for earlier glaciations, notably the pre-Cromer, it is too fragmentary to allow a reconstruction of the extent of the ice. Nevertheless there is abundant evidence from North Sea basin and North Atlantic sediments for alternating warm and cold periods well before the pre-Cromer (Section 2.3). In addition, there are several terrestrial sequences which represent interglacial periods. At Harreskov, for example, in western Jutland, there are freshwater sediments that

Table 2.3 Correlation of Quaternary deposits in Siberia with the Quaternary stages of northern Europe

Glacial complex (Gc) or interglacial (G)	Northern Europe	Siberia		Deposits in the extraglacial zone
			Sartan glacial advance	Yel'tsovka loess
Gc	Weichselian	Zyryanka	Karginsky interstadial complex	Iskitim soil complex Tula soil complex
			Early Zyryanka glacial advance	Loess
IG	Eemian	Kazantsevo	Taz glacial advance	Berd soil
Gc	Saale complex	Bakhta	Shirta interstadial	Suzun loess
			Samarovo glacial advance	
IG	Holstein	Tobol		Lacustrine and fluvial deposits
			Late Shaitan glacial advance	Loess
Gc	Elster complex	Shaitan	Unnamed interstadial	Loess deposition on Ob Plateau
			Early Shaitan glacial advance	
IG	Cromer	Talagaikino		Loess deposition on Ob Plateau

Source: based on Arkhipov *et al.* (1986a, b)

are tentatively ascribed to the Cromerian interglacial which was succeeded by the Elsterian cold stage. There are also freshwater sediments of Holsteinian age and Eeemian age in Jutland, at Vejlby and Hollerup respectively. In the central areas of Scandinavia, however, there is evidence for only one glacial stage. This was the Weichselian, the last glacial stage, which eroded evidence for the earlier warm and cold stages as much of Scandinavia was engulfed in an ice sheet. There are a few sites of Eemian interglacial age such as Leveäniemi in Swedish Lapland and Fjösanger near Bergen, Norway.

In Siberia, Arkhipov *et al.* (1986a) have recognised three periods of ice advance: the Shaitan, Bakhta and Zyrianka. These are given in Table 2.3 and are each characterised by two glacial advances separated by interstadial periods. The Tobol and Kazantsevo interglacial periods are represented by marine facies in northern Siberia that have been correlated with Holsteinian and Eemian facies in northwest Europe on the basis of foraminiferan fossil assemblages. According to Arkhipov *et al.*, these correlations are also justified by a comparison of mammal-bearing deposits in Siberia with those in western Europe. However, not all of Siberia experienced ice cover

during the glacial stages. In those areas peripheral to the ice sheets, periglacial conditions prevailed, characterised by sparse vegetation leaving skeletal soils vulnerable to wind erosion. As a consequence, loess, an aeolian deposit consisting mainly of silt, accumulated in extraglacial regions during cold stages; whereas soil complexes developed in the intervening warm stages. Arkhipov *et al.* (1986b) have related such deposits to the various glacial and interglacial stages (Table 2.3). Lozhkin and Anderson (1995) have recently described last interglacial deposits from northeast Siberia (between the East Siberian Sea and Bering Sea) where they have investigated alluvial, fluvial and organic deposits in river terraces in the valleys of Bolshoii KhomusYuryakh and Bolshaya Kuobatkh-Baga. During this period there was a 600 km northwestward expansion of larch (*Larix dahurica*) forests and a northwestward extension of birch (*Betula ermani*) as compared with the present. The evidence indicates July temperatures 4–8 °C warmer than today.

The records in the Greenland and Antarctic ice cores also reflect environmental change in high latitudes, as discussed in Section 2.4. Funder *et al.* (1994) have summarised extensive evidence from

stream-cut sections and cliffs in Greenland for the
past 240 K years. Within this period there were
three major glacial stages, the last of them being the
Weichselian, during which there is evidence for at
least two interstadials. During the Langelandselv
interglacial (equivalent to the Eemian; see Table 2.2)
botanical evidence and insect remains indicate a
summer temperature some 2–3 °C higher than
during the Holocene optimum. The higher
temperature of this interglacial is also reflected in
the evidence for it from middle latitudes as well as
from ocean-sediment cores and ice cores (Sections
2.3 and 2.4). Evidence for environmental change
during interglacial stages 1, 5, 7, 9 and 11 has been
presented by Eide *et al.* (1996) uing a core from the
Iceland Sea. Their analysis of coccolithophores and
planktonic foraminifera indicate that 5e, the
Eemian, was the warmest in accordance with the
Greenland evidence.

2.8 Environmental change in middle latitudes

Large areas of the temperate zone of the northern
hemisphere have been affected by glacial advances
and retreats during the past 2 to 3×10^6 years. The
same is likely to be true of the equivalent latitudinal
zones in the southern hemisphere, though the
absence of extensive landmasses in this region
means there is only a limited source of information
from terrestrial deposits. Although there is a wealth
of information from middle latitudes in the northern
hemisphere, the volume of information from the
southern hemisphere is increasing substantially,
especially from Australasia.

In terrestrial sequences, one of the most complete
records derives from the Netherlands and this has
recently been reviewed by Zagwijn (1992) in the
context of the European Quaternary period (defined
by Zagwijn as the past 2.3×10^6 years). He points out
that the early interglacial stages which are older than
1×10^6 years are distinct from the later interglacials.
The earlier interglacials do not show a clear succession
of tree species representing the immigration of forest
species; this is because most of the component species
appear to have been present throughout the warm
stage. This probably reflects higher overall
temperatures within a cold–warm cycle before the
amplification of temperature differences. From
ca. 1×10^6 years BP (above oxygen isotope stage 23)

interglacial deposits exhibit an obvious succession of
forest species as they immigrate, spread then decline.
Zagwijn opines that these middle and late Quaternary
interglacials can be categorised as either having an
oceanic climate associated with high sea levels or
having a cooler and relatively continental climate
associated with lowered sea levels (Section 2.10).
Additional evidence for the environment of 2.3 to
3.3×10^6 years ago has been presented by Rousseau
et al. (1995) from La Londe in Normandy, France.
Here pollen analyses from a 15 m core have revealed
initial warming then a general cooling trend,
culminating in a major cooling event ca. 2.4×10^6
years BP that corresponds with evidence for ice rafting
in the North Atlantic (Section 2.3). Emontspohl
(1995) has also provided evidence for vegetation
communities during the Brorup and Odderade
interstadials (Table 2.2) from Watten, Northern
France, where the two intervals are recorded in one
sequence.

Further long pollen sequences from middle
latitudes have been obtained from Greece
(Mommersteeg *et al.*, 1995), Japan (Fuji, 1986) and
Israel (Horowitz, 1989). As is the case elsewhere, the
first phase of major cooling in Israel occurred 2.4 to
2.6×10^6 years BP with subsequent cooling and
warming phases, in general agreement with those of
ocean-sediment cores (Section 2.3). The record from
Tenagi Philippon in Greece (Mommersteeg *et al.*,
1995) provides a ca. 1×10^6 year vegetation and
inferred climatic history with changes corresponding
to Milankovitch's orbital forcing, eccentricity and
obliquity (Section 2.6). Tzedakis (1993) has also
presented evidence from Ioannina, Greece, for a 430
K year record of vegetational and climatic history.
In addition to the parallels with other long records,
Tzedakis uses the data to discuss the refugia concept
and application to the recolonisation of glaciated
and periglaciated regions in Europe after the retreat
of ice sheets. On the basis of the pollen evidence for
the presence of temperate species throughout the
sequence, it is suggested that the southern
populations of trees provide source material for
subsequent expansion. Thus, the traditional concept
of a refugium, i.e. an isolated population of a group
of species, is no longer acceptable.

The Quaternary history of the British Isles has
been summarised by Bowen *et al.* (1986a, b) with an
update by Bowen (1994). The various stages are
given in Table 2.4 (see also Table 2.2). Bowen states
that 'the Early Pleistocene was dominated by ice-
ages with a 41 k year (orbital tilt) frequency. These

Table 2.4 Stages of the British Quaternary in relation to oxygen isotope stages

	STAGE NAMES			GLACIAL EPISODES	INTERGLACIALS AND INTERSTADIALS	^{18}O STAGES	APPROXIMATE DATES (10^3 YEARS)
LATE QUATERNARY	FLANDRIAN					1	10–11
	DEVENSIAN	LATE		LOCH LOMOND GLAC.		2	
					WINDERMERE I.S.		14
				DIMLINGTON GLAC.			25
		MIDDLE				3	50
		EARLY				4	70
					UPTON WARREN I.S.	5a	
						5b	
					CHELFORD I.S.	5c	
						5d	
	IPSWICHIAN				IPSWICHIAN I.G.	5e	122
MIDDLE QUATERNARY	WOLSTONIAN COMPLEX			(RIDGEACRE)		6	128–132
					MARSWORTH I.G.?	7	186
						8	245
	HOXNIAN				HOXNIAN I.G.	9	297–302
						10	440–428
					SWANSCOMBE I.G.	11	
	ANGLIAN			ANGLIAN		12	423
	CROMERIAN				CROMERIAN I.G.	13	478
						14	
					WAVERLEY WOOD I.G.	15	
						16	
						17	
						18	
						19	790
						20	
						21	
						22	
						23	
							900
	BEESTONIAN				C		1600
EARLY QUATERNARY	PASTONIAN				T		
	PREPASTONIAN				C		
	BRAMERTONIAN				T		
	BAVENTIAN				C		
	ANTIAN				T		
	THURNIAN				C		
	LUDHAMIAN				T		
	PRE-LUDHAMIAN						2300

Source: based on Bowen *et al.* (1986a), Bowen (1994) and Šibrava (1986)
I. G. = interglacial
I. S. = interstadial
T = temperate
C = cold

were of relatively high frequency and low amplitude. But what was mainly mountain glaciation was not on the scale of later Middle and Late Pleistocene advances.' Much of the evidence for the early Pleistocene derives from the sediments and terraces of the Thames (e.g. Whiteman and Rose, 1992; Bridgland, 1994). Within the terraces of a large river that flowed from the Midlands to East Anglia, at Waverley Wood, Warwickshire, there are organic deposits which have been ascribed to oxygen isotope stage 15 (Bowen et al., 1989; Shotton et al., 1993), i.e. probably part of the Cromerian complex which encompasses oxygen isotope stages 19 to 13. Evidence for this period is sparse and, as Bowen (1994) points out, there is much dissension as to the course of events during and immediately after the Cromerian complex. Table 2.4 indicates that a major glaciation, the Anglian, corresponds with oxygen isotope stage 12. Deposits of Anglian age are well documented and it is generally considered to have given rise to the most extensive ice coverage of the British Isles. It also caused the Thames to divert from the Vale of St Albans to its present course.

Until recently it was generally considered that two interglacials, including the Holocene, post-dated the Anglian glaciation. These were the Hoxnian and Ipswichian, named after type sites at Hoxne and Ipswich in East Anglia. There is now evidence (Section 2.2) for four interglacials; the two additional interglacials are Swanscombe and Stanton Harcourt, the latter being contemporaneous with the controversial site of Marsworth (Section 2.2). Swanscombe comprises fluvial and brackish water deposits in the Thames Valley; the Stanton Harcourt site has been briefly described in Section 2.2 All contain evidence of flora and fauna that indicate temperate conditions. The pollen assemblages from the sites of interglacial ages reflect the interglacial cycle of vegetation and soil development described in Section 3.2. With some exceptions, notably *Abies* (fir) and *Picea* (spruce) – sometimes present in abundance during the earlier interglacials but which are not components of the present interglacial flora – the tree species of interglacial times were the familiar *Quercus* (oak), *Ullmus* (elm) and *Tilia* (lime), etc. The same is not true of fauna. Remains of animal bones indicate the presence of rhinoceros, horse, hippopotamus, etc. The sites and their flora and fauna have been described in detail by Jones and Keen (1993), as has the evidence for the glacial periods separating these

interglacials and the interstadials of the Devensian, the last glacial period. As is discussed in Section 3.3, there is evidence, not only in Britain and Europe but also in many other parts of the world, for an additional short-lived cold period just before the Holocene and known as the younger Dryas. In high and some middle latitudes this period was characterised by the recrudesence of upland glaciers, etc. (Sections 3.3 to 3.5).

In the southern hemisphere there is evidence for environmental change from a variety of sources including the Vostok ice core (Section 2.4) and a number of ocean-sediment cores (Section 2.3), which provide the longest records, e.g. Heusser and van de Geer (1994) mentioned in Section 2.3. Pillans (1994) has reported on a sedimentary sequence comprising alternating terrestrial and shallow marine sediments from Wanganui Bay, south North Island, New Zealand, which covers the entire Quaternary period. Using a range of techniques to reconstruct palaeoenvironments and a variety of dating methods, Pillans has developed a comprehensive picture of environmental change. This is augmented by correlations, facilitated by tephras, with non-marine sediments including loess, dune sand, and alluvial deposits and ocean sediments. The information is summarised in Table 2.5. This is not only a remarkable record but it also reflects the value of a multidisciplinary approach to establishing the nature of environmental change. The data show that during the interglacial stages high sea levels caused the formation of marine terraces, river terraces and the deposition of dune sand. In addition, Williams (1996) has presented a 230 K year record of glacial and interglacial environmental change from Aurora Cave in New Zealand's Fjordland. During this period there were seven glacial advances, including five in the Otiran, the last major glacial stage.

Elsewhere in the southern hemisphere Colhoun et al. (1994) have compared glacial, terrestrial and marine pollen data that relate to Tasmania's environmental history extending back to oxygen isotope stage 5. This was characterised by wet forest and experienced a relatively warm wet climate prior to ca. 75 K years. In stage 4 there was an ice advance and areas beyond the ice had a herb and shrub vegetation indicating a cold climate. Subalpine woodland and shrubland was dominant in stage 3 whereas in stage 2 the last glacial maximum occurred between 25 and 19 K years BP. Moreover, there is no evidence for a late-glacial readvance of ice here, as there is in the United Kingdom, for

Table 2.5 Evidence for environmental change in New Zealand's Wanganui Basin

Age (K years)	Oxygen isotope stages	S. Taranaki–Wanganui marine terraces	Loess	Rangitikei river terraces	Loess	Coastal dune sand	Rhyolite tephra
0.0	1					Foxton	Taupo
	2		L1	Ohakea	✓		Kawakawa
	3	Rakaupiko	L2	Rata	✓	Huxley	Rotoehu
	4		L3	Porewa	✓		
100	5	Hauriri Inaha Rapanui	L4a L4b	Cliff Greatford	✓	Rapanui (U) Rapanui (L) Mt. Stewart	
200	6		L5 L6	Marton	✓		
	7	Ngarino Waipuna				Brunswick	
	8		L7 L8	Burnand	✓		
300	9	Brunswick Braemore	L9			Mt. Curl	Fordell Griffins Rd (U, M, L)
	10		L10	Aldworth ✓ Waituna? ✓			Rangitawa
400	11	Ararata					
	12	Rangitatau	L11				
500	13	Ball				Rangitatau East	
	14		L12?				
600	15	Piri					Onepuhi
	16						
700	17	Marorau					Kupe
	18						
	19						
800	20						
	21						Kaukatea
	22						
900	23						
	24						
	25						
	26						
	27						Potaka
1000	28						
	29						
	30						

Source: adapted from Pillans (1994)

example. The terrestrial record of environmental change reflected in oxygen isotope signatures for the past 70 K years in a Tasmanian speleothem also mirror those of the Greenland GISP core (Section 2.4), reflecting parallelism (Goede, 1994). On the continent of Australia there are few long records from middle latitudes, though D'Costa and Kershaw (1995) have presented a pollen and charcoal record dating back to ca. 75 K years BP from lake and swamp deposits around Lake Terang in western Victoria. The period ascribed to the last glacial period of ca. 50 K years BP is recorded at the base when the concentration of *Eucalyptus* pollen was high.

2.9 Environmental change in low latitudes

There is a vastly increased volume of information on environmental change from low-latitude regions compared with a decade ago. The most complete continental sequence is from the Funza I borehole (Hooghiemstra 1984, 1988, 1989) and the Funza II borehole, near Bogotá, Colombia (Hooghiemstra *et al.*, 1993; Andriessen *et al.*, 1994; Hooghiemstra and Ran, 1994). A summary of the pollen analyses from, and correlations between, these two sequences is given in Figure 2.7. In Funza I and II, 55 pollen assemblage zones and have been identified, providing a vast volume of information on vegetation changes throughout the Quaternary period. Dating of both cores has involved potassium–argon dating and fission track dating of zircon in volcanic ash horizons. Correlation between the cores is based on pollen assemblage zones and the Funza I stratigraphy has been related to the oxygen isotope stratigraphy of ocean-core sediments. At least 27 climate cycles have been identified based on changes in the height of the tree line. During warm or interglacial stages forest extended to an altitude of ca. 3000 m in the eastern Cordillera of Colombia, the mountains surrounding the lake basin and perennial snow was present only above ca. 4500 m. During cold or glacial stages the vegetation belts were depressed with forest extending to ca. 2000 m and perennial snow present above ca. 3500 m. Superparamo, grassparamo and subparamo, all types of grassland with herbs and shrubs, occurred between the tree line and snow line, and oscillated altitudinally in line with climatic change.

The dating control reflects a periodicity in vegetation change similar to the variations in oxygen isotope stratigraphy of ocean-sediment cores. This provides yet further independent evidence for the operation of astronomical factors in climatic change. An analysis of the pollen of species indicative of changes in water level, i.e. species of deep water and marsh environments, has also been undertaken as a proxy record of lake-level changes. Overall, during cold stages, the lake level was elevated in comparison with warm stages. This could reflect increased precipitation during cold stages or reduced rates of evaporation, though ice-core records, notably from Vostok (Section 2.4), suggest that cold stages were generally characterised by reduced rates of precipitation. The Funza cores suggest there was a ca. 9 °C temperature difference within a glacial–interglacial cycle at an altitude of ca. 2550 m (the height above sea level of the Funza sites), reflecting a change from ca. 6 °C to ca. 15 °C. This is in accordance with temperature reconstructions from the Huascarán ice cores in Peru (Thompson *et al.*, 1995). These reconstructions also show that the atmosphere of the last glacial stage contained 200 times more dust than it does today, reflecting increased continental aridity.

Further evidence for environmental change in low latitudes derives from ocean-core sediments off the northwest coast of Africa (Leroy and Dupont, 1994). A 200 m core from Ocean Drilling Programme site 658 provides a record spanning 3.7 to 1.7×10^6 years, which is correlated with oxygen isotope stages from ODP site 659, another drilling site in the east Atlantic, among others (Section 2.3). Cyclic fluctuations in pollen assemblages parallel oxygen isotope stages. By about 3.26×10^6 years BP, increases in *Ephedra* pollen indicate the development of arid conditions and may represent the commencement of the trade winds. This corresponds to the onset of glaciation in the northern hemisphere and indicators of trade winds increased in three steps until 2.5×10^6 years BP. Humid conditions alternated with arid conditions to generate latitudinal shifts of vegetation zones, notably wooded savannas and desert. Pollen-analytical data from six east Atlantic sites (Dupont, 1993) have also revealed evidence for changes in northwest Africa's vegetation belts during the last 0.7×10^6 years, i.e. during 16 oxygen isotope stages which represent five cold stages with intervening warm stages. During isotope stages 2, 5d, 6, 10 and 12 (Table 2.2) pollen productivity was low and arid conditions prevailed

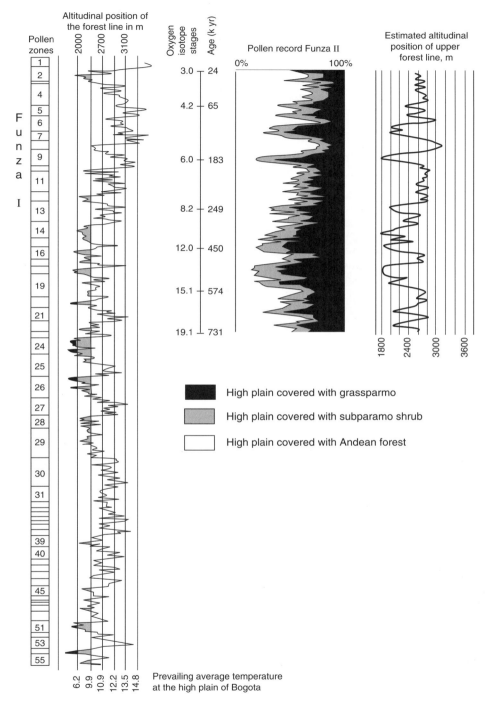

Fig. 2.7 Relationship between climatic change, as represented by marine oxygen isotope stages, and changes in the altitudinal position of the tree line in the High Plain of Bogotá, in the eastern Cordillera of Colombia, deduced from pollen cores Funza I and II (adapted from Hooghiemstra, 1989; Hooghiemstra and Ran, 1994).

as the southern boundary of the Sahara desert expanded, possibly to 14–15° N as compared to its ca. 20° N position today. In contrast, humid conditions developed during stages 8, 11 and 13, and pollen productivity increased substantially. The desert boundary moved north to ca. 20° N or higher, as did the adjacent savanna and rainforest to comprise vegetation belts similar to those of today. Similarly, on the Mediterranean coast of north Africa, cold stages were characterised by an expanded desert whereas Mediterranean-type vegetation was much reduced in extent. Dupont notes that the Sahara–Sahel boundary has been particularly sensitive to climatic change in the past.

Although there are few other continental records that extend beyond the last glacial period, there are numerous lake-sediment or mire records that cover its latter part. van der Hammen and Absy (1994) have examined pollen-analytical evidence from Rondonia, Carayas and Guyana in Amazonia, which shows that during the last glacial maximum (22 to 13 K years BP) savanna vegetation may have replaced the rainforest. They also suggest that the Amazonian forest may have been disjunct with a large area equivalent to today's west Amazonia and a number of other medium-sized areas. This may reflect a decline in rainfall of 500–1000 mm from a total of ca. 1000–2500 mm and a decline in temperatures of 2–6 °C as compared with those of today. In the Yucatan Peninsula, a core from Guatemala's Lake Quexil (Leyden *et al.*, 1994) also reflects a 6.5–8 °C decline in temperature relative to today, along with the development of highly arid conditions during the period ca. 25 to 13 K years BP. In common with the Amazonian data (van der Hammen and Absy, 1994), the Quexil data reflect a series of environmental changes during the time period equated with the last cold stage, i.e. oxygen isotope stages 2 to 5d, which shows that the terrestrial record in the tropics was sensitive to interstadial as well as stadial conditions. Changes of a similar magnitude are reflected in pollen assemblages from Lake Tulane in Florida (Grimm *et al.*, 1993; Watts and Hansen, 1994). In particular, changes in *Pinus* (pine) pollen appear to parallel climatic changes. Abundant *Pinus* reflects a wetter climate than intervening drier phases dominated by *Quercus* (oak) and *Ambrosia*-type (ragweed and marsh elder) pollen. Moreover, peaks in *Pinus* pollen correlate with the Heinrich events recorded in ocean-sediment cores (Section 2.3; see also Fig. 2.3). However, it is not clear how the production of

icebergs in the North Atlantic is related to climatic change, which in turn manifests itself as humid episodes favouring *Pinus*-dominated forests in Florida.

Evidence for glacial-stage environmental change derives from other parts of the high latitudes, including Indonesia and tropical Australia. Hope and Tulip's (1994) pollen analysis of a core from a lowland mire in northern New Guinea covers the past 60 K years. Throughout this time, montane forest has prevailed but there have been marked influxes of pollen from high-altitude taxa at certain times, e.g. 25 to 10.5 K years, approximating to the last glacial maximum. According to Hope and Tulip, a temperature change of 3–4 °C is indicated by such changes in pollen spectra. From nearby Australia, by drawing together evidence from offshore sediment cores and pollen analyses from two volcanic craters in the Atherton Tableland of Queensland, Kershaw (1994) has synthesised the evidence for environmental change during the past ca. 2×10^6 years. He concludes that the composition of the vegetation of the region has changed little during the Quaternary period except for the last ca. 140 K years BP. During this time, extensive moist rainforest was superseded by open *Eucalyptus*-dominated woodland, though there is some evidence that, because it was not synchronous between sites, this process may have been influenced by human activity, e.g. burning by Aboriginal people.

2.10 Sea-level changes

Most of the evidence for environmental change referred to above reflects the operation of astronomical (Milankovitch) factors in climatic change. Atmospheric processes not only affect the distribution and transport of heat energy but also the global hydrological cycle. Alternating cold and warm stages (glacials and interglacials) brought about major shifts in the global reservoirs of water. As ice sheets and glaciers waxed and waned, so too did the volume of water contained within the oceans, bringing about changes in sea level. Moreover, as extensive ice sheets accumulated in high latitudes, landmasses became depressed beneath their weight and subsequently rose as ice melted. Changes in sea level which are a consequence of variations in the amount of water in the oceans are termed eustatic changes, and those due specifically to ice-sheet formation and regression are termed glacio-eustatic, while isostatic movements in sea level are due to the Earth's crustal readjustment to ice accumulation or loss. There is abundant evidence for both eustatic and isostatic sea-level oscillations and, in high latitudes especially, it is often difficult to distinguish between the two; see the review of sea-level changes in Scandinavia in Donner (1995) and comments by Lambeck and Nakada (1992). In relation to ocean-sediment stratigraphy, global ice volume and the establishment of a general framework for environmental change during the Quaternary period, glacio-eustatic sea-level oscillations are particularly significant. Some of the available evidence for such changes is considered below.

There are few data available for sea-level changes during the early and middle Quaternary period. Moreover, much attention has focused on the evidence for, and pattern of, sea-level change in stage 5e because, as the last interglacial period, it is considered to be a useful analogue for the present interglacial, and on the period ca. 18 K years ago to ca. 10 K years ago, when the last ice age came to an end and sea levels rose. In recent years many studies have been undertaken on corals, the dating of which has been made possible by the development of the uranium series method. Examples include Chen *et al.* (1991) for the Bahamas, Bard *et al.* (1990) for Barbados and Ludwig *et al.* (1991) for Hawaii. Uranium series dating can also be used to date calcite speleothems (Section 2.2) from caves in coastal areas now submerged. As Lundberg and Ford (1994) point out, 'the speleothem functions as a recording dipstick' because the speleothem grew when the caves were exposed, but growth ceased when the caves were submerged, creating hiatuses. Using this approach on layers of a flowstone speleothem from Lucayan Caverns, Grand Bahama Island, Lundberg and Ford have presented a sea-level curve that extends back to ca. 300 K years. This is given in Figure 2.8, which shows a strong correlation between the marine isotopic record and sea-level oscillations. In itself this is important because the uranium series dates are independent evidence of sea-level change. High points in sea level occurred at 233, 213, 125 and 100 K years BP, i.e. in oxygen isotope stages 7 (both 233 and 213 K years BP), 5e and 5c which are warm stages. There is, however, no evidence for high sea levels at 195 K years or 80 K years (oxygen isotope stages 7a and 5a). The latter is recorded, for example, by Muhs *et al.* (1994) from the Pacific coast of North America. Here, uranium-series dates from corals indicate the

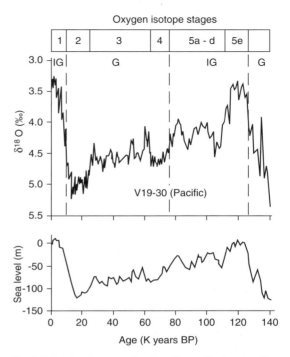

Oxygen isotope stages

Fig. 2.8 Sea-level curve for the Huon Peninsula, New Guinea, derived from marine terrace levels (based on Chappell and Shackleton, 1987). The oxygen isotope stages are from the Pacific deep-sea core V19-30 (Shackleton *et al.*, 1983).

formation of marine terraces in at least five areas extending from Oregon to Baja California. This study has also dated terraces representing the 125 K year sea-level maximum in eight areas.

Another long record is that of Chappell and Shackleton (1986), who have suggested that the sea-level curve derived from studies on the coral reefs of the Huon Peninsula in New Guinea may be representative of global Quaternary sea-level changes for the past ca. 140 K years. The results are given in Figure 2.8, which shows that high sea levels occurred during oxygen isotope stage 5e. Chappell and Shackleton estimate that sea level was ca. 6 m above present sea level during this interglacial and ca. 130 m lower than present sea level during the last glacial maximum ca. 18 K years BP. Additional work by Shackleton (1987), based on oxygen isotope studies of ocean-sediment foraminifera, confirms these overall trends in sea-level changes but suggests they were lower by ca. 20 m during the last glacial period (Figure 2.8). There is also general agreement with sea-level fluctuations that have been determined from sites in the Caribbean, the Pacific, Japan and Timor. Once again, the periodicities of sea-level changes appear to be closely related to astronomical forcing factors.

Crowley and Kim (1994) have examined in detail the relationship between Milankovitch forcing and sea-level change during the last interglacial period. Their comparison of Milankovitch insolation curves and sea-level curves from corals (e.g. Chen *et al.*, 1991) show that the two do not completely correspond. In particular, sea level was as high as it is now some 6000–4000 years before the peak in insolation receipt ca. 126 K years ago. It is suggested that the early sea-level rise was due to a peak in obliquity before precession (Fig. 2.6) with higher precession values than at present before its peak. These data indicate out-of-phase variations that occur in Milankovitch forcing factors and the environmental response they can generate. Burkle (1993) has also addressed the issue of rising sea levels during the Quaternary period in relation to the collapse and melting of the West Antarctic and Greenland ice sheets. This is particularly relevant to the potential impact of global warming on these ice sheets in the future. Burkle reviews the evidence for sea-level change in interglacial stages during the past 500 K years, and especially for oxygen isotope stage 5e. Although stage 11 (Table 2.2) is considered to be the warmest interglacial during this period, there is little evidence to suggest that either of these ice sheets disintegrated during these warm stages. Burkle suggests that the warm stages may not have lasted long enough to cause collapse. However, it is not clear what this means in relation to twenty-first century warming.

There is also detailed evidence for sea-level change during the last cold stage. For example, Richards *et al.* (1994) have reported thorium-230 ages for speleothems in underwater caves in Grand Bahama and South Andros Islands. Overall the results show progressively reducing sea levels as the glacial maximum approached. As the ice sheets subsequently retreated, sea levels rose rapidly. Thus, the development of a major ice sheet is accompanied by a stepwise reduction in sea level of up to ca. 150 m whereas deglaciation is paralleled by rapid rates of sea-level change. In relation to ice volume and sea-level decline, Colhoun *et al.* (1992) have attempted to reconstruct ice-sheet volumes at their margins and the impact they had on sea-level change. Their data show that, during the last glacial maximum, ice margins in Antarctica were thinner than suggested by models, so that ice accumulation

caused a drop in sea level of only 0.5–2.5 m instead of the predicted 25 m. Evidence for this limited sea-level fall derives from altitudinal measurements of raised beaches in the Ross embayment and east Antarctica. This suggests that the build-up of ice before the last glacial maximum had the most important impact on sea-level fall, or that ice accumulation was most significant in the Greenland ice cap. The data of Colhoun *et al.* also show that deglaciation was well advanced by ca. 10 K years BP and was concluded by 6 K years BP. This confirms the results from other studies and justifies the steepness of the curve in Figure 2.8.

2.11 Conclusion

Since the early nineteenth century, ideas relating to environmental change and Earth surface processes have undergone considerable refinement. The inception of the glacial theory and the subsequent acceptance of the occurrence of multiglaciation during the past 2 to 3×10^6 years paved the way for a multidisciplinary arena of research, which can be broadly described as Quaternary studies. The compilation of evidence for environmental change during this particularly dynamic period of Earth history is contributing to an understanding of not only how the Earth's surface has changed but also the underpinning causes. The process of environmental reconstruction is rather like a jigsaw; as new evidence is collected the jigsaw is reassembled, often with considerable difficulty, until the best fit for all the available pieces is achieved. The more evidence that is compiled, the more robust the resulting hypotheses.

Much early work in Quaternary studies concerned terrestrial or continental evidence for environmental change. However, the advent of ocean-sediment core stratigraphy in the 1950s caused the terrestrial record to be reappraised and in particular revealed the exceedingly complex nature of environmental change in the past 2 to 3×10^6 years. The recognition of more than 20 cold–warm or glacial–interglacial cycles reflected this complexity and led to the reinstatement of Milankovitch's (astronomical) theory of climatic change. Since the 1970s polar and a few high-latitude tropical ice caps have provided an additional source of palaeoenvironmental data. These three major sources of evidence, plus long lacustrine sequences, are now being synthesised to provide a framework for global environmental change. The terrestrial and lacustrine records of environmental change based on sedimentary and fossil evidence from high, middle and low latitudes is now providing a means of finely tuning this global record. The ultimate goal involves the use of such data for the prediction of future rates and directions of environmental change, though this is fraught with difficulties. Moreover, the development of new dating techniques, which can be used to obtain precise dates on a wide range of Quaternary situations, and a wider application of dating techniques which are already available, are essential for the advancement of global change studies and for improving the predictive capability of models derived therefrom.

Further reading

Andersen, B.G. and Borns, H.W. (1994). *Ice Age World*. Scandinavian Press, Oslo.
Dawson, A.G. (1992). *Ice Age Earth*. Routledge, London.
Jones, R.L. and Keen, D.H. (1993). *Pleistocene Environments in the British Isles*. Chapman and Hall, London.
Lamb, H.H. (1995). *Climate, History and the Modern World*, 2nd edn. Routledge, London.
Turekian, K.K. (1996). *Global Environmental Change Past, Present and Future*. Prentice Hall, Englewood Cliffs NJ.
Williams, M.A.J., Dunkerley, D.L., DeDekker, P., Kershaw, A.P. and Stokes, T. (1993). *Quaternary Environments*. Edward Arnold, London.

CHAPTER 3

Environmental change in the late- and post-glacial periods

3.1 Introduction

As the last ice sheets waned some 14 to 15 K years ago, environmental change occurred relatively rapidly and varied considerably on a global basis. Overall, climatic amelioration ensued as ice sheets retreated in high and higher-middle latitudes to approximately their present positions. As a result, ecosystems became re-established in regions that had been previously covered in ice. The ecosystems of low-middle and low latitudes simultaneously underwent adjustment as new climatic regimes became established. Some areas such as Britain did not become completely ice-free until 10 K years BP, and the late-glacial period from 14 to 10 K years BP was one of rapid climatic and ecological change. The complexity of environmental change during this period is reflected in many areas beyond northwest Europe and it is possible that the climatic and environmental oscillations of the late-glacial period reflect global, rather than simply local and regional, events. The adjustment to more congenial climatic regimes as the last ice age drew to a close and the present interglacial period commenced involved considerable changes in floral and faunal elements as well as changes in soil genesis. The fossil record indicates that biotic and pedogenic development did not always occur in tandem, and it is the fossil record that provides the most comprehensive evidence for the nature and direction of environmental change during the past 15 K years. This period also witnessed tremendous changes in the development and activities of human communities as *Homo sapiens sapiens* began to emerge as a dominant force in environmental systems. Innovations in resource use, especially the domestication of plants and animals and the establishment of agricultural systems also brought about environmental change. In consequence it is not always possible to distinguish from the available evidence which changes relate to climatic change and which to human activity.

The early part of the post-glacial period witnessed the formation and stabilisation of soils in high-middle and high latitudes as vegetation communities became established and developed in response to both climatic amelioration and ecological factors such as rates of migration from source areas, habitat suitability and competition. Similar environmental changes occurred in low-middle and low latitudes as climate ameliorated. In these regions, in contrast to those which experienced glaciation and periglaciation, climatic amelioration prompted the replacement of existing vegetation communities by others, usually with a significant tree component and hence an increased biomass. In the later part of the post-glacial period there is little evidence for substantial climatic change other than that associated with the 'Little Ice Age'. Nevertheless, some natural changes in global environments did occur in what was generally a period of climatic stability.

3.2 The interglacial cycle

Although this chapter deals with the specific environmental changes that have occurred in the past 15 K years, it is important to realise that this is part of an interglacial period. Although it is by no means certain that another ice age will develop, there is no reason to suppose that the repeated cycle of cold and warm stages, or glacials and interglacials, as reflected in the ocean-sediment ice core and terrestrial stratigraphy (Sections 2.2 to 2.4) will cease. Moreover, if the fossil evidence, particularly pollen assemblages, for the last 15 K years is compared with evidence from previous interglacials, a number of similarities are obvious, which also imply that the climatic cycles so far experienced will continue. First recognised by Jessen and Milthers (1928), on the basis of Danish evidence, Iversen (1958) proposed a simple model of the ecological processes operative during interglacials. A modified version of this scheme is given in Figure 3.1, along with similar schemes for other areas of western Europe, the eastern Mediterranean and Florida. Each cycle comprises four stages which provide a framework not only for the investigation of earlier interglacial deposits but also for the environmental changes of the past 15 K years.

The first phase is termed the cryocratic phase and is characterised by cold, arid continental conditions with skeletal, cryoturbated, base-rich soils supporting hardy herbs that today have an Arctic–Alpine distribution and are shade intolerant. As temperature ameliorates, the unleached base-rich soils are colonised by shade-intolerant herbaceous species, shrubs and pioneer tree species to create grasslands, scrub and open woodlands during the protocratic phase. This is followed by the mesocratic phase during the climatic optimum of the interglacial, when brown earth mull soils support temperate deciduous woodland as shade-intolerant species decline. The tree species characteristic of this period may invade as a direct response to climatic amelioration and/or because skeletal soils have been modified by vegetation communities of the protocratic phase in the process of vegetation succession. The final phase in the interglacial cycle is the telocratic phase and it is retrogressive. It is characterised by podzolised soils and peats that have become nutrient depleted and which carry acidic woodland dominated by conifers and ericaceous heaths. The latter part of the mesocratic phase and the early part of the telocratic phase have been termed the oligocratic phase by Andersen (1966). He suggested that the soil deterioration characteristic of this period may occur before climatic deterioration, due to natural soil processes such as leaching. Ultimately the cryocratic phase is re-established as global cooling occurs and the glacial–interglacial cycle begins again.

Taking the evidence for the interglacials of the British Isles (Godwin, 1975) as an example, it is clear that the actual plant species which become established vary between interglacials. This is why it is possible to distinguish between interglacial sites of different ages. Nevertheless, if the plants are categorised as protocratic, mesocratic, etc., each of the interglacial periods recognised in Britain follows Iversen's (1958) general model. This illustrates parallelism of development and reflects the overriding control of climate on environmental change during an interglacial cycle, while in any specific phase pedogenic characteristics, competition, migration rates and source areas of propagules will also influence the floristic composition.

Although Iversen's (1958) model was designed to provide a framework for the examination of interglacial sequences in northwest Europe, where the optimum stage of the interglacials was characterised by mixed deciduous forest, Birks (1986) has reviewed the evidence for the applicability of this model to other regions (Fig. 3.1) and the species which are characteristic of each phase are given in Table 3.1. Moreover, Hooghiemstra's (1984) work, mentioned in Section 2.9, and the work of others on long lacustrine sequences, suggests that the concept of the interglacial cycle is widely applicable. In the high plain of Bogotá, for example, oscillations in the altitudinal limits of the various vegetation communities (forest and paramo) characterise the interglacial cycle (Fig. 2.7). More recently, Colhoun (1996) has applied the model to the past 20 K years of environmental change in western Tasmania, Australia. Here the cryogenic phase gives way to subalpine shrubs followed by rainforest and eventually moorland or heathland. In all cases the climate is the overriding control on the developmental sequence, and the parallelism of development suggests that the glacial–interglacial cycle does indeed provide a useful framework within which to examine the environmental changes of the past 15 K years.

3.3 Climatic change during the late-glacial period

Evidence from the ocean-sediment cores (Section 2.3) indicates that ice ages ended relatively abruptly, with the change from glacial to interglacial conditions taking place within just a few thousand

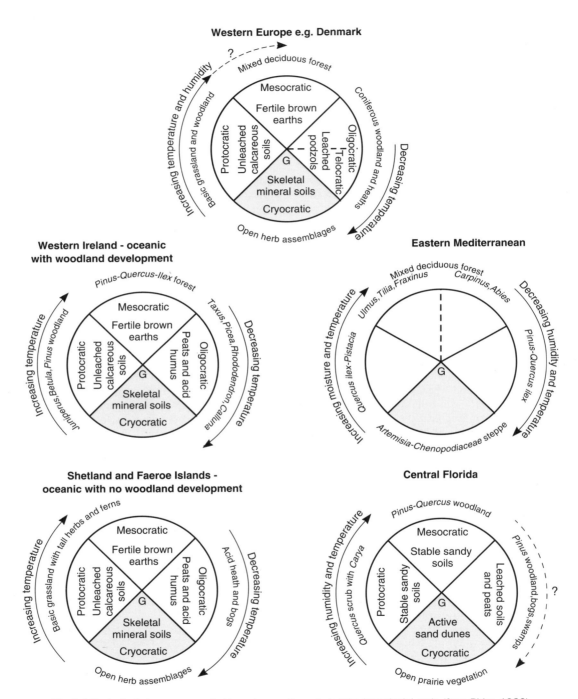

Fig. 3.1 Ecological changes occurring in various regions during the interglacial cycle (from Birks, 1986).

Table 3.1 Characteristic plant assemblages of the interglacial cycle

Phase of interglacial cycle	Faeroe and Shetland Islands	Western Ireland	Eastern Mediterranean	Florida
Cryocratic	Sparce cover of Arctic–Alpine herbs Soils: Skeletal mineral		*Artemisia*–goosefoot steppe	Open prairie vegetation Soils: Active sand dunes
Protocratic	Grassland with abundant tall herbs and ferns Soils: Unleached calcareous	Pine, birch, poplar, juniper woodland Soils: Unleached calcereous	Hollyoak–pistachio woodland Elm, lime, ash	Oak scrub with hickory Soils: Stable sandy
Mesocratic	Grassland with abundant tall herbs and ferns Soils: Fertile Brown Earths	Pine, oak and holly forest Soils: Brown Earth	Elm, lime, ash Hornbeam and fir	Pine, oak woodland Soils: Stable sandy
Oligocratic	Acid heath and bog communities	Fir, spruce, heaths, rhododendron	Pine	Pine woodlands, bogs, swamps
	Soils: Podzols, Peats	Soils: Podzols, Peats	Holly, oak	Soils: Leached soils and peats
Telocratic			and pine	
Original source:	Birks and Peglar (1979)	Watts (1967)	van der Hammen *et al.* (1971)	Watts (1971, 1980a)

Source: based on Birks (1986)

years. Although this is true of the early glacial periods of the Quaternary just as much as it is true of the last glacial period, it is for the last glacial period that most evidence is available. Even before the development of radiocarbon dating in the 1950s, lithological and palaeobotanical evidence from various parts of northwest Europe indicated that the end of the last glaciation was short-lived as well as climatically and ecologically complex. Among the earliest evidence for climatic oscillations during this period (ca. 15 K to 10 K years BP) was that reported by Hartz and Milthers (1901). They found laminated lower and upper clays, containing soliflucted pebbles, separated by an organic mud in a lake basin at Allerød in Denmark. The organic mud contained remains of birches, whereas the clays contained the remains of Arctic–Alpine species and dwarf shrubs. The sequence is interpreted as representing two tundra-like phases separated by a temperate period. Since this early work, other sites have been investigated and much has been written about the regional trends and anomalies that occur in palaeobotanical data from this period throughout Europe. Traditionally, this series of events has been considered to be of significance in northwest Europe

only. There is now evidence for a return to cold conditions before the onset of the Holocene in diverse regions of the world. Thus it is possible that the return to cold conditions, known in Europe as the Younger Dryas, which occurred ca. 11 to 10 K years BP, was a widespread response to an episode of global cooling (Broecker, 1994), possibly associated with a Heinrich event (Section 2.3). In addition there is evidence for a short-lived cold period before the Younger Dryas. This is called the Older Dryas in northwest Europe and is dated to ca. 12.2 K years BP. Yet a third cold period, the Oldest Dryas, occurred ca. 14.5 to 13 K years BP.

Much evidence for the environmental changes that occurred between 15 and 10 K years BP derives from ocean sediments and ice cores (Sections 2.3 and 2.4). In relation to ocean-sediment cores, Broecker (1994) has considered some of the available evidence for the Younger Dryas event, which is characterised by a significant increase in ^{18}O. This reflects an increase in global ice volume and a drop in temperature. That a decline in temperature was widespread is confirmed by evidence for it from a northeast Atlantic core (Duplessy *et al.*, 1981), cores from the Alboran Sea (western Mediterranean) and adjacent Atlantic ocean (Weaver and Pujol, 1988) and cores from the Southern Ocean (Charles and Fairbanks, 1992). It is not, however, recorded in sediments from the South China Sea (Thunell and Miao, 1996). The link between Heinrich events and global cooling has also been investigated by Bond *et al.* (1993), who have demonstrated that a major decline in abundance of the planktonic foraminifera *Neogloboquadrina pachyderma* occurred in two North Atlantic sediment cores (DSDP 609 and V23-81) at the same time (ca. 11 K years BP). Peaks in the concentrations of detrital carbonate also occurred at this time, reflecting a drift from the Labrador coast of sea icebergs carrying sediments rich in detrital carbonate. Bond *et al.* (1993) have also correlated events reflected in North Atlantic sediment cores with the Greenland ice core stratigraphy (Fig. 2.3). There has been much speculation as to the cause of the Younger Dryas event. Berger *et al.* (1987), for example, recognising that none of the Milankovitch cycles (Section 2.6) could generate this relatively short-lived climatic excursion, suggested that it may be due to the effect of glacial meltwater on the production of North Atlantic Deep Water (NADW).

On the basis of ocean-sediment core evidence, Berger *et al.* have suggested that the Younger Dryas event was caused by an influx of meltwater from the Arctic, which temporarily suppressed the production of NADW. Both Berger *et al.* and Overpeck *et al.* (1989) consider that, as the Laurentide ice sheet retreated, meltwater discharge switched from the Mississippi River to the emergent St Lawrence River, causing a disruption to the NADW. This debate has continued, as exemplified by Duplessy *et al.* (1992), whose data from two Atlantic sediment cores on sea-surface temperatures, salinity and density are given in Figure 3.2. This shows that all three parameters vary in unison over the past 18 K years in both cores. Duplessy *et al.* conclude that meltwater entering the North Atlantic caused stratification as surface salinities decreased. Sea-surface temperatures also decreased in response, though it is unlikely these changes alone generated the Younger Dryas event. Duplessy *et al.* suggest that evaporation from the ocean decreased, that there was increased precipitation and a curtailment of the northward movement of salty subtropical water. In combination these mechanisms may have reinforced the shifts in NADW production to generate the Younger Dryas cooling, though

Devernal *et al.* (1996) present evidence for reduced meltwater run-off, obtained from dinoflagellate-cyst assemblages in sediment cores from the Gulf of St Lawrence. Figure 3.2 also shows an earlier period of depressed salinity and sea-surface temperatures. This is the Oldest Dryas, a cold period centred on ca. 14 K years BP which is recognised in many terrestrial sequences (see below). Duplessy *et al.* attribute this to the operation of mechanisms similar to those which caused the Younger Dryas cold period. Other explanations for the occurrence of the Younger Dryas include a depletion of atmospheric carbon dioxide which, because of the diminished greenhouse effect, would have caused cooling. Kudrass *et al.* (1991) present evidence for this from the oxygen isotope record of two cores from the Sulu Sea (South China Sea) and one core from the North Atlantic ocean. They cite additional evidence from a number of other ocean-sediment cores, including those from the Gulf of Mexico, the North Pacific Ocean, the equatorial Atlantic Ocean and the Bengal Fan in support of their hypothesis. Moreover, if a depressed concentration of atmospheric carbon dioxide is a causal factor, then the Younger Dryas must have been a global event. However, a mechanism is required to explain the change in atmospheric composition, a mechanism which could involve changes to NADW production and dissemination.

The recognition of the Younger Dryas event in polar and tropical ice cores lends weight to the suggestion that it was global in extent. For example, at Dye 3 and Camp Century in Greenland, the Younger Dryas–Holocene transition is dated to $10\,720 \pm 150$ years BP (Dansgaard, 1987) and is marked by an abrupt increase in ^{18}O concentrations, as is the earlier warm period, the Allerød. Johnsen *et al.* (1992) have summarised the oxygen isotope data for these cores and for those from Renland and Summit (GRIP). These data are given in Figure 3.2. Not only is the Younger Dryas signal present but there is also evidence for a number of climatic oscillations in the preceding few millennia, as is additionally reflected by the Mayewski *et al.* (1993) analysis of GISP2 glaciochemical records. These correlate with pollen-analytical evidence for climatic excursions from northwest Europe (see below). The Younger Dryas and Allerød events are also reflected in methane concentrations from the GRIP core (Chapellaz *et al.*, 1993), as shown in Figure 2.2, which again reflects a link with greenhouse gas concentrations in the atmosphere. Interestingly,

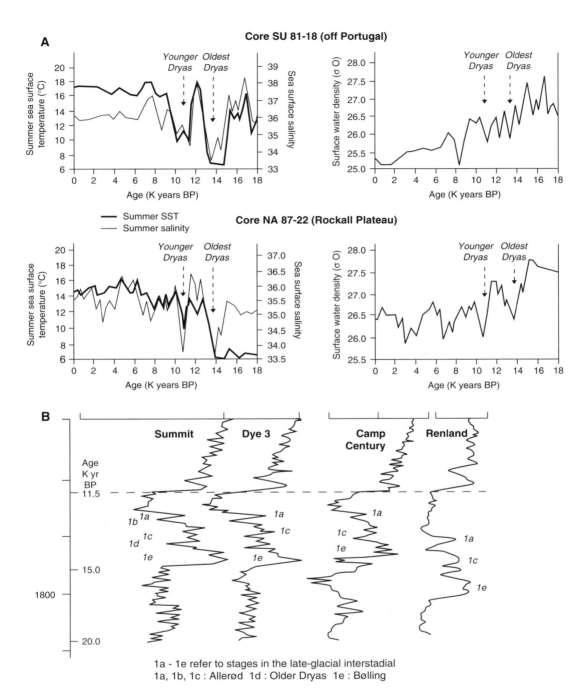

A

Core SU 81-18 (off Portugal)

— Summer SST
— Summer salinity

Core NA 87-22 (Rockall Plateau)

B

Summit   Dye 3   Camp
Century   Renland

1a - 1e refer to stages in the late-glacial interstadial
1a, 1b, 1c : Allerød 1d : Older Dryas 1e : Bølling

Fig. 3.2 (A) Reconstructions of sea-surface temperatures, salinity and water density for ocean-sediment cores NA 87-22, Rockall Plateau, and SU 81-18, off the coast of Portugal (based on Duplessy *et al.*, 1992). (B) A summary of ^{18}O stratigraphies from four Greenland ice cores for the late-glacial period (based on Johnsen *et al.*, 1992).

warming phases at ca. 12.8 K years BP (equivalent to the Bølling oscillation) and 10 K years BP, which is the start of the Holocene, are indicated by atmospheric carbon dioxide concentrations reconstructed from $^{13}C/^{12}C$ ratios in mosses and sedges preserved in peat (White *et al.*, 1994). The analyses of carbon dioxide concentrations from the GRIP and GISP2 (Greenland) cores are not yet available but it seems likely that changes in the concentrations will parallel those of methane. The changes that occur in the Vostok core (Antarctica) during the period between 10 and 15 K years BP are not the same as those in the Greenland cores, as Sowers and Bender (1995) have discussed (see also Jouzel *et al.*, 1987). This reflects non-synchronous trends in the two polar regions. Warming at the end of the last glacial period began in Antarctica some 3300 years before it began in the Arctic, where it is dated at ca. 15 K years BP. A weak trend toward colder conditions between ca. 12.5 and 14 K years BP is recognised from the Vostok and Byrd cores, indicating that the signal for the Younger Dryas in Antarctica is weaker than in the Arctic and that it occurred earlier. Sowers and Bender (1995) have suggested this lack of synchroneity is due to a stationary polar front which insulated the Arctic from global warming. This possibility is reinforced by ocean-sediment core data which indicates that the polar front only migrated northward ca. 15 K years BP (Ruddiman and McIntyre, 1981; see also Fig. 3.4). This makes the evidence for a global cooling event prior to the Holocene less conclusive than some of the other evidence presented above, though it may be that the Arctic region experienced a lagged response to global warming. In the Huascarán ice cap of Peru (Thompson *et al.*, 1995; Section 2.3) there are marked changes in ^{18}O and deuterium which are dated to ca. 12.16 K years BP, a date intermediate between dates from the Arctic cores and those from the Antarctic cores but which is considered to represent the Younger Dryas event.

Certainly there is sufficient evidence to show that it was a fairly widespread event and one which had its greatest expression in the North Atlantic and adjacent landmasses, including the British Isles. Here additional palaeoclimatic reconstructions of the late-glacial period have been facilitated by coleopteran analysis. According to Coope (1986), climatic conditions, particularly temperature, are among the most important factors that determine the geographical range of insects. Atkinson *et al.* (1987) have defined the climatic tolerances of 350

beetle species previously identified from late-glacial deposits relating to the last 22 K years, and on the basis of their mutual climatic range, derived from the intersection of their tolerance ranges, have reconstructed temperature changes. Their results are given in Figure 3.3, which shows that between 22 and 18 K years BP, when ice cover was at its maximum, average summer temperatures were less than 10 °C and the coldest winter month was ca. −16 °C. Between 14.5 and ca. 13 K years BP, when ice retreated from England and Wales then Scotland, climate remained cold and continental, with winter temperatures as low as −20 to −25 °C, giving rise to a temperature range of between 30 and 35 °C, considerably larger than the 14 °C range of today. Such low winter temperatures are attributed to the presence of continuous sea ice on the Atlantic, as has also been suggested by Ruddiman and McIntyre's (1981) investigation of ocean cores from the Atlantic. Just before 13 K years BP, however, the coleopteran evidence suggests there was rapid warming. Between 13.3 and 12.5 K years BP, summer temperatures rose to 17–18 °C and winter temperatures ameliorated to between 0 and 1 °C, but this warm period was short-lived. A cooling trend occurred between 12.5 and 12 K years BP, and although summer temperatures remained constant for some 500 years, winter temperatures declined from ca. −5 °C at 12 K years BP to ca. −17 °C at 10.5 K years BP; from 11.4 K years BP summer temperatures once again declined to ca. 10 °C by 10.5 K years BP and by 9.8 K years BP temperatures had returned to values similar to those at present.

Atkinson *et al.* (1987) have also used the data to calculate rates of temperature change. They estimate that during the warming periods, centred on 13 K years BP and 10 K years BP, changes occurred at the rate of 2.6 °C per century and 1.7 °C per century respectively. Conversely, the rates of cooling during phases centred on 12.5 K years BP and 10.5 K years BP were slower; but even so, temperature changes were still occurring relatively rapidly. Thus it is perhaps not surprising that the palaeobotanical evidence for this period is often at variance from site to site, and often in apparent conflict with coleopteran evidence. This is illustrated if comparisons are made between the palaeobotanical (Pennington, 1977) and palaeocoleopteran evidence (Coope, 1977) from Lake Windermere (Table 3.2). For example, by 13 K years BP, coleopteran evidence suggests that temperatures were similar to those of today, but the pollen assemblages indicate the slightly lagged replacement of pioneer plant species with woodland species. But since Atkinson *et al.* (1987) showed how thermal conditions were ameliorating rapidly, it seems likely that plants were less well able to respond to the climatic stimulus, possibly because habitats remained relatively hostile due to insufficient time for soil development. This possibility becomes even more likely if the habitat requirements of the more thermophilous native British species, such as oak and elm, are considered. Not only do they require climatic conditions similar to the present day, but they also require well-developed soils. It is also possible that the period in question was too short-lived for such thermophilous species to migrate from the refuges in Europe to which they had retreated during the earlier severe cold period that witnessed the maximum extent of Devensian ice in Britain. However, a comparison between the fossil coleopteran (Coope, 1977) and pollen-analytical (Pennington, 1977) data from Lake Windermere for the period between 12 and 10 K years BP suggests that the two sets of indicators are in agreement. Both suggest the onset of cold conditions. The coleopteran data indicate a decline in average summer temperatures to ca. 10 °C, and a decline in winter temperatures to ca. −17 °C. The pollen assemblages indicate a return to Arctic-tundra conditions, with a vegetation dominated by open-habitat grasses and sedges. It is likely that this close agreement is a function of the more gradual cooling that occurred, providing sufficient time in this instance for the vegetation to establish equilibrium with habitat and climate.

There are, however, many sources of error when interpreting both pollen and coleopteran assemblages. There is always the possibility of long-distance transport of some pollen types from far distant source areas and the possibility of reworked grains from older deposits. Nor should the possibility be discounted that fossil coleoptera reflect microclimatic rather than macroclimatic conditions, as discussed by Andersen (1993). Moreover, sites which are suitable for pollen analysis, such as Lake Windermere, may not be particularly suitable for coleopteran analysis and vice versa. For example, a site at Gransmoor in Yorkshire, UK, which has been investigated as part of the North Atlantic Seaboard Programme of the International Geological Correlation Programme 253, is better suited to coleopteran analysis than to pollen analysis as it is a shallow basin. It is also a

lowland site in common with most other UK sites from which coleopteran assemblages have been analysed. Regional, altitudinal and intersite variations will occur which militate against the delineation of overall trends. Care must therefore be exercised in formulating generalities which are not widely applicable, though the level of certainty associated with such generalities will improve as more data become available, or the generalities will be modified. This highlights the need for the continued collection of primary data and, if possible, the validation of general trends by reference to independent data. This approach is exemplified by recent work at Gransmoor, where coleoptera-based palaeotemperature data have been compared with the snow accumulation record from the GISP2 ice core (Section 2.4). The results are given in Figure 3.3, which shows that higher temperatures throughout the late-glacial period correlate with periods of increased snow accumulation (Lowe *et al.*, 1995). The high degree of correlation based on ^{14}C dates is considered by Lowe *et al.* to indicate broad synchroneity of climatic change in Greenland and the British Isles. The reconstructed temperature curve has also been related to the curves for birch (*Betula*) and juniper (*Juniperus*) pollen percentages (Walker *et al.*, 1993). The results are given in Figure 3.3, which shows that the highest temperatures of the late-glacial period occurred at its opening.

Overall, the last protocratic phase was a period of complex climatic and ecological changes. Walker *et al.* (1994) have produced a synthesis of much of the pollen-analytical data covering this period in Britain and northwest Europe, and their work reflects its complexity. The reconstructed events have been attributed by many workers to oscillations of the oceanic polar front. The same conclusion, in relation to British palaeocoleoptera data, is advocated by Atkinson *et al.* (1987), who also refer to Ruddiman and McIntyre's (1981) work on Atlantic ocean cores. These works have suggested that, since the last glacial maximum, the eastern edge of the oceanic polar front has shifted position several times, as shown in Figure 3.4. As the last ice sheet waned, this front rapidly moved northeastwards from a position at the latitude of northern Spain to one north of Iceland, bringing in its wake a rapidly ameliorating temperature regime to the British Isles (and northwest Europe). It then returned ca. 12 K years BP, bringing cold polar waters south and initiated a climatic regime which

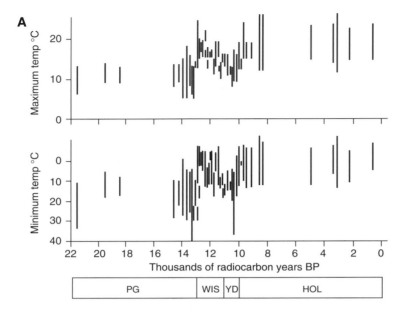

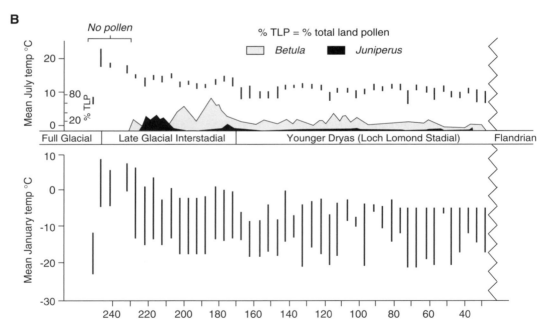

N.B. Late glacial interstadial is defined here as it has been defined by Coope (1977) for Lake Windermere (see Table 3.2).

Fig. 3.3 (A) Reconstructed mean temperatures for the warmest and coldest months of the year, based on coleopteran assemblages from 26 sites in Britian and Ireland, relating to selected periods during the past 22 K years (from Atkinson *et al.*, 1987). (B) Mean July temperature reconstructions based on the mutual climatic range of coleoptera; data are from Gransmoor, east Yorkshire, UK, as are the pollen curves for *Betula* (birch) and *Juniperus* (juniper) (based on Walker *et al.*, 1993).

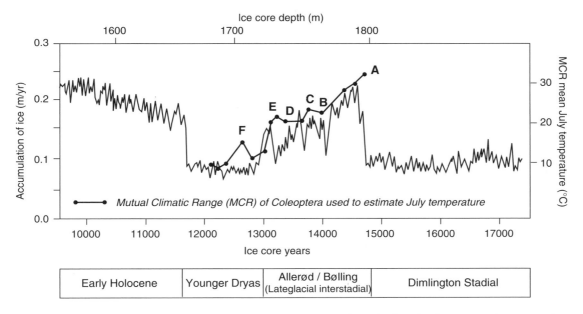

Fig. 3.3 contd. (C) Comparison of GISP2 data on accumulation rate, Greenland, and palaeotemperature reconstructions using coleoptera data from Gransmoor, east Yorkshire, UK (based on Lowe *et al.*, 1995).

in turn prompted the redevelopment of Arctic–Alpine vegetation communities in many parts of Britain. In addition, local ice caps and glaciers became re-established in many upland areas in Britain, where this period is known as the Loch Lomond stadial. As Figure 3.4 shows, by 10 K years BP the polar front had retreated northeast and ameliorated temperatures ensued.

If this hypothesis is correct, the late-glacial movements of the polar front could explain many of the phenomena observed in the palaeoenvironmental record. Firstly, the steep temperature gradient across the front may have been responsible for the rapid temperature changes postulated by Atkinson *et al.* (1987) for the warming periods of 13 K years BP and 10 K years BP. Secondly, since such a movement of the polar front was time-transgressive, the ecological response was probably also time-transgressive, so similar fossil assemblages between sites do not necessarily imply synchroneity. Thirdly, since the movement of the polar front was over the Atlantic Ocean, it may account for the fact that the late-glacial oscillation was recorded in maritime western Europe but is not particularly well marked in pollen assemblages in eastern and southern Europe (Watts, 1980b). Such regional differences will be discussed in Section 3.4.

The growing body of evidence for a widespread expression of the Younger Dryas and earlier events, however, suggests that Ruddiman and McIntyre's (1981) idea of a more or less stationary western margin for the polar front is erroneous. For example, work by Mott *et al.* (1986) and Mott and Stea (1993) from 26 sites in Canada's maritime provinces convincingly demonstrates the operation of an overall warming trend but with local variability from ca. 12.7 K years until ca. 10.8 K years BP when marked changes in the stratigraphy and pollen assemblages indicate the onset of cooling. There is also evidence for the readvance of glaciers in Quebec. This cold period lasted until ca. 10 K years BP, when there was renewed climatic amelioration. In addition, Levesque *et al.* (1994) have presented evidence for a pre-Younger Dryas cold period from the same region, which they call the Killarney oscillation and which may correlate with a minor climatic regression during the Allerød (Table 3.2). For example, a decline in *Betula* (birch) has been noted at several sites in Britain, and a climatic deterioration would explain this. Levesque *et al.* believe the Killarney oscillation lasted some 250 years and that there is evidence for similar oscillations in European pollen diagrams as well as Greenland ice cores and ocean-sediment

Table 3.2 The late-glacial environment as deduced from fossil plant and coleopteran assemblages from Low Wray Bay, Lake Windermere, English Lake District

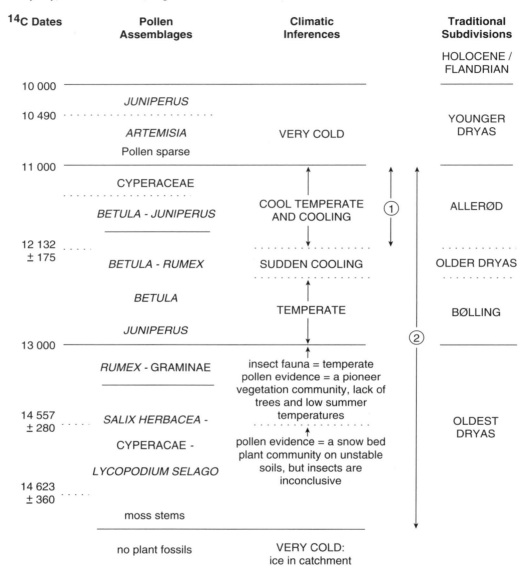

¹⁴C Dates	Pollen Assemblages	Climatic Inferences	Traditional Subdivisions
			HOLOCENE / FLANDRIAN
10 000			
	JUNIPERUS		
10 490			YOUNGER DRYAS
	ARTEMISIA	VERY COLD	
	Pollen sparse		
11 000			
	CYPERACEAE		
	BETULA - JUNIPERUS	COOL TEMPERATE AND COOLING	① ALLERØD
12 132 ± 175	*BETULA - RUMEX*	SUDDEN COOLING	OLDER DRYAS
	BETULA	TEMPERATE	BØLLING
	JUNIPERUS		
13 000			②
	RUMEX - GRAMINAE	insect fauna = temperate pollen evidence = a pioneer vegetation community, lack of trees and low summer temperatures	
14 557 ± 280	*SALIX HERBACEA -*		OLDEST DRYAS
	CYPERACAE -	pollen evidence = a snow bed plant community on unstable soils, but insects are inconclusive	
	LYCOPODIUM SELAGO		
14 623 ± 360			
	moss stems		
	no plant fossils	VERY COLD: ice in catchment	

① Length of interstadial on the basis of plant remains (Pennington, 1977)

② Length of interstadial on the basis of coleopteran assemblages (Coope, 1977)

cores from the North Atlantic Ocean. They therefore propose that it should be known as 'the amphi-Atlantic oscillation'. Its discovery reflects the level of resolution possible using close interval sampling for pollen analysis and adds to the complexity of the late-glacial period. In this context a less tortuous name would perhaps be more appropriate.

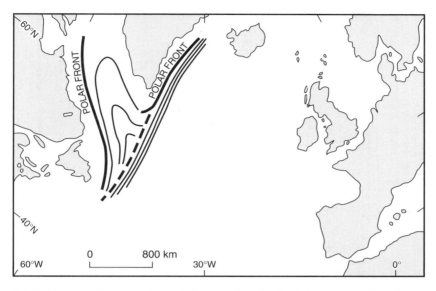

Fig. 3.4 Changes in the position of the oceanic polar front during the early Holocene (based on Ruddiman and McIntyre, 1981).

3.4   Regional expression of changes during the late-glacial period

The climatic changes of the late-glacial period were both rapid and variable, as discussed in Section 3.3. Thus it is not surprising that the ecological response was spatially variable. There are also difficulties associated with interpreting the available data and in establishing correlations within and between regions. For example, the sampling interval used when undertaking pollen analysis is all important. Levesque *et al.* (1994) identified a cool period during the time span equivalent to the Allerød, i.e. 12 to 11 K years BP (Section 3.3); they were able to do this because they used a 1 cm sampling interval. In many pre-1970s pollen diagrams the sampling interval of 5 cm or 10 cm would not necessarily permit the identification of such a relatively short-lived (ca. 250 year) oscillation. This effect may be exaggerated where sedimentation rates have been low. Moreover, the accuracy of dating, notably radiocarbon dating, makes correlation difficult. The error factor is too wide for precise dating, especially within a period that was only ca. 4 K years but which was highly changeable.

Both of these factors may contribute to a greater or lesser degree for the variation that exists between sites in Britain alone. An example of variation is the

Bølling oscillation; not recorded at many sites in Britain, but as Table 3.2 shows, it is clearly registered in Lake Windermere. As stated in Section 3.3, there are the altitudinal and latitudinal variations that would be expected in view of the diversity of the British landscape; this produced considerable regional variation as well as site-specific responses to climatic change. Significant differences also exist between UK sites and those in Europe, and between those of northwest Europe and southern Europe. Nevertheless, there is increasing evidence that the climatic change during the period 14 to 10 K years BP was particularly marked in the northern hemisphere and may have occurred in the southern hemisphere. The changes that are indicated by palaeoecological data from widely dispersed sites reflect the local response to what may well have been a period of global climatic change.

Much of the available information about regional variation in Britain and Ireland has been summarised by Walker *et al.* (1994) and is given in Table 3.3. In addition, data from Gransmoor, Yorkshire (Walker *et al.*, 1993) are given in Figure 3.3 and discussed in Section 3.3. As Table 3.3 shows, open-ground plant communities that characterised the end of the last glacial period, assuming it closed ca. 13 K years BP, were replaced

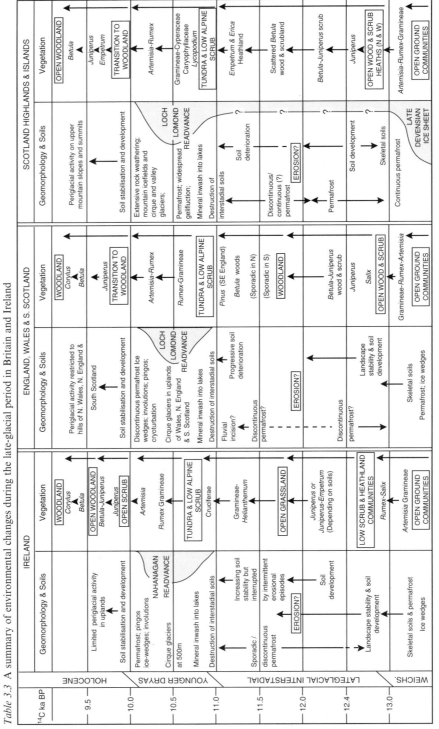

Table 3.3 A summary of environmental changes during the late-glacial period in Britain and Ireland

(Shaded areas represent the period of active glaciation)

Source: based on Walker *et al.,* (1994)

with open woodland and scrub heaths followed by birch – juniper woodland (*Betula* and *Juniperus*), or in Ireland juniper or juniper–crowberry (*Empetrum*) communities. In southeast England pine (*Pinus*) was present along with birch (Godwin, 1975). There is evidence for the operation of soil erosion in the middle of the interstadial, perhaps equivalent to the Older Dryas (Table 3.2), a short-lived period of climatic deterioration. At ca. 11 K years BP there is geomorphological evidence for local glacier readvance in the upland zone and palaeoecological evidence for cooling. Mean July temperatures decreased to ca. 7–10 °C. This is the Younger Dryas, which is clearly recorded in Greenland ice cores (Section 3.3).

Considerable regional variation is also apparent in Ireland, as has been discussed by Watts (1985) and Andrieu *et al.* (1993). Andrieu *et al.* have adopted the suggestion of Watts (1977) that five pollen-assemblage zones provide an adequate framework for describing the many variations that exist in Ireland's late-glacial vegetation history. These zones and their characteristics are given in Table 3.4 in relation to the original pollen zones of Jessen (1949). A comparison of the pollen assemblage zones from Lurga and Illauncoran, and other sites discussed in Andrieu *et al.* (1993), which are situated close together and are 30 m and 47 m above sea level respectively, shows there are many similarities but also significant differences. The significant differences occur in the late interstadial; four substages recognised at Illauncoran are not present at Lurga. Moreover, Fossit's (1994) recent work on three sites in Donegal also reflects local differences in the timing and character of late-glacial vegetation changes. An overall summary of these changes has been included in Table 3.4, which shows the lack of correspondence with the late-glacial pollen zones of Andrieu *et al.* (1993). This may be explained, at least partly by the more northerly location of the Donegal sites, though none are more than 40 m above sea level. Although it is difficult to discern overall trends common to all the sites it is clear that the late-glacial period was complex in terms of the environmental changes that were occurring.

Regional variations also occur in northwest European pollen diagrams, general trends from which have been summarised by Walker *et al.* (1994). In general, the Younger Dryas can be recognised on the basis of pollen assemblages that reflect climatic regression; but for the period before 11 K years BP local site conditions and probably competition between species operated to generate a particularly varied response to a fluctuating climate. The evidence suggests that plant communities reacted in many different ways to the warming trend and its oscillations over the period ca. 14 to 11 K years BP, whereas the subsequent cooling episode of the Younger Dryas brought about a response of similar magnitude in diverse regions. And when examining the late-glacial record in the Netherlands, Bohncke (1993) points out the difficulties associated with comparing pollen records from peat sequences with those from lacustrine deposits, as peat sequences often contain hiatuses. He provides evidence for a decline in mean annual temperatures below 0 °C during the periods 12.1 to 11.9 K years BP, possibly the Older Dryas, (Table 3.2) and 11.3 to 10.9 K years BP, which corresponds with the Younger Dryas. Hammarlund and Buchardt (1996) have presented stable oxygen and carbon isotope data from lake sediments at Graenge, southeast Denmark, which reflect a period of disturbance, possibly caused by forest decline and aeolian activity during the Younger Dryas. The Older Dryas is also recorded elsewhere in Europe, e.g. north Belgium (Walker *et al.*, 1994), western Norway (Birks *et al.*, 1994) and possibly Iceland (Ingólfsson and Norddahl, 1994).

In Iceland, for example, there was a glacier readvance that ended ca. 11.8 K years BP, whereas in Norway there is evidence from Rogaland in the southwest for accelerated soil erosion. From further north, at Bergen and Sunnmøre, there is evidence for a minor ice-sheet advance and the presence of a discontinuous cover of pioneer vegetation types. Thereafter, during the Allerød, birch populations gradually expanded in sheltered areas and willow scrub with dwarf shrub heaths developed whereas open tree-birch vegetation communities with willow scrub characterised the southern sites. There is also evidence from a sedimentary sequence at Krakenes, western Norway, for changing atmospheric carbon dioxide concentrations during the late-glacial period. Beerling *et al.* (1995) have used the stomatal density of *Salix herbacea* leaves to calculate that, during the warm Allerød, carbon dioxide concentrations were ca. 273 ppm then declined to ca. 210 ppm during the cold Younger Dryas. According to Beerling *et al.*, where they are available, suitable plant assemblages can be used to reconstruct atmospheric carbon dioxide histories to complement data from ocean cores (Section 2.4). Amman *et al.* (1994) have presented pollen, coleoptera and oxygen isotope based temperature reconstructions for lowland Switzerland, none of which agree entirely. This

Table 3.4 Interglacial pollen assemblage zones in Ireland

Pollen zones Jessen (1949)	Dates (years BP) (Mangerud et al., 1974)	Pollen zones Watts (1977)	Regional PAZs Watts (1977)	Lurga, Co. Galway Andrieu et al. (1993)	Illauncronan, Co. Clare Andrieu et al. (1993)	Sites in Donegal (Fossitt, 1994)
YOUNGER Salix herbacea (ZONE III)	YOUNGER DRYAS 11000	YOUNGER DRYAS STADIAL (PAZ 5)	Artemisia	Artemisia–Cyperaceae–Thalictrum	Artemisia–Cyperaceae–Thalictrum	Empetrum heath and herb-rich grassland
BIRCH OR ALLERØD (ZONE II)	ALLERØD 11800	LATE INTERSTADIAL (PAZ 4)	GRAMINEAE	Gramineae–Cyperaceae–Helianthemum	Four substages all with Gramineae but with other taxa also dominant	Juniperus–Empetrum heath
	OLDER DRYAS 12000					
	BØLLING 13000	EARLY INTERSTADIAL (PAZ 2,3)	Juniperus–Empetrum	Gramineae–Juniperus–Empetrum	Juniperus–Empetrum–Betula	Pioneer herbaceous vegetation
			Rumex–Salix	Rumex–Salix–Empetrum	Rumex–Salix–Empetrum	Barren deglaciated landscape
OLDER Salix herbacea (ZONE I)		PRE-INTERSTADIAL (PAZ 1)	Pinus–Artemisia–Gramineae			

PAZ = pollen assemblage zone

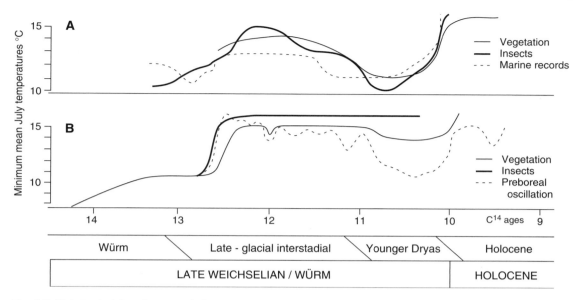

Fig. 3.5 (A) Late-glacial environmental change in south Sweden and Denmark (based on Berglund *et al.*, 1994). (B) Late-glacial environmental change in lowland Switzerland (based on Amman *et al.*, 1994).

probably reflects the varying response of plants, beetles and oxygen isotope ratios to changing climate, particularly changing temperature, and it illustrates why precise palaeoenvironmental or palaeoclimatic reconstructions are so difficult. As Figure 3.5 shows, temperature reconstructions based on the oxygen isotope curve are the most variable in comparison with those based on palaeobotanical and palaeoentomological data. This may reflect the high sensitivity of oxygen isotope ratios to temperature change, and perhaps the less direct response of plants and insects. All three curves rise steeply at the opening of the interstadial (*sensuo* Coope, 1977), reflecting an increase in mean July temperatures of ca. 5 °C between 13.5 and 12.5 K years BP. The curves based on palaeobotanical and palaeoentomological data do not show much variation during the interstadial, as birch woodland then pine–birch woodland developed. The Younger Dryas event is most marked in the oxygen isotope curve, though there is evidence for the reoccurrence of pioneer plant species within the pine–birch woodland and increased erosion. Altitudinal variability in the response of vegetation communities is exemplified by David's (1993) work involving plant macrofossil and pollen assemblages from eight neighbouring sites in the northern French Alps. Climatic changes are recorded for the Oldest Dryas, Allerød with a possible cool period towards its close,

and the Younger Dryas. The response of the vegetation, however, varies with altitude; the strongest response occurs at higher altitudes.

Figure 3.5 also provides data on the environmental changes that occurred during the late-glacial period in southern Sweden and Denmark. As in the case of lowland Switzerland (see above), there is variation in the reconstructed temperature curves based on three indices (the marine curve is based on changes in foraminifera and mollusc records in sediment cores from the Kattegat). However, all show a marked rise at ca. 13 to 12 K years BP at the opening of the interstadial and another marked rise at the opening of the Holocene following the Younger Dryas climatic regression. The curve based on coleoptera data shows the greatest variation during the interstadial and a decline in all curves at ca. 11.5 °C reflects the Older Dryas (Table 3.2), which is followed by a period of relatively stable but cooler climate equivalent to the Allerød. At no time in the late-glacial period is there evidence for main July temperatures as warm as those occurring at the opening of the Holocene or today. This contrasts with the data from lowland Switzerland (Fig. 3.5), which show that temperatures as high as those of the early Holocene and today occurred during a large part of the interstadial. For central Europe, Goslar *et al.* (1993) have presented high-resolution proxy climatic data from Lake Gosciaz, central Poland, in which there are annually laminated sediments. In addition to palaeoenvironmental reconstruction, Goslar *et al.* have determined that the onset of the Younger Dryas occurred within 150 years and that it drew to a close in just 20 years. The rapid end is confirmed by Mayewski *et al.* (1993); on the basis of data from Greenland ice core GISP2, they believe that it also began rapidly, i.e. within 10–20 years.

The conclusions of Watts, made some 15 years ago (Watts, 1980b), in relation to the absence of evidence for late-glacial climatic change in southwestern Europe have not been borne out by recent work from a wide variety of sites; this work has been summarised by de Beaulieu *et al.* (1994). Here the earliest evidence for warming is at 15 K years BP and there is evidence of a widespread warming trend which begins by ca. 13 K years BP as well as for the climatic regression of the Younger Dryas, characterised by increased aridity. There are also several sites in the Massif Central and the western Pyrenees with evidence for a cooling episode, which may be the equivalent of the Older Dryas (Table 3.2). In common with most of the data mentioned in this section, the southern European data reflect a rapid increase in temperature between ca. 13.5 and 13 K years BP, though in some areas the sediments of Younger Dryas age may be absent due to the increased aridity of the period. Changes in vegetation that are contemporary with the Allerød and Younger Dryas of western Europe are certainly recorded at Laghi di Monticchio in southern Italy (Watts *et al.*, 1996). Moreover, Wansard (1996) has presented evidence from ostracods for several late-glacial oscillations from the La Draga sedimentary sequence at Banyoles, north-east Spain. Moreover, mean July temperatures at this time reached ca. 16 °C as they did at the opening of the Holocene (see also Lowe *et al.*, 1994).

During the past decade many new sites have been investigated beyond Britain and Europe. The late-glacial interstadial was once considered to be a phenomenon confined to northwest Europe, but it has now been identified in a wide variety of sites on the Atlantic seaboard of North America (summarised in Lowe *et al.*, 1994). The work of Mott and Stea (1994) has already been mentioned in Section 3.3. Other regional syntheses include those for the northeastern United States (Peteet *et al.*, 1993), New Brunswick (Cwynar *et al.*, 1994), Nova Scotia (Mott, 1994), Newfoundland (Anderson and MacPherson, 1994), Quebec (Richard, 1994) and Baffin Island–Labrador

(Andrews, 1994). The results are summarised in Figure 3.6; there are considerable variations from region to region, with New Brunswick showing the greatest differences in temperature as reflected in reconstructed July temperatures for the surface water of lakes. The Younger Dryas is recorded in all cases. In New Brunswick the warming trend at the opening of the interstadial begins ca. 12.1 K years BP (this and all the other dates for the Atlantic seaboard sites are later than those for Britain and northwest Europe) and reaches a peak ca. 11.3 K years BP. A minor cool period occurs between 11.2 and 10.9 K years BP, considered by Cwynar *et al.* (1994) to represent the amphi-Atlantic oscillation (Section 3.3). A major cooling episode, the Younger Dryas follows a short-lived increase in temperature and occurs before the Holocene. Miller's (1996) coleopteran assemblages from Cape Breton Island, Nova Scotia, confirm the trends indicated by pollen analysis from numerous eastern Canadian sites. In contrast the least variation occurs in New England (Peteet *et al.*, 1993), where the only major change after the onset of the interstadial at ca. 12.4 K years BP is the Younger Dryas decline in temperature. During this period the boreal tree species of spruce (*Picea*), larch (*Larix*), paper birch (*Betula papyrifera*) and alder (*Alnus*) increased at the expense of more thermophilous species, e.g. oak (*Quercus*) and white pine (*Pinus strobus*), which colonised the area in the earlier warm period, i.e. the equivalent of the Bølling–Allerød (Table 3.2).

Although the data presented above reflect the operation of considerable climatic variation in the north Atlantic region as the last ice age drew to a close, there is evidence from elsewhere for an even more widespread expression of climatic change. According to Edwards *et al.* (1993), the Younger Dryas was characterised by a reduced rate of sea-level rise when considered with earlier or later periods. This was due to a reduction in the rate of ice-sheet melting. There is also evidence for lake-level declines associated with the Younger Dryas event, as Magny and Ruffaldi (1995) have discussed in relation to the Jura mountains of France. Mathewes (1993) has reviewed the available data on the Younger Dryas event along the Pacific coast of North America from Oregon to Alaska. There are clear grounds – including peaks in mountain hemlock (*Tsuga mertensiana*) pollen, indicating a cool and moist climate, the replacement of arboreal with non-arboreal pollen and changes in the foraminiferan assemblages of continental shelf cores from warm- to cool-water tolerant species (Mathewes *et al.*, 1993) – to indicate that a cold period

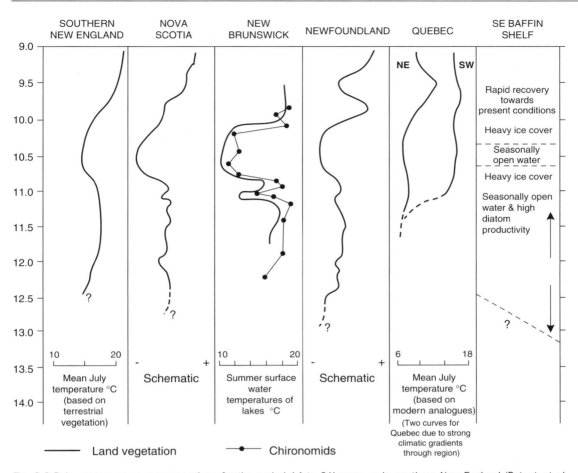

Fig. 3.6 Palaeotemperature reconstructions for the period 14 to 9 K years BP in southern New England (Peteet *et al.*, 1994), Nova Scotia (Mott, 1994), New Brunswick (Cwynar *et al.*, 1994), Newfoundland (Anderson and MacPherson, 1994), Quebec (Richard, 1994) and the southeast Baffin Shelf (Andrews, 1994). The July temperature scale for New Brunswick refers to chironomid data; the curve based on vegetation is schematic only.

occurred between 11 and 10 K years BP. Mathewes also states that the most well-defined expression of this climatic regression is in sites close to the ocean. Additional pollen-analytical data from the Allegheny Plateau indicates that a climate reversal occurred between 11 and 10 K years BP, involving a temperature decline of 2–5 °C and a 10–20 cm decline in precipitation (Shane and Anderson, 1993). Positive evidence of the Younger Dryas cooling event is now required from Japan, the coastal region of China and Kamchatka, in order to demonstrate the operation of an event throughout the northern hemisphere at least.

In South and Central America there is some, though limited, evidence to suggest that there was a cooling episode equivalent to the Younger Dryas

but possibly earlier, e.g. 12.5 to 10 K years BP. Hansen (1995), for example, has drawn attention to the dated site of Laguna Juin in Peru and other sites in Peru and Ecuador which provide some, though limited, evidence for a Younger Dryas cooling. Recent work by Islebe *et al.* (1995, 1996) has refuelled the controversy insofar as they have identified a cooling event in pollen assemblages from the La Chonta bog in Costa Rica (see also Leyden, 1995). This is dated to between 11.07 ± 130 (measured age) and 10.4 K (interpolated age) years BP and is considered to have caused a depression of vegetation belts by some 300–400 m. Reference has already been made to the cool period which centres on 12.16 K years BP and is indicated by oxygen isotope ratios in the Huascarán ice core from Andean Peru (Thompson *et al.*, 1995). There is corroborative evidence from the same region, as Francou *et al.* (1995) have presented evidence for glacier readvance between 11 and 10 K years BP. Moreover, Khury *et al.* (1993) have recognised a cooling event in pollen assemblages in the eastern Cordillera of Colombia, which they have called the El Abra stadial. They demonstrate how this was preceded by a warm interval, termed the Guantiva interstadial, equivalent to the Bölling–Allerød and which ended ca. 11 K years BP. The altitudinal limit of forest during the El Abra stadial was lowered by ca. 400 m from its location in the Guantiva interstadial, reflecting a decline in temperature of 2–3 °C. This and other work in the Andes has been reviewed by Heine (1993), who believes that inadequate dating controls make the correlation of Andean sites with the Younger Dryas of Europe equivocal. The dating of a climatic regression in the Peruvian Andes from the Huascarán ice core (see above) confirms this point of view, though a similar event in the Vostok core (Section 2.4) is dated even earlier, ca. 14 K years BP. This may reflect a lagged response to global warming. Whatever alternative explanations are possible for the lack of dating agreement, it seems likely that some relationship exists between the various warming episodes indicated in Arctic, Antarctic and Andean ice records.

Other sites in southern South America are considered to show little, if any, evidence for a cooling event equivalent to the Younger Dryas. Markgraf (1993) has suggested that the high incidence of charcoal particles during the period 11 to 10 K year BP in sites in Tierra del Fuego reflects the importance of fire rather than cooling in determining the vegetation changes. On the basis of pollen and beetle analyses from South Chilean sites, Ashworth *et al.* (1991) also conclude there is no evidence for a return to cold conditions between 11 and 10 K years BP. If a time lag did operate between the southern and northern hemispheres (see above), Ashworth's evidence would not be surprising; though Markgraf's data from Tierra del Fuego do not reflect an earlier cold period either. In contrast, however, Heusser (1993) interprets the pollen-analytical data from a further six sites in Tierra del Fuego as reflecting two cold periods: the first occurs between 11 and 10 K years BP. Clearly there is considerable scope for reconciling these data. Elsewhere in the southern hemisphere there is some

geomorphological evidence for a climatic regression. In New Zealand, for example, McCalpin (1992) has presented evidence friom the Wairau Valley, near Marlborough, for a glacier advance that is radiocarbon dated to between 9.5 and 10.2 K years BP. Similarly, Denton and Hendy (1994) have reported a radiocarbon age of 11.05 K ± 14 years BP for a moraine deposited by the Franz Joseph Glacier in the southern Alps. Lees *et al.* (1993) have also presented evidence for coastal dune emplacement on the western Cape of northern Australia, which they date to 11.2 K years BP by thermoluminescene. They suggest that the dune was stabilised as a minor marine transgression occurred. This would have occurred as more of the Earth's water was incarcerated in ice as glaciers, etc., once again advanced. Moreover, Goede *et al.* (1996) have presented oxygen isotope analysis of a calcitic stalagmite from a cave in eastern Victoria, which shows clear evidence for a cooling period between 12.3 and 11.4 K years BP. The general lack of palaeoecological data from these regions for a Younger Dryas climatic regression suggests that ecological thresholds were not crossed in the southern hemisphere as they were in the northern hemisphere.

Worldwide, the late-glacial period was complex and was characterised by environmental change rather than stability. It represents the relatively short time period when glacial or cold conditions were replaced by interglacial or warm conditions with all the attendant pedological, ecological and hydrological changes. It has been the focus of much research in the past decade, not least because it provides an insight into rates of change and the way in which plant and animal communities react to climatic stimuli. In this respect it is considered to provide an analogue for a warmer world.

3.5 Faunal changes during the late-glacial period

Much of the foregoing discussion has focused on evidence from fossil pollen and fossil coleopteran assemblages which indicate that the period between 15 and 10 K years BP was one of rapid ecological change. It was also a period that witnessed major changes in vertebrate faunas. For example, Sutcliffe (1985) states that 'the faunal changes of the British Isles, especially those which occurred towards the

Table 3.5 Typical mammalian faunas of Middle and Upper Quaternary deposits in the British Isles

	Stage	Typical mammal fauna	Approximate age (K years BP)	Oxygen isotope stages
HOLOCENE	FLANDRIAN	European elk, giant ox, red deer, wild boar, wolf, beaver		1
			10	
LATE QUATERNARY	DEVENSIAN	Lemming, spotted hyena, cave lion, brown bear, woolly mammoth, horse, woolly rhinoceros, red deer, reindeer, bison, musk ox		2–5d
			120	
	IPSWICHIAN	Hippopotamus, straight-tusked elephant, spotted hyena, narrow-nosed rhinoceros, fallow deer		5c
			130	
MIDDLE QUATERNARY (Wolstonian Complex)	(RIDGEACRE)			6
	(STANTON HARCOURT)	Brown bear, mammoth, horse, extinct horse (*Equus hydruntinus*) a small wolf, a large lion, northern vole	170	7
	?	Vole, lemming, ground squirrel, cave lion, brown bear, mammoth, straight-tusked elephant, horse, Merck's rhinoceros, narrow-nosed rhinoceros and woolly rhinoceros, red deer, giant ox, bison		8
	HOXNIAN	Lemming, horse, red deer, straight-tusked elephant, Merck's rhinoceros and narrow-nosed rhinoceros, fallow deer	300–400	9

Source: based on Sutcliffe (1985)

end of the Pleistocene (by which time most familiar present-day species of mammals had already evolved), are unsurpassed in their contrast.' Although best documented in Europe and North America, similar changes were occurring in other parts of the world.

Table 3.5 provides a summary of the dominant fossil mammal species present in the British Isles during the middle and late Quaternary periods, taking into account the faunal remains from Marsworth (Green *et al.*, 1984; Section 2.2). Although Table 3.5 presents only a fraction of the available evidence – more details are given in Sutcliffe (1985) and Stuart (1982) – it shows that tremendous changes occurred in the distribution of mammals during this period of ca. 1×10^6 years. For example, during the last interglacial in Britain (Ipswichian, *sensuo stricto*) hippopotamus roamed the area of what is now central London, accompanied by cave lions, straight-tusked elephant, narrow-nosed rhinoceros, as well as wild boar, bison

and giant ox. Thus, the interglacial mammal faunas, as evidenced by fossils from the earlier Hoxnian and Cromerian interglacials, as well as the Ipswichian, were very different to those of the current interglacial period. The optimum phase of the Ipswichian was probably warmer than now, since many deposits of this period contain fruits of water chestnut (*Trapa natans*), a thermophilous aquatic that today reaches its northerly distribution in central France. These warmer, possibly more continental conditions, and the presence of extensive mixed-oak forest favoured the spread of species such as hippopotamus up to an altitude of 400 m in the Yorkshire Moors (Sutcliffe, 1985). Sometime during the following glacial period of the Devensian, the hippopotamus and straight-tusked elephant disappeared from the British Isles, never to return.

The advent of the Devensian ice, which reached its maximum extension approximately 18 K years BP, brought great changes in the herbivorous mammalian fauna. Collared lemmings, reindeer and

Table 3.6 Chief mammalian extinctions during the latter part of the late Quaternary

Region	Species
North America	Shasta ground sloth, Jefferson's ground sloth, sabre-toothed cats, scimitar cat, giant beaver, North American camel, mastodont and mammoth Total = 33 genera Percent extinct = 73.3
South America	*Glyptodon* (a relative of the armadillo), a giant ground sloth, a sabre-toothed cat, various horses, a giant capybara Total = 46 genera Percent extinct = 79.6
Australia	Various Protemnodons (wallabies), Procoptodons (large kangaroos), various *Macropus* spp. (kangaroos), *Megalania* (a large lizard), *Thylacoleo* (a marsupial lion) Total = 19 genera Percent extinct = 86.4
Northern Eurasia	Woolly rhinoceros, woolly mammoth, straight-tusked elephant, giant deer, bison, hippopotamus, musk ox Total = 17 genera Percent extinct = ? but low
Africa	*Hipparion* (a large true horse), a species of elephant, giant deer, camel and three bovids Total = 7 genera Percent extinct = 14.3

Source: based on Martin (1984)

musk ox, for example, along with woolly mammoth and woolly rhinoceros occupied the tundra zone to the south of the ice sheet. The carnivorous species, however, remained relatively unaffected, preying upon cold-tolerant herbivores instead of temperate species. However, the closing phase of the Devensian witnessed further considerable changes in mammalian faunas, notably the extinction of many large species. According to Saunders (1987), mammoth, horse, woolly rhinoceros and reindeer all occur in Devensian deposits between 26 and 18 K years BP, when Arctic tundra was prevalent. Until recently, it was generally considered that, of these species, mammoth did not survive into the late Devensian. Coope and Lister (1987) have, however, reported the presence of a mammoth skeleton at Condover, near Shrewsbury, which is radiocarbon dated to ca. 12.8 K years BP, some 5 K years younger than previous finds. This suggests the advent of full-glacial conditions did not initiate the final extinction of this species, as had been thought. Nevertheless, by the opening of the Holocene, 10 K years BP, many of these large mammals had disappeared from the British landscape, including the woolly rhinoceros, woolly mammoth, bison, cave lion and spotten hyena. Moreover, the Irish elk (*Megaloceros giganteus*), so prolific in deposits of late-glacial age in Ireland, is last recorded before the Younger Dryas cold period. According to Barnosky (1994), this extinction can be used to test the role of climatic change, particularly because there is no evidence for human presence in Ireland until ca.

8.5 K years BP (Woodman, 1977). The occurrence of the Younger Dryas cold period caused a reduction in the availability of nutrient-rich plants in spring and early summer and a reduction of plants for browsing during the summer. For the Irish elk it seems likely that the Younger Dryas cooling crossed thresholds of tolerance.

Clearly these extinctions caused a significant alteration in the mammalian faunas (especially megafaunas, whose body weight exceeded 44 kg) in the British Isles. Similar changes occurred in many other parts of the world at the same time, and some of the most notable extinctions are given in Table 3.6. Kurtén and Anderson (1980) present evidence for the extinctions of 35 megafaunal species by the close of the last ice age (Wisconsin) in North America ca. 10 K years BP. Dates from Rancho La Brea, Los Angeles (Marcus and Berger, 1984), one of the most important bone-bearing sites in North America, suggest that extinction occurred ca. 11 K years BP over a relatively short time span. In South America there is evidence for the extinction of 46 genera at about this time, including *Glyptodon*, a mastodon relative of the armadillo. Extinctions also occurred in Australia in the period 26–15 K years BP (Horton, 1984). Losses in northern Eurasia and Africa were not quite so severe, but nevertheless significant.

Marshall (1984) has summarised the dates for extinctions; most fall in the range 12 to 8 K years BP and there is currently much speculation as to the cause (Martin and Klein, 1984). Was climatic change or human activity the underlying factor, or possibly a combination of both? The role of climatic change in causing extinction has been discussed by Barnosky (1994), who points out the difficulties of ascertaining climatic cause and effect relationships. Moreover, there is clear evidence that humans did hunt some of these large animals. In Britain, for example, barbed bone points have been found in association with a skeleton of the elk *Alces alces* at High Furlong near Blackpool (Hallam *et al.*, 1973), which is dated at 11 to 12 K years BP. Martin (1984) has suggested that people were responsible for most of these extinctions, at least in North America. His view is based on the fact that such animals had survived earlier cold periods, including actual ice ages, that many of the extinctions coincide with the arrival of humans in North America (the Clovis culture), that areas such as Madagascar and New Zealand witnessed extinctions as humans arrived or some time after, and that a megafauna similar to the fauna of the late Quaternary survived in Africa. The question of when humans first arrived in the Americas is as much a focus of controversy as the extinction issue. Dillehay and Collins (1988) have presented evidence for human presence at Monte Verde, Chile, ca. 33 K years BP. Even if this is debatable, since it predates other sites by ca. 20 K years, there is evidence for human presence at Taima-Taima in Venezuela (Gruhn and Bryan, 1984) associated with mastodon kill at ca. 13 K years BP, some 1000 years before the major extinction period (Section 4.3). This need not imply the absence of a relationship between human populations and animal extinctions, but the presence of humans considerably earlier than the extinctions weakens the case for cause and effect.

A similar debate exists in relation to the extinction of the Australian megafauna. Horton (1984), for example, has suggested that humans, present in Australia for at least 40 K years, did not play a major role in extinction, but rather that enhanced aridity caused the demise of woodland habitats to which most of the megafaunal species were adapted. In Europe, the Younger Dryas cold event (Section 3.4) may have been responsible for the extinction of at least some species. The giant deer, for example, disappeared from Ireland before a cold stage equivalent to the Younger Dryas, and some 2 K years before human colonisation of the island. It is possible that the relatively rapid climatic and ecological changes that occurred during this period may well have contributed to the demise of

many faunal species. Perhaps, both human populations and climatic change were instrumental in affecting the faunal elements of these dynamic ecological systems.

3.6   Climatic change during the early Holocene

The early Holocene lasted from ca. 10 to 5 K years BP. Temperatures were once again ameliorating by 10 K years BP, as the Younger Dryas cold period ended. The opening period of the Holocene (oxygen isotope stage 1) represents the mesocratic phase of the interglacial cycle (Fig. 3.1). It was one of rapid climatic and environmental change. An extensive forest cover developed in much of northwest Europe with peatlands in the boreal zone. The optimum extent of forest developed by ca. 5 K years BP, the end of the early part of the post-glacial period. In both hemispheres new patterns of vegetation became established as ameliorating temperatures altered precipitation regimes, rising sea levels and changing soil processes provided the backdrop against which various faunal and floral elements migrated, retreated and competed to produce mesocratic landscapes worldwide. As occurred in the preceding protocratic phase, there were considerable attitudinal and latitudinal variations in the response of the biota to climatic change. There is also considerable debate as to the role of humans in environmental change during this period and the role of climatic change in influencing human activity, as will be discussed in Chapter 4.

It is not yet known what brought the Younger Dryas to a close, though Ruddiman and McIntyre's (1981) suggestion that it was due to a northward shift in the oceanic polar front (Fig. 3.4) is one possibility as is changing North Atlantic Deep Water circulation. The demise of the Younger Dryas was rapid, as is reflected in a wide range of proxy climatic data, e.g. fossil coleoptera (Fig. 3.3.) and pollen (Fig. 3.5) as well as oxygen-isotope stratigraphies in ocean sediments from the North Atlantic, where it is recorded as termination 1B (Section 2.3) and ice cores (Section 2.4). Mayewski *et al.* (1993), for example, have estimated on the basis of annual ice layers from the GRIP2 core that the Younger Dryas lasted 1300 ± 70 years and ended abruptly within 10–20 years. Even where a Younger Dryas event is not strongly recorded, evidence points to a rapid shift of climate between the last glacial stage and the present interglacial. The oxygen isotope curve shown in Figure 2.1 is fairly typical of ocean-sediment cores and shows a steep change between oxygen isotope stages 2 and 1. This horizon is dated to ca. 10 K years BP, the generally accepted date for the opening of the Holocene period. However, in Antarctica, the Vostok and other Antarctic ice (Section 2.4) the beginning of warming was at least ca. 1 K years earlier. The discrepancy between Arctic and Antarctic ice cores remains unexplained, though it is likely to be a real difference rather than an artefact of dating. It is possible that events in the northern and southern hemispheres are related and metachronous rather than being unrelated. In terms of overall patterns of climatic change, however, there is a high degree of concurrence between sediment records from North Atlantic ocean cores, Arctic ice cores and palaeoecological evidence from lake sediments and peats. The availability of a wide range of data from the North Atlantic region has facilitated many local palaeoenvironmental reconstructions and is now contributing to regional syntheses such as those cited in Sections 3.3 and 3.4. The amount of evidence from the southern hemisphere has also increased considerably in recent years, though the range of data is not as great as for the northern hemisphere.

According to Greenland ice-core data (Section 2.3), ^{18}O concentrations increase just before 10 K years BP and rise rapidly thereafter until ca. 9.8 K years BP. They then apparently remain more or less constant for the remainder of the Holocene (Fig. 2.3), reflecting a relatively stable climate. The coarseness of the data, however, means that changes of relatively small amplitudes would not be detected. The opening of the Holocene brought an increase in average July temperatures of ca. 5 °C from 10 °C to 15 °C, representing an increase of ca. 1.7 °C per century (Atkinson et al., 1987) and a decline in the annual range of temperatures to ca. 14 °C. It is generally considered that average July temperatures have not fluctuated much more than by ± 1 °C. This increase in temperature was caused by an increase in June insolation (e.g. McManus et al., 1994). This relative stability is, however, untypical of interglacial periods if the record of the last interglacial, as reflected in GRIP, is judged to represent reality (see Section 2.4 for a discussion of this controversy). According to White (1993), three climatic modes operated during this interglacial: a mode like the

present, a cooler mode and a warmer mode. And the shifts between modes were rapid, often a decade in length, whereas the modes themselves lasted decades or millennia. Perhaps the Holocene is unusual in its stability or perhaps the discrepancies between the GISP2 and GRIP cores reflect either an imperfect record, due to ice deformation in the GRIP core, or the real operation of localised climate fluctuations in its vicinity. There is some, though limited, evidence from pollen assemblages in Europe (Field et al., 1994) and from rock magnetism (Thouveny et al., 1994) to support short-lived climatic excursions. This issue reflects some of the difficulties of achieving accurate climate reconstructions. There are equally significant problems when using other forms of proxy climatic data such as plant remains. The importance of plant remains in this context is highly debated. Suffice it to say that the distribution of plants and vegetation communities is largely determined by macroclimate, so it can be used in a general way to reconstruct past temperature and precipitation regimes; nevertheless, most plants will survive under a range of environmental conditions. Thus, fossil or pollen assemblages cannot be used to reconstruct precise palaeoclimatic conditions. Many of the issues relating to this have been discussed by Delcourt and Delcourt (1991).

There has also been much debate as to when optimum temperatures were reached in the Holocene. Inevitably there appears to have been much regional variation. According to Thompson et al. (1995), ice-core oxygen isotopes from Andean Peru (Section 2.4) indicate that climate was warmest between 8.4 and 4.2 K years BP. Optimum temperatures in northwest Europe were probably reached before the development of a climax vegetation community. The so-called climatic optimum, a term still widely used, is applied to the period ca. 6 to 4.5 K years BP. It reflects the optimum expression of the mesocratic phase of the interglacial cycle (Fig. 3.1) when mixed-oak deciduous woodland developed to its maximum extent. The term has historic associations with early Quaternary scientists, though it is now well established that the climax vegetation varied considerably from region to region (Fig. 3.1). Moreover, the term 'hypsithermal' is often used to denote the period ca. 9 to 2.55 K years BP, which clearly includes most of the early Holocene as defined here. Neither term is particularly appropriate because of local and regional variations,

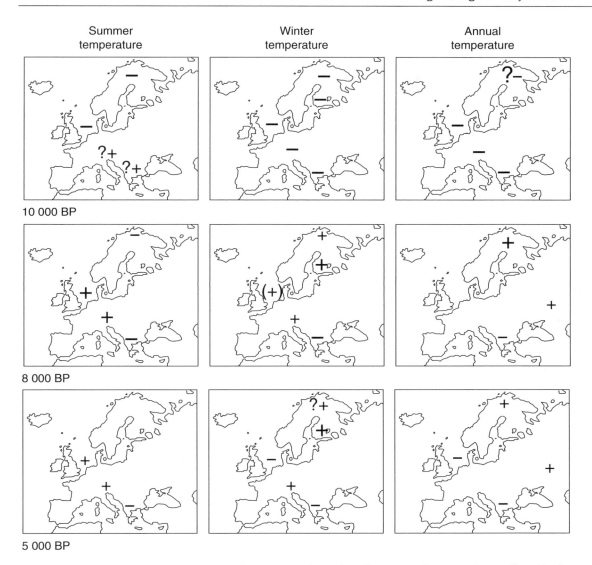

Summer temperature   Winter temperature   Annual temperature

10 000 BP

8 000 BP

5 000 BP

Fig. 3.7 Regional climatic changes from 10 to 5 K years BP inferred from European pollen assemblages (from Huntley, 1990).

the likelihood of mosaic patterns of vegetation communities and the probable lag of vegetation development behind climatic amelioration.

Nevertheless, numerous attempts have been made to reconstruct many facets of early (and late) Holocene environments. Huntley (1990), for example, has reconstructed vegetation patterns from pollen data extracted from a wide range of European sites and from those patterns he has made some general statements about temperature regimes during a series of time slices. They are given in Figure 3.7, which shows that early in the Holocene (8 K years BP) summer temperatures and annual temperatures in higher latitudes were increasing; they were actually warmer than the present day except in southern Europe. By 5 K years BP the North Sea basin was cooling but temperatures in southern Europe remained below their present levels. Temperatures in southern Europe are highlighted by Watts *et al.* (1996), who have calculated that in the early Holocene at Lagi di Monticchio, southern Italy, early Holocene winter conditions were colder (ca. $-5\,°C$) than today ($3.9\,°C$). It is interesting to note that these patterns are similar to those predicted to occur if global warming due to enhanced greenhouse effect were to ensue (Section 8.5)! It is also likely that the impact of the global amelioration of temperatures in the early Holocene was accompanied by the impact on climate of rising sea levels. For example, as Britain became an island ca. 6.5 K years BP climate may have become more oceanic rather than continental. Similarly, the conjunction of the formerly isolated Baltic Sea with the Atlantic Ocean would have had local and regional effects in Scandinavia and northern Europe. Sea-surface temperatures also increased during the early Holocene. Alkenone-derived temperatures (Section 2.3) for the east equatorial South Atlantic (off the coast of Angola) show that mid Holocene sea-surface temperatures were ca. $2\,°C$ higher than they are now (Schneider et al., 1995). Moreover, the increase in salinity of ocean waters that occurred as climate switched from glacial/cold to glacial/warm was accompanied by temperature changes in sea-surface waters. Summer sea-surface temperatures in the North Atlantic Ocean rose rapidly to ca. $14\,°C$ from ca. $10\,°C$ during the transition from the Younger Dryas to the Holocene (Duplessy et al., 1992), an order of magnitude almost compatible with the temperature increases reflected by land biota, e.g. coleoptera (Fig. 3.3). Zhao et al., (1995) have also shown that sea-surface temperatures off the coast of northwest Africa changed as the Holocene opened. Their alkenone stratigraphy (Section 2.3) indicates a temperature of $18\,°C$ during the last glacial, which rose to $21.5\,°C$ in the early Holocene. An even larger increase in sea-surface temperatures occurred in the Aegean Sea. According to Aksu et al., (1995), planktonic foraminiferal, coccolith and dinoflagellate data indicate an increase of $5-10\,°C$ between ca. 14 and 9.6 K years BP.

Geomorphological evidence for an intensified monsoon bringing increased precipitation to parts of the United States is discussed by Forman et al., (1995) and Leigh and Feeney (1995). Forman et al. have examined dune systems on the Piedmont and High Plains region of the United States and have identified several periods when the dunes were reactivated. These periods include 20 K years BP and several late Holocene episodes. Forman et al. suggest that, at the opening of the Holocene,

increased monsoonal precipitation allowed grass and shrubs to grow, stabilising the dune system. There was also a greater difference between land and sea temperatures due to rising insolation values and the cooling effect of meltwater on the Gulf of Mexico. Leigh and Feeney's (1995) evidence focuses on meandering palaeochannels in the floodplain of the Ogeechee River in southeast Georgia. As well as being active during part of the last glacial period, the channels were active during the period 8.5 to 4.5 K years BP. This was probably because flood palaeodischarges were at least double those of today as intensified monsoonal circulation developed, bringing with it increased precipitation.

Moreover, Vance et al. (1995) have synthesised palaeociological data from the Canadian Prairie provinces in order to reconstruct palaeoenvironments. They state that the early Holocene climate of the western foothills of the Rocky Mountains and the prairies was relatively arid. Conditions were warmer and drier than those of today. Tree lines were higher than they are today, alpine glaciers were smaller, lake levels were lower and forest fires were more prevalent. The data indicate that summer temperatures at ca. 6 K years BP may have been $3\,°C$ higher than now, with an increase in mean annual temperature of $0.5-1.5\,°C$ and a reduction of ca. 65 mm of precipitation. Further evidence for a warmer and drier environment in the early Holocene as compared with today comes from Vancouver Island, from where Nagorsen et al. (1995) have identified fossil remains of Ursus americanus. Pollen analyses of sediments clinging to the bones indicates the presence of a mixed coniferous forest, and radiocarbon dates from the bones indicates this had developed by ca. 9.8 K years BP. Additional pollen analysis from British Columbia has led Hebda (1995) to suggest that between 9.5 and 7 K years BP the west coast of Canada enjoyed a 'xerothermic' climate, i.e. warm and dry. Worona and Whitlock (1995) suggest that the early Holocene in the central Coast Range of Oregon was characterised by summer drought and increased fire frequency. These conclusions support the work further north in Canada, discussed above. Instead of decreasing, precipitation increased in other parts of North America, producing pluvial episodes (Davis and Sellers, 1987). The same is true of many low-latitude regions, which were generally wetter during the early Holocene than during the late Holocene. In southern India, for example, Sukumar et al. (1993) have obtained a record of carbon-13 from tropical peats, indicating a wet period between 14 and 8 K years BP. Similar results have been obtained from Tibetan lakes (Fontes *et al.*,1993). The prevalence of aridity at high latitude levels and the wetter climates of low latitudes may have produced relatively low concentrations of methane in the atmosphere during the early Holocene. This is suggested by a detailed analysis of the Holocene section of the GRIP ice core from Greenland (Blunier *et al.*, 1995). Lack of water at high latitudes could have restricted the formation of wetlands hence the production of methane.

3.7   Regional expression of changes during the early Holocene

This topic of immense proportions has been partially reviewed by Tallis (1991), who examined the vegetation history of this period and the late Holocene. However, the following examples should illustrate the magnitude of environmental change that has occurred during this period of 5 K years.

At the opening of the Flandrian (Holocene) Britain witnessed the invasion of tundra vegetation, dominated by grassed, sedges and dwarf shrubs, and by pioneer tree species tolerant of relatively poor soils and little shade. The first of these invaders was the tree birch (*Betula pubescens* and *B. pendula*), which was rapidly followed by pine (*Pinus sylvestris*) under relatively dry continental conditions that were at least as warm as today. These species appear to have become well established by ca. 9.5 K years BP, presumably creating a sufficiently rich soil for the invasion of hazel (*Corylus avellana*) that may have been either a subordinate understorey species in birch and pine woodland, or a woodland dominant. By approximately 8 K years BP, more thermophilous species such as oak (*Quercus*) and elm (*Ulmus*) had become established, and by 6 K years BP they had become abundant over large parts of the British Isles, so that pine (*Pinus*) and birch (*Betula*) were relegated to the more inhospitable areas of upland Scotland which could not support the warmth-demanding species. Lime (*Tilia*) is first recorded in Britain by ca. 7.5 K years BP and is considered to be the most thermophilous of the post-glacial invaders. It spread to its northern limit by ca. 5 K years BP.

On the basis of 12 dated pollen diagrams from various parts of the British Isles, Bennett (1988) has summarised the major components of vegetation communities during the Flandrian. This provides a generalised representation but nevertheless indicates that woodland mosaics, as opposed to single-species woodlands, existed throughout the Flandrian and varied spatially and temporally. For example, in southern and eastern England the dominant lime was accompanied by oak and elm. Alder (*Alnus*) was also a relatively common species that may even have become established during the late-glacial period (Bush and Hall, 1987), subsequently expanding its population to become locally abundant by ca. 7 K years BP, especially in wetter environments. Ash (*Fraxinus*) was one of the later mesocratic species to expand, and populations increased between 6 and 5 K years BP. It may have occupied sites that were too damp for oak and elm, but not wet enough for alder.

Bennett and Humphrey's (1995) calculation of rates of vegetation change in two contrasting situations in Britain has also demonstrated regional variation in the response of vegetation communities to early Holocene warming. The sites in question are Hockham Mere in Norfolk and Loch Lang in Scotland's Western Isles, for which radiocarbon-dated pollen sequences are available. Bearing in mind the problems associated with determining rates of change over limited time periods with ^{14}C dating (Pilcher, 1991), Bennett and Humphrey use a variety of statistical techniques to demonstrate that the rate of change was greatest at both sites as the Holocene opened. However, the rate of change at Hockham Mere was five times higher than at Loch Lang. Among the various reasons for this difference, discussed by Bennett and Humphrey, is the possibility that the continental climate of Norfolk altered more rapidly than the oceanic climate of the Western Isles. This possibility requires much further testing before it can be accepted unequivocally. But Bennett and Humphrey do raise the possibility that if contemporary global warming is as rapid as it was at the opening of the Holocene, it will affect areas with continental climates more acutely than areas with oceanic climates. By comparing pollen diagrams from various parts of the British Isles, it is also apparent that not everywhere became forest-covered during the early Flandrian. Godwin (1975) estimates that there was a 60 per cent forest cover by ca. 5 K years BP, and Rackham (1986) has implied that areas such as northern Scotland and the Outer Hebrides lay beyond the limits for continuous forest cover. It is difficult to refute or confirm either of these statements. However, pollen diagrams for the outer Hebrides (Bennett *et al.*, 1990a) show only low frequencies of arboreal pollen, perhaps attributable to long-distance transport from the mainland, though there are abundant remains of trees preserved beneath the peat deposits of these islands (Wilkins, 1984). Thus the palaeobotanical evidence is conflicting. Indeed the low arboreal pollen concentrations may arise from exposure to Atlantic winds, which could have impaired flowering capacity and hence pollen productivity, rather than the absence of trees. Until recently it has generally been considered that grasslands in Britain were the result of anthropogenic activity, having their origins in the advent of neolithic agricultural practices that developed after ca. 5 K years BP (Chapter 4). However, work in the Yorkshire Dales (Bush, 1993) has shown that grassland communities originating in the late-glacial period persisted well into the Flandrian and may have been the progenitors of modern grasslands in this area today. Grasslands probably also existed in river valleys and coastal areas that were less suitable for trees, and in some areas, especially in upland regions, blanket and ombrogenous bog developed in response to the more oceanic climate that developed after ca. 7 K years BP.

Faunal changes accompanied the floral changes. Following the extinctions that occurred in the late-glacial period (Section 3.5), the development of a forest cover probably encouraged an increase in herbivores, notably aurochs, elk, red deer and roe deer. The woodmouse, bank vole and field vole also arrived in Britain in the early Flandrian, along with the hedgehog, badger wild boar, beaver, mole, wolf and marten. Of particular interest is the record of the European pond tortoise (*Emys orbicularis*), since the northern limit of this species is controlled by temperature and cloudiness during the summer. According to Stuart (1982), reproductive success requires a mean July temperature of 17–18 °C, abundant sunshine and few damp, cloudy or rainy days. Its presence in early Flandrian deposits indicates that summer temperatures were warmer than today by 1–3 °C. It does not occur after 5 K years BP in Britain, though it is recorded much later in Denmark and southern Sweden, reflecting the more continental nature of the climate on the European mainland.

Changes of a similar magnitude to those in Britain were also characteristic of the mesocratic phase of the interglacial cycle elsewhere. In northwest Europe the trends were similar to those in Britain (Birks, 1986).

Khotinsky (1993) has suggested that major vegetation changes occurred in the Russian Plain as continentality diminished. The pre-Holocene vegetation zones were replaced with communities similar to those of today. In the central Russian Plain, for example, birch (*Betula*) and pine (*Pinus*) forests were replaced with oak (*Quercus*), elm (*Ulmus*), lime (*Tilia*) and hazel (*Corylus*) forests. Woody plant species extended into the tundra region of northeastern Russia between 9.5 and 8 K years BP (Lozhkin, 1993). And Kremenetski (1995) provides evidence for a warmer climate than at present in the early Holocene in the southwestern Ukraine. Birch and pine replaced late-glacial shrubland in south Sweden and Denmark, and hazel and elm had become established by 9.5 K years BP (Berglund *et al.*, 1994)

There is also considerable evidence for changes in lake levels worldwide, considered to be a reflection of climatic change. In Europe such work has been carried out by Guiot *et al.* (1993) and Harrison and Digerfeldt (1993). Harrison and Digerfeldt consider changes in lake levels in southern Sweden and in the Mediterranean region. The results are summarised in Figure 3.8, which shows that lake levels were lower than they are now in southern Sweden at ca. 9 K years BP and between ca. 5.5 and 3.5 K years BP. Intermediate lake levels occurred in the Mediterranean ca. 9 K years BP, but by 6 K years BP almost all lakes were high, though a low phase followed after 5 K years BP, occurring rapidly in the west but relatively slowly in the east. This contrast between north and south led Harrison and Digerfeldt to suggest there were major changes in the patterns of atmospheric circulation over Europe during the Holocene. There is increasing evidence for lake-level changes elsewhere, as Grove (1993) has illustrated for Africa. All the closed basins of intertropical Africa were occupied by lakes and many lakes rose more than 100 m above their present levels. This reflects wet (pluvial) conditions, which may have been caused by 15–47 per cent more precipitation than now. The possibility of increased precipitation during the early Holocene is also suggested by pollen and plant macrofossil assemblages from Ahakagyezi Swamp in southwest Uganda (Taylor, 1993). From ca. 11 K years BP dry montane scrub on the surrounding hillsides was replaced by moist montane forest, in which likely successional changes then occurred until ca. 5.8 K years BP. Elenga *et al.* (1994) have presented similar pollen-analytical evidence for the development of humid conditions in the Congo after 13 K years BP,

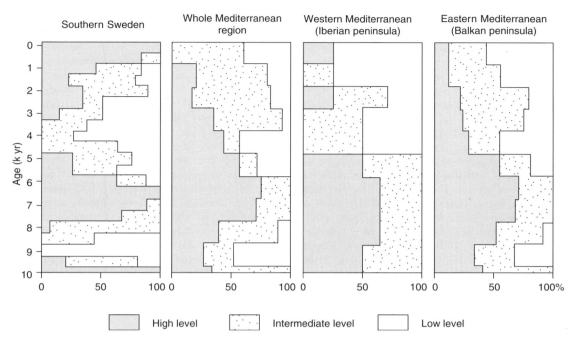

Fig. 3.8 Lake-level changes in southern Sweden and the Mediterranean region during the Holocene (based on Harrison and Digerfeldt, 1993).

when grasslands were replaced by mesic forest. Fang (1991) has surveyed the evidence for lake-level changes during the late Quaternary in China. By ca. 6 K years BP nearly all lakes in southern China, and many in other parts of China, had reached the highest levels of the post-glacial period. This correlates with a palaeoprecipitation study by Maher *et al.* (1994), which indicates that precipitation was higher during the early Holocene with increases being especially great near the northeastern edge of the Tibetan Plateau. These data reinforce those from other sources (see above and Section 3.7) for the prevalence of humid conditions and high precipitation rates in low latitudes during the early Holocene.

Palaeoecological work in Alaska and northern Canada also shows that the early Holocene was a period of rapid change. Anderson and Brubaker's (1994) synthesis of late Quaternary vegetation changes in north central Alaska, for example, reflects the variations that occurred with aspect and altitude. Shrub tundra continued to dominate most upland areas by 9 K years BP whereas poplar (*Populus*) woodlands developed in river valleys and south-facing slopes. Alder (*Alnus*) shrubs then

expanded rapidly throughout north-central Alaska between 8 and 7 K years BP. At the same time spruce (*Picea glauca*) spread north from source areas in northwestern Canada; spruce forests became particularly common along river banks with isolated patches within the shrub tundra. By 6 K years BP spruce (*Picea mariana*) dominated all forests and had invaded areas of shrub tundra. Birch-shrub (*Betula*) tundra and meadow communities expanded in southwestern Alaska ca. 9.8 K years BP (Hu *et al.*, 1995). Alder (*Alnus*) thickets became established ca. 7.4 K years BP but spruce (*Picea glauca*) was a relatively late arrival at 4 K years BP when compared with north-central Alaska. Vegetation change in the Mackenzie Mountains of Canada's Northwest Territories reflect similar controls of aspect and altitude as well as latitude (Szeicz *et al.*, 1995). Pollen-analytical data from three lakes in today's tundra along with zones of forest tundra and open forest have revealed the dynamics of vegetation change as the Laurentide ice sheet retreated. This occurred ca. 11.5 K years BP, when a herb tundra dominated by wormwood (*Artemisia*) and willow (*Salix*) developed. Approximately 10.2 K years BP birch (*Betula glandulosa*) expanded into the herb tundra on a regional basis, and balsam poplar (*Populas balsamifera*) extended beyond the tree line. Spruce (*Picea glauca*) immigrated ca. 8.5 K years BP and together with the later invading alder (*Alnus*) came to dominate the vegetation by ca. 6 K years BP.

Substantial environmental changes ensued in the southern hemisphere as temperatures ameliorated. Many studies from the Andes reflect the importance of altitude and aspect. For example, Hansen *et al.*, (1994) have detailed vegetation change based on pollen records from the central Peruvian Andes at heights above 4000 m. The opening of the Holocene is marked by a rapid increase in Urticales as Compositae and *Polylepsis–Acaena* type pollen declines. Thereafter, Urticales and plantain (*Plantago rigida*, a cushion bog forming species) increase. This is considered a response to rising temperature and enhanced moisture between 11 and 7 K years BP. A similar response of *Plantago rigida* to early Holocene climatic change is recorded in the eastern Cordillera of Colombia, though in a warm humid phase ca. 7.8 and 6.6 K years BP this bog plant declines (Bosman *et al.*, 1994). In addition, Wirrmann and Mourguiart (1995) have shown that Lake Titicaca on the altiplano of Bolivia and Peru experienced a major decline in water level. From ca. 13 K years BP the lake changed from a deep freshwater lake to a shallow lake containing evaporite deposits. These results are in agreement with other lake-level changes in the region.

Still further south, in Australia and New Zealand, the early Holocene climatic changes generated responses in geomorphic systems and ecosystems. In the Lake Eyre basin in Australia's western Simpson Desert, Nanson *et al.* (1995) have shown that illuviation occurred in the early to mid-Holocene. This was a response to increased monsoonal precipitation providing sufficient moisture to bring about floodplain formation. This did not occur in the eastern part of the Lake Eyre Basin, suggesting that the western part of the basin was a key area for registering changes in late Quaternary precipitation patterns. Climatic amelioration at ca. 11 K years BP is also reflected in the pollen assemblages obtained from Lake Bolac in western Victoria (Crowley and Kershaw, 1994), where there occurred a marked increase in casuarina woodland (mainly *Allocasuarina verticillata*). This persisted until ca. 8 to 7 K years BP when eucalyptus (*Eucalyptus camaldulensis*) woodland replaced it, possibly due to increasing precipitation. Similar changes have been

recorded at Lake Terang, also in western Victoria (D'Costa and Kershaw, 1995). During the early Holocene at Taranaki, North Island, New Zealand, McGlone and Neal (1994) have shown that between ca. 12 and 11 K years BP there was a rapid change from grassland to a mixed conifer/broadleaf forest, which gradually lost its cool temperature components by 9.5 K years ago. This led McGlone and Neal to speculate that climatic conditions similar to those of the present were reached fairly rapidly in the Holocene. However, Mildenhall (1994) has suggested that between ca. 9.8 and 6 K years BP temperatures were at least 1 °C higher than they are now, in order to support mangrove communities in the East Cape region of New Zealand's North Island, evidence for which comes from pollen analysis of coastal sediments.

The publication of a wide range of new data from various parts of the southern hemisphere in the last decade has prompted several syntheses to be undertaken. Markgraf *et al.* (1995) have reviewed vegetation history for the last 24×10^6 years. During the early Holocene, areas that are now forested were rapidly colonised by trees as open-habitat species declined. Thomas and Thorp (1995) have reviewed the geomorphological evidence for environmental change in the humid and subhumid tropics. They conclude that such areas in Africa, Amazonia and Australasia experienced an early Holocene pluvial episode between 9.5 and 7.5 K years BP and that lowland rainforests did not become established until after 9 K years BP. The vegetation history of tropical forests in Africa and South America has been discussed by Servant *et al.* (1993), who demonstrate that a dense forest cover had developed by 9 K years BP, after a slow late glacial increase in tree species. In the southern tropical zone of South America, however, there was a marked regression of forest between 6 and 5 K years BP, though there was no evidence of a similar regression in Africa.

In Central America there is charcoal evidence for human disturbance of rainforests from ca. 11 K years BP (Bush *et al.*, 1992); if this pattern is repeated elsewhere in tropical rainforests it is possible that this contributed to the promotion of species diversity, along with other types of disturbance such as changing fluvial regimes causing flooding and deposition (Colinvaux, 1987; Thomas and Thorp, 1995). If this proves to be correct, it may weaken the much disputed refugium hypothesis, which requires that groups of plants (and animals) survived in isolated patches during periods of climatic deterioration and there developed genetically isolated populations which did not intermix when ameliorated conditions facilitated their expansion. Supporters of the refugium hypothesis may, however, take comfort from the work of Joseph *et al.* (1995) on the genetic characteristics of Australian rainforest birds. They have demonstrated that for four bird species present in the now conjunct rainforests of Daintree and the Atherton Tableland, but which were separated during the cold periods of the Quaternary, genetic differences in the two populations indicate the development of two distinct populations. Perhaps isolation is a more appropriate way to express the mechanism of species differentiation rather than the existence of refugia. Moreover, it is possible that geomorphic processes such as those mentioned above also contributed to speciation through isolation. These studies illustrate the dynamic nature of tropical regions during the early Holocene and highlight the fact that the rainforests of today are the products of a changing rather than a stable environment.

3.8   Climatic change during the later Holocene

The later Holocene lasted from ca. 5 K years BP to the present. Palaeoenvironmental evidence from northwest Europe indicates that the oligocratic phase of the interglacial cycle (Fig. 3.1) had begun by ca. 5 K years BP. There is abundant evidence for the replacement of deciduous forests by such vegetation communities as heathlands, moorlands and peatlands though many of these changes were initiated by human activity rather than by climatic change. Indeed, separating those changes due to climatic stimuli and those due to human activity is at best difficult and often impossible. Moreover, some changes during the progression of an interglacial cycle, as proposed by Iversen (1958) do not have to be driven by climatic change. This is particularly so for the oligocratic phase; Iversen proposed that soil deterioration due to progressive leaching and consequent nutrient impoverishment could initiate retrogressive vegetational development.

Polar and tropical ice cores, however, provide direct evidence for climatic change during the late Holocene. Reference has already been made, for example, to apparent discrepancies between last interglacial and Holocene records which suggest that the latter was a period of comparatively stable rather than unstable climate (Section 2.4). Nevertheless some, though minor climatic changes when compared with those of the glacial–interglacial transition, are reflected in ice-core data. There is evidence for a temperature decline of ca. 1 °C per century during the period 5 to 3 K years BP with an ensuing slightly warmer period between 2.5 and 1 K years BP. There is also evidence for a warm period, which is known as the medieval optimum, during the thirteenth century AD when temperatures were comparable with those of the period 6 to 5 K years BP (the so-called climatic optimum). There is increasing evidence from diverse regions to suggest this may have been a global event. Arctic ice cores also record a cold period centred on the fifteenth century AD, which has become known as the Little Ice Age. This is a time-transgressive event but does appear to have been global; during the three to four centuries when it occurred there was also at least one relatively warm period. Detailed analyses of Holocene climate and palaeoenvironments have been produced by Thompson *et al.* (1989) and Thompson *et al.* (1986, 1995) based on ice cores from Tibet and Peru respectively. Both record Little Ice Age cooling, which in the Quelccaya Ice Cap of Peru is dated to between AD 1500 and 1900 (Thompson *et al.*, 1986). Data from the Huascarán ice core (Thompson *et al.*, 1995) indicate that after the warmest phase of the Holocene, between 8.4 and 5.2 K years BP, climate cooled gradually with the Little Ice Age, dated as occurring between AD 1450 and 1750. In agreement with the Tibetan cores and other sources the last few decades have witnessed unprecedented warming. Similar recent warming has been identified in tree-ring sequences (e.g. Jones *et al.*, 1993).

The use of tree rings for climatic reconstruction has been examined in Section 2.5, in which examples have been given from a variety of locations. Lara and Villalba (1993), for example, have presented a 3620 year reconstructed temperature record from alerce trees (*Fitzroya cupressoides*) in southern Chile. The results are given in Figure 3.9, which shows that the most pronounced period with temperatures above average occurred between ca. 80 BC and AD 160. Prolonged intervals with below-average temperatures occurred between AD 300 and 470 and between AD 1490 and 1700. The latter reflects the Little Ice Age and the dates are in comparatively

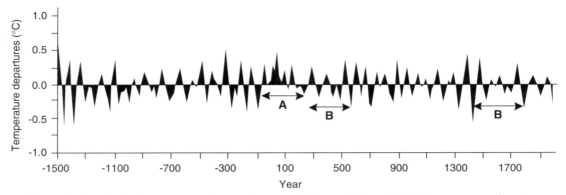

A: longest interval with above average temperatures   **B**: longest interval with below average temperatures

Fig. 3.9 Reconstruction of summer temperature departures from average values (December to March) from a south-central Chile tree-ring record for the past 3.5 K years (based on Lara and Villalba, 1993).

good agreement with those from the Huascarán ice core in Peru (see above). In his synthesis of information on environmental change in South America, Clapperton (1993) suggests that the late Holocene period (i.e. last 5 K years BP) was generally cooler than the early Holocene (Section 3.6). Moreover, there was a diminution in the high rates of monsoon-derived precipitation that characterised the early Holocene in the south-central and southeast regions of the United States; this caused a reduction in flood discharges (Leigh and Feeny, 1995). In addition, the relatively arid phase that characterised the early Holocene in northwestern America (Section 3.6) came to an end after 5 K years BP as precipitation increased (e.g. Vance *et al.*, 1995).

Section 3.6 referred to the value of lake-level changes as a means of reconstructing past climates. In particular, there is abundant evidence for high lake levels in the early Holocene in many tropical regions. During the late Holocene many lake levels declined, probably due to a decrease in precipitation, though there is considerable variation in the response of tropical lakes to late Holocene climates. In tropical Africa, for example, Grove's (1993) review of lake-level changes indicates that there was an arid period during the mid-Holocene (i.e. 4.5 to 3.5 K years BP) when many lakes dried up or became highly saline. After a partial recovery in the early third millennium BP many lakes achieved their present levels by ca. 1.8 K years BP but have never recovered their early Holocene levels. Ely *et al.* (1993) have used another hydrological indicator, the record of palaeofloods, to reconstruct a 5000

year history of flood events in Arizona and Utah in the United States. They have examined flood deposits from 19 rivers and have erected a chronology on the basis of radiocarbon dates. Three distinct periods of extensive floods emerge from this record: 3.6 to 2.2 K years BP, a period centring on 1 K years BP and the period after 0.5 K years BP. Based on pollen and plant macrofossil assemblages from Potato Lake in Arizona, Anderson (1993) has demonstrated that the mid-Holocene was a period of aridity when the lake almost completely dried up. By ca. 3 K years BP, however, the basin was active and lake levels similar to those of today developed. There is some evidence from South Africa for wet periods occurring at the same time as those in the United States. February (1994), for example, on the basis of xylem-cell size in charcoal found at two archaeological sites in Natal Drakensberg has suggested that rainfall was higher at 2.4 K years BP and 0.2 K years BP when compared with present-day values. In addition, Smith (1992) has reviewed the evidence for periods of large floods in the northern and southern hemispheres. A possibly global episode of increased flooding occurred between 3.0 and 1.8 K years BP. As well as recognising a flooding period associated with the Little Ice Age, Smith suggests that the Younger Dryas may also have been a period of above-average flooding.

The period ca. 3.5 K years BP, apart from being characterised by wetter conditions in some parts of the world, may also have been cooler. Dorale *et al.* (1992) have presented oxygen isotope analysis of a speleothem from Cold Water Cave in northeast Iowa for the last 8 K years. The data suggest that after a rapid warming of climate ca. 5.9 K years BP, the so-called climatic optimum, the next major change in average temperatures occurred ca. 3.6 K years BP, when temperature fell by ca. 4 °C. This change was paralleled by a vegetation change from prairie to forest as air from the Arctic penetrated further south than hitherto. Indeed, the movement of the forest–prairie boundary, as registered in pollen and plant macrofossil assemblages, is widely accepted as an indication of climatic change throughout the Holocene in North America (Delcourt and Delcourt, 1991). Wetter and cooler conditions than during the period pre 3.5 K years BP are also indicated by Schäbitz's (1994) work on playa lakes in Patagonia, Argentinia. From ca. 3.5 K years BP there was a gradual increase in lake levels reflecting a change from arid to semiarid conditions with a pronounced seasonality as is the case today. A shift to drier conditions after 3.5 K years BP has also been recorded by Scott *et al.* (1991) from swamp and lake deposits in Namibia and by Zhou *et al.* (1991) from lake sediments in Tibet.

Apart from many localised environmental changes, reflected by a wide variety of palaeoenvironmental indicators, that occurred between ca. 3.5 and 1 K years BP one of the most significant periods of environmental change is the Little Ice Age. The evidence for this has been extensively reviewed by Grove (1988), who points out its time-transgressive nature. There are many parallels that can be drawn between the Little Ice Age and the Younger Dryas (Sections 3.3 and 3.4), not least the debate about their global status. There is now a general belief that both were indeed global, though the Little Ice Age appears not to have been associated with as large a drop in temperature as the Younger Dryas. The available evidence for the Little Ice Age indicates that it was characterised inevitably by regional variation. In high and middle-high latitudes there is evidence for glacier readvance known as the neoglacial. Zhou *et al.* (1991) state that glacier advance occurred in Tibet ca. 0.4 K (i.e. AD 1600) years BP whereas Matthews (1991) gives maximum age estimates for southern Norway of AD 1516, 1450, 1440 and 1424. Clapperton's (1993) synthesis of information on the Little Ice Age in South America indicates that it was under way by the time the Spanish arrived in the early sixteenth century, though it may have begun earlier by ca. AD 800, and it caused a snow-line depression of ca. 300 m in the Ecuadorian Andes. Bradley and Jones (1993) have examined and summarised much of the available

evidence for the Little Ice Age, which includes dendrochronological, ice core, geomorphological and recorded meteorological data, in order to reconstruct climatic variations. They conclude that the coldest conditions occurred between AD 1570 and 1730, and again in the early part of the nineteenth century until warm conditions once again prevailed after the 1920s. Bradley and Jones suggest that summer temperatures declined rapidly after AD 1560, declining by ca. 1 °C in Scandinavia with a much larger drop in Svalbard and northern Canada, reflecting a more severe expression of the cooling trend at high latitudes than in low latitudes, at least in the northern hemisphere. This work illustrates not only the complexity of this period, with cold periods being interrupted by warm periods, but also the difficulties associated with defining what constitutes cool (and warm) trends and the problems of delineating climatic change as opposed to climatic fluctuation. Distinguishing changes from fluctuations is important in the context of whether or not global warming is presently occurring. This will be considered in Section 5.5.

3.9 Regional expression of changes during the later Holocene

Just as there are difficulties in unravelling the record of climatic change during the late Holocene, there are problems associated with distinguishing those environmental changes which were a response to natural rather than cultural agencies. Although human communities were present in almost all parts of the world during the entire Holocene, it was not until the late Holocene that they began to make major impacts on the environment. This is a broad generalisation and a more detailed examination of the cultural impact of Holocene and earlier human communities is given in Chapters 4 and 5; the impact of post-1750, i.e. modern, human communities is considered in subsequent chapters.

In Britain there is evidence for climatic deterioration in the period after 3 K years BP, when there was a trend towards intensified oceanic conditions with increased rainfall and decreased summer temperatures, though possibly by only ca. 1 °C. Ice-core records (Section 2.4) indicate that the Holocene was climatically stable, especially in comparison with the last interglacial. This conclusion was, however, reached on the basis of wide-interval sampling of Arctic and Antarctic ice cores. Evidence from other sources (pollen analysis, glacial moraines, etc.) indicates there were climatic excursions. There is, however, evidence to suggest that some environmental changes occurred as a consequence of natural soil processes during the interglacial cycle (Section 3.2) rather than climatic change. The development of peat in parts of upland Scotland was due to natural soil degradation whereas peat formation in other parts of Britain was the result of deforestation by early agriculturalists. Similar changes occurred elsewhere in Europe as the forest canopy was opened up and moorlands, heathlands and peatlands formed. However, the elm decline is perhaps the most enigmatic event that occurred throughout Europe at the opening of this oligocratic phase. This event is radiocarbon dated to a few centuries before and after 5 K years BP. Despite a wealth of information relating to it, there is still considerable dispute as to its cause(s). The classic interpretation of events at and around the elm decline is that of Iversen (1941, 1944). His pollen analysis of Danish sites shows an initial decline in elm and ivy and a contemporaneous increase in ash. Iversen interpreted these changes as having a climatic cause, since ivy is particularly sensitive to winter temperatures, along with holly and mistletoe. He concluded that an increase in continentality was responsible, with greater seasonal extremes of temperature; this is in general agreement with other northwest European evidence for climatic deterioration at this time (Section 3.8). Stratigraphically above these changes (i.e. later) Iversen (1941) also demonstrated that oak, lime, ash and elm underwent a further though temporary decline as birch, alder and hazel increased in association with charcoal remains. He attributed these changes to clearance or 'Landnam' phases due to human activity. Troels-Smith (1960), however, has questioned the climatic interpretation and has suggested that the earlier elm decline was also anthropogenic as elm, ivy and mistletoe may have been used as fodder plants, analogous to modern practices. Troels-Smith was also one of the first researchers to suggest that Dutch elm disease or a similar pathogen may have played some part in this neolithic elm decline.

Reviewing the evidence for the elm decline in Britain, Smith (1981) emphasises that there is nothing in the palaeobotanical record which necessitates a climatic interpretation. There is, for example, no evidence for a primary elm decline such as Iversen (1941) found in Denmark. Almost all pollen diagrams from the British Isles record an elm decline only in association with other changes that can be interpreted as representing anthropogenic interference in woodland ecosystems. These include the presence of cereal pollen and pollen of ruderal species such as ribwort plantain (*Plantago lanceolata*), the presence of charcoal, increases in grass and herb pollen and occasionally the presence of neolithic artefacts. Indeed, some of these features are characteristic of the Danish sites on which Iversen (1941) initially worked, and are related to his Landnam phases which he envisaged as a three-stage process. Stage 1 included a decline of the dominant arboreal taxa and an increase in grass and herb pollen as clearance was effected. Stage 2 was characterised by the presence of pollen of cereals, ribwort plantain and other ruderals, and was a period of farming, although increasing birch, sometimes alder and then hazel pollen suggested to Iversen that woodland succession and regeneration were also in progress, culminating in stage 3. It is debatable how long these clearance phases lasted. Iversen (1956) suggested that they were probably short-lived, possibly only 50 years or so, since forest regeneration, while initially providing new shoots for animals, would relatively quickly form a closed canopy, and early agriculturalists might then have found it more appropriate to move to a new area and begin the process again.

Subsequent work by Troels-Smith (1956) on the Danish bog Amosen led to some modifications of Iversen's (1941, 1956) ideas. Here Troels-Smith found evidence for farming practices at the level of Iversen's (1941) initial elm decline and below the Landnam phase. These include the presence of cereal and broad-leaved plantain (*P. major*) pollen as well as pollen of ramsons such as wild garlic (*Allium cf. ursinum*), but no evidence of pasture. The presence of species such as wild garlic is particularly interesting since it is sensitive to trampling, and it led Troels-Smith to suggest that the elm decline was caused by the stripping of elm leaves for use as a feed for stalled rather than grazing animals. This impaired the flowering capacity and thus pollen productivity of the elm population. Troels-Smith (1960) later suggested that species such as ivy may have been used for the same purpose.

There is now evidence for opening of the forest canopy, minor clearance phases and cereal cultivation in some parts of Britain and Ireland before the elm decline (Edwards and Hirons, 1984).

These findings imply that there may not be an association between the first agricultural practices and the elm decline, since some arable farming at least clearly predates it. Since a climatic cause for the elm decline in the British Isles has also been rejected, there is renewed debate as to the cause, especially in relation to the possibility of Dutch elm disease. This has been prompted by observations on the recent outbreak of Dutch elm disease, which has caused the serious demise of elm populations in Britain and parts of Europe, and by several recent palaeoecological investigations. The disease results from infection by *Ceratocystis ulmi*, a microscopic fungus transmitted by two elm bark beetles, *Scolytus scolytus* and *S. multistriatus*. For the first time in palaeoentomological studies, Girling and Greig (1985) recorded beetle cases of *S. scolytus* in a horizon some 10 cm below the elm decline, from Flandrian deposits at Hampstead Heath, London. This is not conclusive that elm disease was responsible for the neolithic elm decline, since *S. scolytus* can exist independently of *C. ulmi*, but it is the first palaeoecological evidence to suggest even a vague association between the two. Girling and Greig have postulated that elm populations may have been weakened by disease before neolithic agriculture and that subsequent human impact hastened their decline.

Research by Perry and Moore (1987) has added further, though again not conclusive, evidence that elm disease may have played a part in the neolithic elm decline. They have examined the pollen assemblages from humus profiles in Scords Wood in Kent where the recent elm decline began in 1978 and which lost all adult elms in the following five years. The soil pollen assemblages representing this period show changes that are very similar to those that characterise the neolithic elm decline. Elm pollen itself declines from levels greater than 10 per cent to nil, whereas other tree and shrub proportions are unaffected. However, some other non-tree taxa do increase during or after the elm decline, including wild garlic (*Allium* spp.) and blackberry (*Rubus*). As woodland components, they probably increase their flowering as the canopy is opened up. Taxa that are not woodland components, notably ribwort plantain (*Plantago lanceolata*), daisy (*Bidens type*) and bellflower (*Campanula*), also increase in frequency; Perry and Moore suggest this is also due to the more open nature of the canopy that facilitates improved pollen dispersal. This study therefore provides a rather convincing modern analogue, and the fact that elm disease has a long history in Europe (Rackham 1980) must surely make it a very likely cause of the neolithic elm decline, a conclusion strongly favoured by Rackham (1986). Molloy and O'Connell (1987) have also presented a convincing argument for elm disease as the major cause of the neolithic elm decline in western Ireland, as has Peglar (1993a), whose work on Diss Mere, Norfolk, suggests that woodlands were initially disturbed by agriculturalists who helped to spread the elm disease. From further afield, Davis (1981) has suggested that the decline of the eastern hemlock (*Tsuga canadensis*) at 4.85 K years BP in North America was also pathogen-related. This is further substantiated by Allison *et al.* (1986), who have found many similarities between events during this decline and those during the decline of the American chestnut (*Castanea dentata*), which was almost exterminated by a fungal pathogen in the early 1900s.

There is much information for environmental change in Africa during the late Holocene, some of which is due to natural rather than cultural agencies. The period 3.5 to 3 K years BP appears to have been important insofar as climatic conditions and vegetation zones similar to those of today became established. On the basis of pollen-analytical data and wood identification from charcoal from sites in Egypt and Sudan, Neumann (1991) has shown that aridity increased in the eastern Sahara from 5.3 K years BP. This caused a southward movement of vegetation, notably tropical savannas, by ca. 500–600 km until ca. 3.3 K years BP, when a distribution similar to the present day was achieved. A similar increase in aridity by ca. 3.5 K years BP is indicated by Jolly *et al.* (1994) who analysed a core from Kuruyange in the Burundi Highlands. Increasing aridity is reflected in a decline of arboreal pollen overall and an increase in *Celtis*, a semideciduous tree. In addition, Scott's (1990) synthesis of pollen-analytical results from South Africa indicate that cooling occurred after 6.5 K years BP, giving rise to open vegetation communities; although there appears to have been adequate moisture for most of this period, there is also some evidence for minor arid phases. From the Congo, Elenga *et al.* (1994) report the development of humid forest until ca. 3 K years BP, when precipitation decreased; there is evidence for an increase in grasses and a decline in trees, possibly reflecting a fragmentation of the forest and certainly reflecting a vegetation community similar to that of today. There is some debate that these changes may have been due to human activity, but the record of similar vegetation changes elsewhere in the Congo region suggests they were climatically driven. However, Sowunmi (1991) has also shown that by ca. 3 K years BP vegetation communities similar to those of the present day had become established around Lake Mobutu in Uganda; in particular, montane species disappeared and low-altitude forest species became established in response to increasing temperatures and precipitation. In contrast, a dry period is recorded by Burney et al. (1994) in a Holocene pollen record from the Kalahari Desert. This occurs ca. 5 to 4 K years BP, which is not synchronous with the dry episodes discussed above. Overall the pollen record from this site indicates that the vegetation in this part of Africa has undergone fewer changes than have occurred elsewhere on the continent.

There is evidence for slight climatic deterioration in other parts of the world during the period after 5 K years BP. In Japan, as climate deteriorated, Igarashi (1994) states that oak–alder (*Quercus–Alnus*) and fir–alder (*Abies–Alnus*) assemblages replaced spruce (*Picea*) and fir (*Abies*) forests after 5 K years BP. A short period of increased aridity is recorded in northern China at ca. 3.5 K years BP in Holocene sediments from Lake Manas, Xinjiang Province (Jelinowska et al., 1995). This occurs just after a glacier readvance in Tibet (Zhou et al., 1991), as discussed in Section 3.8, where maximum aridity is recorded in two lake sediments at 4.3 K years BP (Fontes et al., 1993). Indeed, while investigating the stable isotope record of Lake Siling on the Qinghai-Tibetan Plateau, Moringa et al. (1993) have recorded two periods of increased aridity when the lake sediments became desiccated. These occurred between 5 and 4 K years BP and between 3 and 2 K years BP. Zhou et al. (1994) also present evidence for climatic deterioration after 5 K years BP in the Loess Plateau, where neoglacial activity is indicated by the deposition of loess.

In the Yucatan Peninsula, Mexico, oxygen isotope data from ostracod and gastropod shells from Lake Punta indicate that relatively wet conditions prevailed between 3.3 and 1.8 K years BP followed by drought conditions between 1.8 and 0.9 K years BP (Curtis and Hodell, 1996). These environmental changes may have contributed to the cultural evolution of Mayan groups; in particular, the occurrence of episodes of severe drought appears to coincide with cultural hiatuses after 0.9 K years BP (Whitmore et al., 1996). Shulmeister and Lees (1995) have identified a decrease

in effective precipitation between 4 and 3.5 K years BP and its recovery since 2 K years BP, from tropical northern Australia. They suggest that this is evidence for the onset of the El Niño – southern oscillation changes on climates of the Pacific that remain important today. Elsewhere in Australia, notably on the western plains of Victoria, there is little evidence from pollen assemblages for substantial change in vegetation patterns until the advent of Europeans (D'Costa and Kershaw, 1995). But evidence from Lake Eyre in central Australia suggests that a lacustrine period, when the basin was occupied with water, was replaced by an ephemeral regime similar to the present day at ca. 3 to 4 K years BP (Magee et al., 1995). McGlone and Neall (1994) state that pollen sequences for North Island, New Zealand, indicate there have been no major changes of vegetation during the late Holocene, though there is some evidence for summer water deficits since 5 K years BP.

As discussed in Section 3.8, the Little Ice Age was a period of marked climatic deterioration in the recent past at ca. AD 1200 to 1800. Much of the evidence for the occurrence of this event and its environmental and cultural impact has been reviewed by Grove (1988). Further evidence, however, includes that of Tyson and Lindesay (1992) for southern Africa. Synthesising a variety of data, including oxygen isotope ratios of cave speleothems, mollusc remains, pollen assemblages and dendroecology from a variety of sites, they have summarised the nature and the effects of climatic change during the Little Ice Age. In general, cooling occurred between AD 1300 to 1850 with a warm interval between AD 1500 and 1675. This corresponds with data from other parts of the world. In addition to cooling, there was a decline in rainfall in the northeast of southern Africa whereas the amount of rainfall in the southwest increased. Overall lake levels declined, tree cover declined and droughts occurred. On the basis of pollen assemblages from a site in the Arsi Mountains of Ethiopia, Bonnefille and Mohammed (1994) have shown that at ca. 0.56 K years BP a depression of the vegetation belts occurred in response to an average annual cooling of 2 °C. This equates with the Little Ice Age and it was followed by a more recent rise in vegetation belts, adding weight to the possibility of recent global warming, as has been indicated by tree-ring and sea-surface temperature data (Section 2.5). An event of similar magnitude has been recorded by Campbell and McAndrews (1993) from southern Ontario, Canada. Here, after ca. AD 1400 the thermophilous beech

(*Fagus*) was replaced first by oak (*Quercus*) and later by pine (*Pinus strobus*). Campbell and McAndrews consider this represents a 2 °C decline in average annual temperatures, a value identical to Bonnefille and Mohammed's (1994) estimation of temperature change based on changes in forest composition in Ethiopia (see above).

3.10   Conclusion

Diverse and extensive environmental changes have occurred in the past 10 to 12 K years. The evidence relies heavily on palaeoenvironmental data, especially from oceans, ice cores and pollen analysis of limnic sediments and peats. Improvements in analysis and interpretation should lead to an improved understanding of the causes, consequences and rates of environmental change, and ultimately to predictive models.

The transition from full glacial conditions to the present interglacial was indeed complex as ecological systems adjusted to ameliorated climates. The final phase of deglaciation ca. 15 to 10 K years BP witnessed rapid climatic changes, reconstructed in detail using fossil coleopteran assemblages. However, most of this work is confined to the British Isles and its wider application should lead to greater appreciation of thermal regimes, not only during the late-glacial period but also during the Holocene. Whether the late-glacial oscillation was a global event is also open to question, but recent evidence suggests that it was more widespread than hitherto suspected. The evidence from northwest Europe and North Atlantic North America certainly indicates it was a period of rapid climatic and associated ecological changes, most of which have no modern analogues, and which may be a reflection of disequilibrium between climatic and ecological systems.

The period 10 to 5 K years BP saw the establishment of optimum interglacial conditions, characterised in temperate latitudes by the development of mixed deciduous forests. Tree lines reached their maximum heights in many parts of the world, especially in the early part of this period. Palynological and other palaeoenvironmental data should not be used in isolation to infer temperature and precipitation changes, since the migration rates of individual species and edaphic changes must also have affected the vegetation communities and therefore the pollen assemblages. Nevertheless,

palaeoecological data from this period and the later Holocene have a significant role to play in elucidating such problems as the refuge hypothesis in relation to tropical rainforests, and the premise that long periods of stability result in ecological diversity. The present state of knowledge implies that tropical forest diversity is not the result of stability but is more likely to be a consequence of instability.

The later Holocene, especially in northwest Europe, also opens with another controversial datum, the elm decline. Climatic change, human interference and disease have all been implicated and it may be that all three factors contributed to this significant alteration in European temperate forests. The past 5 K years in general have witnessed the development of the later stages of the interglacial cycle and there is abundant evidence for climatic deterioration, especially a cooling trend over the past 2 K years. One of the most notable periods is the Little Ice Age, which was probably a global event, centred on the thirteenth to the nineteenth centuries. It is also possible that the Little Ice Age was the most extreme period of climatic change to have characterised the later Holocene in many parts of the world. In addition there is increasing evidence for warming since the 1970s (Section 2.5).

Throughout much of the later Holocene there is abundant evidence for human influence, particularly the development of prehistoric farming practices considered in Chapter 4. More recent environmental changes related to agricultural and industrial development during the past 300 years will be examined in Chapters 6, 7 and 8.

Further reading

Berglund, B.E., Birks, H.J.B., Ralska-Jasiewiczowa, M. and H.E.Wright (eds) (1996) *Palaeoecological Events During the Last 15 000 years.* J. Wiley, Chichester.

Crowley, T.J. and North, G.R. (1991) *Palaeoclimatology.* Oxford University Press, Oxford.

Elias, S.A. (1994) *Quaternary Insects and Their Environments.* Smithsonian Institution Press, Washington DC.

Goudie, A.S. (1992) *Environmental Change*, 3rd edn. Clarendon Press, Oxford.

Grove, J. (1988) *The Little Ice Age.* Methuen, London.

Lowe, J.J. and Walker, M.J.C. (1997) *Reconstructing Quaternary Environments*, 2nd edn. Longman, Harlow.

CHAPTER 4

Prehistoric communities as agents of environmental change

4.1 Introduction

The hominine ancestors of modern humans (*Homo sapiens sapiens*) probably evolved from an ape-like precursor some 5 M years BP, though the details of the evolutionary relationships between the apes and between the various hominine species are controversial. Reconciling the fossil evidence for human evolution with the biomolecular evidence is generating much debate. The debate is focusing on whether modern humans originated exclusively in Africa or in several different parts of the world. However, the ability of hominines to use and later modify their environment more effectively than their predecessors probably began when they learnt to make tools ca. 2.5×10^6 years BP. The earliest tools were made of stone and it is for this reason that the earliest cultures are known as Palaeolithic (Old Stone Age). The practice of stone tool manufacture has dominated the history of human beings since it persisted until metal-using technology was developed, only ca. 9 K years BP, when the first copper tools were produced. That is not to say the Palaeolithic period was one of stagnation and uniformity. Numerous innovations occurred and food-procurement strategies became increasingly calculated and sophisticated. Deliberate herding of animals during the later Palaeolithic and Mesolithic (Middle Stone Age) periods culminated in the domestication of both plants and animals that led to permanent agriculture during the Neolithic (New Stone Age). In many ways the selection of specific plants and animals and the initiation of agricultural systems were more significant in terms of environmental change than the later innovation of metal technology. The copper, bronze and iron implements that were developed in later prehistory served only to intensify the processes begun during the Neolithic by providing more effective tools.

The beginning of the domestication process, continued today in plant and animal breeding programmes, along with the use of fire and the invention of the wheel, must rank among the most important thresholds of human history, indeed in environmental history. They essentially represent the emergence of *Homo sapiens sapiens* as a controller of the environment, especially of ecosystems, rather than as an integral component. They also provided the means whereby ecosystems could be transposed into agroecosystems, which ultimately led to the agricultural systems of today. Such innovations in food procurement also led to changes in social organisation, as did the advent of metal technology which further enhanced the efficiency of agricultural practices and probably had implications for population growth.

4.2 The evolution of modern humans

This book covers many controversial topics but perhaps the most controversial is the evolution of modern humans (*Homo sapiens sapiens*). This is partly because biomolecular characteristics of modern humans and their primate relatives have entered the debate to augment fossil evidence. However, new finds of ancient skeletal remains have refuelled the debates concerning the hominid lineage and the locus of origin for *Homo sapiens sapiens*.

Since the 1960s molecular biology, particularly the study of human genes, has been applied to the vexed questions surrounding human evolution. This has given rise to a subdiscipline known as molecular anthropology, the principle of which is that the evolutionary history of *Homo sapiens sapiens* is recorded in the genes, as is the case for all living species. The information obtained from the genetic codes of hominines and their relatives, the primates, can be used to ascertain the ancestral relationships between and within these families. It can also be used to determine the length of time that has elapsed since the divergence between families and species. The moment of divergence is particularly controversial because it is underpinned by the assumption that a molecular 'clock' operates, i.e. that after two species have diverged from a common ancestor, itself a product of earlier divergence, genetic mutations accumulate at a constant rate. Consequently, differences in the 'genetic material' of the two species represent a measure of time. This has generated much controversy (Lewin, 1993a,b), especially about the first member of genus *Homo* and when it appeared. The advent of new techniques and innovative ideas in science often prompt controversy, as did Charles Darwin's theory of evolution (Section 1.2) which even now has its dissenters. Such new developments are, however, to be welcomed because they encourage a reappraisal of existing theories.

In molecular biology the invention of techniques known as DNA (deoxyribonucleic acid is the inherited material passed to subsequent generations as genes) annealing or DNA hybridisation has furnished just such a stimulus to hominid palaeontology (this term includes both humans and the African apes). This is because it has revealed the closeness of the relationship between humans and other primates. The method involves the separation of the two strands of the double helix that make up DNA. Single strands from different animals can then be mixed to produce a hybrid double helix, and where there are matching, i.e. complementary, units the two strands coalesce. If heat is then applied, the strands will separate again; higher temperatures are required to separate strands with the greatest degree of complementarity, i.e. those that are most closely related. Although not proven, it is considered that the degree of difference between the two DNA strands represents the passage of time because mutations, i.e. the differentiating genetic characteristics, occur on a regular basis. This and other molecular data, such as interspecies comparisons of blood proteins (Sarich and Wilson, 1967), led molecular anthropologists to propose that the first hominines diverged from the Old World

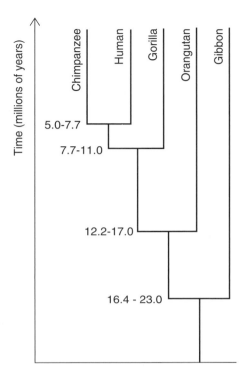

Fig. 4.1 Relationships between the major primate groups based on DNA hybridisation (based on Lewin, 1993a).

monkeys (African apes) only ca. 5×10^6 years ago (Simons, 1964; Pilbeam, 1968). This was highly controversial when it was originally proposed as the divergence had been considered by palaeontologists to have been no earlier than 15 M years BP. However, additional fossils of *Sivapithecus*, a group to which *Ramapithecus* belongs, were discovered in the early 1980s and, as Andrews (1982) has discussed, they produced evidence that this group is more closely related to the orang-utans than to humans (Figs 4.1 and 4.2). Along with further work on the molecular clock theory, this has led to a reconciliation of the molecular and fossil evidence: both palaeontologists and molecular biologists concur that humans and African apes, i.e. the gorillas and chimpanzees, diverged between 5 and 10×10^6 years ago. Indeed the precise relationship between these three groups is still disputed and Figure 4.1 represents only one of several possibilities.

Although this reconciliation occurred in the early 1980s, there have been and continue to be many other contentious issues regarding human evolution.

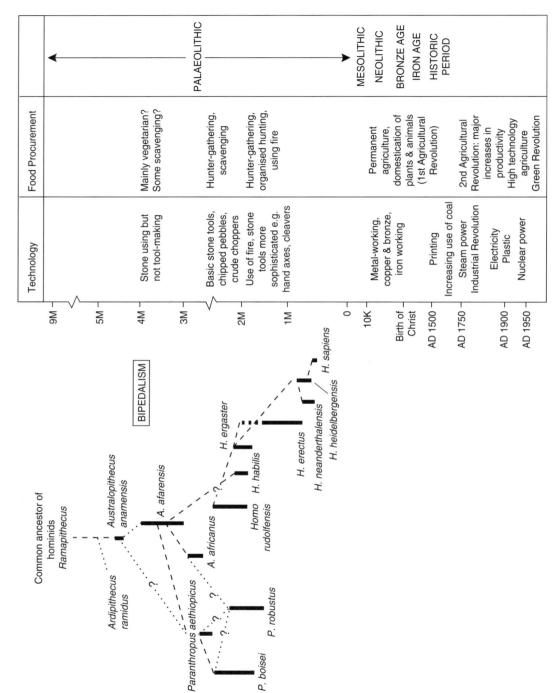

Fig. 4.2 Human evolution in relation to technology and food-procurement strategies (based on Wood, 1994; Mannion, 1995a).

They are prompted by new fossil finds and by changes in nomenclature that sometimes lead to disputes. One such issue is bipedalism and when it developed as an alternative to quadrupedalism. The first evidence for early hominine bipedalism was found at Laetoli in Tanzania by Mary Leakey in the mid-1970s (Leakey, 1979) in the form of footprints in ash layers. These are dated to 3.6×10^6 years BP and are probably footprints of *Australopithecus afarensis*, named in 1979 by Johanson and White. Much of their work was based on fossils, recovered from Hadar in Ethiopia (Kimbel *et al.*, 1994) which included the now famous partial skeleton of the individual known as Lucy. Johanson and White considered this species ancestral to all subsequent hominines and the so-called missing link between the apes and the hominines. Although widely accepted until 1995 (see below) as occupying this important position in the hominid family tree, *A. afarensis* was considered by others to be just one of the two or more species represented in the Hadar collection (Leakey and Lewin, 1992). This was based on size differences in the collection of fossils; Johanson and White believed it represented male and female of the same species but others considered them to be different species. However, *A. afarensis* was bipedal and, as Lewin (1993a) states, 'superficially, *A. afarensis* does indeed appear to be essentially apelike above the neck and essentially humanlike below the neck.' Figure 4.2 shows how the latest ideas on hominine evolution retain *A. afarensis* as an important component but do not give it a pre-eminent position.

This reconsideration is due to new fossil finds reported in 1994 and 1995. The first of them was by White *et al.* (1995) from Aramis in Ethiopia's Afar depression; in a companion paper WoldeGabriel *et al.* (1994) reported on the ecological and temporal context of the 17 fossils comprising dental cranial and post-cranial specimens. In their comparison of the fossils with those of *A. afarensis*, White *et al.* (1995) revealed sufficient differences for the new fossils to be considered as a new species. Initially this was called *A. ramidus*. As in the case of *A. afarensis*, *A. ramidus* was invoked as the new 'root species for the Hominidae'. It has many ape-like characteristics but whether it is bipedal, a character considered to be unique to the hominines, is debatable. It is dated to 4.4×10^6 years BP and occupied a wooded habitat. For various reasons, some of them given by Wood (1994) in a commentary on the *A. ramidus* finds, White *et al.*(1995) have subsequently renamed the fossil *Ardipithecus ramidus*. This is a new genus whose relationship with *Australopithecus afarensis* is unclear, though White *et al*, (1995) consider it to be the ancestor of *A. afarensis*. Another fossil find, this time from Kanapoi, southwest of Lake Turkana in Kenya, has also contributed to the debate (Leakey *et al.*, 1995). Remains of teeth, several jaws and a leg bone, the latter a clear indicator of bipedalism, are considered to be those of *Australopithecus anamensis*, a much more likely direct ancestor of *A. afarensis* than *Ardipithecus ramidus*. As Figure 4.2 shows, all three may be related and much hinges on whether additional bones of *Ardipithecus ramidus* are discovered to prove that it was bipedal. The dating of *Australopithecus anamensis* to between 4.2 and 3.9×10^6 years BP clearly places it as a temporal intermediate between *A. afarensis* and *Ardipithecus ramidus*. Apart from the likelihood that further species of australopithecines and/or *Ardipithecus* species existed and are yet to be discovered, there is the possibility that existing fossils of *Australopithecus afarensis* may represent between one and three species (Groves, 1994). For example, Clarke and Tobias (1995) have recently discovered four articulating hominid foot bones from Sterkfontein, a site near Johannesburg in South Africa. These bones have features that reflect bipedalism but other features which are ape-like. Dated to 3.5×10^6 years, the bones provide evidence for the presence of hominines in South Africa before 3×10^6 years BP. However, the bones have not been ascribed to any particular species. According to Clarke and Tobias, the foot bones 'probably belong to a species of *Australopithecus*, possibly the earliest available *A. africanus*, or to another early hominid species.' The first fossils of *A. afarensis* outside east and southern Africa have recently been reported by Brunet *et al.* (1995) from Chad. This extends the range of the species ca. 2500 km west of the Rift Valley at ca. 3 to 3.5×10^6 years BP.

There are other important considerations besides the questions surrounding the hominine lineage. Firstly, there is the issue of why bipedalism developed, and secondly there is the issue of why hominines diverged from the hominids, i.e. why humans diverged from the African apes. Both questions are probably related. There has been much debate about the reasons for bipedal development, as Lewin (1993a) has reviewed. Several hypotheses implicate food procurement, either improved stalking of prey or improved plant resource gathering and transport (e.g. Lovejoy, 1981). Another possibility is the freeing of hands for toolmaking, though toolmaking is more likely to have been an outcome not a cause of bipedalism. Avoidance of predators, is yet another possibility. Lewin favours the thesis of Rodman and McHenry (1980), which is essentially deterministic. They suggest bipedalism could have developed as a response to environmental change that caused a change in the distribution of food resources. There is certainly abundant evidence for climatic change during the late Miocene, which culminated in the onset of major glaciation at ca. 3.5×10^6 years ago (Sections 2.2 and 2.3). Rodman and Henry propose that habitat fragmentation in the late Miocene meant that hominids were forced to travel more widely than hitherto. Consequently, evolution may have favoured a mode of movement with improved energy efficiency: bipedalism is just such a mode at walking speeds. Although this hypothesis is favoured by Lewin (1993a) and Leakey and Lewin (1992), and is entirely plausible, it means that the distinction between humans and apes was meagre yet ultimately momentous. The evolution of bipedalism was itself not such a major change in anatomy but it generated substantial attributes and advantages that eventually culminated in modern humans. Broadly speaking this could be described as the development of culture. Further evidence, more by analogy than directly, for the importance of environmental change in the course of human evolution has been presented by Kalb (1995), who has examined the fossil record for elephantoids in Africa for the past 22×10^6 years. Kalb has demonstrated that environmental change influenced elephantoid evolution and migration as geology and climate changed. During the past 4 to 6×10^6 years, a crucial period in hominid evolution, there were major changes in elephantoid dentition. These changes reflect an adaptation to a coarse foliage characteristic of an arid to semiarid climate as compared to the soft vegetation with a high water content of earlier times. This desiccation also caused changes in the level and location of lakes in the rift valley between the African and East African plates which shifted position at this time. If elephantoids reacted and adapted to such environmental changes then it is probable that the hominids did so too, along with their immediate ancestors and successors (Vrba, 1996).

This debate has been augmented by the recent discovery of a mandible attributed to *Homo*

rudolfensis in the Malawi Rift (Shrenk *et al.*, 1993; Bromage *et al.*, 1995). The investigation was part of the Hominid Corridor Research Project, established to examine the relationship between early hominids and hominines in east Africa and southern Africa. The Malawi Rift represents a link between the two and provides an opportunity to test the relationship between them in the context of hominine evolution. The report of Schrenk *et al.* (1993) on the hominine mandible is the first unequivocal fossil of a species of *Homo*. The sediments in which the fossil was found have been dated at 2.5 to 2.4×10^6 years BP, by which time there had been a major change from a warm, moist climate to a cold, arid climate (Sections 2.2 and 2.3). The mandible recovered reflects an adaptation to increased chewing, possibly because vegetation had become tougher with only a low moisture content. A similar adaptation is found in *Paranthropus aethiopicus* (Leakey and Lewin, 1992), a temporally parallel lineage (Fig. 4.2). Evidence from other mammalian faunas, notably the presence of many species originating in the south (in the Malawi Rift), indicates that at ca. 2.5×10^6 years BP there was a mammal movement from southern Africa to eastern Africa. This movement may reflect the increasing aridity and the movement north (equatorward) of the continent's vegetation belts. Bromage *et al.* (1995) also interpret these findings as confirmation that the centre of evolution remained in the East African Rift where woodlands still persisted. Bromage and Schrenk (1995) also advocate a climatic basis for the evolution of several early hominids, a possibility advanced by many workers (see above), including Vrba (1992), and one which is essentially deterministic (Section 1.3).

By ca. 2.5×10^6 years BP the first hominine species had evolved (Fig. 4.2). As Groves (1994) and Tudge (1995) discuss, there is also a dispute about the status of the first true *Homo* species, traditionally known as *Homo habilis*. A similar problem exists with what has traditionally been known as *H. erectus*, the successor of *H. habilis*. Wood (1992) has made a case for the reclassification of *H. habilis* fossils into at least two groups: *H. habilis* and *H. rudolfensis* with *H. ergaster* being a possible third. Wood's argument centres on the large variation that exists between *H. habilis* fossils, which he considers unacceptable for a single species. Wood (1992) also makes the point that just before 2×10^6 years BP, a period of global cooling, the emergence of two or three species of *Homo* represents what he describes as a substantial radiation of early hominids that coexisted.

An important difference between the australopithecines and species of the genus *Homo* is brain size; *Homo* species have a large brain with a capacity of ca. 600 cm^3, though this is considerably smaller than modern humans, whose brain size is 1000–2000 cm^3 (Lewin, 1993a). This increase in brain size was accompanied by several anatomical changes in the skull, though the teeth are similar to those of the australopithecines. The pattern of wear on both australopithecine and *Homo habilis* fossils indicates a plant-rich diet, especially of fruit. However, there is also evidence of increased manipulative ability in the hands. This may reflect the capacity of the species to make tools and/or the continued but partial, occupation of trees. The fossil evidence for tool use has been re-examined by Susman (1994) who concludes that by ca. 2 to 2.5 × 10^6 years ago all hominids (i.e. australopithecines and hominines, *H. habilis* and *H. rudolfensis*) were making tools. As Aiello (1994) has discussed, this overturns conventional wisdom which assumes that only hominids with a large brain could be toolmakers. It appears the key is the presence of a broad head of the thumb (first metacarpal bone) in relation to length, indicating the sometime presence of substantial muscle for the generation of the force essential for toolmaking. The implication of this deliberation is that finds of stone tools (such as those attributed to the Oldowan culture; see Section 4.3) cannot be considered unequivocally as the work of the ancestors of modern humans; they may be the work of other primates.

As Groves (1994) has discussed, there is also evidence to suggest that the fossil remains of the successor of *H. habilis* and related hominines may represent more than one species (Wood, 1992). It is now widely accepted that the fossils represent at least two species: the original *Homo erectus*, so named on the basis of a fossil skull from Java, and which includes finds from Zhoukoudian and Hexia in China, and *Homo heidelbergensis* which embraces fossil finds from Africa and other sites in China and Europe. Indeed the recently discovered tibia from Boxgrove in West Sussex, UK (M.B.Roberts *et al.*, 1994) and some 1300 bones from Atapuerca in Spain (Carbonell *et al.*, 1995) are considered to be from this species. There is also some debate about the relationship of *H. ergaster* to *H. habilis* and *H. erectus*. Wood (1992) regards *H. ergaster* as an early African equivalent of *H. erectus*, though in terms of

the age of the fossil at ca. 2 × 10^6 years BP it may be more closely related to *H. habilis*. Yet other authorities do not consider that the African fossils of this age are anything other than *H. erectus*. Whatever the reality, *H. erectus* and counterparts represent an important state in early hominine development. Firstly, there is dental evidence to show that meat had become an important component of hominine diet, and secondly, members of these groups were the first hominines to migrate out of Africa into other parts of the world. There is also evidence for organised hunting.

According to Lewin (1993a), *H. erectus* had evolved by 1.8 × 10^6 years BP, which coincidentally or not, corresponds with the opening of the Quaternary period (Section 2.2). However, the recent report (Gabunia and Vekua, 1995) of a mandible of *H. erectus* from the archaeological site of Dmanisi in the Republic of Georgia (Fig. 4.3), dated at 1.8 to 1.6 × 10^6 years BP, means that the species evolved earlier than this, probably more than 2 × 10^6 years BP. In comparison with its predecessors, *H. erectus* had an increased brain size of 850–1100 cm^3. It also had an increased body size and different cranial features, notably prominent brow ridges. The presence of distinctive stone tool assemblages often found in association with *H. erectus* and related fossils (Fig. 4.3), though not in east and southeast Asia, reflects systematic toolmaking, especially the development of first Oldowan and later Acheulian tool assemblages. (Both are named after type sites where characteristic tools have been found and are referred to again in Section 4.3.) The considerable evidence that hunting had superseded scavenging and that stone tool production was becoming increasingly sophisticated reflect major changes in the status of hominines in relation to their environment. This evidence may be interpreted as Gaian evolution, i.e. reciprocal evolution between life and its environment, insofar as it is highly likely the climatic changes of ca. 2 × 10^6 years ago helped stimulate the evolution of *H. erectus*, *H. heidelbergensis*, etc., and the new species change the environment as they became more technologically adept. In the reconstructed lifestyles of *H. erectus*, etc., it is possible to see the emergence of hominines as dominants within ecosystems. Moreover, *H. erectus* eventually gave rise to *H. sapiens sapiens*, i.e. modern humans, which are a distinctly powerful force on the global landscape. The significance, in terms of cultural and environmental change, of these developments could be regarded as momentous.

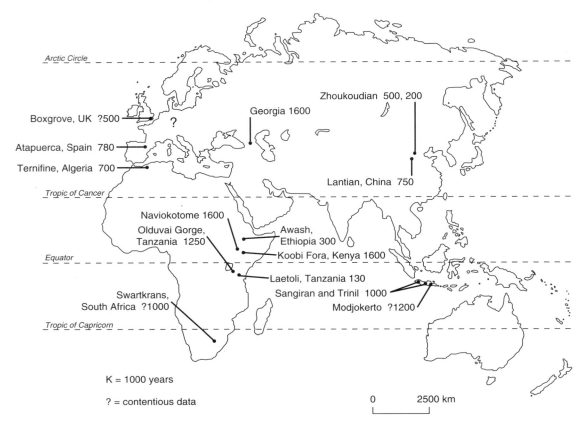

Fig. 4.3 Sites at which *Homo erectus* and *H. heidelbergensis* have been found (based on Lewin, 1993a). Dates are in thousands of years BP.

However, among the many questions that surround *H. erectus*, etc., two are especially significant. Why did these early hominines migrate within and beyond Africa? Did they communicate with each other using language? According to Cavalli-Sforza and Cavalli-Sforza (1995), *H. habilis*, etc., probably used a primitive language. This view is based on cranial structure, which in *H. habilis* and later hominines is slightly asymmetrical to accommodate the somewhat larger left hemisphere of the brain which deals with language. Such an accommodation of anatomy is not present in earlier hominid fossils, i.e. those of *Australopithecus afarensis*. The capacity for language probably increased thereafter in line with brain size and intelligence until the evolution of modern humans ca. 300 to 200 K years ago. Even then, it is unlikely these early modern humans had the same capacity for language as humans enjoy today. Cavalli-Sforza and Cavalli-Sforza suggest that language is a relatively recent development on the grounds that all languages spoken today are similar in relation to their degree of complexity, as is the ability of different groups to learn language. However, there is little that can be stated with certainty about the role of language in the lifestyle of *H. erectus* except that it probably did play a significant role in food acquisition and migration. There is a little more evidence on which to speculate about the migrations of *H. erectus*. The species began to move from its centre of origin in east Africa soon after it first evolved. Even its evolution is a matter for speculation, as Maslin (1994) has recently suggested that renewed volcanic activity in east Africa's rift valley prompted this event. Moreover, the change from a vegetarian to an omnivorous diet may have played a role in this. According to Walker and Shipman, quoted in Lewin (1993a), such a change in diet requires either a change to a smaller body size or an expansion of foraging range. As *H. erectus* was bigger than its predecessors, it had to expand its range. Moreover, change to a high-protein diet encourages the movement of a species, insofar as individuals can graze on many different types of herbivore. Each type of herbivore is adapted to certain plant communities, so their movements are more constrained. However, the recent discovery of archaic hominid teeth at Longgupo Cave in Sichuan Province, central China, considered to belong to *H. habilis* or *H. ergaster* and dated to ca. 1.9×10^6 years BP (Wanpo *et al.*, 1995), displaces *H. erectus* as the first hominine to migrate beyond Africa. As Wood and Turner (1995) point out, the new find could reflect the presence of a centre of evolution in Asia.

The ability of *H. erectus* to use fire, evidenced from several sites including Zhoukoudian (formerly Choukoutien) near Beijing in China (Fig. 4.3), may have been used to help attract and trap animals. At many archaeological sites where the remains of *H. erectus* have been found (Fig. 4.3) there is evidence for communal living, fire and crude dwelling structures, as well as Acheulian tool assemblages. However, there is nothing in the archaeological record to explain why *H. erectus* became a meat-eater. Did the increased sophistication of the Acheulian tool assemblages provide the means to butcher animal carcasses? Or did the scavenging of dead animals provide an incentive for more sophisticated toolmaking? Whatever the answer, Brand-Miller and Colagiuri (1994) have suggested the adoption of a low-carbohydrate, high-protein diet may have been prompted by the advent of an intensified glacial–interglacial climate that favoured large grazing animals. They also suggest this change required alterations to metabolism, notably the development of insulin resistance, which means the insulin produced has a reduced effectiveness. Rates of reproduction are impaired in animals and modern humans with insulin resistance. Consequently, it is likely that natural selection has favoured genetically determined non-insulin-resistant individuals. Interestingly, perhaps, insulin resistance is considered to be an early sign that some people will develop non-insulin-dependent *diabetes mellitus*. The reason why the incidence of diabetes is relatively low in Europeans may be because they are among the first to adopt permanent agriculture which focused on cereals (Section 4.6). It is likely that other modern-day diseases are a product of our evolutionary past.

The next stage in human evolution, the emergence of *H. sapiens* and *H. neanderthalensis* (Fig. 4.2), is the most controversial issue in palaeoanthropology. It hinges on whether archaic modern humans, i.e. *H. sapiens*, evolved from disparate *H. erectus* populations in various parts of the world, or whether they evolved only in Africa from where they migrated, like *H. erectus*, into Europe, Asia, etc. (Fig. 4.4 shows the location of sites.) These two hypotheses, respectively the multiregion and single-region hypotheses, have occupied archaeologists, palaeontologists, anthropologists and molecular biologists for almost a decade (Sussman, 1993; Klein, 1994). They are also, known as the candelabra and Noah's Ark replacement model (Lewin, 1993a). Lewin (1987) and Aiello (1993) have also suggested that a combination of these possibilities may have occurred, i.e. a single geographic origin might have been followed by migrations with interbreeding between the migrants and established populations of archaic *sapiens*. In the light of recent fossil discoveries and advances in molecular biology, the single-region (Noah's Ark) hypothesis is now generally favoured. However, it has not been proven unequivocally and some workers, notably Milford Wolpoff (Wolpoff *et al.*, 1994; Frayer *et al.*, 1993) continue to advance the multiregion hypothesis.

One of the major protagonists of the single-region hypothesis is Stringer (1992, 1994), who points out that it satisfies both the fossil and genetic evidence. The genetic evidence focuses on mitochondrial DNA; this was first proposed as a reliable indicator of human evolution by Cann *et al.* (1987). Mitochondria are organelles within cells and are concerned with energy metabolism. Unlike nuclear DNA, mitochondrial DNA (mtDNA) accumulates mutations at a relatively fast rate, usually 10 times faster than nuclear DNA. Moreover, mtDNA is inherited only from the mother, so its history is easier to unravel than that of nuclear DNA. Cann *et al.* examined the mtDNA of 147 people of different ethnic origins and showed there were strong relationships between them, indicating a common ancestor. This had to be a woman, and on the basis of a regular accumulation of mutations, Cann *et al.* suggested she lived ca. 200 K years ago. This was highly controversial but more recent evidence from mtDNA and nuclear DNA reinforces the earlier conclusions (Cann, 1993; Cann *et al.*, 1994; Bowcock *et al.*, 1994). It remains debatable how or why these archaic modern humans replaced existing

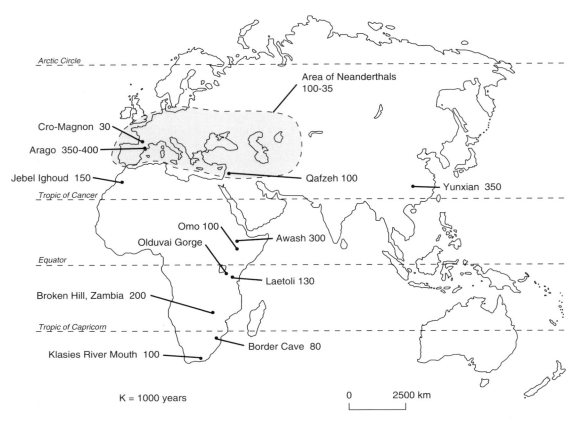

Fig. 4.4 Archaeological sites revealing the earliest evidence of modern humans (*Homo sapiens sapiens*) (based on Lewin, 1993a). Dates are in thousands of years BP.

hominines, mostly *H. erectus* but also *H. neanderthalensis*. As well as confirming the work of Cann *et al.* (1994), Penny *et al.* (1995) used mtDNA to support an African origin for modern humans and they suggested a two-phase model separating the movement of a common ancestor out of Africa from the major expansion in human populations. Described as the 'cultural explosion' model, support has recently come from Zaire, where tools associated with modern humans have been dated to 90 K years BP (Yellen *et al.*, 1995). No tools of a similar age have been found outside Africa, and those of a similar type which have been found outside Africa are considerably younger. Additional work on the morphological characteristics of fossils of *H. sapiens* (Waddle, 1994) also supports the single-region hypothesis with a centre of origin in Africa and/or the Middle East. Why archaic *sapiens* should have moved out of Africa is open to

question. Perhaps climatic amelioration during an interglacial or interstadial encouraged a radiation; this may have been a response to animal migrations. Equally important is the question of how archaic *sapiens* replaced *H. erectus* and *H. neanderthalensis* and what is the phylogenetic relationship between the three.

According to Wood (1994), both *H. sapiens* and *H. neanderthalensis* probably evolved from *H. heidelbergensis*, as shown in Figure 4.2, though until recently it had been assumed that *H. neanderthalensis* was the direct ancestor of *H. sapiens*. In 1988 a new date from Jebel Qafzeh, a site near Nazareth in Israel, where fossils of modern humans have been found, was given as 100 K years BP (Valladas *et al.*, 1988; see also Bar-Yosef, 1994) based on thermoluminescence. This implies that *H. sapiens sapiens* had evolved before 100 K years BP, not ca. 45 K years as had been thought. The new date also makes it highly unlikely that *H. sapiens sapiens* had evolved from *H. neanderthalensis*; instead it may have replaced *H. neanderthalensis*, which became extinct 35 to 45 K years ago. Leakey and Lewin (1992) warn: 'These arguments rest on dates of fossils, not on anatomical analysis. Change the dates and you may change the evolutionary progression.' However, a further date of 119 ± 18 K years BP has been reported by Mercier *et al.* (1993) from another site where *H. sapiens* remains have been found, Es Skhul in Israel. Acceptance of this date means that the two species coexisted, at least on a temporal basis, for ca. 40 K years, though there is no evidence for interbreeding. Leakey and Lewin have suggested the two species may have had quite different distributions, so they were spatially separate. According to Leakey and Lewin, the Neanderthals were well adapted to cold environments whereas modern humans were adapted to warm environments, at least initially. As climate cooled during the stadials of the last ice age (Sections 2.2 to 2.4), *H. sapiens sapiens* moved south leaving the Middle East to the southward-moving Neanderthals. As the modern humans developed skills and technologies to migrate out of Africa, and eventually into the temperate zone and high latitudes, they became more successful than the Neanderthals, which became extinct. This is just one possibility; extinction may have occurred for other reasons and competition with modern humans may not have been a factor. It is also sobering to consider the relatively short time span that this close relative of *H. sapiens sapiens* spent on Earth. The hominine line is clearly not immune to extinction. This fact has recently been highlighted by Gott (1993) and is inherent in the Gaia hypothesis (Section 1.3), though it was first suggested by Copernicus (1473–1543) and again by Darwin (1859).

Whatever the reason, following its evolution, *H. sapiens* migrated out of Africa at least 200 K years ago and perhaps as much as 400 K years ago. Modern humans (*H. sapiens sapiens*) appeared in Europe ca. 40 K years ago, but apart from Qafzeh (see above), the earliest dates for its appearance are from Africa (Fig.4.4), as would be expected under the single-region hypothesis and if population increase occurred some considerable time after its initial emergence. These dates are 74 to 130 K from Border Caves and 74 to 115 K from Klasies River Mouth (Klein, 1992). Humans had spread into most parts of the world by about 60 to 70 K years ago. For

example, R.G.Roberts *et al.* (1994) report a date of 60 K years BP from different dating techniques for the initial occupation of Australia. These dates are from a rock shelter in Deaf Adder Gorge in Australia's Northern Territory at the foot of the Arnhem Land Plateau and ca. 100 km from the coast. As Flood (1995) has discussed, there are many archaeological sites close to the coast as well as other groups of sites in central and southern Australia. Indeed, until recently, the oldest dates were those from one site discovered in the late 1970s in the Upper Swan River Valley near Perth in the southwest of Western Australia. This is radiocarbon dated to 38 K years BP and comprises an open-air campsite with worked flake tools. Though there are some reservations about the dates from Deaf Adder Gorge (e.g. Bowdler, 1993), they do lend weight to the thesis that *Homo sapiens sapiens* entered Australia via the north (there is no evidence for earlier species of hominines in Australia). However, this is rather simplistic and somewhat misleading; Australia's northern coastline is thousands of kilometres long. Moreover, there is a major biogeographic divide between Asia and Australasia (which includes New Guinea). This has been in existence since long before the evolution of modern humans or their immediate ancestors and existed as an ocean barrier even in times of lowered sea level during the Quaternary. The modern humans who reached Arnhem Land, or elsewhere on Australia's northern coast, had to be seafarers. Cases for the advantages of low (Butlin, 1993) and high (Chappell, quoted in Flood, 1995) sea levels are inconclusive. It may be that migration occurred using raft-like structures of bamboo under either set of conditions. It is likely that new sites and new dates will, in the future, alter the story of Australia's colonisation but they are unlikely to change the story dramatically.

When the first humans entered the Americas is disputed. It may have been as early as 30 K years BP or as late as 14 K years BP. The oldest undisputed evidence of human presence in the Americas is dated at ca. 12 K years and relates to projectile points manufactured by the so-called Clovis people. It is considered that these hunter-gatherers migrated from Asia across the Bering land bridge, which existed between Siberia and Alaska when sea levels were lower. They then moved south via an ice-free corridor between the Laurentide ice sheet, which covered the centre of North America, and the Cordilleran ice sheet, which covered the Rocky Mountains (Macdonald and McLeod, 1996).

Table 4.1 Archaeological sites in North and South America which may represent pre-Clovis cultures and which are disputed

Site	Location	Date (K years BP)	Comments
Old Crow Flats	N.W. Canada	40.0	Carved bones
Bow River Valley	Alberta, Canada	> 21.0	Stone artefacts
American Falls	N.W. USA	> 43.0	Artefacts
Santa Rosa Island	S. California	> 40.0	Hearths
Meadowcroft	Nr. Pittsburgh	20.0	Rockshelter: artefacts
Coopertown	Texas	20.0	Butchered bones
El Cedral	N. Mexico	31.0	Traces of fire; bones
Tlapacoya	N. Mexico	21.0	Charred animal bones
Caulapan	N. Mexico	21.0	Scraper tool
El Bosque	Nicaragua	20.0	Stone objects
Taima-Taima	Venezuela	16.0	Artefacts
El Abra	Colombia	12.4	Artefacts; hearths
Guitarrero	Peru	12.5	Artefacts; hearths
Pikimachay	Peru	18.0	Settlement; artefacts
Toca do Boqueirão	E. Brazil	49.0–14.3	Fire; lithic industry
Monte Verde	Chile	32.0	Occupation layer?
Monte Verde	Chile	13.0	Settlement remains
Los Toldos	Argentina	12.6	Lithic industry
Cueva Fell	Chile	11.0	Settlement remains

Source: based on Lorenzo (1994) and Meltzer (1993)

Alternatively, they may have moved south via an emerged continental shelf. Clovis people hunted large herbivores, notably mammoth and mastodon, and their arrival and spread in North America is often considered to be the main reason for the extinction of these creatures, and possibly other large mammals (Section 3.5). The Clovis people were replaced by the Folsom people by ca. 11 K years BP. However, there are several archaeological sites in both North and South America with disputed evidence of human presence and/or disputed dates which place the peopling of the Americas much earlier than suggested by the Clovis sites. These controversial sites are listed in Table 4.1. The site at Monte Verde in Chile is particularly controversial because there are two potential habitation levels with radiocarbon dates of ca. 33 K years and 13 K years BP (Dillehay and Collins, 1988; Quivira and Dillehay, 1988). And the disputed site of Pedra Furada in Brazil, data from which have been reviewed by Meltzer *et al.* (1994), has radiocarbon dates ranging from 49 to 14.3 K years BP. Additional evidence for a pre-Clovis human presence in the Americas has been provided by Chlachula (1996), who reports on pebble-tool industries present in fluvial gravels of the Bow River, Alberta, Canada. These industries are considered to be older than 21 K years BP on the basis of their stratigraphic position. Although recognised as controversial, some researchers, e.g.

Lorenzo (1994), believe these sites represent the material culture of an initial wave of human migration through the Americas, a migration which may have begun some 65 K years ago.

However, a recent investigation of sites in Beringia (Hoffecker *et al.*, 1993) has put the earliest dates for settlement in the traditional window of 12 to 11 K years BP, the implication being that the sites listed in Table 4.1 are indeed spurious. Kunz and Reanier (1994) have reported dates of 11.66 to 9.730 K years BP for hearths and sediments containing Clovis implements from Mesa in Alaska, in line with Hoffecker *et al.* (1993). As well as confirming the settlement of the Clovis people in Alaska, rather than their passage through the region, Kunz and Reanier (1994) record a hiatus in the record of human occupation between 11 and 10.3 K years BP. This they ascribe to climatic deterioration during the late-glacial period (Section 3.4), which they suggest may have prompted migration south. An analysis of mtDNA from various South and Central American natives by Horai *et al.* (1993) presents a slightly different possibility. On the one hand, the data suggest that four ancestral populations entered the Americas in four separate waves of migrations, a theory also suggested by analysis of Native American languages (Gibbons, 1993). On the other hand, Horai *et al.* (1993) infer that the first wave of migration across the Bering land bridge occurred 21 to 14 K years BP.

Nor is there any evidence for a 'bottleneck' in migration across the Panama isthmus, as is also indicated from further research on mtDNA by Monsalve *et al.* (1994). Dates considerably older than 12 K years BP for human presence in nearby Siberia and Japan also add fuel to this controversy (Bonnichsen and Schneider, 1995). Additional views have been expressed by Heusser (1994), who has used the occurrence of charcoal as a proxy record for human presence in South America during the past 50 K years. Based on the charcoal counts from 10 sites in southern Chile and Argentina, Heusser has suggested that fire was used as a hunting strategy and the charcoal record may indicate human presence since at least 33 K years BP. But Heusser believes the absence of charcoal during the last glacial maximum (25 to 14 K years BP) reflects the inhospitable environment and glaciations in the high Andes. Palaeoindians then returned as the ice sheets and glaciers melted. In view of the much earlier peopling of Australia (see above), where there is a biogeographic divide to cross, it is surprising that the peopling of the Americas occurred so late, relatively speaking. Although the vagaries of ice-sheet movements may have been major deterrents, the much lower sea levels associated with ca. 100 K years of the last ice age (Section 2.10) provided an opportunity for migration from Asia though this was apparently seized only at the last minute.

Modern humans were adapting to a rapidly warming climate by the end of the last ice age ca. 15 K years BP. They were still stone-users, i.e. Palaeolithic, but had become sophisticated hunter-gatherers. Within another four millennia the first plants and animals had been domesticated and settled agriculture had developed.

4.3 The relationship between environment and Palaeolithic groups

The term 'Palaeolithic' refers to the Old Stone Age, which lasted from the evolution of the first stone-tool makers (Section 4.2) until the end of the last ice age ca. 14 K years ago. During this time many changes occurred in the technology and subsistence base of Palaeolithic groups, as shown in Figure 4.2. Although the evidence for Palaeolithic resource exploitation is relatively meagre, it indicates that major changes took place as new hominines evolved

and developed. The interpretation of Palaeolithic archaeology is also hampered by inadequate dating controls and a less than perfect understanding of palaeoenvironmental evidence. Moreover, heavy reliance on stone artefacts and human and animal bones, which survive well in comparison with wooden artefacts and plant remains, means that the evidence is biased. As the range and accuracy of dating techniques improve, and as additional archaeological sites are discovered, an improved insight into Palaeolithic resource use and environmental impact will emerge.

The terminology associated with the Palaeolithic period varies on a regional basis; strictly speaking, the term itself is a reference to a cultural type, but there is a high degree of interregional correspondence in a temporal context. The various terms are given in Table 4.2 along with approximate dates. As this shows, and as Lewin (1993a) has discussed, the earliest stone artefacts were produced by *Homo habilis* about 2.5×10^6 years ago, though the recent suggestion (Susman, 1994; see Section 4.3) that all hominids (i.e australopithecines and hominines) could make tools means that artefacts at archaeological sites may not have been produced by species of *Homo*. Evidence from those early artefacts derives from numerous African sites, the most well known being the Olduvai Gorge. The earliest artefacts from here are dated to ca. 1.9×10^6 years BP; these and similar tools from elsewhere are called Oldowan (Table 4.2), which represents the culture of the time. However, the oldest Oldowan tools, ca. 2.5 M years old, have been found in the Hadar region and the Omo Valley of Ethiopia and from Lake Turkana in Kenya (Coppens and Geraads, 1994; Chavaillon, 1994). They consist of flake tools and chopping tools. At some riverside sites there is evidence for permanent settlement, and the animal bones accompanying the tools, etc., indicate a bias towards certain parts of the animals. It is likely that some of the tools were used to dismember animal carcasses where the animals died, either through hunting or more likely scavenging, and then selected parts were taken back to the base camp. The most commonly exploited animals were antelope, hippopotamus, zebra and giraffe. As stated in Section 4.2, *H. habilis* used language. In combination with the factors above, this all points to the operation of social interactions at the levels of family and extended family.

The tool assemblage belongs to mode I (Table 4.2) which represents a crude type of tool manufacture. Of the tools themselves, the flakes may have been the most important, especially as

Table 4.2 The terminology associated with the Stone Age (Africa) and the Palaeolithic period (Eurasia) with approximate dates (based on Lewin, 1993a)

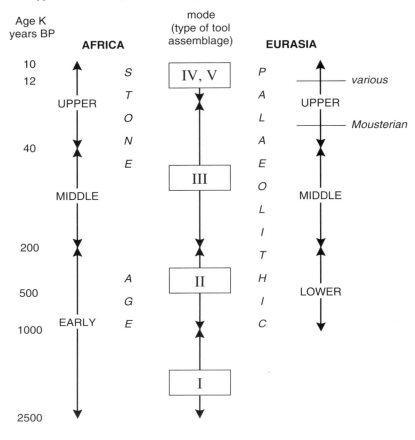

Mode I   : Chopping tool   Mode IV : Blade-based tools
Mode II   : Bifacial tool   Mode V : Microlithic technologies
Mode III : Prepared core

they allowed these early hominines to slice through hide to obtain the meat. The choppers may have been the remains of stones after flakes had been produced, but may themselves have been used to carve an animal carcass or for breaking bone in order to obtain marrow. These tools were probably also used to make implements from bone and wood, notably digging sticks. There is no evidence to enable the identification of the plant types exploited by *H. habilis*, though plant foods must have played a major role in the diet of this hominine species. According to Leakey (1979) the Oldowan industry underwent some marked changes around 1.6×10^6

years BP. In particular, the number of choppers declined, the number of small scrapers increased and the first, comparatively simple, hand axes and cleavers appeared; Mary Leakey calls this assemblage the Developed Oldowan. It represents a transitional phase between the Oldowan and later Acheulian (Table 4.2). According to Gowlett's (1988) reappraisal of these three tool assemblages, they are not discrete entities but represent a continuum of tool development. They represent a major change in the hominine/environment relationship, as reflected in Chavaillon's (1994) statement that 'the first true hand axes and cleavers had appeared and humankind was on the threshold of a technical and cultural transformation.' Chavaillon also points out that Oldowan industries exist elsewhere in Africa: in the Maghreb, Angola and South Africa.

By ca. 1.8×10^6 years BP, at about the same time as *Homo erectus* appeared, hand axes typical of the Acheulean tradition were being produced. Whether the two are related remains to be determined. As *H. erectus* populations migrated into Europe and India they took with them the Acheulian tradition of toolmaking. They also took with them the Oldowan tradition of toolmaking to Europe, China and Java. Precisely when the hominines began to use fire is conjectural but *H. erectus* is likely to have been the first species to do so. As stated in Section 4.2, *H. erectus* is considered to be a major innovator and developer of 'human' characteristics, reflecting a more complex lifestyle than the ancestral *H. habilis* or *H. rudolfiensis*. This increasingly complex lifestyle is paralled by an increase in brain size (Section 4.2). Brain and Sillen (1988) have reported the earliest use of fire at between 1 and 1.5×10^6 years BP from Swartkrans Cave in South Africa, though it is not clear which hominine species was responsible.

This use of fire occurred in the early Palaeolithic, when *H. erectus* migrated out of Africa and into Asia and Europe. Figure 4.5 summarises the development of Palaeolithic traditions in north China, where there are several significant archaeological sites. Unfortunately, the dating of some of them is questionable (Xinzhi and Linghong, 1985) and until further advances are made the dates given in Figure 4.5 must remain tentative. This is particularly so in relation to Xihondu, Shaanxi Province (Lanpo, 1985), part of a complex of Palaeolithic localities near Kehe. Cores, flakes, choppers and scrapers have been found at this site in association with deer antlers that bear signs of chopping and scraping, and mammalian ribs that appear to have been burnt. Lanpo suggests the site is at least 1×10^6 years old, and palaeomagnetic dating suggests it is 1.8×10^6 years BP. However, this is controversial and there is some evidence to indicate the artefacts may have been transported in water. The oldest hominid fossil-bearing site in China is Yuanmou, Yunnan Province (Rukang and Xingren, 1985), where two teeth of *H. erectus* have been found and are thought to date from 700–500 K years BP, though this dating is controversial. Remains of *H. erectus* have also been found at two localities in Lantian County in Shaanxi Province

(Rukang and Xingren, 1985). The oldest is a cranium, found at Gongwangling, which predates a mandible discovered at Chenjiawo. The age range for these fossils is probably 800 to 650 K years BP and both are associated with cores, flakes and scrapers as well as vertebrate remains.

Perhaps the most well-known archaeological site in China is Zhoukoudian (formerly Choukoutien), near Beijiing; its stratigraphy and fossils are summarised in Table 4.3. According to Senshui (1985), the lithic assemblages show a typological development, and the presence of numerous ash layers and burnt bone attest to human use of fire, though whether this was manipulation of naturally occurring fire is open to question. Ho and Li (1987) have suggested that evidence from this and other sites in north China indicates two different subsistence strategies. At Zhoukoudian, the lithic assemblages below level 6 (see Table 4.3) consist of large rather crude artefacts whereas above level 6, which is dated to ca. 370 K years BP, smaller and more intricate tools predominate. There is also a change in faunal assemblages; xerophytic grassland herbivores are dominant in the younger horizons whereas forest species dominate below level 6. Ho and Li postulate that the earlier subsistence strategies were based on resource gathering whereas the small tool and grassland herbivore remains represent a change to hunting. A further Palaeolithic site at Xujiayao (Lanpo and Weiwen, 1985) has also yielded implements made of bone and antler. This site is dated to 100 K years BP and indicates that Palaeolithic groups were exploiting more than just stone to produce a toolkit. Rukang and Lanpo (1994) have also reported on plant and animal remains from Zhoukoudian. Pollen-analytical data from nearby indicate that the hominine fossils were deposited during an interglacial when temperate deciduous forests and grasslands occupied the plains and valleys and coniferous forests occurred on the mountains. Plant remains in the cave itself, notably charred seeds, indicate that the Chinese hackberry, walnut, hazelnut and possibly the fruits of pine, elm and rambler rose were sources of food. The cave also contains an abundance of fossil bones, including bones from ca. 3000 deer comprising two species: the thick-jaw bone deer (*Megaloceros pachyosteus*) and the sika deer (*Pseudaxis grayi*). There is plentiful evidence for toolmaking (Rukang and Lin, 1983).

Similar broadening of resource use and the manufacture of stylistic implement types occurred

1   Choukoutien, Beijing *(E,M,L)*
2   Xihoudu, Shanxi *(E)*
3   Dingcun, Shanxi *(M)*
4   Xiachuan, Shanxi *(L)*
5   Shiyu, Shanxi *(M,L)*
6   Salawusu, Inner Mongolia *(L)*
7   Sanmenxia, Henan *(E)*
8   Xiananhai, Henan *(L)*
9   Lantian, Shaanxi *(E)*
10 Shuidonggou, Ningxia *(L)*

E   Early Palaeolithic
M   Middle Palaeolithic
L   Later Palaeolithic

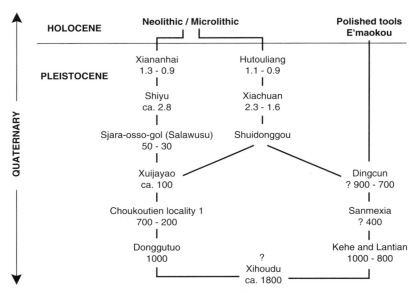

Fig. 4.5 Development of Palaeolithic traditions in north China (based on Lanpo and Weiwen, 1985; the map is from Rukang and Olsen, 1985). Dates are in thousands of years BP.

Table 4.3 The chief characteristics of the Choukoutien (Locality 1) archeological site near Beijing, China (based on Senshui, 1985)

	Layers	Stratigraphy	Evidence for use of fire	Lithic industry	*H. erectus* fossils	Approx dates (K years BP)
Late stage of the Early Palaeolithic	1-2	Coarse breccia with stalagmites		Vein quartz remains dominant raw material but milky quartz nodules and flint were also used.		
	3	Coarse breccia	✓	Flake tools increase in frequency and heavy cave tools decrease in frequency.	✓	
	4	Upper ash layer	✓	Stone awls appear for the first time.	✓	300 to 200
	5	Black & grey clay stalagmite crust				
Middle stage of the Early Palaeolithic	6	Hard breccia	✓	Quartz is the dominant raw material but sandstone and rock crystal were also used. Flake tools predominate over core tools.	✓	400 to 300
	7	Cross-bedded sands				
Early stage of the Early Palaeolithic	8-9	Coarse breccia with ash lenses	✓	Quartz and sandstone were used as raw materials. The lithic assemblages are typologically simple choppers and scrapers with a predominance of core tools.	✓	
	10	Red clay with weakly brecciated limestone blocks	✓		✓	660 to 400
	11	Reddish breccia	✓		✓	

elsewhere. Joshi (1994) has discussed changes in southern Asia, i.e India, Pakistan, Afghanistan, Nepal, Bhutan, Myanmar and Sri Lanka, where there is no evidence for *Homo habilis* (Section 4.2) and evidence for *H. erectus* is confined to Pakistan, India and Myanmar. The tools in Myanmar were made by placing stone on wood, and although they reflect the chopper-tool tradition they are quite distinct from the Lower Palaeolithic tools of India and Pakistan. These artefacts can be classified as Soan or Sohanian pebble tools (Misra, 1987), a complex which had developed in India by ca. 300 K years BP, or hand axe industries of the Acheulian tradition. Core and flake tools of this type were being manufactured by 200 K years. There is some, but limited evidence, for resource exploitation. For example, in India's Karnataka region (in south India) excavations in the Hunsgi Valley have highlighted the use of limestone as a source material for artefacts.

There is much more evidence for Lower Palaeolithic activity in the Near East (southwest Asia) and Europe. This reflects the intensity of research activity in these areas and the relative paucity elsewhere. However, the Near East is particularly important because of its geographical position; it links Africa, the acknowledged centre of hominine evolution, and the rest of the world. A review of Lower Palaeolithic archaeology in this region has been presented by Hours (1994). Despite its strategic location, few human fossils have been found, though there are many sites at which stone tool assemblages have been discovered. One of the oldest sites is the Ubeidiya Formation in the central Jordan Valley, the lowest sequence of which is considered to be more that 800 K years old and possibly 1.4×10^6 years old (Tchernov et al., 1994). The contained implements are Oldowan and are made of flint, basalt and limestone. According to Hours (1994), hominines (*Homo erectus*) probably lived on the shores of what was Lake Ubeidiya and hunted big game animals. The nearby site of Evron Quarry may be contemporaneous with Ubeidiya (Tchernov et al., 1994) and is located directly east on the Mediterranean coast, though there are no absolute dates available. The lithic assemblages contain quartz and limestone pebbles, and flint cores in combination with faunal remains. The faunal remains include 12 species from a variety of biogeographic provinces i.e. Africa, the Palaearctic and the Orient.

Numerous sites elsewhere in the Near East, especially the Levantine Rift and along the Mediterranean coast, have been described by Hours (1994) as middle Lower Palaeolithic. Most of them are characterised by Acheulian lithic industries and are present in sediments of the Latamneh formation in the rift. They closely resemble sites of a similar age in Africa and reflect a technological advancement on the Ubeidiya assemblage. However, by the late Lower Palaeolithic, bifacial tools had become common and are described by Hours as late Acheulian or advanced late Acheulian. The assemblages at Birket Ram in southwest Syria are sandwiched between two basalt flows dated to 230 K and 800 K years BP; the assemblage itself is considered to be 300 K years old. It comprises many small tools fashioned using the Levallois technique. This involves the creation of a core with an upper flat surface and a lower convex surface. To produce a flake, the core must be hammered onto an anvil stone at an angle. These tools were relatively commonplace during the latter stages of the Lower Palaeolithic and reflect advanced technology and improved 'finishing'.

In Europe it is likely that *Homo erectus* arrived ca. 1×10^6 years BP (Rolland, 1992), though some researchers, e.g. Lambert (1987) and Delson (1989), have suggested it may have been much earlier, possibly as early as 2×10^6 years BP. Yet others, e.g. Roebroeks and Kolfschoten (1994), consider that a date of 500 K years is more appropriate. However, one of the earliest dates for human occupation is from Soleihac in the Massif Central of France and is between 970 and 900 K years BP on the basis of palaeomagnetism, though the basalt and flint artefacts may not be contemporaneous with the sediment in which the artefacts are found (Villa, 1994). This makes the site equivocal as the earliest European location for *Homo erectus*. The oldest hominine fossils are from Atapuerca, Spain (Section 4.2) and are dated at 780 K years BP (Parés and Pérez-González, 1995) on the basis of palaeomagnetism. At Isernia in central Italy (Coltori et al., 1982), a further site has yielded thousands of stone artefacts and animal bones and has been potassium–argon dated to 736 K years BP. According to Giusberti and Peretto (1991) there is evidence for systematic breakage of bison bones by early hominines, probably to obtain marrow.

Villa's (1994) synthesis of evidence on the Lower Palaeolithic of Europe indicates two interesting factors. Firstly, there is little if any evidence for the use of fire until after 500 K years BP, which is surprising in view of its much earlier use in Africa (see above). Secondly, the movement of *H. erectus* into the northern and central regions of Europe coincides with interglacial periods, when temperate climates prevailed and ice sheets had retreated poleward. Currant (1986) has examined this issue in relation to the mesocratic phase of British interglacials characterised by mixed deciduous woodland (Fig. 3.1). Insofar as interglacials were relatively short-lived periods (typically ca. 20 K years) in comparison with the cold stages (typically ca. 100 K years), it is possible that early hominine groups did not adapt to such forested environments because of familiarity with more open habitats and/or because of the lack of pressure to exploit such environments. This view is endorsed by Gamble (1986), who has suggested that hominine survival strategies were more in tune with the long cold stages of the Quaternary (Sections 2.2 to 2.4), when open habitats prevailed, rather than with forested interglacial environments. Moreover, Foley (1987) has suggested that the movement of hominines into Eurasia was a component of a major faunal shift that involved migratory animals, especially herbivorous ungulates. It is also likely that animals and hominines expanded poleward in tune with the change in vegetation belts that occurred as ice sheets waned, with the reverse occurring as ice sheets expanded (see the discussion in Section 4.2 in relation to the occupation of the Near East by *H. sapiens sapiens* and *H. sapiens neanderthalensis*). This movement of hominine communities reflects the operation of climatic or environmental determinism (Section 1.2). And this factor is also expressed in Gamble's (1986) assessment that the length of the growing season was all-important in influencing the early hominine colonisation of Eurasia.

The Early/Lower Palaeolithic ended by at least 200 K years BP or possibly as early as 350 to 300 K years BP, according to the re-examination by Mercier *et al.* (1995) based on new dates of lithic industries in the Near East. By the time the transition had occurred changes had taken place in modes of tool production (Table 4.2), notably the introduction of bifacial tools which characterise the Acheulian industry. The Oldowan and Acheulian assemblages endured relatively unchanged and with relatively few components for more that 1×10^6 years each. New and shorter-lived technologies developed with the onset of the Middle Palaeolithic (Figure 4.6). In particular, prepared core or Levallois tools dominate Middle Palaeolithic tool assemblages (see descriptions above). This is mode III production (Table 4.2) and tools made using this technique are generally called Mousterian. This name reflects the name of a Neanderthal site at Le Moustier in the Dordogne region of France. In addition, the advent of Middle Palaeolithic industries paralleled evolutionary changes in the hominine line (Fig.4.2) as *Homo heidelbergensis* was undergoing changes that led eventually to *H. sapiens* but with *H. neanderthalensis* evolving slightly earlier (ca. 180 K years BP). On the basis of thermolumines-cence dating, Mercier *et al.* (1995) indicate that

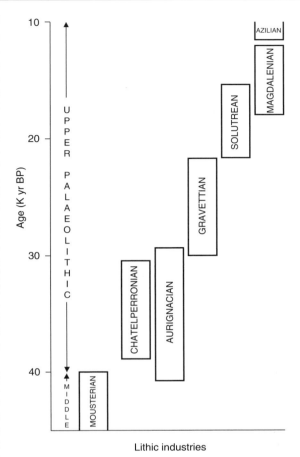

Fig. 4.6 Lithic industries during the Middle and Upper Palaeolithic.

Neanderthals were present in the Near East possibly as early as 170 K years BP (oxygen isotope stage 6). According to Valoch (1994a), the emergence of Middle Palaeolithic industries from those of the early Lower Palaeolithic may have been influenced by the onset of a major glacial period, i.e. the Riss (Table 2.2). In combination with increasing brain size as the hominine lineage evolved, the new industries allowed hominines to spread into regions they had never before occupied, including periglacial areas. The most important tools were side-scrapers and retouched flakes that were probably used as knives when attached to bone or antler hafts. The numerous points found at Middle Palaeolithic sites may reflect the use of spears which were created when the points were attached to wooden shafts. Variations on this and early Lower Palaeolithic industries occurred throughout Europe. At Wallertheim near Mainz, in Germany, a Middle Palaeolithic site has yielded flake tools and a range of animal bones, though butchery marks on the bones occur only for a species of steppe bison, *Bison priscus* (Gaudzinski, 1995). This may reflect specialised rather than indiscriminate hunting. During the late Middle Palaeolithic, a Mousterian tradition, based on the earlier Acheulian, developed into the Châtelperronian (Fig. 4.6) in western Europe.

Valloch (1994b) has reviewed the evidence for the use of resources during the Middle Palaeolithic, especially in relation to the Neanderthals who, like their ancestor *Homo erectus* were hunter gatherers. In addition to the hunting of large herbivores and omnivores, some of which was specialised with targets of specific animals, and organised on a group basis, there is evidence for river fishing. This derives from many spatially disparate sites. For example, Valloch states that 'Seventy-five per cent of the 26 000 bones found in the Kudaro cave in the Great Caucasus belong to the salmon. At Ogzi-Kitchik in Kazakhstan, 13 600 bones out of a total of 15 000 are the remains of steppe turtles.' Neanderthals occupied a range of habitation sites, including caves, many of which had earlier been occupied by *Homo erectus*, open-air sites which required the construction of shelters, and rock shelters. Many such sites (see Fig. 4.3 for the area that Neanderthals occupied) were on river banks and often close to stone suitable for tool manufacture. There is some evidence for tool exchange and clear evidence for the use of fire in hearths. It is likely they lived in small groups of 20–30 people, representing a few families within which there was strong bonding. Traces of healed wounds, for example, indicate mutual care and there is substantial evidence for ritual burial practices, as discussed by Leakey and Lewin (1992). One of the most well known is that of Shanidar in the Zagros Mountains of what is now Iraq (Trinkaus, 1983). The evidence points to a burial, 60 K years BP, of an old man who was laid to rest on a bed of plant material, with 'grave goods' that included many spring flowers, some of which did not grow near to the burial site and may have been used for medicinal purposes. In addition, Hublin *et al.* (1996) have reported on recent finds of Neanderthal remains from Arcy-sur-Cure, Yonne, France. The remains are dated at ca. 34 K years BP and represent the youngest Neanderthal remains discovered. This reinforces the view that Neanderthals and modern humans coexisted, the abundant non-human bone assemblage reflect a rich bone industry and the use of bones as personal adornments.

The transition between the Middle Palaeolithic and the late Upper Palaeolithic occurred ca. 40 K years BP, by which time modes IV and V (Table 4.2) traditions of tool making had become established. Blade manufacture replaced flake manufacture; blades are types of flake which are half as wide as they are long. Moreover, the Upper Palaeolithic is characterised by a wide range of regional and stylistic types, each with a large number of tools including projectile points and needles. Lewin (1993a) states that 'for the first time in human prehistory, there is a strong sense of directed design and elaborate use.' The transition to the Upper Palaeolithic also coincides with the development of modern human activities, i.e. activities of *Homo sapiens sapiens* as opposed to *Homo sapiens*; see Section 4.2. Gamble (1986) attributes this development in lithic industries to changes in food procurement akin to Wymer's (1982) distinction between savagery and barbarism. Savagery involves directed hunting and gathering with deliberate and planned manipulation of animal herds, and possibly some plant communities; barbarism is indiscriminate and a more opportunistic manipulation of resources. Whether or not this is true is open to question, but the advent of cave and rock art by ca. 30 K years BP, which usually depicts animals, may reflect a more considered and strategic approach to hunting. Mobile art, i.e. carved statues, also originated at this time.

The switch to Upper Palaeolithic traditions occurred during the last ice age (Table 2.2) and lasted until ca. 12 K years BP, though when this occurred varies regionally. The movement of ice sheets and climatic fluctuations must have had some effect on hominine populations and there is much debate as to the relationship between *H. neanderthalensis* and *H. sapiens sapiens*, and why *H. neanderthalensis* became extinct (see above and Section 4.2). *Homo sapiens sapiens* certainly became more successful, colonising most of the Earth's land surface relatively rapidly. The species was able to adapt to a wide variety of climatic regimes ranging from near polar to arid desert. Nevertheless, the use of stone remained dominant, and albeit in an increasingly sophisticated form, hunting and gathering remained as the means of food procurement. According to Klima (1994), there was increased innovation in tool construction. For example, there is evidence for harpoon and bow and arrow construction. In the lithic assemblages it is possible to distinguish between hunting implements, those used for butchering animals and preparing skins and implements used for working wood or bone. Klima also suggests that the production of blades represented a more economical use of raw materials than production of flakes in the Middle Palaeolithic. The various lithic assemblages that characterise western Europe are given in Figure 4.6 and the details relating to each of these industries are given in Otte (1994).

There is abundant evidence from elsewhere to attest to Upper Palaeolithic activities. From Siberia, for example, Goebal and Arsenov (1995) have reported new accelerator radiocarbon dates from Makarovo and Varvarina Gora, which are Upper Palaeolithic sites near Lake Baikal. Dates of ca. 38 to 39 K years BP indicate that the Upper Palaeolithic began earlier in this region than previously thought, and the settlement type and material culture was quite varied. Both are open-air sites with an abundance of lithic fragments. At Makarovo there are 4119 lithic pieces, including 113 cores and some 1700 blades and flake blades; at Varvarina Gora there are 1451 artefacts, including 226 tools comprising flakes, blades, scrapers and knives and a sharpened fragment of ivory tusk. There are hearths as well as animal bones at both sites, which may represent temporary hunting camps. There is no evidence for the plant component of the food resource This is rarely documented anywhere for the Palaeolithic in general, but it must have been very important. This issue has recently been considered by Mason et al. (1994) in relation to Europe, where the evidence is particularly sparse. They suggest the analysis of occupation layers could indeed provide some insight into the plant resources used by Upper Palaeolithic groups. Initial work on the archaeological site of Dolni Věstonice II in the Czech Republic, involving pollen analysis and the extraction of plant remains from a hearth, reflects the presence of a broad-based plant resource and the likely consumption of roots of species of Asteraceae or Compositae. Many species of these families are edible and may have been growing locally. There is much additional work to be undertaken on the role of plant resources including the analysis of plant residues on stone tools. Such work has been reported by Loy et al. (1992) on stone artefacts from the Solomon Islands. The residues include starch grains, identified as being derived from species of taro (*Colocasia* spp.), which

are cultivated today for their edible roots in many Pacific islands. The use of plants in the Near East during the Upper Palaeolithic is discussed by Hillman et al. (1989) in relation to Wadi Kubbaniya in Egypt and is reviewed generally by Miller (1991). Legumes, chenopods, sedges and large- and small-seeded grasses were components of the food base. Some of them eventually gave rise to the earliest domesticated plant species, as examined in Section 4.5.

4.4 The relationship between environment and Mesolithic groups

As Figure 4.6 illustrates, the late Palaeolithic industry known as the Magdalenian underwent several changes in western Europe, leading to a new industry known as the Azilian. This is named after the Mas d'Azil site in Ariège, France. According to Otte (1994), this transition occurred ca. 11.8 K years BP and involved the production of microlithic projectile points, short blades and scrapers, all of which would have been hafted in wood. There was also a significant increase in implements made from antler, including harpoons, as the spread of deciduous forests also encouraged the spread of species such as red deer. As the cold period of the Youngest Dryas (Table 3.2) gave way to the opening of the Holocene, the Azilian gave way to the Mesolithic. Similar changes occurred in other parts of Europe, as is illustrated in Figure 4.7, to produce distinctive Mesolithic cultures from pre-existing Palaeolithic cultures. Northwest Europe, for example, was dominated by the Maglemosian. Otte (1994) also distinguishes between the early and late Mesolithic; the early Mesolithic begins ca. 8 K years ago, when there were further changes in lithic industries. Such a generalisation is not necessarily appropriate throughout northwest Europe because there is increasing evidence for substantial changes in resource exploitation before 8 K years BP. In addition, the existence of a 'radiocarbon plateau' ca. 9.6 K years BP (Kramer and Becker, 1993; Stuiver and Braziunas, 1993) means that precise dating is problematic. For example, Kramer and Becker (1993) obtained constant radiocarbon ages at 9.6 K and 10 K years BP for pine remains in central European river terrace deposits. Such problems occur because of fluctuations in the concentration of ^{14}C in the atmosphere. The plateau occurs over a 400 year period in the case of the early Holocene.

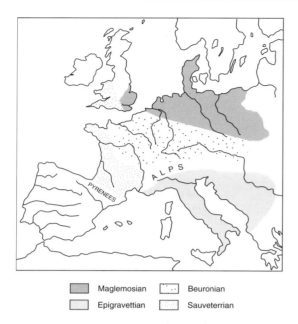

	Maglemosian		Beuronian
	Epigravettian		Sauveterrian

Fig. 4.7 European Mesolithic cultures: Maglemosian in northern Europe, Beuronian in central Europe, Epigravettian in the Mediterranean zone and Sauveterrian in western Europe (based on Otte, 1994).

The only way this problem can be overcome is to use an independent means of dating, if there is one available. In the case of the early Mesolithic (i.e. early Holocene) it is possible to use the tree-ring chronology of Becker (1993) as has been undertaken for Star Carr by Day and Mellars (1994).

The Mesolithic was short-lived compared with the Palaeolithic, lasting from ca.10 to 5 K years BP. It is a transitional period between the Old Stone Age and the New Stone Age, and it has its widest expression in Europe, where it coincides with the protocratic and mesocratic phases of the interglacial cycle (Fig. 3.1), during which temperate forest replaced vast tracts of tundra. As Zvelbil (1986) has pointed out, the use of the term 'Mesolithic' is somewhat spurious since it is often used to delimit a particular chronological period. Although this is loosely correct, it must be borne in mind that it is a term applied to early post-glacial temperate and boreal zone industries that developed in response to a forested environment and which did not occur elsewhere in the Old World where Palaeolithic traditions gave way to those of the Neolithic.

As stated in Section 4.3, there is little evidence for

exploitation of forested environments in the earlier European interglacials. Why then should the forest biome of the current interglacial have been the focus of human activity? Presumably there were unprecedented ecological and/or social pressures that initiated exploitation of these expanding forested environments. Population pressures may have ensued as the flora and fauna to which the Palaeolithic groups had adapted began to diminish. As discussed in Section 3.5, human pressures and/or environmental change may have brought about the extinction of many large herbivores between 12 and 10 K years BP. In the Near East these influences were likely to have been prime factors in the initiation of permanent agriculture (Section 4.5). Whatever the reason for the change in habitat preference, deposits of the early post-glacial period have yielded implements that are different in character to earlier deposits. They are characterised by microliths, many of which were used as tips for arrows, as well as axes and adzes that were attached to wooden hafts and which may have been used for creating small forest clearances and for making canoes and paddles.

One of the most significant Mesolithic sites in Europe is at Star Carr in Yorkshire, Britain, where there are two local phases of human activity. A wide range of materials have been preserved in an anaerobic peaty environment that developed as an ancient lake silted up. The site has been excavated by Clark (1954, 1972), who has reconstructed the activities of the occupants. The habitation site consists of a platform of cut birch brushwood that was placed in the reed swamp at the edge of the lake and which may have been covered by animal skins to form tent-like dwellings. A variety of animal bones reflect the regional fauna, e.g. badger, fox, various deer and numerous birds, as well as other animals that were used as food. Of these other animals, red deer were most abundant, followed by roe deer, elk, aurochs and pig. Remains of domesticated dog, the earliest domesticated animal found in Britain, are also present at Star Carr and at the nearby site of Seamer Carr as well as Kongemose in Sjaelland, Denmark (Clutton-Brock and Noe-Nygaard, 1990). Clark also examined the remains of deer antlers from Star Carr and has suggested that the site was seasonally rather than permanently occupied. The preponderance of antlers broken from the animals' skulls suggests they were hunted during a period from March to October when stags carried their antlers. Further analysis of deer-teeth characteristics from this site by Legge and

Rowley-Conwy (1989) suggests that Star Carr was inhabited during the summer. However, Clutton-Brock and Noe-Nygaard (1990) have suggested it was a hunting camp that was visited throughout the year by Mesolithic groups who also exploited coastal resources. There is also the possibility that stag frontlets with the antlers still in place may have been used in ritual activities, and there is evidence that antlers of red deer were used for making barbed points. The presence of stone burins, awls and scrapers indicate that animal skins were processed, possibly for the provision of shelter, clothing and/or containers. As attested by the birch platform, wood was also used in the Mesolithic economy of this site. The remains of a wooden paddle indicates the inhabitants made boats and the presence of rolls of birch-bark may relate to the extraction of resin, possibly for fixing arrowheads and spearheads.

Recent work by Day and Mellars (1994) involves charcoal analysis of the Star Carr sediments including the occurrence of macrocharcoal particles. These are particles larger than 600 mm, considered to represent local as opposed to regional fires, though it is not possible to ascribe with certainty such burning events to deliberate burning by human communities because it is feasible that natural fires occurred. In the first phase of burning, reflected by 'large' charcoal particles, there are three peaks. Based on a new chronology to accommodate the ^{14}C plateau, this occurred at ca. 10.7 K years BP (calibrated). The second phase is dated 10.55 K years BP (calibrated). The charcoal record may reflect several burning events over a 120 year period. And the charcoal derives from the reed *Phragmites* spp., which suggests the local wetland vegetation was the fuel source. Although there are no other concentrations of macrocharcoal, the continuous presence of microcharcoal throughout the profile attests to the significance of fire in the vicinity of the camp. Overall there is more evidence for Mesolithic disturbance at upland sites than from lowland sites. However, charcoal and pollen analysis of two sites in Oxfordshire, Cothill Fen and Sidlings Copse (Day, 1991), reflect the significance of fire in the early Holocene. Similar evidence has been found at Lashford Lane Fen in Oxfordshire, close to Cothill (Benton and Mannion, 1995) and at Tidmarsh Wood, west of Reading (Mannion and Benton, 1997). The charcoal counts from these sites are shown in Figure 4.8. There is no evidence to suggest that Mesolithic activity caused these fires, but the

presence of Mesolithic artefacts and Mesolithic campsites in south-central England means that it cannot be excluded.

Whatever the reality of these inferences, the remains from Mesolithic sites indicate that Mesolithic people were using a wide range of resources, evidence for which is widespread in Europe. From Oronsay, Inner Hebrides, Mellars (1987) has excavated several shell-midden sites dated at ca. 5 K years BP. The abundant remains of molluscs, especially limpets, seals, birds and fish, especially the coalfish (*Pollachius virens*), indicate the inhabitants relied heavily on marine resources in an environment that probably contained very limited terrestrial animal resources. Similarly, research in the Netherlands, Denmark and Germany (Barker 1985) points to Mesolithic exploitation of red deer, roe deer and aurochs as wells as coastal resources. Indeed, Price (1987) has suggested that marine resources may have contributed up to 90 per cent of the diet of some north European Mesolithic groups and that carbon isotope analysis of human bone reflects a change towards terrestrial food sources only in the Neolithic period.

This is perhaps a little exaggerated as there is increasing evidence for forest and woodland exploitation, as discussed below. Nevertheless, Smith's (1992) review of Mesolithic marine resource exploitation in the British Isles indicates that a wide range of sources of marine protein were utilised and that water craft were in use by ca. 9 K years BP. Such craft may have been similar to dugout canoes, involving a wooden frame and a skin covering. Sea-level was rising at this time in response to ice-sheet melting; early Mesolithic encampments on what was then the coast were abandoned as they were inundated, thus depriving palaeoecologists and archaeologists of valuable insights into how they were used by Mesolithic people. Evidence for the Mesolithic exploitation of large mammals in Europe is substantial. At many Mesolithic lakeside archaeological sites in Denmark, Rowley-Conwy (1993) has demonstrated that five species are consistently represented: *Bos primigenius* (aurochs), *Alces alces* (elk/moose), *Cervus elephas* (red deer), *Sus scrofa* (wild pig) and *Capreolus capreolus* (roe deer); all of them were hunted. At one of the sites, Holmgaard on the island of Zealand, a bow made of *Ulmus glabra* (narrow-grained elm) has been found and dated at ca. 6 K years BP (Bergman, 1993). Elsewhere in Scandinavia Mesolithic bows made from *Taxus baccata* (yew) and *Sorbus*

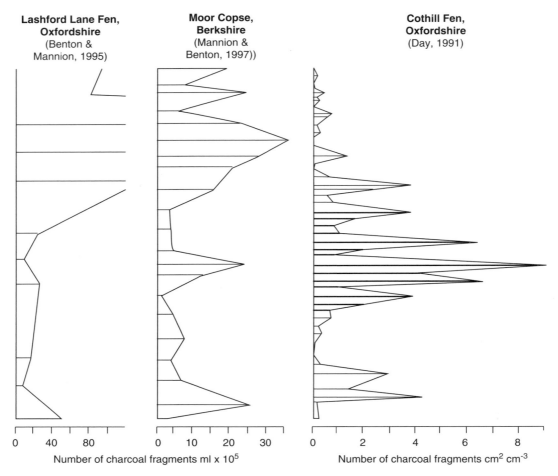

Fig. 4.8 Evidence for the significance of fire in early Holocene environments (ca. 6.5 to 10 K years BP) in south-central England. Correlations between sites are not implied.

aucuparia (rowan) have been found and reflect skilled woodworking. According to Otte (1994) and Bandi (1994), there is a range of art types for Mesolithic settlement sites. For the period between 10 and 7 K years BP there are animal figures made of amber in the Baltic region, and carved ivory figures, human figurines and rock engravings from many parts of Europe.

In common with the Upper Palaeolithic, the evidence from Mesolithic sites in Europe reflects exploitation of animal resources. Evidence for exploitation of plant resources is much less widespread, although it has already been suggested in Section 3.7 that Mesolithic communities may have played a significant role in environmental change during the early Holocene. The abundance

of hazel in British early post-glacial pollen diagrams, for example, may have been a response of the species to burning and/or coppicing (Smith, 1970), though recent work on the relationship between fire and vegetation in East Anglia (Bennett *et al.*, 1990b) has rather disproved this hypothesis. A survey of possible plant use and evidence for apparatus associated with food production and preparation during the Mesolithic in Europe has been undertaken by Zvelbil (1994). His survey suggests that plant use was extensive throughout Europe with clear evidence for the consumption of nuts, especially hazel, acorns and water chestnuts. At several archaeological site in Latvia, Lithuania and northwest Russia the remains of water chestnuts are abundant and in some cases they are associated with plant-processing equipment such as wooden hoes and mattocks (soil-loosening pickaxes). Kubiakmartens (1996) has also reported on the presence of root and tuber tissues ascribed to *Sagittaria sagittifolia* (arrowhead) and *Polygonum* spp. (knotgrasses and bindweeds) from Calowanie, an Upper Palaeolithic/Mesolithic site in the Polish Plain. This is the first site in the North European Plain to produce evidence of plant foods other than water chestnuts. Zvelbil (1994) also suggests that the axes, adzes and picks, often made of antler, from Mesolithic settlement sites may have been used for digging rather than for woodworking, as is generally accepted. Although Zvelbil's survey cannot be considered as unequivocal evidence for intensive and widespread manipulation of wild plant resources and incipient agriculture, it indicates that systematic research is needed to clarify the role of plants in Mesolithic *genres de vie*.

Zvelbil's survey also included information on pollen-analytical evidence for small-scale clearances during the early Holocene, i.e. between 10 and 5 K years BP, which may be attributed to the activities of Mesolithic communities. More than two decades ago Simmons (1975) drew attention to the many pollen diagrams from upland Britain with evidence for clearance often associated with charcoal remains. More recent research, summarised in Simmons and Innes (1987, 1996a, b, c), Simmons (1993c, 1995, c) and Simmons and Innes (1996a, b, c) adds support to the idea that Mesolithic communities were exploiting the uplands of northern England in a highly organised fashion, including the controlled use of fire. Work on North Gill in North Yorkshire indicates that burning of the woodland may have been undertaken on a patch basis, with each burnt patch comprising tens of metres rather than hundreds of metres (Turner *et al.*, 1993). As Clark and Robinson (1993) have discussed, there is abundant evidence for the occurrence of fire in Quaternary sediments, including those of the early Holocene in various types of forests worldwide, but it can rarely be attributed unequivocally to human rather than natural sources. (See also Heusser's (1994) work on evidence for fire in South America in relation to the peopling of the Americas, mentioned in Section 4.2.) Fire is nowadays a major ecosystem management technique (Archibold, 1995) and there is no reason to suppose it was less important in the early Holocene. Irrevocable changes in vegetation communities occurred in some cases of woodland disturbance. For example, the woodland may have given way to peat or moor formation (Simmons and Innes, 1985).

Pollen-analytical and/or charcoal evidence for disturbance of early Holocene woodland in Scotland and the Scottish islands is also abundant. For example, Bunting (1994) reports that Mesolithic activity in west Mainland of Orkney affected birch–hazel woodland, which eventually disappeared ca. 5 K years BP as a consequence of Neolithic activity. From Islay, Edwards and Berridge (1994) have presented pollen, charcoal and sedimentary data for likely Mesolithic environmental change and suggest that hunter-gatherer communities were present on Islay from ca. 10 K years BP. Possible Mesolithic interference in woodlands on the island of South Uist is suggested by Bennett *et al.* (1990a). Numerous other sites in the British Isles where there is evidence of disturbance or the presence of edible plants are listed in Zvelbil (1994). The majority are from coastal and upland regions with a paucity of sites in the English Midlands and the Southeast. This partly reflects the paucity of work in these areas, though there is evidence for fire in the early Holocene at several sites in Oxfordshire and Berkshire, as illustrated in Figure 4.8. Moreover, Waller (1993) has suggested that openings in the forest canopy in the vicinity of Pannel Bridge, East Sussex, before ca. 5 K years BP were due to Mesolithic activity. Evans' (1993) review of molluscan data from the English chalklands indicates that the presence of Mesolithic camps did not necessarily result in woodland disturbance, but the increasing archaeological evidence for Mesolithic presence in the region implies the need for a systematic search for sites that will yield information about Mesolithic and later human impact.

The record of intense manipulation of environment in the British Isles and Europe during the Mesolithic, especially when compared with earlier periods (Section 4.3), may reflect the establishment of permanent settlements. Even if not the case, the sophisticated infrastructure of advanced hunter-gatherer groups which developed during the early Holocene would have provided a sound base on which permanent agriculture could become established as the Neolithic culture developed. The introduction of domesticated plants and animals from Europe to a well-organised proto-agricultural society would not have required such climactic adjustments as are often conjectured. Simmons and Innes (1987) suggest that the development of Neolithic economies heralded a change in the scale of food production rather than a fundamental change in community structure. They conclude that 'Holocene economic adaptations may be viewed as a progressive intensification of food production; early Mesolithic foraging followed by advanced foraging with environmental manipulation which culminated in a control of food resources can amount to horticulture and herding of ungulates.' This emphasises continuity between Mesolithic and Neolithic cultures and reflects a continuum of development in the people/environment relationship.

4.5 Domestication of plants and animals: the beginnings of agriculture

It is a matter of fact that agriculture is one of the most significant cultural agents of environmental change. Its impact has been wide-reaching temporally and spatially (Mannion, 1995a), as is considered in Chapters 7 and 8. Moreover the issues of when, where, how and why agriculture began have been the foci of considerable research in the past century. Although many advances have been made, the answers to the questions about why agriculture developed remain elusive. Nevertheless, the advent of agriculture set in train many cultural and environmental processes, processes that were often irreversible and certainly consequential. It is accepted that agriculture would not have developed without the precursor of organised hunter-gatherer societies and that it represents the culmination of earlier events. Nevertheless, its emergence signified a turning-point in cultural and environmental history through the changed relationship between people and environment. In this context there are parallels between the emergence of agriculture and the Industrial Revolution of the eighteenth century. Both also had profound effects on global population growth, the division of labour, energy flows and biogeochemical cycles. Moreover, the advent of agriculture reflected the intensifying ability of humans to manipulate their environment.

In 1936 the archaeologist Vere Gordon Childe coined the term 'Neolithic Revolution', since it was during the archaeological period known as the Neolithic (New Stone Age) that permanent agriculture began. However, in archaeological terms it is not the emergence of agriculture that defines the Neolithic, but the presence of pottery (ceramics) that reflects a change in the material culture of human groups. In reality the advent of agriculture predates the first pottery. The beginning of agriculture occurred in a period that is often known as the aceramic Neolithic. Although the emergence of agriculture was not inherently revolutionary, its representation in the archaeological record is abrupt. This is because the presence of domesticated, as opposed to wild, plants and animals represents the culmination of developmental stages between hunting/gathering and permanent food production. As Smith (1995) states, 'Debate continues today on the fine points, but there is considerable agreement on a good starting definition: domestication is the human creation of a new form of plant or animal – one that is identifiably different from its wild ancestors and extant wild relatives.' How long a period of time or how many generations of species are required between selecting a wild species and transforming it into a domesticate is highly conjectural. Moreover, it is important to note that the term 'Neolithic' refers to a cultural period that is not applicable on a worldwide basis; even where it is applicable it is not necessarily synchronous. For example, it is a term which is applied only to Old World archaeology because in the New World, the various pre-Columbian or Aboriginal cultures are described on the basis of local or regional names based on type-site localities. Within the Old World the earliest Neolithic cultures emerged in the Near East ca. 10 K years BP but did not develop in Europe's periphery until ca. 5 K years BP.

Of all the questions that surround the beginning of agriculture, the most perplexing concerns why it was brought about. In the 60 years since Childe's (1936) discourse on the subject, the considerable volume of research that has been undertaken in various parts of the world has not provided an uncontestable answer. Although the archaeological and palaeoenvironmental records provide insights into the issues of where and when they are unable to reveal the motives of innovators. Overall, there are two groups of stimuli: environmental stimuli and cultural stimuli, the former a type of environmental determinism. These two groups are not necessarily mutually exclusive but have given rise to two polarised views that are summarised in Table 4.4. According to MacNeish's (1992) synthesis of research on plant domestication, this dichotomy arose in the 1930s when Childe (1936) advocated environmental change as a major stimulus to the inception of agriculture, whereas Ivan Vavilov, a

Table 4.4 Two possible models for the emergence of agriculture

	INPUTS		OUTPUTS
MATERIALISM OR CULTURAL MATERIALISM	Need through population increase ⟶	A	
	Greed through a desire to produce a surplus ⟶	G	Continued population increase, permanent settlements
		R	
	Need or greed through a shortage of food created by environmental change ⟶	I	Trade, bringing an increased awareness of resources
	A sedentary existence ⟶	C	Division of labour, facilitating diverse activities
		U	
ENVIRONMENTALISM OR CULTURAL ECOLOGY	Climatic change at the end of the last ice age ⟶	L ⟶	Food security, political pre-eminence
	This concentrated plants, animals and people around oases in the Near East, the cradle of civilisation where desiccation was occurring ⟶	T U R	Improvements and innovations in food production
	An overall shift of the resource base due to climatic/ecological change ⟶	E	Development of ceramics

Source: Mannion (1995a)

well-known Russian botanist, proposed that the most important stimuli were cultural (Vavilov (1992) is a translation of Vavilov's 1920–1940 papers by D.Löve) e.g. population growth and/or the desire to produce a surplus and reliable food supply. Since climatic change may have created uncertain ecological conditions, it may have encouraged plant and animal domestication to allow human communities to increase their control on food supplies. As discussed in Section 4.4, the highly organised hunter-gatherer societies of the late Upper Palaeolithic (Section 4.3) and the Mesolithic (Section 4.4) may have begun permanent agriculture via relatively low-key innovations.

However, there is growing evidence that environmental change did indeed prompt the emergence of agriculture (McCorriston and Hole, 1991; and Byrd, 1994). As the last ice age drew to a close, many faunal extinctions occurred in different parts of the world (Section 3.5); the debate as to why this occurred focuses on whether it was environmental or cultural, i.e. there are many parallels with the debate surrounding the emergence of agriculture. As examined in Section 3.5, this issue remains unresolved but increasing evidence for rapid global climatic and ecological change (Sections 3.3 and 3.4) between 14 and 10 K years certainly points to a dynamic environment. Moore and Hillman (1992), reviewing

existing palynological data from the Near East, have pointed out that it reflects climatic and ecological instability. It is unlikely this would have had no impact on human communities. At Abu Hureyra, an archaeological site in modern-day Syria occupied from 11.5 to 7 K years BP, a change in climate and vegetation is dated to ca. 10.6 K years BP, just before the inception of agriculture (Moore, 1992). Perhaps it is no coincidence that the earliest domesticated species, which derive from the Near East, coincide with the end of the late-glacial period and the early Holocene. As Harlan (1992) states: 'The question must be raised. Why farm? Why give up the 20-h work week and the fun of hunting in order to toil in the sun? Why work harder for food less nutritious and a supply more capricious? Why invite famine, plague, pestilence and crowded living conditions? Why abandon the Golden Age and take up the burden?' Why indeed if it were not for existential necessity?

It is possible that human populations had become sedentary, causing a depletion in food resources so that agriculture was the only option. This is the population-led and materialist approach to the beginning of agriculture advanced by Boserup (1965) and based on analogies with the process operative in modern hunter-gatherer societies. Similar events would occur if a human community wanted to produce a food surplus for barter and to ensure food

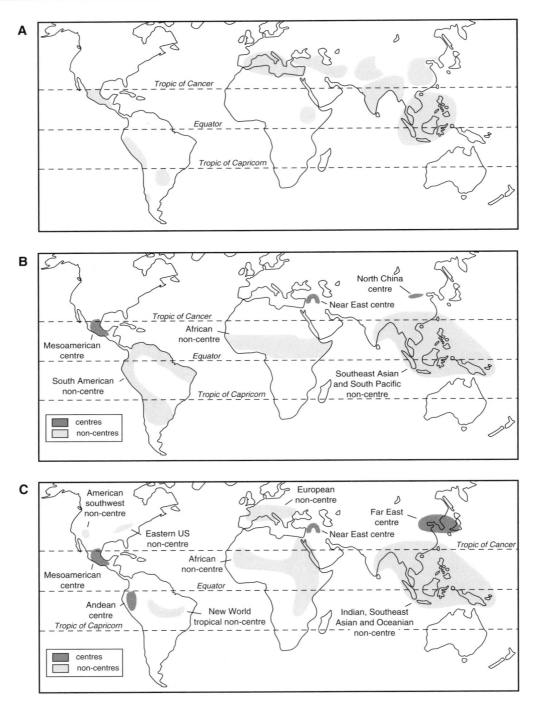

Fig. 4.9 Centres of origin for crops: (A) suggested by Vavilov in 1926 (based on Harlan, 1992); (B) suggested by Harlan in 1971 (based on Harlan, 1992); (C) suggested by MacNeish (reviewed in MacNeish, 1992);

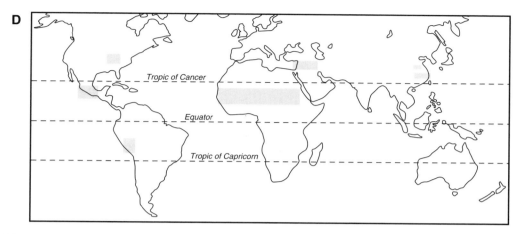

Fig. 4.9 contd. (D) suggested by Smith (1995).

security. Such events are not borne out by archaeological evidence from the Near East. For example, Roosevelt (1984) has compared the palaeopathology of Palaeolithic and Neolithic skeletons of the region. The Palaeolithic skeletons show little evidence of severe or chronic stress or malnutrition, which is not the case for the Neolithic specimens. Both Roosevelt and Layton *et al.* (1991) conclude that the inception of agriculture placed more strains, not less, on human communities, so its inception was unlikely to be due to population pressure. Further palaeopathological evidence has been reported by Molleson (1994) from Abu Hureyra. The human skeletal remains from the period just after the inception of agriculture reflect the effects of hard physical labour and a comparatively poor diet. Molleson has suggested that deformities of the spinal cord could have resulted from the porterage of heavy loads as well as protracted kneeling necessitated by the grinding of cereals in pestles and mortars. Protracted kneeling is also suggested by Wright (1994), whose examination of grinding stones and mortars from the Near East attest to the labour required in grinding cereals. She also suggests that such plants would only have been adopted to provide a major food source if other wild foods became scarce. Interestingly, perhaps, climatic deterioration between 11.5 and 10.5 K years would have favoured wild grasses at the expense of other gathered plants in the Near East. Wright's analysis shows that it was shortly after this phase (Youngest Dryas; see Tables 3.2 and 3.3) that domesticated forms of wheat and barley appeared at several Near Eastern sites. At the same time there was an increase in the abundance of food

processing equipment. The implications of these results are that the Youngest Dryas, which was probably a global event (Sections 3.3 and 3.4), at least accelerated the process of plant domestication and thus represented a cultural as well as an environmental threshold. However, it must be noted that the short-lived nature of the environmental changes that occurred between 14 and 10 K years BP (the late-glacial period discussed in Sections 3.3 and 3.4) makes them difficult to date precisely, because of the errors associated with radiocarbon dating, so the precise nature of the relationship between environmental change and the emergence of agriculture remains enigmatic.

There is more certainty about where the domestication of plants and animals took place than about why it took place. The oldest known domesticated plant and animal remains derive from archaeological sites in the Near East but there are also other centres of *in situ* domestication, notably the northern Andes, Mesoamerica, north and south China, sub-Saharan Africa and the eastern United States. The identification of these regions as centres of domestication owes much to the work of Vavilov (1992), who suggested the centres of plant domestication were likely to have been in regions containing high biological diversity (biodiversity), i.e. regions in which many different potential sources of plant foods were available. Such an abundance of wild foods could have been responsible for the adoption of sedentary lifestyles by hunter-gatherers who developed techniques, e.g. burning strategies, to enhance the supply of these resources. Subsequent environmental change may then have caused people to plant and reap; a similar outcome would have occurred if human population levels had increased to exhaust the supply of wild food. In 1940 Vavilov proposed there were eight centres, or hearths, of domestication (Vavilov, 1992), as illustrated in Figure 4.9. Subsequent workers, e.g. Sauer (1952), adopted Vavilov's views though with some modifications and redefinitions. In 1971 Harlan proposed there were three centres, i.e. definable geographical areas within which a number of plant species were domesticated and then diffused into neighbouring areas, and three non-centres, i.e. large regions in which plant domestication occurred widely (Harlan, 1992). These centres and non-centres are shown in Figure 4.9. A further adaptation has been proposed by MacNeish (1992), also illustrated in Figure 4.9. This comprises four centres and six non-centres. The non-centres differ from those of Harlan because they are well-defined areas which received their initial domesticates from the centres before *in situ* domestication of indigenous species took place. Smith (1995) defines seven centres in which the domestication of indigenous plants occurred independently; these too are illustrated in Figure 4.9, which shows a degree of concurrence between the various schemes.

Table 4.5 gives the earliest dates quoted in the literature for plant domestication in the various centres (and sometimes non-centres). The earliest overall dates are from the Near East and place the first domesticated species, i.e. wheat and barley, as emerging just after the Youngest Dryas cold period (Sections 3.3 and 3.4). In all the other centres the first domestications were a little later, though this does not necessarily imply a lack of a cause and effect relationship between environmental change and the emergence of agriculture, because the onset of the Holocene was also a period of rapid environmental change. When compared with their wild counterparts, the recorded species have in all cases been altered through the influence of human activity. As Blumler and Byrne (1991) have pointed out, it is possible that plant (and probably animal) domestication may have occurred before actual cultivation, at least partially. Moreover, there is considerable debate as to the rate at which a species may become domesticated. In relation to wild wheats and barley, Hillman and Davies (1990) have suggested that once the species were being planted they became domesticated in as little as 20–30 years under ideal conditions and intense management; at most the process for these species would be no more

Table 4.5 Some of the world's most important crop plants and their approximate dates of origin

Crop	Common name	Approximate date (K years BP)
A. The Near East		
Avena sativa	oats	9.0
Hordeum vulgare	barley	10.2
Secale cereale	rye	9.0
Triticum aestivum	bread wheat	7.8
T. dicoccum	emmer wheat	9.5
T. monocccum	einkorn wheat	9.5
Lens esculenta	lentil	9.5
Vicia faba	broadbean	8.5
Olea europea	olive	7.0
Cannabis sativa	hemp	9.5
B. Africa		
Sorghum bicolor	sorghum	8.0
Eleusine coracana	finger millet	?
Oryza glaberrima	African rice	?
Vigna linguiculata	cowpea	3.4
Dioscorea cayenensis	yam	10.0
Coffea arabica	coffee	?
C. Far East		
Oryza sativa	rice	9.0
Glycine max	soybean	3.0
Juglans regia	walnut	?
Catanea henryi	Chinese chestnut	?
D. Southeast Asia and Pacific Islands		
Panicum miliare	slender millet	?
Cajanus cajan	pigeonpea	?
Colocasia esculenta	taro	9.0
Cocos nucifera	coconut	5.0
Mangifera indica	mango	9.2
E. The Americas		
Zea mays	maize	5.0
Phaseolus lunatus	Lima bean	5.0?
Manihot esculenta	cassava	4.5
Ipomea batatus	sweet potato	4.5
Solanum tuberosum	potato	5.0
Capsicum annuum	pepper	8.5
Cucurbita spp.	various squashes	10.7?
Gossypium spp.	cotton	5.5

Source: based on Evans (1993) and work cited in the text

than two centuries. This view is generally in line with that of Ladizinsky (1987), who has shown that the domestication of lentils may have been taken only ca. 25 years. Galinat (1992), however, believes that the domestication of maize required 100–400 years, though, as Smith (1995) has reviewed, the domestication of maize is a controversial issue (see also Benz and Iltis, 1990). Insofar as it seems possible for the domestication of at least some staple species to have proceeded in a matter of decades, it would have been necessary for only a few generations of farmers to establish them as farm crops, thus strengthening the relationship between plants, people and possibly power, since a food surplus would have provided advantage. Today that relationship represents wealth generation in many forms (Mannion, 1995a).

Smith (1995) has reviewed the evidence for early agriculture in the centres and non-centres shown in Figure 4.9. The major crops domesticated in each centre are given in Table 4.5; a more detailed discussion of plant domestication is given in Evans (1993).

4.5.1 Centres of plant domestication: The Near East

Two of the world's most important cereals, wheat and barley, originated in the Near East (Zohary and Hopf, 1993). According to Lagudah and Appels (1992), modern bread wheat (*Triticum aestivum* L.) originated from Emmer (*T. turgidum* subsp. *dicoccum*) and *T. tanschii*, a hybridisation that Harlan (1992) suggests occurred in the region southeast of the Caspian Sea. From its centre of domestication in what is now Israel/Syria/Jordan, bread wheat spread into Europe and Asia along with emmer (*T. dicoccum* subsp. *dicoccum*) itself and einkorn (*T. monococcum*). Several pulses (Table 4.5), e.g. lentils, which were also domesticated in the Near East, often accompanied the wheats and barley (*Hordeum vulgare*) as crop complexes were introduced into areas beyond the centre of domestication. Like the wheats mentioned above, the wild ancestors of barley are widely distributed in the Near East and Western Asia; they are

Table 4.6 Neolithic cultures of the Near East

Date (K years BP)	Culture
11.7–11.0	Late Natufian (Palaeolithic)
11.0–10.5	Pre-pottery Neolithic A
10.5–9.25	Early pre-pottery Neolithic B
9.25–8.25	Late pre-pottery Neolithic B
8.25	Beginning of Halaf, Hassuna, Samarra (painted pottery cultures)
7.25–7.0	End of Halaf and beginning of Ubaid 3 (end of painted pottery and beginning of plain pottery)
6.0	Beginning of city states, e.g. Uruk, Akkad and Sumer

Source: based on Mellaart (1994)

collectively known as *Hordeum vulgare* L. subsp. *spontaneum*. As Table 4.6 shows, many other species were also domesticated in the Near East, especially between 10 and 7 K years BP, including fibre and oil plants, i.e. hemp and flax. The domestication of flax has been discussed by Diederichsen and Hammer (1995).

4.5.2 Centres of the Far East

The Far East has two centres of domestication, both of them in China (Fig. 4.9). In southern China there is the Yangtze River corridor, where the first domestications occurred ca. 8.5 K years ago, and further north there is a centre in the Yellow River basin, where the first domesticates occurred ca. 7.8 K years BP (Smith, 1995). The crops developed in these two regions include broomcorn millet (*Panicum miliaceum*), water chestnut (*Trapa natans*) and rice (*Oryza sativa*). The origins of rice are particularly contentious. Traditionally (e.g. Chang, 1989) no single focus of rice domestication has been discovered and it is generally considered to have occurred somewhere within a broad belt extending from India to China, some considerable distance south of the Yangtze River. This belt corresponds to the modern-day distribution of wild rice. However, there is growing evidence that this is incorrect. For example, the geographical range of wild rice has been extended due to recent finds in the Yangtze River basin; Wenming (1991) has reported the discovery of the remains of domesticated rice dated to ca. 8 K years BP from Pengtoushan in the mid-basin of the Yangtze River. Along with Smith (1995), Wenming believes that domestication of rice and certainly the origin of rice cultivation will eventually be traced to the area of the Yangtze Delta. Millet originated in the northern centre, where the Yellow River leaves the western highlands and enters the plains. Excavations of several villages have produced two domesticated species of millet: broomcorn millet (*Panicum miliaceum*) and foxtail millet (*Seraria italica* sp. *italica*), which were domesticated by c. 8 K years BP and possibly earlier.

4.5.3 The sub-Saharan centre

The other species of domesticated rice, African rice (*Oryza glaberrima*), originated in the savanna zone of west Africa from an annual grass adapted to a seasonally distributed rainfall (Harlan, 1994). In fact, Harlan states: 'A complete suite of cultivated plants was domesticated in sub-Saharan Africa .... Some of the most important include sorghum, pearl millet, finger millet, teff, fonio, cowpea, Bamara groundnut, African rice, African yams, watermelon, okra, cola nuts, coffee and oil palm.' Harlan believes that the entire sub-Saharan zone produced these domesticated species, though the spatial and temporal pattern of domestication cannot be determined from currently available archaeological data. This has led Harlan (1992) to advocate the non-centric origin of agriculture, at least in this region, where the variation in ecology and the diverse range of naturally occurring plant species would have provided many potential domesticates. In terms of modern agriculture, sorghum (*Sorghum bicolor*) is the most important crop to be produced from this region. Referring to Harlan's work, Smith (1995), recounts that the earliest evidence for domesticated sorghum is dated at ca. 4 K years BP and comprises a single grain impression on pottery. In addition, he reports that its ancestor is *S. verticilliflorum*, whose habitat is tall-grass savanna in the Chad–Sudan region. Today sorghum is a mainstay of African agricultural systems, especially in regions where there is drought stress and regions unsuitable for other cereals.

4.5.4 Centres of the Americas

The three remaining centres of plant domestication are in the Americas (Fig. 4.9), where the first domestications occurred at approximately the same time at ca. 4.5 K years BP, though until recently the domestication of maize was considered to be much earlier at ca. 8 K years BP (MacNeish, 1992). The revised dates of ca. 4.7 K years BP result from recent radiocarbon dates, using accelerator mass spectrometry (AMS), on the actual fossil maize cobs rather than the conventional radiocarbon dating method which used material such as charcoal from the matrix of deposition (Fritz, 1995). The locus for maize domestication is also disputed because there is evidence to suggest it may have been in the Mexican centre (MacNeish, 1992) or the Andean centre (Bonavia and Grobman, 1989). However, it is also possible that it was domesticated in both centres at different times, and possibly in other areas as well. The new AMS dates also cast doubt on Pearsall and Piperno's (1990) research on

phytoliths (deposits of silica with a particular shape within a plant that are species specific) reflecting cultivated maize in sediments 5.3 K years BP from Ecuador's Amazon Basin. Moreover, Van der Merwe (1982) has undertaken carbon isotope analysis of several human skeletal bones from a range of sites in Central and South America, analysis which reflects the type of plants consumed. His results show that a maize-dominated diet had developed in the lowlands of the Orinoco River in Venezuela only 1.6 K years ago. However, it also adds to the debate surrounding the AMS dates for Mexican maize because the carbon isotope ratios of bones of skeletons from Tehuacán, from where early maize cobs have been obtained, indicate that by 6 K years BP maize was of prime importance in the diet.

Apart from the controversy discussed above, there has been much debate about the wild ancestry of maize. Galinat (1992) points out that the problem arises because the cob structure of the cultivated species is very different from wild maize types. The predominant theory today, supported by molecular evidence (Doebley, 1990), is that teosinte (*Zea mays* L. subsp. *mexicana* Iltis and subsp. *parviglumis* Iltis and Doebley), a grass, is the ancestor of maize. However, quite when the first maize was domesticated is open to question, as discussed above. Maize spread north and south from Mexico. Its spread into South America has already been mentioned briefly. In addition, Smith (1995) quotes evidence for the presence of domesticated cobs at two sites along the Orinoco River, La Ponga and Nueva Eva. These sites are dated at 2.8 to 2.4 K years BP. At Panean a further site in Andean Peru, the earliest maize in a sequence reflecting its development is dated at 1.5 K years BP.

The Mesoamerican centre also gave rise to several other domesticated species, including the common bean (*Phaseolus vulgaris*) and lima bean (*P. lunatus*), though according to Gepts (1990), both of them were probably domesticated in two separate centres, Mexico and the southern Andes. Cotton (*Gossypium hirsutum*) is an upland New World species whose earliest remains to be associated with human activity occur in Mexico's Tehuacán Valley and are dated to ca. 5.5 K years BP. However, it is not clear precisely where domestication occurred, and the ancestral species remains unidentified. On the basis of the genetic structures of cultural and wild cottons in Mexico, Brubaker and Wendel (1994) suggest it may have been domesticated initially in the Yucatan peninsula, from where it spread into southern

Mexico and Guatemala. This region became a second centre of diversity but was not the primary centre of origin, as previously thought. *G. hirsutum* is one of the four species of domesticated cotton; the other two are *G. barbadense*, *G. herbaceum* and *G. arboreum*. *G. barbadense* is sea island cotton and is the most important species cultivated for commercial purposes today; the other two species originated in Sudan–Ethiopia (Harlan, 1992). Squash (*Cucurbita* sp.) is the other important crop plant to be domesticated in Mesoamerica. According to Smith (1995), five species of squash were domesticated in the Americas; the most common species today is *Cucurbita pepo*. Taxonomic studies (Decker-Walters *et al.*, 1993) indicate this species was domesticated independently in the eastern United States and in Mexico. Original work on the archaeological site of Guilá Naquitz in the Oaxaca Valley of central Mexico (south of Tehuacán) produced *C. pepo* seeds dated to 9.3 K years BP (Flannery, 1986). As is the case for several of the species from Mesomericean sites, these seeds are being redated using AMS. Like the other plant remains, direct dating rather than contextual dating may yield quite different results. Although *C. pepo* may have been one of the earliest domesticated species in Mesoamerica, its domestication seems unlikely to have been before 7 K years BP. Its wild ancestor has not yet been identified.

In the centre of domestication that is now eastern North America (Fig. 4.9), *C. pepo* was domesticated from its wild ancestor, *C. pepo* subsp. *ovifera* var. *ozarkana*, which as its name suggests, grows near streams in the Arkansas Ozarks. Another crop domesticated in this region is goosefoot or lamb's-quarter (*Chenopodium berlandieri*), the oldest remains of which are 3.5 K years old at least. Smith (1995) believes that actual domestication may have been 5.5 K years ago. Marsh elder (*Iva annua* var. *macroparpa*) and sunflower (*Helianthus annuus* var. *macrocarpus*) were amongst the species domesticated. In the remaining centre of domestication, the Peruvian/Bolivian/Ecuadorian Andes, one of the most important crop plants is the potato, of which *Solanum tuberosum* is the most common of the four species domesticated; Hosaka (1995) examines the genetic characteristics. According to Hawkes (1991), the oldest remains of domesticated potatoes so far discovered are from the Chilca Canyon in Peru and are dated at 7.0 K years BP. Both Hawkes (1991) and Smith (1995) suggest that the most likely wild relative of *S. tuberosum* is *S. stenotomum*, which currently grows in

the region between northern and central Bolivia and southern Peru. As well as producing the domesticated species *S. tuberosum* subsp. *tuberosum*, *S. stenotomum* may have interbred with *S. sparsipilum*, a wild species of Peru and Bolivia, to produce *S. tuberosum andigena*, another domesticate. It is this species that Hawkes (1990) believes was first introduced into Europe. Moreover, it may have crossed with yet another Andean species to produce *S. tuberosum* subsp. *tuberosum*, a native to southern Chile that was introduced into Europe after the devastation of *S. tubersosum* subsp. *andigena* by late blight (*Phytophthora infestans*). Another crop of importance, and which is now being reintroduced to the Andean centre, is quinoa (*Chenopodium quinoa*). This may have been domesticated in the Andes of Peru and Bolivia from its wild relative *C. hircinum* ca. 5 K years BP (Smith, 1995). The seeds of the species can be used in much the same way as rice and the leaves are rather like spinach. According to Tohme *et al.* (1995), nuna beans (a species of *Phaseolus vulgaris*) may also have originated in the Andes of Peru and Bolivia.

4.5.5 The domestication of animals

Far fewer animal species than plant species have been domesticated. Nevertheless all of the issues relating to plant domestication i.e. where, how and why apply equally to animal domestication. In addition, the centres of plant domestication, notably the Near East and the northern Andes, were also centres of animal domestication, as shown in Figure 4.10.

The first animal to be domesticated was the dog (*Canis lupus*). According to Clutton-Brock (1995), the oldest fossil of domesticated dog is dated at 14 K years BP and consists of a mandible from a late Palaeolithic grave at Oberkassel in Germany. Davis (1987) has suggested that it was first domesticated in the Near East, though the Oberkassel find predates dog remains in the Near East by ca. 2 K years. There is little evidence for butchering of the species, so it is likely the dog was prized for its ability to herd, hunt and guard, much as it is today. The precise locus of domestication, though there may have been many separate centres, has not been identified, nor has the subspecies of wolf (Morey, 1992 and Dayan, 1994).

As Figure 4.10 shows, the Near East was a major centre of animal domestication. It is from here that many species originated, species which nowadays have become the most economically important livestock.

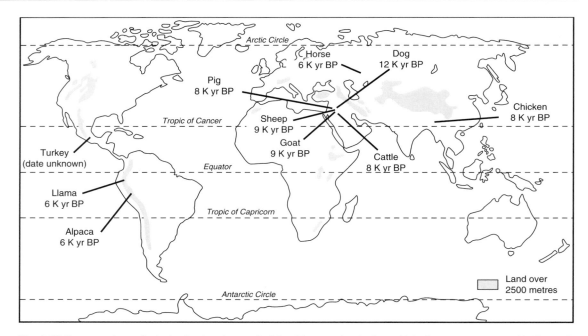

Fig. 4.10 Common domesticated animals: places of origin with approximate dates (data from sources quoted in the text).

The domestication of sheep (*Ovis aries*), goat (*Capra hircus*), cattle (*Bos taurus*) and pig (*Sus scrofa*) all occurred between 9 and 7 K years BP (Davis 1987; Smith, 1995). The ancestor of the domestic sheep is the mouflon (*Ovis orientalis*), though it did not play an important role in the food resources of the western parts of the Near East (the area around Jericho) as its remains are rarely found. This is not the case further north and east, where in the foothills of the Zagros Mountains it was probably domesticated, as shown in Figure 4.11. Once domesticated its range expanded. Goats were also domesticated a considerable distance from the Mediterranean coast in the region southwest of the Caspian Sea (Fig. 4.11). The beozar goat (*Capra aegagrus*) is the ancestor of the domesticated goat (*Capra hircus*) and before domestication it was widely hunted in its native region. Smith states that, along with sheep herding, goat herding had become a major activity in the Near East some 300–800 years after initial domestication. Cattle in the Near East were domesticated slightly later than sheep and goats, though there is bimolecular evidence, i.e. mitochondrial DNA, for a second centre of domestication in India (Loftus *et al.*, 1994; Bradley *et al.*, 1996). The ancestors of the domesticated species (*Bos taurus*) are the aurochs (*Bos primigenius*). These were widely hunted before domestication. The locus of

domestication was probably in what is now Turkey and Syria (Fig. 4.11). Similar to cattle it is highly likely that the pig was domesticated in more than one area, notably in parts of Europe and China as well as the Near East. In the Near East domestication occurred in the area that is modern-day Syria and Turkey (Fig. 4.11). It was definitely domesticated by ca. 8 K years BP but Smith (1995) quotes limited evidence for domestication ca. 10 K years BP. Before their husbandry pigs were hunted in large numbers in the northern region of domestication.

Although all the species mentioned above were hunted as wild animals before domestication, it is interesting that several other animals which were extensively hunted, and therefore sought-after as a food source, were not subsequently domesticated. One of the most important of these is the gazelle. There is considerable evidence for gazelle hunting in the Near East (e.g. Henry, 1989; Cope, 1991). Dayan and Simberloff (1995) state, 'The mountain gazelle (*Gazella gazella*) was one of the most frequently hunted ungulate species during the Upper Pleistocene of the southern Levant.' This occurred during the Natufian period i.e. 12.5 to 10.2 K years BP. Moreover, Cope's (1991) analysis of gazelle assemblages from pre-Natufian and Natufian sites suggests that the Natufian sites are distinct insofar

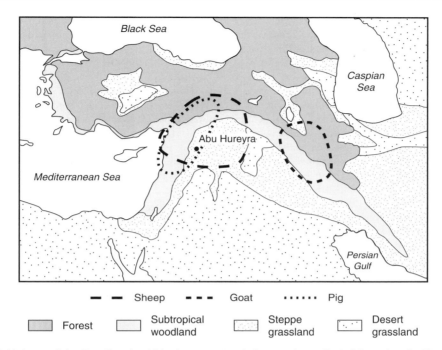

Fig. 4.11 Areas of the Near East in which sheep, goat and pig were domesticated (based on Smith, 1995).

as they show evidence for dwarfism, which may be the result of intensive human manipulation in a form of 'protodomestication'. Whether or not this is true, and a reanalysis of Cole's data by Dayan and Simberloff (1995) does not confirm the original conclusion, there still remains the question of why this species was not domesticated when it was clearly sought-after as a source of meat. The issue here is why species like the wild ancestors of sheep, goats, cattle and pig, not generally well represented in archaeological sites, were domesticated, whereas species like the gazelle, an extensively hunted animal, was not domesticated. This issue is pertinent to other parts if the world besides the Near East. Interestingly, perhaps, it is only recently that the farming of deer has begun, as opposed to their hunting. It is also curious that so few of the animal kingdom have been domesticated when compared with the plant kingdom. These issues have been addressed by Clutton-Brock (1992), who believes it may be a result of the animals' patterns of social behaviour. She also discusses why a tamed animal is not necessarily a domesticated animal. What is noteworthy is that comparatively many species were domesticated in the Near East.

The chicken is the earliest domesticated bird, and

according to West and Zhou (1988) there is good evidence for its earliest domestication in southeast Asia from the red jungle fowl (*Gallus gallus*). This initial domestication may have occurred in what is now Thailand (Fumihoto *et al.*, 1994). From its centre of domestication the chicken was introduced to China. By 8 K years BP it had become well established in China. West and Zhou (1988) report that abundant chicken bones have been found at archaeological sites in the Lower Yellow River. By 5 K years BP the domesticated chicken had been introduced to Europe. Other animals domesticated in Asia include the water buffalo (*Bubalus bubalis*), which may have been domesticated separately in several places in southeast Asia (Tanaka *et al.* 1995). There may have been one centre in China and there is evidence for its presence at the archaeological site of Ho-mu-tu, just south of the mouth of the Yangtze River (Smith, 1995). This is the earliest evidence for water buffalo in China and it is dated at 6.5 K years BP.

A few animals were also domesticated in the Americas. Although it is not known when the turkey (*Meleagris* sp.) was domesticated, its locus is thought to have been Mexico. The vicuña and guanaco were extensively hunted in Andean South

America during the early Holocene (Fiedel, 1992). By ca. 6 K years BP llamas had been domesticated (Bahn, 1994) from the guanaco whereas the alpaca's wild ancestor is the vicuña, though as Wheeler *et al.* (1995) point out, hybridisation between the two groups after the Spanish conquest means the genetic distinctions between the two are blurred. Data from Telarmachay Rockshelter indicate that general hunting of guanaco and vicuña occurred before 7.2 K years BP, there was specialised hunting of these species between 7.2 and 6 K years BP, domesticated species originated ca. 6 K years BP, and a herding economy had developed by 5.5 K years BP (Wheeler, 1995). Smith (1995) has discussed the domestication of the guinea pig (*Cavia porcellus*); the earliest evidence for its domestication dates to 4.5 K years BP. Bones of domesticated guinea pig have been found at several sites in the Ayacucho Valley in the Andes of southern Peru, where other guinea pig bones are associated with hunter-gatherers of the early Holocene.

Finally, mention must be made of the horse which was domesticated in the Ukraine. According to Anthony *et al.* (1991), herds of wild horses roamed the steppelands of what is modern-day Ukraine in the first half of the Holocene. On the basis of evidence from several archaeological sites in the Ukraine domestication occurred ca. 6 K years BP. Indeed Anthony *et al.* believe the horse was exploited as wild game, as a domesticated source of meat and as a mount. Mounts provided a source of energy for transporting people and goods; they would also have been advantageous for hunting. At Dereivka, an archaeological site 250 km south of Kiev, remains of domesticated horse occur alongside those of dogs, cattle, sheep and pigs. Both horses and dogs may have been used in ritual or religious activities as well being part of the subsistence economy. Moreover, analyses of horse teeth indicate that bits were probably being used. In addition to the advantages that horses would have provided in the form of protein and transport, Anthony *et al.* have suggested they would have helped during warfare and territorial expansion.

4.6   The Neolithic period

Neolithic culture emerged from the Upper Palaeolithic in the Near East ca. 11.5 K years BP. What distinguishes it from the Palaeolithic is the production of pottery and the practice of agriculture (Section 4.5), though the Neolithic actually began in the Near East before either of these innovations occurred. In the Old World, for which the term 'Neolithic' is widely used, its appearance was not spatially synchronous. Instead it occurred metachronously as people, ideas and technology spread beyond their centres of origin.

The cultural sequence that occurred in the Near East is given in Table 4.6, which shows that the Natufian, an Upper Palaeolithic culture, gave way to the pre-pottery Neolithic ca. 11 K years BP. Moreover, Mellaart (1994) suggests that the eventual emergence of agriculture was not necessarily a product of population increase or environmental change (Section 4.5) but represented the culmination of increasingly intensive and planned biotic resource exploitation, evidence for which dates back to ca. 16 K years BP. This early Neolithic period was characterised by permanent dwellings in nucleated settlements, e.g. Jericho, constructed from sunbaked bricks, stone axes, bone implements, stone vessels and vessels of wood and hide. Mellaart quotes evidence for trade and fortification at Jericho. Obsidian from central Anatolia was imported; plant seeds and seed processing equipment were probably the main exports. From the aceramic site of Çayönü in southeast Anatolia, van Zeist and de Roller (1991/1992) report the presence of einkorn and emmer wheat, field pea, lentil and bitter vetch. The pulses dominate the plant assemblages and there is evidence for the collection of wild pistachio (*Pistacia atlantica/khinjuk*) fruits. Willcox (1996) has reported similar plant remains from three aceramic sites in the middle Euphrates valley, which date from 9.8 to 7.8 K years BP. This period of the early Neolithic (PPNA) gave way to the second pre-pottery Neolithic period (PPNB), when several animal species were domesticated to add to the range of domesticated biota hitherto dominated by plant species (10.5 to 8.25 K years BP; see Table 4.6). The weaving of wool and flax was commonplace, as were clay figurines along with stone vessels in various styles. Rectangular houses with small rooms, and possibly two storeys, were the predominant architectural style.

The material culture of PPNB is more widespread, varied and sophisticated than its PPNA predecessor. By the close of PPNB the first pottery was being produced which Mellaart described as coil-built in the manner of basket work with flat bottoms and little decoration. Eventually ca. 8 K years BP (Table 4.6) pottery production increased substantially in the Near East and elaborately painted wares were widely used, as compared with the relatively plain pottery of PPNA. At the same time there was an increase in the number of permanent settlements throughout the region (Fig. 4.11). Sickles and hoes were added to the armoury of agricultural tools; ovens and kilns became widespread and there is some evidence for copper working. House structure diversified and included rectangular, round and T-shaped dwellings. There is also evidence for irrigation in the valleys of the Tigris and Euphrates, representing an increasing manipulation of the environment. A great deal of local variation in material culture occurred during PPNA and PPNB. But there is little evidence for the nature of land subdivision or the ways in which labour and food distribution were organised. The origins of what were to become the great civilisations of Sumer and Akkad lie in this period (see below). By 7.5 K years BP there was a change from painted to plain pottery though the reason is not obvious. Trade continued within the Near East and diversified to include stamp seals and metal ornaments; architectural styles also diversified. By the time the Uruk culture developed ca. 6 K years BP metalworking was widespread (Section 4.7). According to Roberts (1992), the cultures of the Near East had become sufficiently complex by this time that many historians consider it to represent the emergence of civilisation. Perhaps this is naive; the diversification of material culture occurred much earlier, though as far as archaeological investigations allow, the identity of cohesive units, i.e. states, clearly dates from this period.

The foregoing paragraphs briefly describe the characteristics of Neolithic communities in the Near East. People, material culture and technology disseminated from the Near East into Europe and Asia, where Palaeolithic (Section 4.3) and Mesolithic (Section 4.4) cultures were transformed. There has been much debate as to how the practice of agriculture and the technology associated with pottery production spread beyond their centres of origin. Were migrations of people involved? Or were the technologies promulgated through trade contacts which facilitated the spread of ideas? By ca. 5.5 K years BP both technologies were evident in peripheral regions of Europe such as Britain, Ireland and Scandinavia. For the intervening 5 K years there is increasing evidence that migration was the most important way in which technology spread,

though no doubt the diffusion of ideas also played a role, if only a minor one. Evidence for the migrationist process comes from the genetic patterns of modern-day people. Sokal *et al.* (1991) have compared observed genetic patterns with predicted patterns produced by the demic (migrationist) expansion view. The genetic differences, derived from blood group data collected from people at 3000 locations in Europe, reflect genetic distances caused by the separation of groups over time. As Figure 4.12 shows, there is a significant correlation between genetic distances and the spread of agriculture based on radiocarbon-dated archaeological evidence. Further work by Cavalli-Sforza *et al.* (1994) and Piazza *et al.* (1995) has confirmed these results.

The operation of demic diffusion does not necessarily mean a return to the discarded views of archaeologists in the early part of the twentieth century, which focused on waves of aggressive migrants conquering indigenous groups in a form of prehistoric colonialism. The demic diffusion that occurred to disseminate Neolithic culture was probably gradual and non-violent. What prompted demic diffusion is, however, no more clear than the reasons why *Homo erectus* or *Homo sapiens* migrated out of Africa several millennia earlier (Section 4.3). Population pressure is often invoked as a reason, especially as the expansion of early Neolithic groups came just after the environmental changes at the end of the last ice age. Whatever the reason, Neolithic people spread into Europe and Asia where they interbred with indigenous people to register their genetic markers, which are still in evidence today. According to Jones (1991), 'The hunter-gatherers of Mesolithic Europe suffered a process of gentrification – or even yuppification – from the east.' The genetic data, reviewed in Cavalli-Sforza *et al.* (1994), indicate an expansion of agriculture into Europe of 1 km per year on average. This research also suggests a similar pattern and rate of spread of agriculture into Asia from the Near East.

Moreover, it is possible these early Neolithic migrants spread into uninhabited areas, as Chapman (1989) has suggested for much of southeastern Europe. This issue has been reconsidered by van Andel and Runnels (1995) in the light of several archaeological surveys of ancient sites in Greece and the rest of the Balkans. Their data show most Neolithic settlements occur in areas of the Balkans that have little evidence for earlier Mesolithic or Palaeolithic occupation and they tend

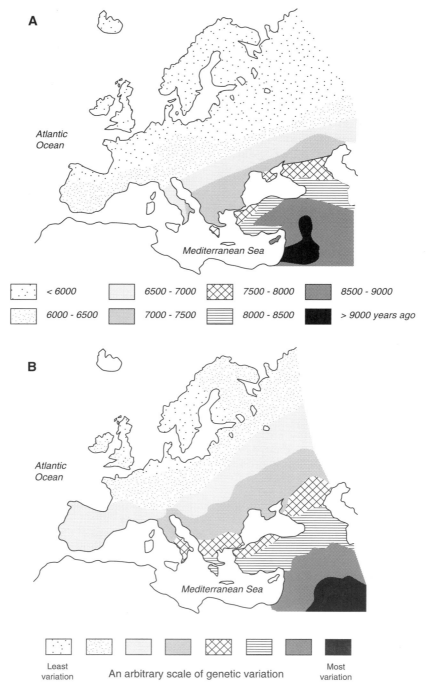

Fig. 4.12 (A) The spread of agriculture in Europe as indicated by radiocarbon dating of Neolithic agricultural sites. (B) The genetic landscape of Europe showing the relationships between regions. Note the correspondence between the two maps, implying a link between them (based on Cavalli-Sforza and Cavalli-Sforza, 1995).

to occur in active river floodplains. According to van Andel and Runnels, active river floodplains were particularly sought after by Neolithic farmers because early post-glacial European rivers were depositing volumes of glacially derived sediments, thereby creating dynamic levée and channel systems. Those levées which became inactive and were no longer submerged provided acceptable sites for dwelling construction and were close to a reliable water source as well as silt-fed fertile patches of land. This type of landscape may also have been favoured by the earliest agriculturalists in that part of the Near East known as Anatolia (in modern Turkey). Coupled with the existence of geographic barriers, this preference for active floodplains probably meant the demic expansion of the Neolithic was somewhat more complicated than suggested by the model of Cavalli-Sforza *et al.* (1994). At the local and regional scales it is inevitable these factors, and others such as woodland density and soil fertility, would influence the choice of agricultural land and settlement patterns, but the more general demic-diffusion model remains applicable at the continental scale. However, by pinpointing factors such as appropriate landscapes which may have prompted migration, van Andel and Runnels (1995) demonstrate that population pressure does not have to be the only stimulus for the movement of human groups. Moreover, these studies emphasise the operation of environmental determinism (Section 1.2). Further studies on the early agriculture of the Balkans have been undertaken by Chapman and Müller (1990) and Chapman and Shiel (1993).

The genetic data cited above and in Figure 4.12 have also been used to trace the origin and spread of European languages (e.g. Cavalli-Sforza *et al.*, 1993; 1994; Piazza *et al.*, 1995). Issues that surround the origin and spread of languages plus relationships between languages, though controversial and complex, they do reflect the development of culture. This is an emotive but important factor which sets humans apart from other animals and initially at least it is a product of environmental characteristics, but it also produces environmental change. Equally, the use of language and the ability to communicate within and between human groups would have facilitated the spread of ideas, including the technological aspects of Neolithic material culture. The genetic analysis indicates that the spread of agriculturalists from the Near East may have given rise to the language variation which occurs between

Asia, Africa and Europe (Cavalli-Sforza and Cavalli-Sforza, 1995) with additional complications caused by the differentiation of the Uralic language groups from the Indo-European language and the inspread of Kurgan people from central Asia. The former involves the development of a distinct linguistic family, Uralic, in western Siberia but which extended west of the Urals to produce the Ugro-Finnic group of languages. Today such languages include Finnish, Hungarian, some Lapp languages and Estonian. The Kurgan influence derives from nomadic pastoralists of the Russian plains, who domesticated the horse (Section 4.5) and who infiltrated peripheral regions to the east and west.

The inception and spread of agriculture is often cited as the basis for the rise of the great civilisations of prehistory, including Sumeria and Akkad in Mesopotamia, notably in the river valleys of the Tigris and the Euphrates, along with the ancient Egyptians. All had arisen by ca. 5.5 K years BP and were characterised by organised religion, skilled craftspeople, extensive urban centres, the existence of writing, temples and monuments. They were centres of high population concentrations characterised by a ruling elite and a division of labour that could only have been made possible by productive and reliable agricultural systems. Indeed there is evidence that all three civilisations employed various forms of irrigation to benefit from the fertile silt as well as the water carried by the great rivers which flowed through these regions. Even today the labours of the ancient Egyptians provide a substantial income via the tourism that focuses on some of the greatest and most well-known ancient monuments in the world.

The spread and character of early Neolithic food gathering and agriculture in the Western Desert has been discussed by Wasylikowa *et al.* (1993) and Wendorf *et al.* (1992); the Nile Valley has been addressed by Wetterstrom (1993). The former have discovered grains of sorghum at Nabta Playa, an early Neolithic site, 100 km west of Abu Simbel in Egypt's Western Desert. Since sorghum is not a member of the crop complex originally domesticated in the Near East (Section 4.5) and even though they have not been confirmed as those of a domesticated species, the finds at Nabta Playa indicate there may have been another centre of plant domestication that eventually affected Egypt's Neolithic agriculture. According to Wetterstrom (1993), the earliest farming villages in the Nile Valley developed ca. 7 K years BP in the north; by 6.5 K years BP the earliest farming villages were developing in the southern Nile Valley. The crop complex that formed the basis of these economies was derived from the Near East (Section 4.5), though the Western Desert, and possibly other parts of north Africa may have also contributed domesticates. The initial introduction of domesticates probably predated the establishment of villages by ca. 1 K years. According to Stanley and Warne (1993), the increased deposition of silt in the Nile Delta, caused by a decline in the rate of post-glacial sea-level rise, provided favourable conditions between Egypt and the Near East for the expansion of the Near East's agricultural practices.

By the time Egyptian civilisation began to achieve pre-eminence ca. 5 K years BP, with the unification of various subunits into a single nation state, the food resource consisted of floodplain-grown wheat and barley supplemented by hunting and some gathering. According to Hassan (1993), the towns that emerged along the length of the Nile Valley were probably centres for the exchange of food and other resources and may have been fortified for protection against desert nomads as well as being religious centres. Their role as centres of food exchange could have developed because of the uncertainty associated with grain production, brought about through flood failure and pest outbreaks. The coordination and integration of resources within a group of villages by chiefs or priests not only diminished the risk of famine but facilitated ritual activities associated with the links between food production and a religion based on Nature. Hassan suggests that the basic aim was to ensure an adequate food supply from year to year rather than to generate a profit. In the Roman empire, for example, some four millennia later, the lowest classes known as the headcount, were each guaranteed a quantity of grain at a fixed, relatively low price. This was considered the most appropriate way of avoiding rebellion and disruption. It is possible that such relative stability within the dynastic state of Egypt facilitated the many cultural and architectural achievements of this ancient civilisation.

Hughes (1992) has suggested the ancient Egyptians employed a form of sustainable agriculture which may have been the reason why their civilisation lasted for ca. 4 K years, a period longer than any other ancient civilisation. He also believes the likely integration between urban and rural activities and populations contributed to this,

as did the Egyptian's reverence for Nature which is implicit in their religious practices. Moreover, the sharp distinction between Egypt's desert and the fertile Nile Valley probably made rulers and slaves alike very conscious of where to grow crops and where not to. That land was at a premium is also reflected in the fact the tombs of the kings and queens are located beyond the Nile Valley in the desert so as not to use fertile land. Whether or not the decline of the ancient Egyptian civilisation was associated with diminishing food production is open to question, but for the demise of the Akkadian Empire there is increasing evidence that this is precisely what occurred. Weiss *et al.* (1994) have suggested the climatic change caused the onset of desertification regionally. In turn this led to a movement of people from the northern state of Subur, which depended on rain-fed cereal cultivation into the irrigated lands of Akkad and Sumer. The twin pressures of rapid immigration and increased aridity caused a collapse of these empires ca. 2.3 K years BP. Neolithic cultures in western Europe were also highly organised with cultural hierarchies and diverse material cultures reflecting the skills of specialised craftspeople. Like the Egyptians, many of these early agricultural societies left behind a legacy that endures today. This legacy comprises the megalithic (large stone) monuments which occur in Ireland, Britain, France and the Iberian peninsula, the most famous being Newgrange in Ireland's Boyne Valley and Stonehenge in Wiltshire, England.

In what is modern-day China there were flourishing Neolithic cultures. According to An (1994), four regional groups can be distinguished: the Yellow (Huange) River valley, the northern steppe, the middle and lower reaches of the Yangtze (Changjiang) River and the southern hilly land. As discussed in Section 4.5, there is some debate as to where agriculture in China first emerged, with increasing evidence for early rice cultivation in the Yangtze River basin, but with a well-established archaeological record for Neolithic activity in the Yellow River valley. Excavations at the village of Banpo, near Xi'an in Shaanxi Province (Fig. 4.5) have yielded some of the earliest Neolithic artefacts dated to 7 to 6 K years BP, when agriculture was already well established. The artefacts and dwellings at this site are attributed to the Yangshao culture which flourished in the Yellow River valley and Loess Plateau between 7 and 5 K years BP. Stone implements, e.g. adzes and axes, digging implements and querns were prevalent along with bone implements, e.g. arrowheads, harpoons and fish hooks. In addition, there are remains of domesticated dog and pig, foxtail millet and broomcorn millet. A variety of pottery kilns have been found with the potential of producing firing temperatures of up to 1050 °C and the pottery itself was varied in shape but with painted red and black patterns, sometimes with human and animal motifs.

The predominant culture in the Yangtze River valley is called the Hemdu culture, An states, and it dates back to ca. 7 K years BP. Like the Yangshao culture, mentioned above, it has its roots in earlier Mesolithic cultures. Woodcraft was widespread, stone and bone artefacts dominate archaeological sites and there is some evidence for the lacquering of a wooden bowl. The pottery is distinct from the Yangshao type insofar as it is black charcoal ware which was lightweight and probably fired at relatively low temperatures (800–850 °C) in bonfires rather than in kilns. Rice cultivation in paddy fields was the predominant agricultural activity and there was some contact with the Yellow River valley cultures, as is evidenced by the presence of painted pottery in archaeological sites in the middle Yangtze River. The Neolithic cultures of the Yangtze River also influenced developments further to the southeast in the coastal region. The material culture suggests a less well-developed tradition of pottery making, which produced red, coarse sandy ware at low temperatures of ca. 680 °C in fires. Agriculture developed much later in this forested tropical zone than in the Yangtze or Yellow River valleys. The Neolithic cultures of the northern steppes may also have been influenced by developments in the Yellow River basin. The steppe regions maintained a tradition of hunting and gathering by nomadic peoples. This Mesolithic tradition was maintained until farming was eventually established.

A synthesis of the Neolithic in northeast China (Dongbei) has recently been produced in various papers in Nelson (1995) with emphasis on the Hongshang culture. This emerged ca. 6 K years BP from the Chahai and Xinglongwa cultures that were characterised by hunting, fishing, animal husbandry, crude pottery types and what Guo (1995) calls a primitive form of agriculture by sedentary people. In contrast, the Hongshang culture, centred on several river valleys in Dongbei, was characterised by a more diverse material culture, notably a wide variety of stone tools, including jade artefacts, along with pottery types including red clay pottery, sandy

grey pottery, polished black pottery and painted pottery produced at high temperatures in kilns. The many dwelling sites that have been found in the region occur on terraces or mounds at least 10 m above the river level and comprise square, subterranean floors with hearths. The Hongshang culture was also characterised by sacrificial sites, including the so-called Goddess Temple at Niuheliang at the centre of a group of sites from which have been recovered vast numbers of human figurines. The tool assemblages from Hongshang sites are made of stone, including polished and chipped artefacts. There are known ploughshares, adzes, chisels, hoes, axes and grinding stones. The abundance of stone ploughs may imply intensive, or at least expanding, agricultural practices. Stock rearing was also important, as is evidenced by abundant pig and sheep bones. According to Guo, such a mixed agricultural economy is a reflection of Hongshang's transitional locus between Mongolian grassland economies based on herding and the cropping agriculture of the Yellow River valley.

A region where the development of a Neolithic culture did not involve the practice of permanent agriculture occurs around Lake Baikal, in modern-day Russia. A recent review by Weber (1995) demonstrates that the Neolithic commenced ca. 5.8 K years BP from the late Mesolithic Kitoi culture which relied heavily on fishing for subsistence and which covered its dead in red ochre before interment in graves. The factor which distinguishes the Mesolithic culture from the Neolithic culture is the introduction of pottery by the Kitoi people, evidence for which derives from excavated living sites. Subsequently, Weber indicates there was a decline in Kitoi populations due to an as yet unidentified cultural or physiological stress. Whether environmental stress was a factor in this decline is not clear. Whatever caused the decline of this ancient Siberian population, it produced a cultural discontinuity that represents the division between the early and middle Neolithic which occurred ca. 4.2 K years BP. This was characterised by Serovo people who may have assimilated the surviving Kitoi or who may have migrated into the completely depopulated catchment of Lake Baikal. The Serovo people practised a more sophisticated foraging subsistence than their predecessors; the bows, arrows and spears from Serovo sites provide evidence for active hunting. The Serovo people also practised different burial rituals involving graves lined and filled with strong stones, and they had a varied range of pottery styles. The middle Neolithic ended ca. 3 K years BP, when copper metallurgy developed with the onset of the Chalcolithic (Copper Age), the forerunner of the Bronze Age. Copperware dates from ca. 3.4 K years BP and is associated with the Glazkovo culture, which developed from the Serovo culture. The work of Kuzmin and Chernuk (1995) in southern Primorye in far-eastern Russia reflects a similar economy to that around Lake Baikal until ca. 4.1 K years BP, by which time millet had been domesticated and cattle were being reared.

4.7   The Bronze Age

Precisely when and how metalworking technologies were originally developed is far from clear. Moreover, there may have been several centres of innovation rather than a single centre. The earliest evidence for metalworking comes from the Near East, where hammered copper objects have been recovered from pre-pottery Neolithic period B sites (Section 4.6). According to De Laet (1994b), they are dated to ca. 9.5 K years BP. Thus, the first use of metal occurred not long after the inception of agriculture. It was, however, ca. 8 K years BP before metal smelting began (Mellaart, 1967), evidence for which comes from Çatal Hüyük in central Anatolia. As Figure 4.13 shows, this region was a major centre of copper working for which, as in many parts of Europe, the term Chalcolithic, i.e. Copper Age, is often applied. This signifies the divergence from stone use to copper use and distinguishes between copper working and bronze working, since bronze working requires a more sophisticated technology. This divergence of resource use enhanced the ability of humankind to manipulate their environment and to bring about environmental change.

Copper oxide ores in Anatolia provided the basis for the extraction of copper, a metal that could be easily hammered and polished but which was not very durable insofar as it is soft and rapidly loses cutting-edge sharpness. For these reasons the Chalcolithic was relatively short lived. Nevertheless, copper exploitation led to the production of bronze, an alloy of copper and arsenic (arsenic bronze) or copper and tin (tin bronze). How the initial production of bronze came about is unknown, but perhaps it was accidental because many of Anatolia's copper ores contain arsenic. Smelting to obtain copper may thus have produced a robust alloy serendipitously. Early metal use also occurred elsewhere in the Near East, as discussed by Shalev (1994) in relation to Israel and Jordan (see also Engel and Frey, 1996). Here, the Chalcolithic was under way by at least 6 K years BP and possibly earlier. Many different types of metal objects have been found, including many jewellery items of gold and silver. This indicates that a range of metal ores were being exploited, not only copper. A similar situation occurred during the early Bronze Age, when unalloyed copper continued to be used for tools and blade weapons despite a marked change in modes of implement production. Chalcolithic villages in the northern Negev processed mined ores into finished products using the same methods; during the early Bronze Age there was much more specialisation with different methods applied to different products. These developments reflect craft specialisation in Chalcolithic and Bronze Age societies.

Bronze production eventually spread throughout Europe not, as hitherto believed, from developments in Anatolia but due to metalworking in independent centres. As Figure 4.13 shows, there were several independent centres, including the Balkans, Italy, Spain and the South of France. According to De Laet (1994b) the exploitation of copper ores at Rudna Glava, former Yugoslavia, and Aibunar in Bulgaria began as early as ca. 8 K years BP. Tools and weapons were made from the copper which was exported along with the implements. By 6.5 K years BP the Chalcolithic Age had begun in the Balkans and it brought with it a degree of unrest due to the competition created by the copper resources (Garašanin, 1994). This in turn led to the emergence of a warrior class, illustrating the influence that resources can have on social organisation and hierarchies. It is possible this Balkan centre and/or the Anatolian centre influenced the development of copper use and bronze production in central Asia, where the Chalcolithic Age began ca. 4.5 K years BP and lasted until ca. 3 K years BP, when the Bronze Age began. According to P'yankova (1994), the earliest village settlements of central Asia were in the foothills of the Kopet Dag Mountains, south of the Aral Sea. These settlements were supported by irrigation-fed agriculture. As in the case of the Balkans, the end of the Chalcolithic was a period of social unrest with the fall of many villages, the rise to prominence of others and the emergence of social

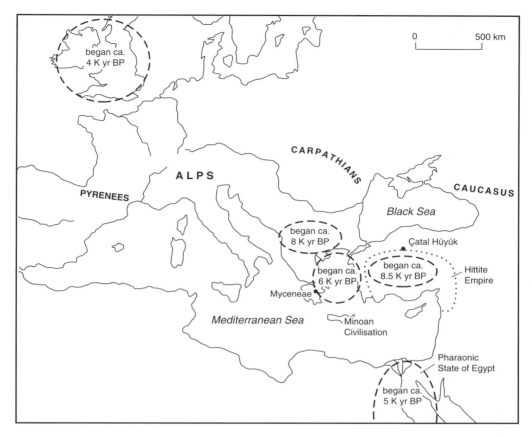

Fig. 4.13 Centres of copper and bronze working in Europe ca. 8 to 3 K years BP and places mentioned in Section 4.7 (based on sources quoted in the text).

stratification. Such changes may have been the result of increasing craft specialisation and the establishment of centres of bronze production in certain Kopet Dag villages. Certainly, P'yankova describes the substantial increase in size that occurred as some of the villages developed into urban centres. Once such centre was Alkyn Depe, the dwellings and burials of which reflect a high degree of social stratification. This and many other centres in the region declined markedly in the Late Bronze Age as the nearby Margiana and Bactria cultures, centred on the Murgab River, rose to prominence. There are conflicting views on the origins of this culture, ranging between local development from pre-existing Chalcolithic groups, which may have originated in the Kopet Dag foothills, to the inspread of a nomadic group from further north in what is modern-day Iran. The presence of copper ores in the steppelands north

and east of the Kopet Dag foothills also meant that the nomadic steppe cultures were producing bronze implements from ca. 4 K years BP.

As Figure 4.13 shows, a centre of bronze production had become established in the Aegean region by ca. 6 K years BP. This was centred on what is modern-day western Turkey and eastern Greece. According to Sherratt (1994a), the early use of hammered copper did not generate profound changes in ways of life: 'Copper, like gold, was a medium of display rather than as a means of changing the material world.' Changes did, however, occur after ca. 5.5 K years BP as metal exploitation increased and bronze was produced; trade between regions increased in both frequency and volume. Sherratt considers this marked the beginning of a new era for the region. Excavations of the city of Troy, for example, have revealed a rich material culture, especially when compared with cultures of the earlier Copper Age. As well as gold and silver jewellery and vessels, and bronze bowls, reflecting a comparatively rich and powerful society, a high degree of craft is reflected in the finely detailed ornamentation. This wealth accrued during a millennium between 5.5 and 4.5 K years BP. A merchant class developed, urban centres expanded, new settlement arose and fortifications were employed. Such wealth generation occurred because of innovations in metallurgy and agriculture, and enhanced trade with other parts of Europe, which brought new products and new ideas. Metallurgical innovations were new methods for extracting silver and new methods of casting copper, including alloying it with tin to produce bronze; agricultural innovations included the domestication of several types of fruit tree.

This wealth and expertise was what underpinned the first truly urban civilisations in Europe: the palace states of the Aegean, such as Minoan Crete and Mycenaean Greece. They enjoyed ca. 800 years of influence before their collapse ca. 3.2 K years BP, when the Bronze Age world of the Aegean and the Near East underwent profound changes. To a certain extent, the palace states influenced the lands of the central Mediterranean, i.e. southern Italy, Sicily and Sardinia, whereas western and central Europe experienced changes and developed largely independently of the Aegean influence (Sherratt, 1994b). According to Wardle (1994), the palace civilisations can be subdivided into several stages as detailed in Table 4.7. Wardle (1994) also states that 'the foundations of the Cretan "Palace Economy" lie in the stable and prosperous development of Early Bronze Age society'. The palaces of this phase in Crete are found at Knossos, Mallia and Phaistos with a likely further palace at Chania. Today these palaces are attractions for Crete's important tourism industry. Each palace was the focus of a large settlement, estimated to house between 15 000 and 50 000 people, and an agricultural hinterland from which produce was transported by pack animal. It is possible food storage facilities housed stores of sustenance that could be distributed in lean times, as occurred in ancient Egypt (Section 4.6). The early Palace society was hierarchical with specialised craftspeople, including potters and metalworkers. There is also evidence for cult worship. The demise of these palaces was caused by a substantial earthquake ca. 3.6 K years BP. In the southern Argolid area of Greece (Fig. 4.13), between the southeast of Corinth and Argos, Jameson *et al.* (1994) have shown that middle Helladic sites, contemporaneous with the early Palace period of Crete, reflect smaller, mainly agricultural communities. Archaeological data indicate an increasingly diverse material culture but a reliance on mixed farming, including new domesticated species such as the olive and the vine. As in Crete, there appears to have been a period of stability and growing prosperity in the southern Argolid but without the palaces, and probably the social hierarchy, that characterised Crete.

The New Palace period in Crete was relatively short-lived but the rebuilding of the palaces after the earthquake on just as grand a scale attests to the prosperity and success of this early Bronze Age civilisation. Wardle (1994) considers that the architects and builders employed considerable skill to rebuild the palaces; the settlements around the palaces were also rebuilt, and script types known as Linear A and B were used to record agricultural produce and possibly other details. Extensive wall paintings, elaborately styled and decorated pottery, simple unpainted pottery, bronze implements and bronze ingots, and evidence of cloth production, all attest to the skills of the Minoans. Indeed Wardle suggests they may have specialised in luxury goods. The decorations on pottery and carved figures also attest to cultural and religious activities. However, the ravages of extensive fires destroyed this civilisation, though the cause of this catastrophe remains unknown. It may have been due to natural fires related to volcanic or earthquake activity, or it may have been cultural as invaders from the nearby mainland annexed the island.

As the New Palace people flourished in Greece, another powerful civilisation arose on the Greek mainland centred on Mycenae (Fig. 4.13). Evidence for the material culture of this early Mycenaean civilisation comes from distinctive shaft graves

Table 4.7 Chronology of the palace civilisations of Crete and Greece

	Approximate date (K years BP)	
Mycenaean palaces and fortresses	3.4–4.2	Mycenaean civilisation
Early Mycenaean civilisation	3.6–3.4	Mycenaean civilisation
New Palace period in Crete	3.6–3.42	Minoan civilisation
Old Palace period in Crete	4.0–3.6	Minoan civilisation

Source: based on Wardle (1994)

which contained amber from the Baltic, finely crafted gold jewellery, and a range of bronze weapons including rapiers, short swords and spears, gold death masks and vessels made from precious metals. As Wardle (1994) points out, there is no obvious reason why Mycenae should have accrued such wealth since it was not a port, nor was it a centre of a rich agricultural area. It may have capitalised on the activities of the prosperous Minoan Crete (see above) in some way or even controlled the Minoans. Whatever the reason, the early Mycenaean people operated a hierarchical structure, as evidenced by grave goods belonging to different classes. The second phase of this civilisation began ca. 3.4 K years BP, corresponding with the destruction of the Minoan palaces (see above). The influence of Mycenae in the Aegean and the western Mediterranean increased, and Mycenae itself began to be fortified along with other towns under its control. Wardle points out that the fortresses and palaces of this period may have been built to reflect wealth, and possibly political power, instead of to meet an immediate need for defence. Defence came later, as is discussed below. Via a series of banks and channels, drainage was undertaken of the marshland around Lake Copais. This was probably to increase the amount of land that could be cultivated. It also allowed the diversion of streams and the avoidance of floods. Like those of the Minoans, the palaces of Mycenae acted as centres of religion and wealth accumulation through agriculture. Crops such as barley, wheat, lentils, grapes and bitter vetch along with cattle, pigs, sheep and goats were taken to the palace centres and stored there or traded. At its apogee, this later Mycenaean civilisation influenced and/or controlled much of what is today eastern Greece, i.e. Thessalonika, the Cyclades Islands and the southern Peloponnese. Both the Minoans and the Mycenaeans had regular trade contact with Egypt and the Near East as well as with westerly parts of the Mediterranean, e.g. Sicily and southern Italy.

According to Popham (1994), there was major upheaval in the lands bordering the eastern Mediterranean ca. 3.2 K years BP. The Mycenaean palaces were destroyed at the same time as the Egyptians and the Hittites, centred in what is modern-day Anatolia in Turkey. There is also evidence for destabilisation elsewhere in the region, which was generally devastated by war. The overall result was the demise of late Bronze Age civilisations despite the preparations, i.e. the fortresses and

strongholds of Mycenae and Crete. However, it is not clear which group or groups posed the threats, or why. Popham gives two possibilities: a threat external to Crete and interstate rivalry. Rivalry may have been brought about by a burgeoning population and the resulting competition for agricultural, especially arable, land. Another possibility, able to account for the problems of upheaval encountered elsewhere in the eastern Mediterranean, is an external force from the eastern Mediterranean which attacked by sea. They may have come from Sardinia, Sicily or Etruria in Italy (Popham, 1994) and even settled peacefully in Crete, becoming part of the Minoan civilisation before rising again to usurp some of the wealth of Crete, Mycenae, and possibly Egypt and the Hittites. Moreover, Lamb (1995) emphasises the possibility of regional drought in lands adjacent to the eastern Mediterranean, which would have caused social unrest and migrations as crops failed. Whatever the reason, a major upheaval occurred in late Bronze Age Greece causing migrations and the regrouping of populations but with some continued use of settlements, such as the city of Mycenae itself, and of language.

Centres of metallurgical innovation existed independently of Anatolia, the Balkans and the Aegean. Egypt, for example, flourished as a unified Pharaonic state; between ca. 5.1 and 0.65 K years BP, which O'Connor (1993) equates more or less with the Bronze Age. There were three kingdoms and three intermediary periods. Each brought change but there was an underpinning continuity via the agriculture of the Nile Valley, and moves towards urbanisation. This also occurred, though to a more limited extent, in Iberia (Chapman, 1995). Here one of the most important archaeological sites is Los Millares near Almería in southeast Spain. It is fortified, associated with megalithic tombs and has other fortified sites in its hinterland. It remained important for a ca. 2.5 K years during a period of agricultural and metallurgical innovation which encouraged the formation of larger settlements than during the Neolithic. This Iberian bronze-producing region also traded with other Mediterranean and north European groups.

Britain is included in north European groups. By ca. 4 K years BP copper was being mined in the British Isles and bronze implements were being produced, as were gold ornaments. Until recently it was generally accepted that Ireland was the only source of copper, from which tin bronze was produced for all of the British Isles, even though no tin source has been identified in Ireland (Budd *et al.*, 1992, 1994). Although it is possible that tin was imported from Cornwall, it is just as possible that tin bronze was produced in Britain. Moreover, Budd *et al.* (1994) state that methods of tin-bronze production may not be conducive to identifying ore locations and types with certainty. Nevertheless, an abundance of bronze implements has been found in Ireland, implying there were indeed important bronze-producing centres there, just as there were significant gold-producing centres. For example, The Mount Gabriel mines of the southwest were an important source of copper (O'Brien, 1990). The importance of Ireland as a copper producer has also been emphasised through the lack of evidence for Bronze Age copper exploitation in Britain. Although there is evidence for metal smelting, it has been assumed the raw materials and finished implements were imported from elsewhere, including Ireland. This view is challenged by a recent report of Dutton *et al.* (1994) on prehistoric copper mining at the Great Orme near Llandudno, Wales. The archaeological excavations have revealed that mining occurred over a 1000 year period between ca. 3.8 and 2.8 K years BP, involving an area of some 24 000 m^2 with 5 km of passages extending 70 m below ground. This was clearly a major production centre; Dutton *et al.* consider it to be of international significance. Other possible mines have been found elsewhere in Wales, e.g. at Copa Hill, Cwmystwyth (Timberlake, 1988), Parys Mountain and Alderley Edge (Budd *et al.*, 1992).

The Bronze Age was generally characterised by increasing agglomerations of people and increasing craft specialisation. Moreover, hierarchical societies became widespread in the Old World, as is evidenced by burial practices and the character of grave goods. There is also evidence for the intensification of agriculture in many parts of Europe between 4.5 and 2.5 K years BP. This caused environmental change, notably deforestation and soil erosion. For example, Dearing's (1994) review of past rates of soil erosion in Britain indicates an acceleration in alluvial sedimentation and colluvial accumulation in the English Midlands and the chalk downlands, respectively, in the Bronze Age. Deforestation occurred in the chalk downlands earlier in the Neolithic, but as agricultural activity expanded it had a detrimental effect on rates of soil erosion. Day (1993) records a period of major woodland disruption ca. 3.8 K years BP at Sidlings

Copse, Oxfordshire, and attributes it to an expansion of pasture and arable cultivation in the nearby area. At several sites in the Upper Kennet Valley, Wiltshire, there is evidence for late Neolithic or early Bronze Age deforestation and soil instability, whereas late Bronze Age settlements and field systems have been recognised on Marlborough Downs and Bishops Cannings Down (Evans *et al.*, 1993).

There is also palaeoenvironmental evidence from other parts of the Old World for culturally induced environmental change. Lagerås and Sandgren (1994) have identified periods of deforestation, using pollen analysis from Bråtamossen Bog in Småland, southern Sweden, which begin in the late Neolithic and continued into the Bronze Age. These phases of deforestation are accompanied by high charcoal counts, indicating that fire was used to clear the forest. And magnetic analyses of the bog sediments reflect periods of soil erosion following forest clearance when soil particles were carried onto the bog surface by wind or by run-off. A synthesis of the landscape history of the Argolid peninsula of Greece by Jameson *et al.* (1994) also indicates that environmental change occurred during the Bronze Age, notably the acceleration of alluviation in the late Bronze Age. Rosch (1996) has shown that shifting cultivation which characterised southwestern Germany during the Neolithic developed into permanent arable agriculture during the Bronze Age. Rosch goes as far as to to say that by this stage the landscape was already similar to Medieval times.

4.8   The Iron Age

By the time bronze working appeared in Britain ca. 4.5 K years BP ironworking was already under way in the Near East. According to Maisels (1990), iron was known and used in the earlier Bronze Age but on a limited basis. The earliest authenticated find of an iron object is from Alaça Hüyük (situated near the modern city of Ankara in Turkey) and is dated to ca. 4.4. K years BP (Collis, 1984). It is a dagger found in association with other grave goods made of bronze. There has been much speculation as to why ironworking was not developed earlier, especially as iron ores are much more widespread and locally available than ores of copper and tin. The late development of ironworking may have been because iron has a higher melting point than copper and tin, and bronze furnaces and pottery kilns could not reach sufficiently high temperatures. Moreover, iron implements only become durable and lose their brittleness through the addition of carbon. There is also the possibility the upheavals that occurred in the western Mediterranean as the Mycenaean civilisation collapsed (Section 4.7) caused major disruptions in the movement of copper, and especially tin. Such shortages may have caused metallurgists to innovate, and once the technology was mastered, the advantages of a cheaper and readily available raw material would have become apparent rapidly. Maisels (1990) certainly subscribes to the importance of this upheaval in technological innovation.

The more general use of iron had come about by ca. 3.5 K years BP in the Near East and by 3.1 K years BP in Greece, though bronze continued to be used for many implements, especially weaponry and armour. The precise origins of iron technology remain obscure. Although at least one centre must have been in Anatolia, probably in the Hittite Empire, it is likely there were other centres of origin for iron production, similar to bronze production (Section 4.7). There is a consensus that this was a period of confusion as readjustments to warfare, migration and destruction were made. For example, Roberts (1992) states, 'we have entered an age too complex and too obscure for straightforward narration'. Apart from the Hittites, iron metallurgy is also associated with the Phoenicians, a seafaring people based in what is modern-day Lebanon. The diffusionist school of thought on iron smelting in Africa believes the Phoenicians introduced the metallurgy into north Africa, possibly via Carthage. However, as Okafor (1993) discusses, there are opposing views: the indigenous and 'cautious' schools of thought. The indigenous school considers that iron smelting developed independently in Africa whereas the cautious school believes that more evidence is required before deciding between diffusion and indigent. The diffusionist view is also espoused by Miller and van der Merwe (1994) for the establishment of iron and copper metallurgy in sub-Saharan Africa. Ironworking was evident throughout Europe by 2.5 K years BP. In view of the abundance of iron ores it is likely that iron metallurgy did indeed develop in several independent centres, as is also likely for India and the Far East. Although they were durable, iron implements did not entirely supersede bronze

articles or items made of stone and bone; instead they added diversity to the material culture and further means by which the environment could be manipulated.

Not all environmental changes were the result of iron technology. The Iron Age witnessed an intensification of environmental change, and in some areas an adjustment to climatic change, of trends established several millennia earlier. The changes in agriculture, social organisation and technology are interrelated and created the wherewithal for, first, the Greek Classical Age and, second, the development of the Roman empire (Section 5.2). These were Mediterranean-based civilisations; elsewhere in Europe the Iron Age saw the emergence of many cultural groups, including the Celts, with various religious practices. Indeed Lamb (1995) has argued that this is true of other parts of the world. In China the period of so-called warring states occurred in what was equivalent to Britain's middle Iron Age. Lamb suggests the unrest may have been caused by a cooling climate and its adverse impact on crop production. He also points out that a few centuries earlier Buddha (563–483 BC) and Confucius (551–479 BC) respectively sowed the seeds of a new religion and advocated social reform. It was the era of the Greek philosophers and, just a little later, Christianity and Islam were born.

Cunliffe (1994) has recounted the rise to pre-eminence of the Classical World centred in the eastern Mediterranean which began in ca. 2.8 K years BP. In particular, the city states of Ancient Greece (includes modern-day Greece and the Mediterranean coasts of modern-day Turkey) were emerging, reflecting an intensifying process of urbanisation. As discussed in Section 5.2, pressures on food supplies increased, causing various diasporas to establish 'colonies' elsewhere in the Mediterranean shorelands, including those to the west. Simultaneously, the Phoenicians (of modern-day Lebanon, see above) and Etruscans (of modern-day central Italy) were establishing themselves as seafaring and trading nations. Competition and some conflict ensued (Section 5.2). Although this Classical World has had a profound effect on western intellectualism, significant developments were also taking place elsewhere in Europe. On the one hand, the trading nations of the Mediterranean had contact with central and peripheral parts of Europe. For example, there was contact with the emerging Hallstatt culture of Austria and central Europe via the Rhône and Saône Rivers. The

Hallstatt culture in turn influenced areas of Europe further north, west and east via rivers such as the Danube and the Rhine. Finds of Mediterranean wares, especially in grave goods, across the region attest to this trade and it is considered that such commodities were much sought-after luxury goods. The Hallstatt people had a clear identity and established specific burial practices for their elite that involved the interment of the vehicle (e.g. chariot) used to carry the deceased to the grave. Hillforts were, however, the preserve of the periphery of northwest Europe (see below in relation to Britain); further north still, encompassing the North European Plain and southern Scandinavia, the landscape was characterised by villages. Many of them originated several millennia earlier as Neolithic constructs.

The influence that the Hallstatt people exerted on north and central Europe was considerable but eventually proved to be fragile. Their power was based on their role as 'go-betweens', importing prestige goods from the Mediterranean then exchanging them with areas to the north, east and west for commodities such as hides and gold, and probably slaves (Fig. 4.14). As the Rhône corridor was closed, disruption of this trade to the passage of Mediterranean goods forced the Etruscans to find other ways of exporting their wares to northern Europe. This they did by finding new routes to make direct contact with the Marne-Moselle region, where a warrior class rapidly developed into an elite. The new contact generated internal changes, including the emergence of a new art style known as Celtic or La Tène art. The influence of this new art form was felt throughout Europe through diffusion but also via a series of migrations. As Figure 4.14 illustrates, the Celts moved south and east in aggressive waves taking with them the La Téne culture.

Hedeager's (1992) study of the Iron Age in Denmark provides an example of a north European village-based society and the interplay between economy and ecology. The settlements were placed so they were surrounded by a radius of 1 km of arable land as well as having access to water meadows, etc., which could provide winter fodder. The density of settlements, which is similar to today, probably reflects a high population, so that villages exploited their hinterlands to the full, i.e. the land was being used at its maximum carrying capacity. Hedeager states 'When the land was so heavily exploited that new areas could no longer be found,

agricultural and technological changes were the only answer to population pressure and soil exhaustion'. The settlement itself comprised several longhouses and outbuildings used for storage. Wattle and daub, turf or timber were the main construction materials and each longhouse would have family dwelling space as well as animal stalls. Fixed field systems, with fields separated by boundary banks (also known as Celtic fields) first appear at the onset of the Iron Age (ca. 2.5 K years BP). This too may reflect high population concentrations which were established in the Bronze Age. Hedeager's research indicates that Iron Age communities were maintaining nutrient levels in the soils by manuring and fallow periods. The keeping of animals in stalls would have facilitated dung collection, and fields were probably cultivated on a rotation basis. Soil exhaustion in some areas of Denmark caused arable land to be abandoned about the same time as the birth of Christ.

According to Lamb (1995), climatic deterioration occurred during the Iron Age, i.e. 3 K years BP to 0 BC, in Europe. In Britain there is evidence for increased wetness in the west and increased dryness, followed by a return to wet conditions, in the east. Lamb suggests that such conditions reflect the dominance of westerly winds and a cooling Arctic region. Glaciers readvanced in Alpine regions and high latitudes and coasts of the North Sea rim were subject to flooding and intense storms. A wet climate prevailed by 2.5 K years BP, and the influence of the westerlies declined. Increased wetness is reflected in the construction of wooden trackways in areas like the Somerset Levels (Coles and Coles, 1986), where there was a marine incursion between 2.8 and 2.5 K years BP (Housley, 1988). In some parts of Britain there is palaeoecological evidence for woodland clearance, especially in Wales, northern England and southern Scotland. Some of this evidence has been summarised by Barber et al. (1994). At Walton Moss in Cumbria, for example, a substantial reduction in arboreal pollen and an increase in grass pollen, with other indicators of disturbance, is dated to 45 (calendar) years BC. It continued until ca. AD 64 (calendar) years and the presence of cereal pollen suggests that woodland was cleared for arable agriculture. However, at nearby Bolton Fell Moss there is no evidence for extensive woodland clearance until Roman times (Section 5.2). Walker (1993) has reviewed the evidence for woodland clearance in upland mid-Wales, which began earlier in the

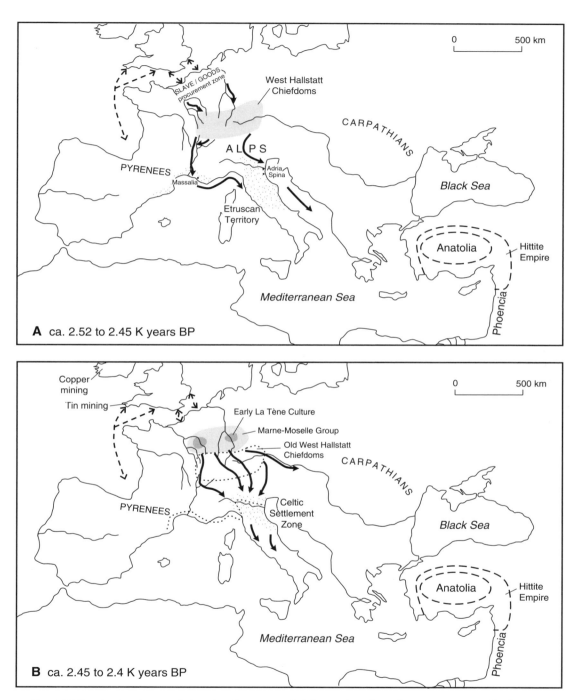

Fig. 4.14 The dominant groups in Iron Age Europe and their interactions (based on sources quoted in the text).

Bronze Age and which continued into the later Romano-British period when it intensified.

Hill (1995) has recently reviewed the evidence for pre-Roman Iron Age (PRIA) conditions in Britain and Ireland, a chronology for which is given in Table 4.8. The period spans the first millennium BC. Apparently the advent of the Romans did not substantially alter Iron Age communities but added a new dimension (Section 5.2). Hill advocates that PRIA developments in Britain and Ireland were characteristic of the European temperate zone in general, and were not indicative of a purely indigenous isolationism as has often been considered. One factor which distinguishes the PRIA from the Bronze Age is the rise to prominence of many upland areas as centres of power, though lowland areas continued to enjoy a high level of economic activity, especially in southern Britain.

In Halstatt Britain and Ireland, which encompasses the late Bronze Age and early Iron Age, bronze and iron working occurred alongside each other, reflecting continuity between the Bronze

Table 4.8 A chronology for the Pre-Roman Iron-Age (PRIA) in Britain and Ireland (based on Hill, 1995)

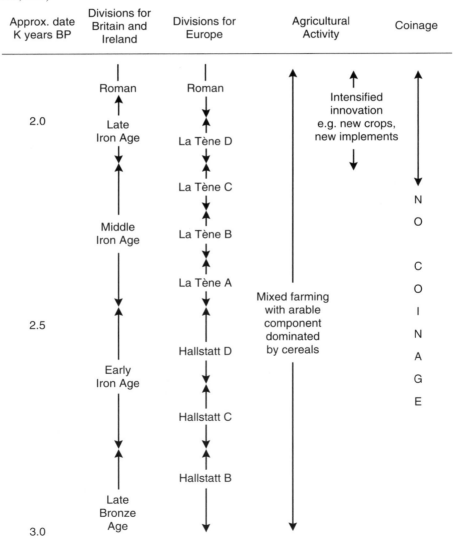

and Iron Ages. The cessation of metal deposition in wetlands and rivers, along with the abandonment of the practice of hoarding metal objects and the appearance of so-called prestige objects, e.g. swords, axes in iron, rather than in bronze, marked the onset of the Iron Age. Nevertheless, the settlement patterns and established centres of pottery production persisted while agricultural activity intensified. Jones (1996) believes the production of cereal crops, within a mainly mixed farming economy, became more important than in the Bronze Age, and the hunting and gathering of wild plants and animals was minimal. This trend continued throughout the Iron Age. There were also changes in ritual practices with offerings of food, household objects and human bones becoming commonplace; burials and cremations diminished considerably in number. The first hillforts also appear in significant numbers in the early PRIA and today they remain as the most visual evidence for Iron Age cultures. Most hillforts occupy commanding positions and are surrounded by stone walls or earth banks. As Hill (1995) points out, few generalisations can be made about their function within an area though they are often considered to be focal points for defence, trade and exchange, centres for collection and dissemination of agricultural produce and for craft production, i.e. metalware and pottery, and foci of ritual religious activities. Many functions are characteristic of some hillforts, including Danebury in Hampshire (Cunliffe, 1995), one of several defended hillforts in the region. Constructed by c. 2.4 K years BP, each may have controlled a specific territory within which numerous smaller settlements were established. Evidence from Danebury indicates the operation of an intense mixed-farming system in its catchment with open pasture on the downland and arable cultivation on gentle slopes and the edges of water meadows. The water meadows supported cattle herds in the spring and autumn. There are grain storage pits and evidence for ritual practices both here and at other hillforts, including several in Scotland (Hingley, 1993).

Settlement patterns and ritual practice in the middle PRIA (MPRIA) reflect continuity with the early PRIA (EPRIA); see Table 4.8. Hill (1995) opines that the features distinguishing the MPRIA from the EPRIA are changes in mortuary practices and metalwork styles. In Europe this coincides with the transition between the La Tène and Halstatt cultures, which occurred ca. 2.5 K years BP. The La Tène culture developed in what is modern-day Switzerland and spread into temperate Europe. Finds of metalwork and some burial practices found in Britain reflect continued but limited contact with Europe; in southern Britain there was a shift in ceramic production, reflecting improved firing control. Agriculture inevitably continued as a primary activity and underpinned (conjectured) population growth. Jones (1995) indicates that innovations in agriculture continued but accelerated in the late PRIA (LPRIA).

As Table 4.8 illustrates, the LPRIA was relatively short-lived. It is characterised by new pottery forms, reflecting renewed vigour in contact with Europe and the introduction of a new potter's wheel in southern England. Many changes occurred in this region involving most aspects of life. Hill (1995) states that 'culturally, this part of England became increasingly like that of neighbouring areas in France and Belgium. This situation continued, if not intensified, after the Roman conquest'. One particularly important development was the introduction of coinage. Coinage underwent several changes during the LPRIA, including the use of silver and gold coins; see Haselgrove quoted in Hill (1995). The influence of Rome was also felt indirectly in Britain before its annexation in AD 43. Grave goods in southern Britain reflect contact with Europe and the erection of shrines became part of local rituals, again reflecting continental influence. The importance of trade is reflected in grave goods and material culture, and there is a marked difference between the LPRIA and the earlier PRIA periods. New constructions also appeared in southern Britain. Known as oppida, they represent a wide range of enclosed sites and may represent urban development, similar to hillforts. Allthough this is controversial, many oppida did indeed provide the sites for later Roman cities. It has been suggested that an even more hierarchical society was developing during the LPRIA with a tribal structure throughout most of Britain. Regions beyond the immediate influence of Europe may have developed in their own ways to generate local identities that involved distinctive buildings and material cultures. Jones (1996) suggests that many innovations, including new crops as well as new implements, were occurring in many areas in Britain, and not necessarily in a way explained by diffusionist theory. According to Raftery's (1994) synthesis of Iron Age Ireland, the LPRIA witnessed the construction of so-called royal sites. They are located mainly in the

Irish Midlands and are large circular enclosures, probably used for ceremonial or ritual purposes.

The Iron Age was a period of substantial changes in Europe. It seems likely there was a link between climate and social organisation and change. Moreover, the available evidence indicates the presence of high population densities and periods of social destabilisation, including warfare. And the land reached its carrying capacity, at least in some parts of Europe, possibly prompting innovation in agricultural practices to prevent or reverse soil deterioration and/or to cope with growing populations.

4.9   Conclusion

This chapter has reflected on the emergence of *Homo sapiens sapiens* as a manipulator and controller of environmental processes. It is perhaps ironic that those very forces of ecological systems which helped to stimulate hominid development should eventually come under hominid control. Stone, wild animals and plants were among the first resources to be used, and by the Upper Palaeolithic, if not earlier, humans were using controlled fire to manipulate flora and fauna in well-organised hunter-gatherer food-procuring strategies.

A major, though probably gradual, turning-point in human and environmental history was the domestication of plants and animals some 10 K years BP. This process, which allowed humans to control the fundamental genetic characteristics of plants and animals, was an essential precursor to the development of agricultural systems that in turn led to major changes in social organisation and underpinned some of the most powerful ancient civilisations. These may well have risen and fallen on the strength of their ability to manipulate food production and the new resources of copper, tin and iron on which metallurgical industries were dependent and which facilitated agricultural intensification. The development of irrigation systems in more arid regions provided a further means by which humans learnt to manipulate their natural resources and to exert increased control over water supplies, an essential ingredient for efficient crop production.

But the ability to manipulate the environment did not develop trouble-free. The archaeological record clearly attests to the problems of environmental management such as overgrazing, declining soil fertility, soil erosion and salinisation. Such problems are now among the world's most important environmental issues, as will be discussed in Chapters 7 and 8.

Further reading

Cavalli-Sforza, L.L., Menozzi, P. and Piazza, A. (1994) *The History and Geography of Human Genes*. Princeton University Press, Princeton, NJ.

Cunliffe, B. (ed.) (1994) *The Oxford Illustrated History of Europe*. Oxford University Press, Oxford.

De Laet, S.J., Dani, A.H., Lorenzo, J.L. and Nunoo, R.B. (eds) (1994) *History of Humanity, Vol. I: Prehistory and the Beginnings of Civilization*. Routledge, London, and UNESCO, Paris.

Evans, L.T. (1993) *Crop Evolution, Adaptation and Yield*. Cambridge University Press, Cambridge.

Flood, J. (1995) *Archaeology of the Dreamtime*, 3rd edn. Angus and Robertson, Sydney.

Nitecki, M.H. and Nitecki, D.V. (eds) (1994) *Origins of Anatomically Modern Humans*. Plenum, New York.

Smith, B.D. (1995) *The Emergence of Agriculture*. W.H. Freeman, New York.

CHAPTER 5

Environmental change in the historic period

5.1 Introduction

The moment when prehistory ended and history began is an arbitrary distinction from region to region and country to country. Many of the technological innovations that enabled human communities to manipulate their environment were developed in Europe. Consequently, history as defined here begins with the Greek and Roman empires. These civilisations wrought considerable environmental change, and the Romans especially influenced the people/environment relationship throughout Europe, as will be discussed below. An appraisal of global environmental change throughout the historic period is beyond the scope of this text, so emphasis will be placed on developments in Britain with some reference to Europe.

Before the eighteenth century, Britain was predominantly rural with an agriculturally based economy. Agricultural improvements in the seventeenth century, however, provided a sound base for industrial development and from the mid-1700s the Industrial Revolution occurred along with the rise of urban centres. Such innovations laid the foundations for modern society but were not without cost in terms of environmental change. The increased use of coal, and subsequently other fossil fuels, as well as the production of artificial fertilisers, were to have far-reaching environmental effects, some of which are only just coming to light today.

Along with other European countries, Britain has significantly influenced environmental change in many other parts of the world. This began in the fifteenth century, a period of exploration when the New World was discovered and trade links were established with the Far East. The initial impact of Europeans in these new-found lands was relatively slight, but new territories were annexed as empires were consolidated and resources exploited. The eighteenth and nineteenth centuries witnessed the large-scale expansion of European interests in their colonies.

5.2 The impact of the Greeks and Romans

After the fall of the Mycenaean civilisation ca. 3.3 K years BP (Section 4.7) much of southeastern Europe entered a period known as the 'Greek Dark Age', which lasted for some 400 years until the emergence of the city states ca. ninth century BC. These states developed into Classical Greece of the fifth century BC, when Athens developed her empire. The abundance of archaeological remains and documentary material attests to the considerable influence of Classical Greece and the sophistication of its society, though there is only limited evidence for its impact on the landscape. In general, agriculture expanded as the olive was brought into widespread cultivation. Isager and Skydsgaard (1992) have illustrated the value of ancient texts as a source of evidence for agricultural practices of the time. Many such texts reflect the intensive production of cereals and pulses while cultivation of the vine and olive generated products that could be traded. The importance of trade plus the fact that Athens' empire was built essentially on naval power (Hornblower, 1994) meant there was a substantial demand for wood for shipbuilding. There was also a demand for precious metal for coinage. The power of Athens resulted from its command of resources from within its own boundaries, from its many island territories and from western Turkey. In particular, it relied heavily on grain imports from south Russia, transported through the all-important Hellespont, the narrow strait joining the Sea of Marmara with the Aegean.

This grain requirement was the result of the limited agricultural potential of Athens' hinterland, Attica, where only relatively small pockets of cultivated land existed and which were more suited to olive than to grain production (Murray, 1994). According to Sallares (1991), as little as 20 per cent of Attica may have been cultivated, and certainly no more than 50 per cent. Many fruit crops were produced besides the olive, including peas, pomegranates, apples, almonds and figs, all of which are mentioned in ancient texts (Isager and Skydsgaard, 1992) as is the practice of manuring the soil. Pastoral activities predominated in the uplands of Attica to generate wool and milk from sheep and goats. Meat was probably only consumed during festivals. Murray (1994) also points out that a 'new deep vein of silver was discovered in the Laurium hills' in the early fifth century BC. This was worked for several centuries by slave labour which was important in Classical Greece because it freed Greeks to become involved in cultural and educational activities. Slave labour was commonplace and existed in a variety of forms from the tethered slaves of the silver mines to the freelance craftspeople who paid a fee to their owners. According to Murray, it is probable there were 100 000 slaves in Attica, approximately the same number as free inhabitants. Human resources thus played a major role in shaping the character of classical Greece; without slave labour the familiar monuments, religion, philosophy, mathematics, literature and drama which have so profoundly influenced Western intellectualism may never have come into existence.

Ancient texts also provide evidence for deforestation. According to McNeil (1992), Plato referred to deforestation and soil erosion in Attica, probably of the mountain Hymettus, which did indeed lose its arboreal mantle (Harrison, 1992). Although most other upland areas of Attica retained their tree cover, even up to today, there was demand for Macedonian timber during the Peloponnesian War, 460–446 BC (McNeil, 1992). Either the Athenians could not or would not use their local resources. According to Hughes (1982), a great deal

of forest did indeed remain unscathed in much of ancient Europe; this was especially so in mountainous regions, which offered protection through remoteness and unsuitability for agriculture. Gerasimidis and Athanasiadis (1995) have recently presented several pollen diagrams from the northern mountains of Greece, which were in the hinterland of Athens' north Aegean coastal lands during the Classical era. At some sites there is evidence for woodland clearance and the spread of open habitat species, some of which may have occurred in the Classical period. In the southern Argolid, a peninsula south of Athens but which was not part of the Athenian empire, there is evidence for soil erosion and deposition during the past 5000 years. According to Runnels (1995) and Jameson et al. (1994), a major phase of erosion corresponds with the spread of Bronze Age agriculture with a further series of soil erosion events occurring between 400 BC and AD 600. They state that 'the landscape of Classical times was not much richer in soil than it is now'. In the early part of this period (the Hellenistic Period) population comprised ca. 12 000, one of the highest population levels ever recorded in this region. There was, however, sufficient cultivable land to support these people, with 1 ha per person available. Soil depletion occurred despite the innovations of terracing, fallow, intercropping, etc. Soil fertility, however, was probably maintained through manuring, as Alcock et al. (1994) have discussed in relation to Classical Greek artefact distributions. The manure was derived from a wide variety of sources, including animal and human wastes, road sweepings, urban waste and domestic waste. These practices continued in the Hellenistic period ca. 323–31 BC. McNeill (1992) has provided information on the Pindus Mountains to the west of modern-day Greece and on Epirus, the plain between the mountains and the Mediterranean Sea. He states that sheep farming was the mainstay of the region, relying on transhumance to utilise the resources of uplands and lowlands. Terraced agriculture provided arable crops but forest remained widespread until the Roman invasion in 167 BC.

The environmental impact of Classical and Hellenistic Greece may not have been confined to the city states and their hinterlands. S.M. Hong et al. (1994) have undertaken the analysis of lead concentrations in a Greenland ice core representing the last ca. 2.5 K years. They found that natural background concentrations, deriving from rock

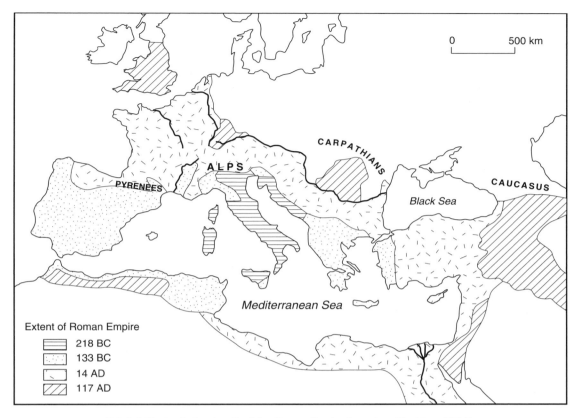

Fig. 5.1 Stages in the growth of the Roman Empire (based on Stockton, 1994).

weathering, soils and volcanic eruptions, was ca. 0.5 pg/g of ice. In the section of the core dating to the time of Christ the concentration of lead reached 2.0 pg/g, falling again to the background level in the core section dated to AD 500. S.M. Hong *et al.* attribute the start of the increase to the intensive silver mining in the Laurium Hills of Attica (see above). The mineral extracted from these mines was galena, lead sulphide; it also contains a small amount of silver. The silver was recovered by heating the ore and then cupelling, i.e. the molten metal was heated in a container with hot air to oxidise the lead, which was skimmed off to leave the silver. Lead itself became commercially important when the Roman empire was established, so its production increased (see below), thus adding to the amount of lead in the atmosphere and thus to that recorded in the Greenland ice core. S.M. Hong *et al.* suggest that the increased lead concentrations represent the deposition of ca. 400 tonnes in Greenland over an 800 year period. This is equivalent to 15 per cent of the lead pollution caused in the past 100 years by lead in petrol.

There is also much written and archaeological evidence for the impact of the Romans. Rome itself was founded in 753 BC though it was not until ca. 264 BC that it came to control the entire Italian peninsula; thereafter its control expanded to much of the Mediterranean world and beyond into central and western Europe (Rawson, 1994). The growth of Rome's empire is illustrated in Figure 5.1. By AD 400 the empire was disintegrating and in AD 410 Rome itself was sacked by Alaric the Visigoth, and Britain was formally renounced. The Romans were responsible for a great deal of environmental change both at home and within their empire. The remains of Roman buildings in many parts of the world bear witness to the imperial presence and whatever exploitation of the conquered peoples Rome was responsible for, it has left a money-spinning legacy in the shape of tourist attractions. Two examples are the great city of Ephesus, in modern-day

Turkey, and the amphitheatre of Verona, the venue of many splendid opera performances. Potter (1987) has documented many of the changes that occurred in Roman Italy. In relation to landscape changes, the most significant developments in Italy were centred on land tenure and the spread of villas and farms. The farms employed slave labour derived from conquered lands of Europe and were engaged mainly with vine and cereal production. The character of land tenure also changed as smallholdings became amalgamated into *latifundia*. *Latifundia* were large estates owned by Roman nobles but usually managed by overseers. According to Pounds' (1990) review of Roman Italy, settlement intensified, especially in Etruria (Fig. 5.1), and further south the area around Naples produced wine, olive oil and various types of fruit for urban markets. However, Pounds also notes that the highly organised agricultural production system of the Romans is particularly well illustrated by the centuriated landscape of parts of northern Italy. In the Apulian Tavoliere, for example, aerial photography has revealed vestiges of this system of land subdivision involving the creation of regular, square fields separated by roads and paths. Lines of pits were also constructed aligned with the roads and were used for storing the olive crop.

The nature of agricultural practices themselves has been examined by Rees (1987), who states that the two major subdivisions of the empire in relation to climate gave rise to two distinct types of farming techniques. 'Dry farming' predominated in the Mediterranean zone, with hot dry summers and warm wet winters. This involved arable crop production in lowland, mainly coastal areas and alluvial plains, with vine and olive cultivation on less productive slopes and summer pasture, mainly for sheep, in the uplands. In contrast, the northern provinces, with a cool, wet climate were suited to 'wet farming'. In particular, arable and mixed farming characteristics of Iron Age communities (Section 4.8) dominated lowland areas; stock rearing on pasture prevailed in the upland zone. Agriculture in the Mediterranean was based on a crop–fallow system with one crop being obtained every two years in order to take advantage of rainfall over two years. Water conservation measures were also employed, including the maintenance of a fine soil tilth to reduce evaporation from the soil and to encourage the deep percolation of winter rains, and frequent weeding to reduce transpiration. Nutrient levels were maintained through the employment of

fallow and stubble burning. It is likely that 'manuring' was also employed (see above), involving household and city waste or animal waste, depending on what was available. The Romans also instituted drainage networks, the remains of which are still in evidence. In the Po Valley, for example, large-scale drainage was undertaken to increase the amount of land suitable for agriculture. Elsewhere in the empire, such as southern Spain, irrigation systems provided water for the cultivation of fruit, olives and vines. In addition to the production of wheat and spelt, millet was grown in the Middle East and Gaul whereas legumes, e.g. beans, turnip, carrots and peas, along with herbs, were produced in irrigated market gardens and orchards. Wacher (1987) also provides evidence for a wide range of secondary animal products such as cheese, glue and lard. A survey of the major crops and domesticated livestock is given by Randsborg (1991).

Alcock (1993) has examined the evidence for landscape and social change in Greece when it was occupied by the Romans. Both archaeological and documentary evidence point to a widespread decline in rural habitation sites which began in the second century BC and continued until the third or fourth century AD. Small sites in particular disappeared, possibly reflecting the amalgamation of small holdings into larger ones akin to the *latifundia* (see above). The landowners themselves may have been Roman nobles and there is evidence for material wealth and social stratification. Documentary evidence may reflect an intensification of agriculture with emphasis on exports of grain and olive oil to Rome. Alcock's survey of palaeoenvironmental evidence for landscape change includes reference to erosion and an expansion of pollen of maquis species in the southern Argolid (Section 4.7). Changes in the pollen spectra may reflect a shift from arable to pastoral agriculture but this would not normally be associated with erosion. It is also feasible that pastoral agriculture replaced arable cultivation because of rural depopulation. There are many references to labour shortages on Italian estates due to the continuous raising of armies to defend the empire. Rome's colonies beyond Italy were not exempt from this process.

The agricultural practices, settlements, mining and smelting operations and communication networks of the Romans changed the character of landscapes. Iron ores were most widely extracted and smelted; in Italy there was a major iron industry in Tuscany which obtained its ore from the island of Elba.

However, the province of Noricum, mostly in present-day Austria, was the most important iron-producing region and continued as such well after the demise of the Roman empire. Lead was also extensively mined and was required by the hydraulic engineering works that characterised most Roman cities. The major sources of lead were Britain, Spain and the Balkans; smelting was sufficiently extensive to increase the concentrations of atmospheric lead as recorded in a Greenland ice core (see above). Hong *et al.* (1996) report that copper concentrations in the atmosphere also increased substantially in Roman times. Their analysis of a Greenland ice core shows that copper production and smelting was at a maximum ca. 2 K years BP. Production may have been as much as 15 000 tonnes and was used for a range of purposes but particularly coinage. All these smelting activities required charcoal. Consequently, woodlands must have been seriously exploited in smelting centres, though the evidence for this is limited. It is generally considered that one of the reasons the Romans were keen to annex Britain was to enhance their supply of cereals. Accordingly, there is some evidence to suggest that arable cultivation intensified during the period of occupation (AD 43–410). This was made possible by the adoption of Roman technology, including improved ploughs, scythes, sickles and watermills. Such an expansion of agriculture would also have increased pressure on woodlands. Rackham (1990) states: 'Among their many activities in the Weald the Romans had military ironworks, whose output has been estimated by De Henry Cleere at 550 tons a year (Cleere, 1976). From this figure I calculate that these ironworks could have been sustained permanently by 23 000 acres of coppice wood. There are many other ironworks in the Weald, whose total influence on the landscape could hardly have been surpassed even in the seventeenth-century hey day of the iron industry'. Similarly, the establishment of pottery-producing areas, which also required wood for kilns, as in the New Forest and Nene Valley, exerted a toll on native woods. Rackham points out that little is known about Roman woodcraft as practised in Britain, though documentary evidence attests to its widespread use in Italy. Substantial quantities of wood would also have been necessary to support the cottage industries, i.e. metalworking, pottery, etc., in the numerous small towns that had developed in England and Wales (Potter and Johns, 1992)

There is also evidence for the extensive use of timber and wood for construction purposes, as has been pointed out by McCarthy (1995) in relation to Carlisle in Cumbria. Timber was usually cut from the tree trunks and used for major structural elements in buildings, such as posts and beams, whereas wood derived from coppiced stools was used for fencing and fuel. Coppiced stools were also used for wattle and daub, which involved the intertwining of coppiced poles that were then covered with mud. Furthermore, many forts were constructed by the Romans, especially in northern Britain along defensive structures such as Hadrian's Wall, of which Carlisle was a major fort, and later the Antonine Wall. All but the Antonine Wall, constructed from turf, required considerable volumes of wood. Until recently there has been little palaeoecological evidence for the Roman impact on British vegetation communities, the major exception being that of Turner (1979) for northeast England. This showed that woodland clearance did occur during the period of Roman occupation. More recent work in northern Britain, e.g. Dumayne (1993) and Dumayne and Barber (1994), has also established that clearance occurred, though it is not clear why this occurred. Was it to increase grain production through arable cultivation which necessitated woodland clearance? Or was it to satisfy the *Machina Romana's* desire for wood for the other purposes mentioned above? It may have been a combination of both. However, Dumayne (1993) argues that pollen-analytical data from several sites in the region between Hadrian's Wall and the Antonine Wall, i.e. southern Scotland, show major increases in grass pollen and decreases in arboreal pollen in horizons dated to the Roman occupation of the region that began when Hadrian became emperor in AD 117; the wall itself was constructed between AD 122 and AD 130. However, because the increases in grass pollen, reflecting increases in open habitats, are not accompanied by substantial increases in other indicators of agricultural activity, Dumayne has suggested that much of the woodland clearance was due to a demand for wood rather than for an extension of agricultural land. If this is correct, it appears the Romans were operating in an exploitative and unsustainable way in relation to a natural resource. In some respects this is surprising in view of their native woodcraft skills and because of their heavy reliance on wood for most of their enterprises. In other respects, such an exploitative attitude may have been typical of a colonial power.

There are pollen-analytical assemblages from other parts of Britain that also reflect woodland clearance, or continued woodland clearance from the earlier Iron Age, during the period of Roman occupation. Bartley and Chambers (1992) record woodland clearance on Extwistle Moor, Lancashire, which is associated with agriculture. Pollen assemblages from Cromwell Green, Westminster, in the heart of London, also reflect a decline in forest cover from the earlier Iron Age. Greig (1992) reports that the amount of tree pollen declined from 24 per cent to 12 per cent of total pollen. Interestingly, perhaps, there was a recovery to 28 per cent in deposits attributed to the post-Roman period. The woodland in the vicinity of Cromwell Green comprised oak, elm and lime with pine, birch and maple. The Roman deforestation may have been for timber but the increased incidence of cereal pollen also suggests an intensification of arable agriculture. At this time London was a thriving Roman city and port, though *Londinium* itself was located east of Westminster in what is today the City of London. As Britain's foremost port at this time, a wide range of Mediterranean goods and crops were imported into London, including olive oil, dates and figs.

5.3 The Middle Ages (ca. 400–1400)

The early part of this period is often called the 'Dark Ages', i.e. the fifth to the ninth century in Europe. This period is so called because, traditionally, historians have considered it to be a period of stagnation, both intellectually and technologically, following the alleged large-scale disruption caused by the demise of the Roman empire. Now, however, a consensus is growing that believes there was considerable continuity with the earlier Roman-dominated period and that the epithet 'Dark Age' is more a consequence of lack of information than of real social retrogression and absence of innovation. Certainly there is little information available on the role of human activity in environmental change.

The decline of Roman influence in Britain and Europe was a gradual process. Why such a major power should have declined is a matter for much speculation. Although the reasons are complex, it is likely that within a turbulent political structure, increasingly difficult to maintain, the issues involved were environmental as well as economic. The fall of Rome and its outposts, in common with a number of contemporary civilisations such as Han China

and Parthia (the Persian Empire), was at least in part brought about by continuous raiding of the empire's margins by nomadic tribespeople such as the Huns, a tartar people from central Eurasian steppelands (Fig. 5.2). On the one hand, Rome had to allocate much resource in terms of money and food to maintaining its legions in peripheral and vulnerable regions and, on the other hand, these so-called barbarians may themselves have suffered crop shortages that motivated them to seek resources elsewhere. Pounds (1990) states there is little evidence for the latter, so possibly greed and power were the motivating factors in their invasive policies. However, in the Mediterranean region, the heartland of the Roman Empire, there was a major period of drought in the period AD 300–400 (Lamb, 1995). Perhaps this left the Roman empire particularly vulnerable to attack from outside. Such an analysis invokes environmental determinism (Section 1.5), itself controversial but nevertheless relevant to such discussions. Whatever brought the demise of the Roman Empire, it paved the way for numerous migrations within Europe and between Europe and Asia. To Chadwick's (1994) statement that 'the Mediterranean was no longer a Roman lake' it could be added that the Romanisation of much of western Europe left, in the wake of Rome's demise, well-organised indigenous peoples and highly productive farmlands.

In Britain one of the most significant changes that occurred as Roman influence diminished was the decline of the many urban centres that had been established, except London. Centres such as Cirencester, Silchester and Gloucester fell into decline as villages came to dominate rural landscapes. In some scholarly accounts of the immediate post-Roman period, e.g. Bede and Gildas, there is the suggestion of a demographic decline due to hostilities from tribes based on the European side of the North Sea, and possibly famine and plague. Higham (1992) has examined these possibilities, including the evidence from pollen diagrams in many parts of Britain. He concludes that evidence for woodland regeneration is restricted to marginal areas in both upland and lowland zones. This could have been caused by the loss of the Roman market at home and abroad. Higham states: 'The end of Roman government in Britain appears to have occurred without any discernible fracture in the processes of land-use and the maintenance of clearance. Recent interpretations of faunal remains are similarly unsympathetic to a

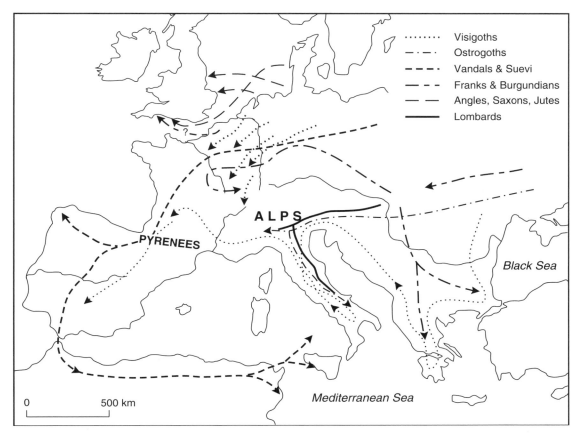

Fig. 5.2 (A) Movement of people in Europe in the first century AD (based on *Atlas of World History*, Times Books, 1993).

major reforestation in the fifth century, suggesting a continuum of pastoral farming in a comparatively unaltered landscape from late prehistory into the mid Saxon period.' This would not have occurred if there had been a substantial population decline. Moreover, there were several waves of migration from Europe to compensate for the loss of the Romans. The view of England, especially, with an agricultural landscape within which intensively managed woodland played an important role while having an extent similar to under Roman management, is supported by O. Rackham (1994a).

Continuity in the countryside is also reflected in a detailed study based on pollen analysis of landscape changes around Diss Mere in Norfolk by Peglar (1993b). The name Diss itself is considered to be Anglo-Saxon in origin, i.e. *dice* or *disce*, which means 'a pool of standing water', and its East Anglian location places it in a prime position for

incoming migrants from Europe. The pollen analysis shows that deforestation began in the Neolithic and there was a marked decline in tree cover between 500 BC and AD 200, i.e. the Iron Age and Roman times. The land use was mainly pasture with some cultivation of cereals. The subsequent period AD 200–1500 was characterised by pollen assemblages, indicating the continued presence of meadows and pastures. The increased presence of mineral matter is taken by Peglar to reflect increased erosion, perhaps due to increased arable cultivation and building associated with the expansion of the town itself. The continued presence of charcoal attests to the use of fire. In the sediments representing the period AD 200–1000 there are pollens of *Secale cereale* (rye), *Cannabis sativa* (hemp), *Linum* spp. (flax), *Brassica* type (cabbages, etc.) and *Vicia* cf. *faba* (broad bean) and other cereals which reflect the crops being grown in the

Fig. 5.2 (B) Subdivisions in Britain during the first century AD (adapted from Hall, 1994).

catchment. Crop diversity increased during the period AD 1000–1500. Thus the Middle Ages around Diss Mere witnessed declining woodland and increasing agricultural activity. There is no evidence for stagnation or a decline in human activity. Moreover, a pollen-analytical study in County Down, Northern Ireland, indicates that cereal cultivation began in the area between the third and fifth centuries AD, though deforestation and pastoral agriculture began earlier (Hall, 1990). In another study on three additional sites in lowland Northern Ireland, Hall et al. (1993) have demonstrated that by AD 860 woodland had been largely cleared for agriculture. This subsequently declined and woodland regenerated to produce scrub by AD 1100, though there is no obvious reason why this should have occurred. Bell (1989) has also cited palaeoecological evidence for continuity of cultivation in other parts of Britain. As Figure 5.2 shows, there were many significant migrations of peoples within Europe and from Asia into Europe. Although these early invaders contributed to the demise of the Roman empire (Section 4.2), the invasions continued in the post-Roman period of the fifth and sixth centuries. The Angles, Saxons, and Jutes were the Germanic and Danish tribes that crossed the North Sea to colonise parts of England. The Vikings arrived in the ninth century. Despite such invasions, settlement evidence tends to reinforce the view of continuity rather than disruption in the English countryside. Although there is some evidence for the desertion of villages, there is more evidence for continued occupation. Eagles (1994) has presented evidence from archaeological sites in Wessex for a moulding together of British and Angle (English) traditions. In relation to place names, there is a mixture relating to British, English and Jutish; interestingly, perhaps, English place names are most numerous, probably reflecting substantial numbers of invaders. Moreover, in his review of Anglo-Saxon Oxfordshire, Blair (1994) reports that Saxons were settling in the Upper Thames region by ca. 550 as evidenced by cemeteries and grave goods. Here, too, English place names predominate.

Further evidence is also given by Murphy (1994), who has examined palaeoecological evidence from several sites in three different landscapes comprising the coastal marshes of Essex, the river valleys of central Essex and the Breckland of west Suffolk. In common with the results from Diss Mere (see above), the palaeoecological data show no evidence for woodland regeneration in the post-Roman period. If there was any regeneration it was local rather than regional. However, at Micklemere (near Pakenham) in West Suffolk there is pollen-analytical evidence for a major change in land use. Murphy (1994) has shown that cereal pollen is substantially reduced but the lack of evidence for woodland regeneration, and the continued presence of pollen of open habitat species, may indicate a shift from arable to pastoral farming. However, at other sites investigated by Murphy, notably in the river valleys of central Essex, arable farming continued into the Anglo-Saxon era whereas sheep farming probably

continued along the Essex coast. Crabtree's (1994) analysis of patterns of animal kills, as revealed by assemblages of animal bones, from several village sites in East Anglia also shows a degree of continuity in agricultural practices from Roman to Anglo-Saxon times. There were, however, marked differences in patterns of trade and exchange. Sheep were predominant at the village of West Stow, though cattle and pigs were also reared. Crabtree indicates that these results are in agreement with the Domesday Book, which reports Stow as an important sheep-rearing region in the eleventh century. At Wicken Bonhunt, in contrast, the abundance of pig bones suggests this village was the focus of pig-rearing activities.

Green (1994) has undertaken a review of the evidence for cereals and plant foods in Anglo-Saxon Wessex. In common with the East Anglian evidence presented above, there are substantial differences from site to site. The cereals, unsurprisingly, dominate the plant assemblages investigated. Six-rowed barley (*Hordeum vulgare*), wheat (*Triticum aestivum/compactum*), oats (*Avena sativa*) and rye (*Secale cereale*) are the predominant types recorded from urban centres, e.g. Winchester, Newbury, Romsey, Andover and Trowbridge. Barley and wheat were most important throughout the Anglo-Saxon period with rye becoming significant during the eleventh century. A wide range of other crops were also cultivated in Wessex, including a variety of legumes and fruit species. Included in the latter is *Vitis vinifera* (vine), which Green suggests may have been widespread, having been introduced by the Romans. The remains of many different types of vegetables have also been identified from urban centres. Examples include carrot *(Daucus carrota)*, celery (*Apium graveolens*), various *Brassica* spp. and beet *(Beta vulgaris)*. Plants were grown for food flavourings, e.g. *(Thymus* sp.), sage *(Salvia* sp.), oregano (*Oreganum* sp.) and nuts were also produced, e.g. hazelnuts *(Corylus avellana)* and walnut *(Juglans regia)*. This diversity of plants for consumption was even greater in France (Ruas, 1992), though the only additional cereal being cultivated was millet *(Panicum miliaceum)* as a minor crop. To these crops referred to above must be added imported exotic species. Green (1994) refers to fig (*Ficus* sp.) recorded from the Wessex site of Hamwic (early Southampton); it has also been recorded in deposits of Anglo-Saxon age in London (Jones *et al.*, 1991), where the cultivation of some food types and medicinal herbs may have occurred *in situ* in garden plots (e.g, J. Rackham, 1994). There is also evidence for the establishment of many additional iron-smelting centres in Wessex during Saxon times (Hinton, 1994).

By 866 there was a Viking presence in Britain. In that year the Vikings captured York and remained in control until 927, when the English gained control. The English ruled until 1069, when William the Conqueror prevailed. Besides York, Dublin was another centre of Viking power, and excavations in both cities during the past two decades have revealed information on their resource bases and the trade they enjoyed. Pounds (1990) recounts how these invaders came from Norway and Denmark and how these cities originated as temporary raiding camps. Moreover, the influence of the Vikings extended beyond the boundaries of York and Dublin. In England, where Danish law and custom prevailed, the Dane law extended south from Northumbria to include East Anglia (Fig. 5.2) though the Vikings were unable to conquer Wessex. Why the Vikings raided and eventually settled in parts of Britain, Ireland and northern Europe is a matter for conjecture. Lamb (1995) suggests that it was a result of their seafaring skills; Pounds (1990) admits that many factors, notably internal political strife, overpopulation and adverse environmental change could have been important. Hall's (1994) review of archaeological work at York paints a picture of a thriving city that relied on the food resources of its hinterland and which became a major manufacturing and commercial centre. York's importance is clearly reflected in the *Anglo-Saxon Chronicles*. Its role as a trading centre is reflected in the wealth of coins from many different sources and the rich material culture including amber, steatite, textiles and jewellery imported from northwest Europe and the Baltic region. However, air quality in the city may have been poor, as has been suggested by Lewis *et al.* (1995), whose analysis of skeletons from a cemetery at St Helen-on-the-Walls reveals a high incidence of maxillary sinusitis when compared with skeletons from a nearby rural area. When the Normans arrived in England in 1066 they found a landscape that O. Rackham (1994a) suggests was less than 15 per cent wooded and which experienced a further 5 per cent reduction by the time of the Black Death in 1348. The Domesday Book of 1089 provides much information on the character of the medieval English Landscape. Woodland apart, 35 per cent of England was arable land and a further 30 per cent was pasture, the remainder consisting of non-agricultural land such as moorlands and heathlands as well as settlement. Many changes occurred in the ensuing 400 years before the first Tudor monarch, including the rise to power of the monasteries and their involvement in sheep farming as well as substantial population increases in the twelfth and thirteenth centuries. The village became the most common form of settlement, though additional urban centres began to rise to prominence. Additional land was brought into production, often under the guidance of Cistercian monks, who first came to Britain in 1128. Many of their monasteries were situated in remote upland regions which they used for cattle and especially sheep grazing. Sheep provided the much-needed raw materials for the developing wool industry. As Bechmann (1990) points out, the Cistercians were not the only stockbreeders. The Templars, soldier-monks who employed estate managers, also cleared land in the vicinity of Chartres, France, creating meadows to grow winter feed for animals. The Cistercians also favoured the maintenance of some forest cover to provide a tripartite system comprising fields for arable cultivation, stockbreeding and forests. The forests also provided grazing opportunities on a controlled basis.

The heightened pressure on the landscape as population increased and agriculture expanded caused legislation to be passed to protect certain land areas from encroachment by agriculture; the protected land included many forested areas. Examples are the royal forests which were designated as such by William the Conqueror and his successors. Such areas, one notable example being the New Forest, were not necessarily well wooded. Bond (1994) states 'The medieval *Forest* was an area circumscribed by defined boundaries and administered by special officials, within which the king reserved for himself the right of hunting animals, particularly deer and wild boar'. Moreover, a forest would not necessarily have been entirely Crown land; it may have included land owned by a variety of individuals and estates. Bond affirms that Hampshire was the most extensively forested county in England, including as it did some 40 000 hectares of the New Forest, probably the most important of the Wessex royal forests. He implies that its large extent was due to the relatively low agricultural potential of its Tertiary sand, gravel and clay soils. There were at least 10 other forests in Hampshire, including Melchet, Pamber and Alice Holt.

The New Forest exemplifies many of the characteristics of medieval royal forests. It comprised extensive common land, open woodland and coppice enclosures which are still in evidence, as are the peatlands of the heavy clay soils and the heathlands of the sand and gravel areas that were cleared during the Bronze Age. Some areas may have experienced a continuity of woodland cover since the wildwood of 6 K years BP (Section 3.7), though the character of the woodland has changed immensely. In particular, lime *(Tilia cordata)* has declined as an important component. O. Rackham (1994b) points out that place names like Lyndhurst are a reminder of lime's former pre-eminence. This is the case for many woods in the south and east of England. Moreover, the sweet chestnut *(Castanea sativa),* introduced into Britain by the Romans (Section 5.2), has proliferated. Changes in the occurrence of hazel *(Corylus avellana)* and beech *(Fagus sylvatica)* have also taken place in response to management, notably coppicing and grazing, and to competition. In the majority of Britain's forests, including the royal forests of Scotland e.g. Ettrick and Darraway (Marren, 1990), management was intensive as both the Crown and the people derived revenues from the woodlands, which were coppiced and pollarded to provide a regular harvest, and from the pannage and herbage they provided for pigs and other domesticated animals (Bond, 1994; Marren, 1990). The crown also derived income in the rent from tenant farmers, stud farms and various metal industries. Marren asserts that the iron-smelting industry of the Forest of Dean produced 50 000 horseshoes for Richard the Lionheart's crusades in the early thirteenth century. These crusades in the Holy Land also caused substantial exploitation of woodlands elsewhere in Catholic Europe as the need for ships and charcoal to produce ironwork increased (Bechmann, 1990). Woodlands were therefore valuable resources that were intensively managed and subject to many conflicts of interest, conflicts which caused a substantial reduction in woodland cover between 500 and 1500.

Other parts of Britain also changed in character as additional land for agriculture was sought. Many wetland areas, for example, became the focus of reclamation schemes. Before ca. 800 settlement had avoided low-lying marshy areas despite the fact they contributed to local economies through the provision of pasture for cattle and sheep. However, the situation changed in the period post 800, leaving

a legacy that persists to the present day. O. Rackham (1986, 1994b) considers that Anglo-Saxon and Norman times witnessed the second period of East Anglian fen drainage from ca. 700, the first occurred during the Roman occupation (Section 5.2). Sea banks and fen banks were constructed to curtail flooding and protect villages, land was subdivided to reflect ownership, inland lakes were important sources of fish and canals and ditches were dug as drainage channels. Rackham states that similar enterprises were being undertaken in other fenland areas such as Romney Marsh, the Somerset Levels and Canvey Island. After the Norman Conquest settlement towards the sea into saltmarsh and inland into peatlands intensified in all these regions. In particular, East Anglia's fenlands rapidly became prosperous as a result of these efforts, financed by fenland abbeys, the Bishop of Ely, private landowners and parishes. The extensive excavation of peat from the fenlands, in the period 1100–1300, mainly for domestic fuel in the absence of major woodlands, produced large pits which subsequently became waterlogged to produce the Norfolk Broads. The prosperity of the region is reflected in the Lay Subsidy Agreement of 1334, which states that 31 of the 106 richest places in England were situated in the seaward fens (Rackham, 1986). The economy was based on animal products, notably meat, butter and wool. Reeds and sedges were used for thatching and fuel.

Prosperity, however, was to be short-lived for the British countryside, despite the fact there was an extensive road network and considerable trade with Europe. The abandonment of villages and hamlets in the fourteenth century has been attributed to a complex set of factors, including the Black Death of 1348 and the conversion of arable land into sheep pasture for increased wool production, leading to evictions. It is also likely the onset of the Little Ice Age (Sections 3.8 and 3.9), which brought a reduction in the growing season and increased crop failures, played a significant role in village abandonment. Whatever the cause, an economic recession ensued.

5.4   The period 1400–1750

By the time Henry VII ascended to the English throne in 1485, to become the first Tudor monarch, an economic recovery had taken place. The widespread decrease in population that occurred throughout Europe due to the Black Death reduced pressure on the land and created labour shortages. Surviving peasants found themselves in demand and their wealth increased as they capitalised on vacant tenancies, etc. Many grew sufficiently wealthy to establish a new yeoman farmer class. Together with entrepreneurial activities by the Church and the Crown, they fuelled an economic recovery. The yeoman class emerged as a new landowning and land-leasing elite in which lawyers and merchants were just as prominent as those who were royally connected. Many architecturally impressive residential buildings were constructed to house this new elite and many of these buildings persist in the modern landscape as a reminder of this affluence.

New urban centres of importance also emerged, including Norwich, Bury St Edmunds and Exeter which were associated with cloth manufacture. This was a major growth industry; it developed as a result of the established wool production of the earlier period (Section 5.2). Pounds (1990) reports that the industry began on a family basis in rural communities but by 1100 it was established in small urban centres throughout western Europe. By the 1400s wool had become Britain's most important export. Both course homespun cloth and fine wear were produced, and linen and silk weaving were beginning to appear in parts of Europe, notably Italy. Cloth replaced wool as Britain's major export and a merchant class was well established by the late fourteenth century (Unwin, 1990). The West Country, East Anglia and the West Riding of Yorkshire emerged as primary centres of cloth production. Ramsay's (1982) analysis of the wool industry shows that London and the Crown derived considerable revenue from an export tax on cloth and from import taxes on commodities such as dyestuffs. Some of this revenue was reinvested to improve infrastructure, notably road and canal construction. For example, the rivers Aire and Calder were deepened in the early 1700s to provide for the transport needs of the Yorkshire wool industry. The importance of the trade to Britain is also reflected in the fact that even during the Civil War of 1642–1645 it was not appreciably disrupted.

Further industrial innovations occurred during the fifteenth and sixteenth centuries. Many were based on metallurgy and most were rurally based, taking advantage of water power, the availability of charcoal from woodlands and the still largely dispersed rural population. Such innovations

occurred throughout much of Europe which, as Pounds (1990) points out, benefited from technical advances. Among the most significant was the replacement of the wind furnace, which had to be positioned to receive a draft, by the hearth or bloomery which relied on bellows for draft. The bellows were worked by waterwheels, so iron production centres came to be located where water power was available. A further innovation was the modification of the bloomery to produce the blast furnace which facilitated the production of increased quantities of smelted iron. This innovation intensified pressure on woodlands because furnaces required large volumes of charcoal. Bechmann (1990) states: 'With the techniques of the Middle Ages, eight to ten kilograms of wood were necessary to obtain one kilogram of charcoal that provided as many calories as 1.67 kilograms of wood ... in the Middle Ages, using wood in the form of charcoal led to consuming five times more wood than if it had been burned directly. To make one ton of refined iron, over 65 cubic meters of wood were needed. A good coppice produced on one hectare about 84 cubic meters every sixteen years'. Bechmann also points out that this reliance on wood often prompted forge masters to purchase forests so they did not have to buy in charcoal, and so incur transport costs.

The increasingly technological production of iron resulted in the establishment of specialist centres, e.g. the Weald. Its pre-eminence was assured by resources comprising abundant ironstone, incised rivers that could easily be dammed to generate water power for bellows and hammers, and abundant charcoal, along with the region's proximity to London. The Forest of Dean, a significant Romano-British centre (Section 5.2), and the Cleveland Hills, also became major centres of iron production. Other metals were also exploited on an increasingly commercial basis. For example, in Derbyshire the lead mining, which began in Roman times, intensified; the history of this industry is reflected in the chemical record of local peat deposits (e.g. Livett et al., 1979), which show an increase in lead deposition from ca. 1400 onwards and huge increases from 1600 to 1700. Similarly, the tin industry in Cornwall, glass-making in the Midlands and salt production in Cheshire became established. Although such activities were on a relatively small scale, they contributed towards the industrial base that eventually gave rise to the Industrial Revolution. Analagous developments occurred throughout much of Europe, notably in France, Germany, Italy and Switzerland.

Bechmann (1990) makes the observation that ashes, the by-product of wood combustion including charcoal production, were not considered as waste products. Ashes contain substances including potash, phosphates and alkalis and may be used for many purposes, including soap or detergent manufacture, in glass-making and even for pastry making. Phosphates are also valuable as fertiliser. Like the production of charcoal, ash collection was a lucrative business and provided a source of employment. Although not widely practised in Britain, probably because of the dearth of woodland by this time, the management of some woodlands in Europe was directed specifically at potash production. One example is the forests of the middle Volga Basin in Russia, reported by Berdnikov (1995). Here the state treasury controlled potash production which generated considerable income. Specific trees were used, namely oak, elm, lime and maple. The significance of the industry is reflected in the fact that in 1620 the forest extent in Volga District amounted to 77.4 per cent but as this declined, the profits of the potash business declined. This caused several inventories to be made, as maps, which reflect the intensity of woodland exploitation for what appears to be an extravagant use of wood.

However, the Industrial Revolution of the eighteenth century would not have been possible without the development of coal mining. Deforestation over the preceding millennia and the ineffective attempts of the Norman and later monarchs to establish conservation measures had resulted in a dearth of fuel resources that were so essential to the developing manufacturing base. Glennie (1990) reports that by the 1550s the northeast of England was the only region producing coal to any great extent. Thereafter annual coal production expanded substantially from 40 000 tons to 800 000 tons by the early 1600s. By the early 1700s a massive 1 400 000 tons were being produced. It was transported by river to the coast and then by sea to London. Indeed by 1600 most of the coalfields worked today were in use, though on a localised scale. This new source of fuel allowed metallurgical industries to shift from the traditional areas like the Weald to areas such as south Wales, Shropshire and the Midlands. Consequently, many local landscapes became 'industrialised' and the population shifted increasingly from agricultural to industrial activities, though still on a local scale in a predominantly rural

environment. Glennie states: 'At micro-scale, much industry took place in small domestic units alongside agricultural pursuits. Interactions between agriculture and industry within households continued to influence industrial change'. Coal was rapidly adopted as a fuel for a wide range of industries, notably iron production. Those coalfields with easy access to cheap water transport underwent the most rapid development. In addition to northwest England, mentioned above, the Shropshire coalfields developed on the basis of their access to the River Severn, those in Nottinghamshire benefited from proximity to the River Trent, and the coalfields of south Wales had access to sea transport. The excavation of deeper mines and the requirement for increasingly sophisticated equipment, notably pumping equipment, necessitated increased investment of capital. This was raised by groups or cartels of entrepreneurs who became the controllers of the industry, which became divorced from agriculture as coal-extractive industries on a large scale were created. Similar developments occurred in Scotland as industry began to emerge from agriculture to become a separate entity and to congregate in specific regions such as the area between Glasgow and Edinburgh. In general, the most important emerging industries throughout Britain were those concerned with coal, iron, textiles and salt.

Agricultural practices changed between 1400 and 1750 in England (Campbell and Overton, 1993) and Europe. Agriculture was affected by the Little Ice Age (Sections 3.8 and 3.9), especially by the occurrence of severe winters. Pounds (1990) asserts that wine yields were reduced in Provence and the altitudinal level of cultivation in northern Europe was lowered, as was the tree line. Grove's (1988) review of the Little Ice Age refers to farm abandonment in southwest Scotland and Dartmoor. She suggests that by 1500 summers were ca. 0.7 °C cooler than during the previous 200 years and caused a reduction in the length of the growing season. This would have been crucial in marginal upland areas but in lowland England the growing season may have been reduced by as much as five weeks. In addition, Lamb (1995) refers to the records of famines and food shortages in Scotland which occurred regularly between 1550 and 1600 and which prompted migrations to Ulster and England; in 1550, 1551 and 1554 harvests were poor and those of 1555 and 1556 were worse.

Thirsk (1987) presents evidence for this period as one of significant agricultural innovations which underpinned the so-called Agricultural Revolution of

Table 5.1 Wheat and barley yields 1600–1800 in selected areas

Approximate date	Norfolk/Suffolk	Lincs	Herts	Oxon	Hants	Midlands
Wheat yield (Bushels per acre)						
1600	11.4	11.8	9.7	15.0	11.1	NA
1700	17.6	15.6	14.6	21.6	10.5	NA
1800	22.7	24.3	20.7	21.9	21.5	20.7 (ave.)
Barley yield (bushels per acre)						
1600	NA	NA	15.5	14.5	15.4	NA
1700	NA	NA	28.0	18.3	16.9	NA
1800	NA	NA	36.0	30.0	30.0	NA

Source: data from Allen (1991), Clark (1991) and Glennie (1991)
NA = not available
ave. = average
1 bushel per acre = $90 \, dm^3 \, ha^{-1}$

1750–1850. Agricultural efficiency and productivity increased overall, especially after the Little Ice Age, and facilitated an increase in population from ca. 3 million in 1550 to 5.5 million in 1700 (Wrigley and Schofield, 1988). Moreover, Clark (1991) has demonstrated that grain yields were beginning to rise by 1550, increasing by ca. 50 per cent during the following 200 years as shown in Table 5.1. One major change which brought about this increase in productivity was the adoption of a three-year or four-year crop rotation, e.g the Norfolk four-course rotation, followed by grass leys of between 6 and 12 years (Overton, 1991). The Norfolk rotation involved wheat, turnips, barley and clover which were grown in succession. The inclusion of clover, a legume, increased the availability of nitrate in the soil through the symbiotic relationship between clover and nitrogen-fixing bacteria. The production of turnips, introduced originally from Holland, increased the availability of animal fodder, and increased numbers of animals produced much-needed manure. Other innovations that contributed to increasing productivity included the increased use of a varied range of fertilisers such as marl, lime and seaweed, the ponding of streams so the nutrients of water meadows were replenished by silt deposition, the introduction of new crops such as rape, lucerne and potato (Section 5.8), improved breeds of animals and the reclamation of wetlands to provide additional arable land.

Landscape change was also brought about in Britain by changes in land tenure, of which enclosure was particularly significant. This involved the enclosing of agricultural units by hedges or walls that replaced the open field systems that were so characteristic of the earlier Anglo-Saxon and Norman periods. The reasons for enclosure, as Yelling (1990) has discussed, were varied and included economic factors such as growing markets as population increased, changes in land use and coalescence of small landholdings into large landholdings. The rate and timing of enclosure varied considerably during the period 1500–1800 and much of it was enforced by Act of Parliament. Overall the enclosures reduced the amount of common land and led to a decline of the small farmer, a process that continued in the period post 1800. According to Rackham (1986), the hedges planted between 1750 and 1850 were at least equal in extent to all those planted in the previous 500 years, adding considerably to an already common feature of the British landscape.

The need for more arable land, as population expanded, also prompted the reclamation of wetlands. Once again interest turned to the fens as agricultural entrepreneurs recognised the potential of the area for crop production. Vermuyden, a Dutch engineer, was responsible for designing and constructing a new drainage network. This included diversion of the Great Ouse River which created the old and new Bedford Rivers in 1637 and 1651. As Rackham (1986) recounts, the scheme was not particularly successful either in terms of drainage or local popularity since it severely limited common rights and was constructed to benefit the investors rather than the indigenous populace. It also initiated environmental degradation, especially soil and peat erosion as the land surface shrank. Nevertheless, between 1637 and 1725 some 2500 km^2 of the Fens were drained and successfully converted into agricultural land. Further drain construction was also necessary to ensure efficiency, often in conjunction with windmills to pump drainage water into channels. The system is maintained today by electric pumps.

The changes in agriculture that were characteristic of Britain and Europe may have influenced atmospheric composition. There is global evidence from polar ice cores for increased methane production in the sixteenth century. According to Subak (1994) this was probably due to an increase in domesticated animal populations, biomass burning and irrigaton, though irrigation was most likely due to rice cultivation in the Far East. The fifteenth and sixteenth centuries were also periods of extensive exploration. Columbus, Vasco da Gama and Magellan are but a few of the notables who established the existence of hitherto unexplored lands. By 1700 several European nations had established colonies in Africa, southeast Asia and the Americas, where they had profound effects on the indigenous people and their landscapes, as will be discussed in Sections 5.7, 5.8 and 5.9.

5.5 Immediate consequences of industrialisation, 1750–1914

Between 1700 and 1800 the population of Britain increased from 5 million to more than 10 million and gross national product increased by an average of 2 per cent per annum between 1760 and 1830. Moreover, Pounds (1990) reports that the population of Europe in the early 1500s was ca. 68 million and by 1815 it had risen to 175 million, the increase occurring mostly in the 1500s (Section 5.4) and the late 1700s. Innovations in agriculture and industry accelerated. Britain became known as 'the workshop of world' as industry became the mainstay of the country's economy. Large-scale urbanisation began. This was a period of rapid change in terms of both social organisation and the landscape. The growth of urban centres, wherein industry was concentrated, encroached on the rural landscape and encouraged the movement of people from the countryside into the towns and cities. Britain was becoming an industrial and urban society instead of an agricultural and rural society.

Chapter 1 referred to Odum's (1975, 1993) classification of ecosystems based on energy-flow characteristics (Table 1.2) The final category of Odum's scheme is the fuel-powered urban-industrial ecosystem; the product of the Industrial Revolution, such ecosystems now dominate the developed world. On the one hand they are wealth-generating systems but on the other hand they generate waste and pollutants which take their toll on the wider environment. The population of urban-industrial systems is not directly engaged in food production, so it relies on solar-powered and subsidised solar-powered ecosystems for its food energy. Solar-powered ecosystems had therefore to become increasingly efficient in order to provide food in excess of the requirements of the food producers. All of these factors contributed to the shaping of British and European landscapes in the nineteenth and twentieth centuries.

The close relationship that existed between agriculture and industry is well illustrated by developments in Britain during the period 1700–1820. The rise of large-scale capitalist farming on a landlord–tenant basis, one outcome of the enclosure movement (Section 5.4), facilitated the development of urban centres because such agricultural systems were able to produce food surpluses as a result of the innovations described in Section 5.4. However, the relationship was mutual; industrial developments with their scientific underpinning had repercussions for agriculture. One of the most notable was the production of artificial fertilisers, which superseded the use of natural nitrates (guano) imported from Chile between 1820 and 1850. Agriculture began to become industrialised and the urban environment suffered a decline in quality through overcrowding and pollution due to coal combustion domestically and industrially.

In terms of the direct impact of industry on the landscape, it was the exploitation of natural resources that wrought some of the most extensive changes. In 1840 there was 250 active tin mines in Cornwall, and although the industry was to decline in significance during the next 30 years, the mining and smelting activities, and associated workers' houses, have left the vestiges of an industrial landscape in what today is primarily a rural area noted for its tourist industry. In fact, Cornwall in the middle of the nineteenth century had one of the most industrialised landscapes in Britain and the mines employed about 30 000 people. In the 1860s industrial decline had set in and, as Blunden and Turner (1985) comment, many of the industrial relics of this period are a testimony to the heyday of the tin industry. For example, the engine houses and tall chimney flues of two mine complexes at Botallack and Levant in West Penwith, which allowed tin to be recovered from beneath the seabed, remain although many of the waste tips have since become overgrown. However, as the tin mining industry declined, attention turned to Cornwall's china clay as the pottery industry, centred on Stoke-on-Trent in Staffordshire, expanded and required raw materials.

Changes in energy provision also brought about landscape changes, especially during the Victorian era (1837–1901) when there was a massive switch from water and wind power to coal. Between 1750 and 1850, water power sustained the early years of industrial development. Thompson (1985) states that in 1850 water power, facilitated by mill-races, weirs, dams and reservoirs, provided about one-third of all industrial power, declining to about 10 or 12 per cent by 1870 as coal-generated steam became the major source of energy. The demand for coal thus increased, creating a very different landscape to that which was harnessed for water power. Thompson describes these new developments as 'awesome, grimy and distinctive colliery landscapes of pithead gear, coal heaps, and a maze of tramroads and colliery railways usually accompanied by colliery villages and always by a thick layering of the surrounding country with coal dust'. As coal output increased from 27 to 230 million tonnes per year, landscapes such as this proliferated as existing coalfields, such as the Northeast, expanded and new fields were developed, such as south Yorkshire and Kent. Thompson suggests that some 60 701 ha of land were exploited for coal during the nineteenth century and were thus lost to agriculture. Data on the growth of coal production in Britain between 1750 and 1950 are presented in Table 5.2 and reflect the pre-eminence of the Northeast, Yorkshire, the Midlands, south Wales and Scotland.

These coal-rich areas were also the focus of industrial development, so industry, along with the people, moved out of the countryside. The population of Birmingham trebled during Victoria's reign to reach half a million by the first census of 1801; the largest rates of population increase in the period 1780-1831 occurred in the southeast and the industrial North, notably Lancashire and Yorkshire (Lawton, 1990). Overall, Britain's population increased by ca. 86 per cent. Land was annexed for urban spread and railway construction. Railways replaced canals as the major form of transport and provided an extensive communications network in all but the most inaccessible areas. Good communications and the increasing mechanisation of agriculture also led to the expansion of towns outside the coalfields. Many of them, e.g. Peterborough, Lincoln and Ipswich, were already important market towns

Table 5.2 Coal production by region, 1750–1938

	Northeast		Yorkshire		Midlands		Lancashire		Staffordshire		South Wales		Scotland		Cumberland		North Wales		Salop		Rest		Total (10⁶ tons)
	T	P	T	P	T	P	T	P	T	P	T	P	T	P	T	P	T	P	T	P	T	P	
1750-55	1.6	35.5	0.5	11.1	0.2	4.4	0.2	4.4	0.3	6.7	0.1	2.2	0.9	20.0	0.1	2.2	0.1	2.2	0.3	6.8	0.2	4.4	4.4
1771-75	2.0	30.3	0.8	12.1	0.3	4.5	0.3	4.5	0.5	7.5	0.1	3.0	1.3	19.7	0.3	4.5	0.2	3.0	0.4	6.1	0.3	4.5	6.6
1791-95	2.8	27.5	1.0	9.8	0.5	4.9	0.8	7.8	0.9	8.8	0.8	7.8	1.7	16.7	0.5	4.9	0.3	2.9	0.4	4.0	0.4	4.0	10.2
1811-15	3.7	21.3	1.8	10.3	1.2	6.9	1.6	9.2	2.4	13.8	1.8	103	2.5	14.4	0.6	3.4	0.4	2.3	0.7	4.0	0.7	4.0	17.4
1831-35	6.1	19.3	3.6	11.4	2.2	7.0	4.5	13.9	4.0	12.6	4.0	12.6	4.0	12.6	0.7	2.2	0.7	2.2	0.8	2.5	1.0	3.7	31.6
1854	15.4	23.8	7.3	11.3	3.9	6.0	9.9	15.3	7.5	11.6	8.5	13.1	7.4	11.4							4.7	7.3	64.7
1875	32.3	24.2	15.9	11.9	12.4	9.3	21.0	15.8	14.5	109	14.2	107	18.6	14.0							4.4	3.3	133.3
1913	56.4	19.6	43.7	15.2	38.8	13.5	28.1	9.8	14.9	19.8	56.8	19.8	42.5	14.8							6.2	2.2	287.4
1938	44.7	19.7	42.4	18.7	37.8	16.7	14.3	6.3	13.4	15.5	35.3	15.5	30.3	13.3							8.8	3.9	227.0

Source: based on data compiled from various sources in Lawton and Pooley (1992)
T = total regional production (10⁶ tons)
P = percentage of British production

and increasingly became involved with the provision of agricultural implements. Moreover, the increases in urban populations created environmental change within the urban centres themselves, where pollution due to coal burning and mineral processing was developing on a large scale. Clapp (1994) has reviewed the environmental impact of industrialisation in Britain, which he reports was at its worst in the late 1800s. He also reports that London received much less winter sunshine than in county towns in southern and eastern England and that vast volumes of soot were deposited in industrialised towns. Some examples are given in Table 5.3, which shows that Sheffield and London were particularly badly affected. Clapp comments: 'It is not easy to grasp the meaning of 55 tons of solids falling per square mile per month. To the inhabitants of Attercliffe it meant eight ounces of dust falling every year on every square yard of the district'.

The increases in population necessitated large-scale building programmes, which were not controlled as they are now by local planning authorities. Haphazard housing and industrial development meant that a high proportion of urban dwellers lived in squalor in overcrowded accommodation with little sanitation. Hare (1954) cites the situation of one street in Leeds where only three toilets were available for 368 people and the plight of 200 000 people living in Liverpool cellars to whom no sanitation was available. Although running water, baths and water closets were available by the 1830s, they were only affordable by the upper-middle and upper classes. Consequently, both excreta and domestic rubbish accumulated in the streets, generating a considerable health hazard. This, Boyden (1987) indicates, contributed to a much higher mortality rate in urban centres than in rural areas and to a high incidence of typhus that

Table 5.3 Mean monthly deposits in four locations of Britain for the period 1914–1916

| | Mean monthly deposit (tons per square mile) | | | | |
	Soot	SO_3	Cl	NH_3	Total solids
Malvern	0.4	1.1	0.5	0.04	5
Attercliffe, Sheffield	9.6	7.2	4.0	0.20	55
London (8 stations)	5.9	5.5	1.7	0.30	38
Manchester	4.3	4.8	1.3	0.13	32

Source: based on data from Shaw and Owens dated 1925, quoted in Clapp (1994)

eventually precipitated the Public Health Movement of nineteenth-century Britain, prompted also by several outbreaks of cholera between 1830 and 1862. The 1875 Public Health Act established a government Department of Health to oversee legislation relating to sanitation.

5.6 Rural changes after 1750

The generation of fuel-powered urban-industrial ecosystems led to an increasing dependency of a large proportion of the population on less-populated food-producing systems (Section 5.5). Consequently, subsistence farming declined and agriculture became more commercialised or industrialised, thus emphasising the close link between agriculture and industry. Although the urban centres relied on rural society for their food, farmers became increasingly reliant on industry to produce the necessary components to increase the efficiency of food production, dissemination and marketing.

In 1851 the size of the agricultural labour force had reached a peak. In conjunction with horticulture and forestry, it employed more than 2 million people, more than 20 per cent of Britain's population (Mingay 1981). By 1896, however, only 1.5 million people were engaged in the food and wood production, representing less than 10 per cent of the population. There were also a sustained population growth in general between these years, from 17.9 million in 1851 to 32.5 million in 1901, although this expansion in population began considerably earlier, from 6.7 million in 1761 to 15.9 million in 1841 as industrial development occurred (Lawton, 1990). Thus the development of the fuel-powered urban ecosystems of Britain prompted considerable changes in agricultural practices. During the eighteenth and early nineteenth centuries there was a marked increase of land under cultivation with a concomitant reduction in wasteland, some of which was used to expand arable production and the remainder for pasture. The introduction of new crops and new rotations, etc., that began in the middle of the seventeenth century (Section 5.4) provided a base for agricultural intensification during the period 1750-1850, though Grigg (1989, 1992) suggests that British agriculture was under pressure during this period, especially during the Napoleonic Wars of 1793–1815, when wheat harvests were poor. It was during this period that the potato, grown earlier in limited amounts for fodder, became a significant food crop. Moreover, between 1780 and 1801 the price of wheat trebled, which also implies that British agriculture was unable to increase productivity in line with the rapidly increasing population. Poverty rather than affluence appears to have been prevalent in the worker classes of rural societies and industrial output, rather than agricultural output, was responsible for the increased wealth of Britain, though much of the finance came from the landowning elite.

Nevertheless, British farmers were producing sufficient carbohydrate and protein to support ca. 80 per cent of the population in 1860. The increased production was stimulated by mechanisation, which began in the 1800s, and guano imports for fertiliser. According to Walton (1990), only 4 per cent of the meat consumed in Britain during the period 1851–1860 was imported. Imports of food were, however, growing; the repeal of the Corn Laws in 1846, which had hitherto protected British agriculture from cheap imports, later resulted in free trade in foodstuffs. Despite a decline in the rate of population increase between 1850 and 1900, the absolute increase in population produced a considerable expansion in food imports between 1860 and 1910. By 1910, 85 per cent of grain for breadmaking was imported, along with 45 per cent of meat, much of which came from North America and Australia. Moreover, during Victoria's reign the amount of wool supplied to the wool industry from abroad, especially Australia, increased to 80 per cent by ca. 1900. There was also a general trend towards importing the basic raw materials which underpinned British industry, especially jute, rubber and leather. Both the needs of the population and industry itself helped to diminish the significance of British agriculture. The increasingly affluent standard of living afforded by industrial occupations led to a demand for animal products such as butter and cheese, and although British farmers recognised this growing market, imports of dairy produce from Ireland, Europe and North America were necessary to satisfy it. Grigg (1995) considers this change in diet as a nutritional transition, i.e protein-rich food replaced starch-rich food. Industry itself also contributed to agricultural change; for example, butter substitutes based on imported oils and fats began to replace butter in the diet of many, especially the poor.

The 1880s witnessed a fundamental change in agriculture. Before then wheat was the most important food crop and output was enhanced by increasing manuring and applying artificial fertilisers. Bad harvests in the late 1870s and the growing importation of cereals led to a change in policy. Increased emphasis was placed on animals and animal products; grazing became predominant in western Britain, as arable lands were set to permanent grass, whereas eastern Britain remained reliant on cereal production. Although the increasing use of cereals for animal feed also meant that many cereal farmers focused their efforts on animal products (Grigg, 1987), the quality of life diminished for cereal growers. The extensive railway network of the Victorian era also facilitated the transportation and rapid dissemination of animal products, especially milk, to the people of towns and suburbs.

The period 1750–1900 also witnessed tremendous changes in the distribution of population. This has been discussed by Lawton and Pooley (1992), who cite various regional trends and implicate rural poverty, the demand for labour from the developing industrial base (Section 5.5) as well as changes in mortality rates. Overall there was a general movement of people out of the countryside into the towns. The agricultural labour force was at a peak by 1850, and although the number of actual farmers changed little in the ensuing 60 years, the number of labourers fell by 23 per cent by 1911, representing only 8.5 per cent of the total occupied labour force in comparison to 21.5 per cent in 1850. The low wages, long working hours and the increasing use of machinery all contributed to the decline in rural populations.

The latter half of the nineteenth century also witnessed a closer liaison between agriculture and science. Plant and animal breeding were well established before the First World War, crop protection chemicals were being produced and artificial fertilisers were being used. All of these innovations were designed to increase the output of plant and animal products as they are today. Mechanisation also became more sophisticated; the first tractors appeared in the United States in the 1890s and were subsequently employed in Britain during the First World War. All these developments have provided the base for modern British agriculture and for the developed world in general. Although agricultural systems may vary considerably from region to region and nation to nation, the common denominator is a much enhanced input of fossil fuel energy both directly and indirectly. Thus coal and oil have become the mainstay of these agricultural systems, replacing human and animal manuring, just as coal and oil underpin much of industry. As will be discussed in Chapter 7, the 'industrialisation' of agriculture, so effectively adopted by the Victorians, has also brought about many indirect environmental changes which are now developing into environmental problems.

Agricultural practices have also altered the character of the countryside, especially since the Second World War. This prompted increased intervention by the state in order to ensure self-sufficiency in food production and was manifested in the 1947 Agriculture Act which guaranteed specific prices for certain crops, especially cereals. In consequence, the arable acreage increased. This support for agriculture is continued today under the auspices of the EU's Common Agricultural Policy. This involves the purchase of produce by intervention agencies to maintain prices, as well as the provision of grants for the reclamation of marginal land and for mechanisation. Thus agriculture has become commercialised and innovations to increase yields have become legion. Hedgerows have been decimated in may parts of Britain, especially in eastern England, to create large fields that are more suited to highly mechanised farming that has also become more monocultural. Mercer and Puttnam (1988) point out that approximately half of Norfolk's hedgerows disappeared between 1946 and 1970, and in East Anglia in general hedgerows disappeared most rapidly between 1969 and 1985, more rapidly than in any other part of England (Section 7.2). Virtually nowhere in Britain has escaped changes in the environment as a result of postwar agriculture. Moreover, such changes have created a considerable conflict of interest between conservationists and agribusiness interests, as will be discussed in Chapter 7.

5.7   Changes in Africa following European settlement

The annexation by the Portuguese of the fortress town of Ceuta from the Moroccans in 1415 marked the beginning of European colonialism in Africa. Assisted by the geographical and navigational expertise developed by Islamic scholars and the production of improved maps, the fifteenth century became an era of exploration dominated by Europeans, particularly the Portuguese, Spanish, Venetians and Genoese. Contact was made with India and China as well as Africa and trade routes were expanded. By the end of the nineteenth century, Europeans were in control of almost the entire continent of Africa. The Portuguese were joined by the Spanish, Dutch and British, whose primary interest in Africa lay in the slave trade that supplied labour to European colonies, particularly North America and the Caribbean. Despite the demise of slavery by the middle of the nineteenth century, there was still a considerable European interest in Africa, and the latter half of that century saw European countries formally annex African territories in their desire for political as well as commercial influence. Since then the majority of these colonies have become independent but the legacy of European influence remains.

Land degradation, a major problem in Africa today, has often been considered a direct outcome of colonialism. Vail (1983), for example, argues that before the 1880s there was an overall balance between African populations and their environment, a balance that was disrupted by colonial influences such as introduced disease, warfare and especially the imposition of unfamiliar people/land relationships. Unfamiliar relationships include the introduction of new crops, new agricultural techniques and the direction of profits into Europe. However, Blaikie (1986), equating colonialism with capitalism, has argued against this by suggesting it is too easy to use colonialism as a scapegoat for all of Africa's environmental problems, old and new. He also draws attention to the fact that such problems can be just as symptomatic of Marxist and Communist agricultural systems and, more importantly, he highlights the social factors which in many cases underpin land degradation problems and which must be considered in tandem with environmental measures if effective amelioration programmes are to be successful. Such comments are intended not to exonerate colonialism as significant in Africa's land degradation problems but to suggest it is just one of many causative factors. Moreover, it is important to point out that many colonial governments were instrumental in initiating conservation programmes, with varying degrees of success, the primary aim of which was to safeguard the environment. The examples which follow provide a basic introduction to some of the problems involved.

Although much environmental change is associated with colonial policies in the late 1800s, there is evidence for land transformation considerably earlier than this period. For example, Lindskog (1995) has compiled evidence for the degradation of the Cape Verde Islands off the coast of Senegal. He has examined written accounts by the explorers who first discovered the islands ca. 1455–1460. The reports of two captains, Cadamosto and Gomes, imply they found well-wooded islands with fresh water. Today, however, the Cape Verde Islands are desertified and degraded. The onset, and possibly the majority of this degradation, may have been the result of late fifteenth and early sixteenth century Portuguese settlers who were engaged in the slave trade. They introduced goats, donkeys, rabbits and cattle which rapidly became uncontrolled. Goat populations in particular increased rapidly and were encouraged to do so because they provided skins for export to the home market. Reports of travellers reflect the significance of goats: one such report in 1566 asserts that 40 000 goat skins were exported in only one year from only one island. This reflects a high-density population and probably reflects a major cause of degradation in an environment with unstable slopes of friable volcanic rock. Moreover, Lindskog refers to the system of land tenure at the time which, through the inheritance system, brought pressure on the land. Land could only be passed on to the eldest son which eventually created a landless class. The farmers were in any case only tenants of a landowning elite who had no responsibilities to their tenants or their land so their approach was largely exploitative rather than sustainable.

However, the period post 1880 was particularly significant in Africa's history because there was a major expansion of European influence. This was prompted by the opening of the Suez Canal in 1869 and the simultaneous 'diamond rush' in South Africa. The ensuing political subdivisions within Africa are given in Figure 5.3, which highlights the significance of many European countries as imperialist and colonial powers. Many of the changes that ensued in the next 50 years have been examined by Blakie and Brookfield (1987), who suggest that a major reason for land degradation was the social disruption caused by the annexation of land from native farmers by European planters and ranchers. The production and export of cash crops such as tea, coffee, sugar, rubber and cocoa were the linchpins of colonial economies. Griffiths (1994) points out that the need to generate profits and to guarantee supplies of tropical products were the major stimuli to the establishment of plantation agriculture. Even today this type of agriculture is an important source of foreign exchange. However, the annexation of land for plantations caused local problems, including the withdrawal of land for traditional agriculture and land shortages. This was particularly the case in Kenya and Zimbabwe. Palmer (1977) reports that 6 million hectares of Zimbabwe, nearly 17 per cent of the entire country, were managed by Europeans after 1890. Moreover, native reserves were created in order to limit indigenous groups to specific, often poorer, areas. Even today, the so-called General Lands, commercial farmlands often owned or managed by white farmers, are less susceptible to erosion than the Communal Lands, where subsistence farming is practised and population densities are high (Whitlow, 1988). Additional causes of social disruption included the spread of diseases such as rinderpest, a debilitating and often fatal disease of cattle which was initially introduced into Ethiopia during the 1890s in infected cattle from Italy.

The creation of overpopulation in specific areas caused changes in traditional agricultural practices, which in some cases accelerated land degradation and eventually caused a decline in crop productivity. Sometimes the length of the fallow period was reduced and nutrient reserves were reduced as a result. This problem is exemplified by the impact of colonialism on the Swaka people of Zambia (Robinson, 1978). Moreover, the imposition of taxes on native people, a rare occurrence in indigenous cultures, led subsistence farmers to grow cash crops in order to raise cash at the expense of local grain crops, etc. In Tanzania the production of coffee by smallholders was encouraged by the British administration in the early 1900s (Berry *et al.*, 1990). Similar practices were encouraged elsewhere as Berry *et al.* have reported for Ruanda and Burundi. Here the Belgian administration not only advocated cash cropping but also introduced so-called *paysannat* schemes involving forest clearance and rural resettlement to make way for coffee plantations. The vegetation cover, rural population distribution and indigenous agriculture in parts of Nigeria were also affected by the decisions of the British colonial government. Udo *et al.* (1990) point out that, although forest reserves were created to conserve Nigeria's overexploited and rapidly diminishing wild-rubber resources, the denial of land to rural populations in some areas gave rise to overpopulation which caused land degradation. The

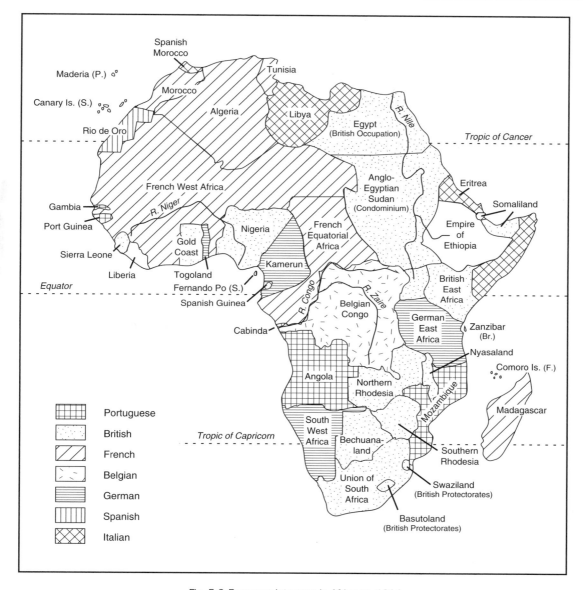

Fig. 5.3 European interests in Africa ca. 1914.

introduction of cash cropping caused additional environmental change. Udo *et al.* state: 'Paradoxically, while the colonial government created forest reserves, it also initiated certain policies and measures that were to accelerate the destruction of unreserved forests. These measures included the introduction and promotion of tree-crop cultivation of cocoa, rubber and oil palm for export, the granting of timber-felling concessions to foreign firms, and the establishment of forest plantations of exotic tree species'.

Otieno and Rowntree (1987) also present evidence for degradation brought about by British colonialism on the Machakos and Baringo Districts of Kenya. Here, before the British arrived in 1890, shifting agriculture and pastoralism were the means of subsistence for the Kamba people of the Machakos Hills; a varied land use predominated in Baringo, from cultivation near the southern shores of Lake Baringo to pastoral subsistence in the bush country north of the lake, and in the Tugen Hills a system similar to the Kamba people's shifting pastoralism. Between 1896 and 1902 the British were responsible for constructing the Uganda railway, which benefited the Kamba and allowed their populations to expand as trade increased and food relief was brought in during times of drought. The agricultural innovations had far-reaching effects, including an expansion of the cultivated area. This was facilitated by the introduction of the hoe and plough to replace the digging-stick, the introduction of cash crops and improved seed strains to increase productivity. Reservations, or native land units, were introduced in 1906 to make way for scheduled areas which were set aside for European settlers and Crown lands. As population numbers increased in the native land units, greater environmental pressure was exerted and exacerbated by the expansion of cultivated areas and the introduction of a taxation system. The outcome of all these changes was a reduction in grazing land and increased exposure of cultivated land to sheet-wash erosion.

In Machakos District the impact of erosion was obvious by 1917. Non-vegetated land areas and exposed soils became common, despite the discouragement of shifting cultivation and an embargo on the cultivation of steep slopes. Although afforestation programmes were initiated in the 1920s, as well as land enclosure and terracing to combat erosion, degradation remained a stark reality affecting between 37 and 75 per cent of the land. Similar problems developed in the Baringo District, especially in the Tugen Hills, which Otieno and Rowntree (1987) believe contributed to the movement of people into the south Baringo plains in the 1920s. Both social and environmental problems were exacerbated by drought periods and pest outbreaks. Conservation measures in the 1930s, including the exclusion of cattle and goats from certain areas and the clearance of native bushland and grass planting, were ineffective at combating the problems and later attempts have not been overly successful either. Despite attempts to rectify the problems created by colonial superimposition, the colonial regime wrought not only environmental change but also environmental degradation and social upheaval. However, many improvements have been made in recent years and erosion has been halted in some areas accompanied by increased productivity (Section 8.4).

Colonialism also had a major impact on conservation policies in Africa. Conservation policies have been intrinsic to colonial attitudes since the 1890s and were developed in response to a growing awareness of environmental problems, especially in relation to soil erosion and forest resources. Grove (1987) has examined the importance of the colonial context in conservation policies in the Cape Colony. These involved forest protection and game reservation but failed to address the environmental degradation. These and similar policies were ineffectual because they failed to acknowledge the significance and the ecological basis of traditional shifting agricultural systems. Indeed Richards (1985) has suggested that improvements in these systems may well be the key to an indigenous agricultural revolution, as exemplified in west Africa today. Such factors went unrecognised by colonial governments, and they have remained largely unrecognised by post-colonial governments.

Where forest conservation in Africa is concerned, colonial regimes have achieved both successes and failures. In common with many conservation programmes, some forest conservation projects have at best achieved only partial success because of their inability to recognise and reconcile conflicts of interest, and their failure to take into account social factors such as native land rights. Anderson (1987) has examined the significance of these factors in the conservation history of the Lembus Forest in Kenya. Conflicts arose here because of three different sets of interests. Social indigenous groups wanted to continue their traditional exploitation, but commercial exploitation and the development of a timber industry was the prime concern of a business consortium via a concession which was granted with little restriction by the High Commissioner for East Africa in 1904. Moreover, the Forestry Department wanted to maintain both the forest and a financial return. Conflicts arose between the timber company and the Forestry Department because the timber company failed to supply plans for exploitation and because their policy to avoid clear-felling hampered reafforestation plans. At the same time, the generous terms of the concession enabled the timber company to avoid paying high royalties, so income to the government also remained low. Growing public concern about the exploitation and the improved financial status of the Forestry Department eventually led to increased official control by 1945.

As conservation measures became more effective,

other conflicts arose involving native rights. These were initially assured by the Coryndon Definition, a document drawn up by Governor Coryndon to define grazing, gathering and cultivation rights within the Lembus Forest. Although policies elsewhere were aimed at confining groups in reservations, the Coryndon Definition remained an anathema. Local groups ensured they used their rights to the full and the cultivation of Lembus Forest glades increased to such an extent that conservation activities were jeopardised. In 1956 the Forestry Department was forced to concede to the claims of the Tugen people, in a political climate that warranted conciliation between natives and colonial government. Thus it was agreed that control of the forest should revert to the Baringo African District Council when the timber company's agreement was terminated in 1959.

Although these examples of colonial policies are limited in number, they illustrate the conflicts of interest that existed in Africa in the late nineteenth and early twentieth centuries in relation to environmental resources. Despite the introduction of conservation programmes, land degradation was exacerbated by a disregard for traditional land-use practices and indigenous social factors. Today, much of Africa retains the legacies of these events but, as will be discussed in Chapter 8, modern resource exploitation and agricultural systems have generally failed to address these problems adequately.

5.8   Changes in the Americas following European settlement

It now seems certain that Viking seafarers were the first Europeans to discover the continent of America, probably in the tenth century. However, it was not until the late fifteenth century that European influence became significant following the four voyages of Christopher Columbus in the years 1492, 1493, 1498 and 1502; the details of these voyages, the associated inland explorations and the European political climate are discussed in Phillips and Phillips (1992). It was in the islands of the Caribbean which Columbus encountered on the first two voyages that European settlement was initially established. Here, in 1493 the first European colony of La Isabella was founded in what is now the Dominican Republic. It was not until 1517 that

attention was directed to mainland Central and South America. By this time mineral resources in the Caribbean Islands had become depleted and the indigenous Caribbean Amerindian cultures had been subjected by their Spanish masters to a system of forced labour. The populations of native indians, notably the Arawaks and Caribs, were also ravaged by diseases brought by the colonists. These factors prompted the Spanish to obtain their forced labour from peripheral islands and the mainland even before European settlement became widespread. This was the precursor to the slave trade that later extended to Africa.

The Spanish colonists increased their interests in the region by exploiting Mexican silver and bringing to an end the indigenous Aztec Empire in 1521 (Nostrand, 1987). A need for further trading and strategic bases prompted the Spanish to establish colonies in other Caribbean islands and in Florida. By this time the French made inroads into Atlantic Canada, with settlements in the maritime provinces that were based primarily on the fishing industry and fur trade, but which were relatively low key in comparison with the Spanish settlements. Harris (1987) states that even by 1700 there were only ca. 20 000 people of French origin who were widely scattered from Newfoundland to the Mississippi. Two thousand of them lived in Quebec, with a further 1200 in Montreal. According to Mitchell (1987) the British entered the North American arena relatively late, with some 250 000 settlers established by 1700 in small settlements dispersed along the Atlantic seaboard from southern Maine to South Carolina. The Spanish, French and British all brought distinctive traditions, social organisation and agricultural systems which, as populations expanded, impinged on the native Indian traditions, displacing traditional land use and social organisations. The environmental and cultural changes prompted by the colonisation of the Americas by Europeans were profound. However, the Americas had a significant affect in Europe, especially in agriculture, because of the introduction of American crops, e.g. maize and potatoes. Axtell (1992) summarises the exchanges between the Old and New Worlds as follows: 'After 1492 the world became a different place. Western Europe's *discovery* of and imperial thrust into the equally old world of the Americas set in rapid motion the final stages of the human and biological exposure of the Earth's constituent parts to each other and the tying of those parts together with nautical lines'.

Pre-Columbian cultures, as the indigenous peoples of the Americas are known, practised varied food-procurement strategies (see Section 4.6 for details of early plant and animal domestication) in regions that were as varied as the Arctic tundra of Alaska, the humid swamplands bordering the Caribbean and the high Andes. Many were agriculturists engaged in the production of crop complexes (Mannion, 1995a). Maize, beans and squash were the major crops being produced in Mexico, southwest North America and the Mississippi Basin (Smith, 1992, 1995; Doolittle, 1992). The environmental impact of many pre-Columbian agricultural systems is reflected in the palaeoecological record for parts of North America (Delcourt and Delcourt, 1991). The character of pre-Columbian agriculture and the environment in the Caribbean islands first encountered by the Spanish has been examined by Richardson (1992). The chief crops were cassava and sweet potatoes with subsidiary crops of peanuts, maize, beans and cotton. All were grown on temporary plots cleared of forest by burning, as occurs in to present-day 'slash and burn' cultivation. The removal of native farmers by the Spanish for enslavement altered traditional agricultural practices; food shortages ensued. Moreover, the enslaved were employed in the search for gold on Hispaniola. Overcrowding in the mines accelerated the death rate and forced the Spanish to find a labour force from the neighbouring islands of Jamaica and Puerto Rico and even further afield from the Bahamas. Food shortages and the ravages of disease resulted in high mortality rates and a rapid decline in aboriginal populations. Introduced cattle, sheep, pigs and goats rapidly increased in numbers with some becoming feral. The feral animals reduced the natural vegetation cover and trampled soils, and vast herds of domesticated cattle, and later horses, occupied the subtropical grasslands. Richardson relates that, although the domesticated animals introduced by the Spanish thrived, many of the crops they introduced were much less successful, e.g. wheat, grape and olive. Consequently, the Spanish exacerbated the problems of food shortage by demanding tributes from the remaining natives.

Of particular note is the introduction of sugarcane to Hispaniola by Columbus from the Canary Islands. Its rapid success encouraged extensive deforestation within a few decades. Today it is a mainstay of many economies of Caribbean islands. The demand for sugar in Europe was a major stimulus to this plantation agriculture, which necessitated large-scale deforestation. The environmental impact of this deforestation is summarised by Richardson: 'The abruptness of this change was unprecedented, and it represented a sharp ecological discontinuity with the past, local reports of soil erosion and similar environmental stress occurred almost simultaneously with the clearing, but Caribbean environmental decisions were no longer being made in the Caribbean itself. The islands suddenly had been absorbed into an expanding European-centred commodity exchange of trans-Atlantic scope'. The introduction of sugarcane into other parts of mainland America had similar environmental repercussions and Europe's American colonies all suffered from distant, often ill-advised and ill-informed decisions.

Not long after the initial discovery of the Caribbean Islands, the Spanish turned their attention to the mainland. By 1520 Hernán Cortés had annexed Mexico for Spain, subjugating the native Aztecs in the process, and by 1532 Francisco Pizarro did the same for Peru and the Incas. In both countries the lure was gold and/or silver; in both countries aboriginal populations were exploited to provide labour and were considerably reduced in number by European diseases, e.g. smallpox. The annexation of land by the Spanish was accompanied by the requirement of a tribute from remaining native farmers. The impact of the Spanish colonisers on parts of Mexico has been described by Whitmore and Turner (1992), who point out that 'hybrid' landscapes developed as a result. There is now palaeoenvironmental evidence to show that pre-Columbian land use was not always environmentally sound. In several Mexican lake sequences Metcalfe *et al.* (1994) and Metcalfe (1995) have identified disturbance episodes that predate the advent of the Spanish. On the other hand there is sufficient evidence to show the environmental impact of the Spanish was considerable.

In the area of Mexico bordering the Caribbean and adjacent Piedmont rain-fed agriculture, involving terraces and earth embankments, was practised by the indigenous Totonac people (Whitmore and Turner, 1992), who also kept orchard-gardens to produce fruit and vegetables. In the Mesa Central, with its semiarid volcanic basins, rain-fed terraced cultivation was practised on slopes below the tree line; in the basins themselves, impaired drainage produced extensive wetlands, parts of which were cultivated in many ways to produce a range of crops, including maize, squash, amaranth, beans, sage, chilli and tomato (Sluyter, 1995). These agricultural systems and land ownership patterns were altered substantially with the arrival of the Spanish. European crops were introduced and indigenous populations were considerably reduced for the same reasons as those of the Caribbean Islands (see above). Notably the decline in population caused terracing, etc., to fall into disrepair and agricultural land to be abandoned. Whitmore and Turner (1992) point out these developments actually assisted the Spanish in their annexation of land. The introduction of livestock and its subsequent population explosion caused widespread environmental degradation and contributed to a global increase in atmospheric methane (Subak, 1994). The introduction of European technology, e.g. the plough, along with draft animals increased cultivation on flat land in valley bottoms. Whitmore and Turner (1992) also report that the settlers' demand for wheat bread led to widespread wheat cultivation in the Mesa Central and further north to the silver mines, e.g. Taxco near Mexico City. Some plantation crops were also grown, e.g. sugar-cane, cacao, cotton and tobacco, much of which was destined for the European market. The commerce was organised by merchants and by government officials, whose objective was maximum profit for minimum expenditure (Patch, 1994).

Comparable changes occurred in Peru. Assadourian (1992) has noted that in Peru's Andean region the demise of the Inca State (1532–33) and the concomitant decline in indigenous populations caused a transformation of the agrarian landscape. Terrace and irrigation systems were abandoned in the wake of population decline. The need for labour generally, but especially in the silver mines of Potosí, led to a redistribution of population from scattered villages to a few large towns. Moreover, an *encomienda* system became established as Spanish settlers entered Peru (and elsewhere in South America) in the wake of the Conquistadors. This system consisted of a group of villages controlled by a Spanish overseer to whom natives pledged their labour and taxes and who in turn afforded them protection. The increase in Spanish immigration caused settlements to expand and centres of control to expand. Many such centres are today the capitals of many Central and South American countries. Examples include Mexico City, Panama City, Bogotá, Quito, Lima and Buenos Aires, all of which

were established in the period 1520–1650. Land subdivision comprised *estancias*, i.e. grazing land, *characas*, i.e. arable land, for which labour was provided through the *encomienda* system as well as paid Indian workers. Land abandoned as native populations declined was absorbed into the *haciendas* that were owned by Spanish settlers or controlled by ecclesiastical groups such as the Jesuits and Augustinians. The labour to operate these estates was increasingly provided by slaves brought in from Africa. According to Mörner (1987), 5224 slaves were employed by the Jesuit estates along Peru's coastal plain in the year 1767; the estates were situated where rivers entered the Pacific Ocean and 62 per cent of the slaves were used in sugar plantations, with a further 30 per cent in vineyards.

Portuguese colonists also had a considerable impact in Brazil. Between 1500 and 1549, Johnson (1987) states, there were four stages of increasing Portuguese involvement in the development of Brazil, from early land leases to the final establishment of a royal administration through a governor. According to Schwartz (1987) this period was one of considerable change, principally witnessing the establishment of sugar-cane agriculture. As in Spanish South America, native populations declined and imported African slaves augmented local labour resources. Initially the dark red massapé soils of the northeast coastal region were favoured for sugarcane production, organised on an *engheno* or plantation basis. Since the produce was intended primarily for export, mills were situated either along river valleys or on the coast, where most settlement was concentrated. Sugar-cane cultivation gradually extended south along the coast to Rio de Janeiro and the industry rapidly replaced the wood-export industry in economic terms, as between 1580 and 1680 Brazil became the world's largest producer and exporter of sugar (Schwartz, 1987).

Dyewood was one of the first resources of Brazil to be exploited by early Portuguese colonists, but attention soon turned to other trees which could be used for shipbuilding and furniture-making, especially brazilwood. However, despite the fact that rubber was utilised by native peoples, its successful cultivation was not achieved by nineteenth-century colonists. The major contribution made by Brazil to the rubber industry was the supply of seeds to Kew Gardens that ultimately led to the establishment of plantations in southeast Asia. Cash crops other than sugar also became important, notably tobacco, and horses and cattle imported from Europe formed the basis for the development of ranching, especially outside the sugar- and tobacco-producing regions. Schwartz states there were more than one million head of cattle in the northeast region by 1710; cattle supplied the transport needs of the sugar industry as well as beef and hides to the growing coastal settlements.

The exploitation of the native labour force and African slaves was not confined to agricultural systems. In fact it was the lure and discovery of precious metals that prompted both the discovery and subsequent exploitation of South America. Reference has already been made to the silver mines of Potosí in Peru; these were first exploited in the 1540s (Tandeter, 1987) and worked by Indian labour drafted in from as far away as 1200 km. The Spanish search for gold and silver was mainly responsible for rapid colonisation and the establishment of settlements, and subsequent agricultural development to support them. Extraction and processing created landscape changes as waste material was discarded and housing for machinery and the labour force was constructed. Bakewell (1987) has reviewed the major developments in the mining industry that occurred during the colonial period in Spanish South America, based mainly on silver, gold and mercury, and has also pointed out their repercussions on the indigenous Indians and the colonial society. Apart from wealth generation, the mining industries prompted both external and internal trade and determined where the power bases of colonial South America were situated. The discovery of precious metals on a large scale came much later in Brazil than in Spanish South America. According to Russell-Wood (1987), major gold strikes occurred some two centuries after colonisation in many areas but especially in the Minas Geris region, north of Rio de Janeiro. Townships became established, African slaves were imported to provide labour, and agriculture was stimulated following a recession due to falling tobacco and sugar prices.

The major impact of a rapidly increasing population was not felt in North America until the mid-nineteenth century, but early settlers of the seventeenth and eighteenth centuries also left their mark on the environment. Cronon (1983) has reconstructed the early colonial history of New England from an ecological perspective. He emphasises the patchwork nature of the varied habitats of the region, including many different forest mosaics, coastal salt marshes and well-grassed valleys. The indigenous Indian groups exploited these habitats via a mobile existence even where crop growing was practised, and fire was extensively used to clear forest for agriculture as well as to maintain an open parkland that was favoured by wildlife which was hunted. A similar picture is presented by Cowell (1995) for Central Georgia and Delcourt and Delcourt (1991), who have examined the impact of indigenous groups in the Little Tennessee River Valley using palaeoenvironmental techniques. Here crop cultivation and the manipulation of forest resources and forest boundaries were just as important as Cronon (1983) indicates for New England. The British colonists of the seventeenth century created permanent settlements that infringed on the Indian idea of communal territory. Thus the landscape changed in character as boundaries were constructed around colonial farms and farming practices prevented forest regeneration. At the same time, the introduction of European diseases caused Indian populations to decline, so their land-management practices, especially firing, also declined. Cronon believes this profoundly altered the wildlife as well as the vegetation. Williams (1990) estimates that approximately 50 per cent of the United State's original forest cover has been cleared since initial colonisation. He states: 'Before 1850 (when accurate figures became available), it is probable that over 113.7 million acres had been cleared. In the ten years between 1850 and 1859 there was a big upswing in clearing when a remarkable 39.7 million acres were affected'. After the Civil War of the 1860s, when clearance declined to 19.5 million acres, the 1870s witnessed the clearance of another 49.3 million acres. Much of this occurred along the east coast and immediately south of the Great Lakes.

Whitney (1994) considers that the destruction of the forest occurred because of the establishment of permanent agriculture (and associated establishment of personal landholdings), the development of a lumber industry and the provision of fuelwood. All three are, however, interrelated (Williams, 1989). Many of the early settlers were lumberjacks as well as farmers and they used wood for many purposes on their farms, including domestic fuel and fencing. Their exploitation of the forest was very different from the native Indians. White pine (*Pinus strobus*) was especially prized for the construction of ship masts, and because of the shortage of timber in England (Sections 5.3 and 5.4) an export trade based on the lumber industry had become well established by the mid-1600s. Whitney asserts that white pine helped to maintain the naval pre-eminence of the English fleet throughout its annexation of parts of North America. The lumber industry expanded rapidly after the Civil War in response to increased population and the establishment of new settlements, each of which constructed a sawmill as a basic necessity. Agriculture continued to expand, causing forest to be removed by felling, burning or girdling (this involved bark stripping which prevented leafing and caused death). Although demands for fuelwood and construction material were the major factors influencing forest demise, the production of charcoal for iron smelting was also important. Whitney reports that wood was also essential for the establishment of many other primary industries, notably salt, glass, lime and brick production. One of England's noted seventeenth century arboricul-turalists, John Evelyn (1664) stated, 'Twere better to purchase all our iron out of America, than thus to exhaust our woods at home'. This probably reflected general views on the exploitation of colonial resources by the imperialist power!

Agriculture, population expansion and lumbering all caused environmental changes; some can be inferred from old maps, documentary records and contemporary accounts, as reflected in the works of Cronon (1983), the contributions in Conzen (1990), the contributions in Dilsaver and Colten (1992) and Whitney (1994). Further evidence is available from the palaeoecological record in lakes, peats and soils. Kelso (1994) has reconstructed the pre- and post-vegetation cover in the Great Meadows area of Pennsylvania using pollen analysis. The western hillside was covered by hardwood forest dominated by oak in the preclearance era of the 1700s, with marsh in a clearing in which General Washington ordered a fort to be built, and shrub vegetation between the two. Forest was cleared during the nineteenth century for arable cultivation, indicated by the presence of cereal and weed pollens. There is also palaeoecological evidence from lake sediments for changes in forest composition and extent as a result of colonial exploitation. Gajewski et al. (1985) have presented such data from three lakes in northwest Wisconsin, which show that after 1850 the proportions of non-tree pollen increased markedly, representing deforestation. At the same time, disturbance indicators and grass pollen increased, reflecting the spread of agriculture.

Further data for forest disturbance caused by European settlement in the eastern United States have been summarised by Davis (1984). Data from Linsley Pond in Connecticut show that in 1700 there was a decrease in hemlock (*Tsuga canadensis*) and an increase in herbs such as grasses, ragweed (*Ambrosia*) and docks (*Rumex* spp) as a farm was established in the vicinity of the lake (Brugam, 1978). Several sites in the western United States also contain evidence for deforestation and logging, which caused a decrease in conifer pollen and an increase in grasses (Baker, 1984).

Palaeoenvironmental indicators in lake sediments also provide insights into the direction and timing of the impacts on lake catchments. Much of this information has been reviewed by Mannion (1989, 1995b), especially in relation to the process of cultural eutrophication, i.e the nutrient enrichment of lake waters. This will be discussed further in Chapter 7, but it merits mention here because the onset of cultural eutrophication can be traced to the colonial period. At Harvey's Lake in Vermont, for example, Engstrom et al. (1985) have shown that both enhanced soil erosion and inputs of human and animal wastes caused changes in the Lake biota ca. 1780 as a consequence of European settlement and the establishment of permanent agriculture. A large-scale study by Bradbury (1975) has also demonstrated that cultural eutrophication in Minnesota and Dakota was a widespread event associated with European settlement in the late 1800s and early 1900s. Two studies on Frains Lake in Michigan also illustrate the effects of European settlement in its catchment. Davis' (1976) examination of erosion rates has shown that just after 1826, when European settlement was established, there was a change in sediment type, reflecting accelerated erosion due to deforestation. Davis also shows that sedimentation was occurring at a low rate before this disturbance but increased markedly after 1830. Before disturbance ca. 9 t km^{-2} of mineral matter per year were lost from the watershed; after a short interval when erosion rates were 30–80 times greater than in presettlement times, erosion was removing ca. 90 t km^{-2} of mineral material per year by 1900. In a related study, Carney (1982) has shown that lake biota reacted to changing inputs. Increases in the accumulation rates of two groups of algae, diatoms and chrysophytes, immediately after the 1830 horizon, and variations in the range of species present indicate a change from oligotrophic or nutrient-poor to eutrophic or nutrient-rich waters. This Carney attributes to clear-felling of the forest and the establishment of intensive agriculture.

There is also palaeoecological evidence for the impact of European settlement on the Great Lakes. In Lake Ontario, Stoermer *et al.* (1993) have shown that there was a major disturbance in the period 1831–1847, when eutrophication occurred due to nutrient loading caused by deforestation for agriculture. As in the case of Frains Lake, the subsequent period to ca. 1900 was a time of apparent stabilisation when there was a reduction in nutrient loading as land-use patterns in the catchment stabilised after the initial ploughing-up period. A period of enrichment is recorded in the sediments of Lake Superior. This dates to ca. 1850–1900 and is attributed by Stoermer *et al.* to similar catchment perturbations. Reavie *et al.* (1995) have similarly identified trends towards eutrophication in several British Columbian lakes; such trends are associated with the advent of European settlement ca. 1850.

The North American environment was subject to increased deforestation, logging and urbanisation in the period 1840–1920, when there was a major increase in the population. According to Ward (1987), the population of the United States in this period increased from 17 million to more than 105 million, and Wynn (1987) states that the population of Canada increased from ca. 3.5 million in 1871 to ca. 8.8 million in 1921. These increases were due to mass migration from Europe and high birth rates in established colonial settlements. By this time the Industrial Revolution was well under way and the technological innovations that resulted provided further means to modify the environment. Many of the palaeoenvironmental studies mentioned above reflect the effects of intensified human activity. The sediments of Frains Lake (Carney, 1982) contain evidence for enhanced cultural eutrophication caused by increased inputs of nutrients as pig farming and road networks were constructed in the 1930s. The escalation of phosphate loading that occurred in Lake Ontario after 1900 also accelerated cultural eutrophication (Stoermer *et al.,* 1993), which in Lake Minnetonka, Minnesota, is reflected in an increase in organic pigments in the period post 1853 (Engstrom and Swain, 1986). The onset of acid rain caused by industrialisation in the 1800s is also recorded in the sediments of many lakes, especially in the northeast United States and the adjacent area of Canada (e.g. Mannion, 1992a; Dixit and Smol, 1994).

Documentary evidence relating to environmental change during this expansionary period includes the work of Trimble (1992) on the Alcovy River swamps of Georgia, most of which were created after the first European settlement in 1814. The Alcovy River drains the Georgia Piedmont, from which soil erosion was accelerated as European settlers arrived to establish settlements and crop cultivation. Early maps and accounts by surveyors of the region show that few swamplands existed in the early 1800s but the subsequent erosion caused the bed of the Alcovy to aggrade; even by the time of the Civil War (1860s) much swamp development on the bottomlands had occurred with subsequent substantial increases. Historical evidence for the post-colonial developments in the Great Lakes forest has been documented by Whitney (1987, 1994). Here both logging and settlement intensified in the late 1880s. After white pine (see above) was exhausted, attention turned to hemlock, which provided lumber and a source of tannin for the growing leather industry. By 1912 sugar maple (*Acer saccharum*) was also being heavily exploited. This was facilitated by the advent of efficient saws, drying kilns and railway developments. The old-growth forests were replaced by second-growth forests in which sugar maple was dominant. Moreover, the increased incidence of both accidental and deliberate forces caused major changes in the mixed-pine forest, encouraging oak *(Quercus alba* and *Q. rubra)* and aspen (*Populus tremuloides* and *P. tridentata*) which had been only minor constituents of the old-growth forests. Between 1920 and 1940 the effects of firing were limited by improved firefighting policies and eventually by reforestation programmes established in the 1930s. These measures allowed the oak, aspen and jack pine *(Pinus banksiana)* populations to mature and since the 1950s they have provided the raw materials for the pulpwood industry in the region. Agriculture, lumbering and settlement were not the only causes of environmental change, pointed out by Francaviglia (1992), who highlights the role of mining in landscape change in numerous parts of the United States. His comparison of maps for 1881 and 1912 showing the Warren copper mining district of Arizona indicate that topography changes with increasing technology. In particular, waste materials increased in volume as ores with 25 per cent copper were exhausted and replaced by lower-grade ores with less than 5 per cent copper by ca. 1915.

Although agricultural development and the establishment of logging and mining industries fundamentally altered large parts of North America, there were also efforts to preserve the landscape as concern mounted about the environmental damage. One direct outcome of this was the establishment of national parks. Six national parks were set up by 1900, the first being Yellowstone National Park, designated in 1872. Forest policy will be discussed in Chapter 9 and problems associated with agricultural practices will be examined in Chapter 7.

5.9   Changes in Australia and New Zealand following European settlement

Environmental problems, especially environmental degradation, have been associated with Australia's development since Europeans first colonised the continent in 1788. The same can be said of New Zealand, initially colonised in the early 1880s. Overall the distribution of natural vegetation communities in Australia, described by Barlow (1994), relate to gradients of effective precipitation which decline markedly towards the arid interior of the continent. These vegetation communities, dominated by eucalypts, acacias and spinifex grasses, underpin the indigenous ecosystems which were more or less intact before European colonisation. There is, however, evidence from palaeoecological studies on lake sediments in many parts of Australia for frequent firing, some of which was undoubtedly due to burning by Aboriginal groups. Kershaw (1994) has suggested that replacement of rainforest by eucalyptus woodland in northern Queensland, during the past ca. 140 K years has been a response to increased burning, by Aboriginals. Palaeoenvironmental research elsewhere in Australia (e.g. Kershaw and Nanson, 1993; Dodson et al., 1994) indicates widespread influence of pre-European Aboriginals through burning. Pyne (1991) has also emphasised the value of fire in Aboriginal food procurement before and during the early days of European settlement. Many of his observations are based on accounts written by early settlers and explorers such as Edward Eyre and Sir George Gray. Nevertheless, such influence on Australian vegetation has been minor in comparison to the changes wrought by European colonists. Rickard (1988) has drawn attention to the

way in which early settlers perceived their new homeland, scotching the myth that they saw Australia as an aesthetically hostile environment. One of the most significant impacts of European settlement on the Australian environment has been the introduction of extensive livestock grazing to some 60 per cent of the total land area. In addition, mining activities, urban settlement and the introduction of a wide range of exotic plants and animals have contributed to environmental change during Australia's 200 years of European occupation. Although the Aboriginals used burning as a technique for manipulating natural resources, the Europeans instituted new patterns of burning that had far more severe environmental effects.

The importation of growing volumes of both meat and wool into Britain from Australia during the nineteenth and early twentieth centuries was made possible by the expansion of ranching into the more arid rangelands of Australia. Heathcote (1983) has described the period between the 1830s and 1920s as the golden age of ranching, stimulated by the growing European market as well as the developing home market as the numbers of migrants from Europe increased. By 1888 there were 80 million sheep, mainly confined to the semiarid grasslands south of the Tropic of Capricorn and especially in New South Wales, as well as 8 million cattle and 1 million horses (Heathcote, 1987). At this stage in Australia's history the cattlemen were the most adventurous, pushing their ranches into the arid centre of Australia, the Northern Territory and even into the Kimberley region of the northwest. By this time pigs, donkeys, goats and camels had been introduced, all contributing to environmental degradation. Meanwhile, 2.8 million hectares, about 0.5 per cent of the continent, had been converted to arable land, notably in the coastal areas and the red soil plains of South Australia and Victoria.

Thus, within the first century of European settlement considerable environmental change had occurred and even in the early nineteenth century there was growing concern about environmental degradation. Messer (1987) has drawn attention to government intervention in 1803 to reduce indiscriminate clearance of vegetation. Nevertheless, sheep and cattle herds continued to increase and expand into unoccupied areas of the vast continent, increasing the extent of damaged land and ousting the Aboriginal populations from their homelands. The environmental problems relating to such large-scale ranching are much the same now as they were in 1888, due to the mismanagement of that most valuable natural resource, the land and its soil, and an incomplete understanding of environmental processes and interactions. No doubt the lure of big profits motivated resource exploitation to a greater or lesser degree as the demand for animal products increased and as ranchers based their exploitation strategies upon unsuitable European farming practices.

The most widespread environmental problem consequent on large-scale cattle and sheep ranching was erosion. This relates to the imperfectly understood operation of the indigenous ecosystems and the way in which the Australian landscape developed before European colonisation. Ranching also contributed considerably to the process of desertification. Of particular significance is the impact of hard-hooved ungulate animals, i.e. the introduced sheep, cattle, horses and camels, on the soil in contrast to the much reduced impact of the soft-footed animals such as the kangaroos and wallabies. According to Heathcote (1987), a sheep treads six times more heavily in relation to its hoof area, and a bullock seventeen times more heavily, than a kangaroo. Hence the poorly consolidated red soils of Australia's heartland were rapidly disturbed by herds of extensively grazed animals, rendering them susceptible to erosion. Such arid areas are also sparsely vegetated due to low and sporadic rainfall, so that in seasons of drought, winds of even moderate speed can blow vast clouds of red soil from their places of origin to bank against fences and farm buildings. Heathcote (1987) states:

'By 1888 some inland pastures consisted of 'cane swamps' – level stretches of hard, white clay thinly covered with coarse grass – with only tiny islands of the original surface soil still standing in the midst. To enable the land to carry more stock, pastoralists had rainwater tanks dug at low spots on their runs and at the inlet they usually put in a smaller tank to act as a silt trap filter. Yet so severe were the effects of soil erosion that, even with this protection, tanks meters deep silted up within five years'.

Even in non-arid areas where grazing is widespread, soil erosion is also a problem due chiefly to the removal of vegetation cover and overstocking. In the humid regions, sheet erosion is a frequent occurrence as the soil is exposed to enhanced overland flow, and wind erosion is a major problem in semiarid areas. In all these examples, not only is the soil cover diminished but

biogeochemical cycles are impaired as the nutrient stores are depleted. Heathcote (1994) asserts that soil erosion was giving cause for concern by the 1890s, though official action was not prompted until the drought (and depression) years of the 1930s, which created a 'dust-bowl' equivalent to its counterpart in the United States. Several state soil conservation agencies were established in the late 1930s. Conacher and Conacher (1986) state that of 130 million hectares of land used for non-arid grazing in Australia, some 37 per cent was considered by Australia's Department of the Environment in 1975 to require treatment. The state most affected is Queensland, with 17.2 million hectares of eroded land, followed by New South Wales with 10.5 million hectares. The magnitude of the erosion problem has been highlighted by Pickard (1994) who has examined the extent and rates of erosion that have occurred over the past 150 years in the semiarid rangelands of Polpah Station, New South Wales. Here the deposition of sediment in creeks reflect the degree of erosion and its impact on creek migration. Since 1883 creeks have migrated by 1–2 m per year, often due to damming and diversion. Pickard also states that up to 1 m of sediments has been deposited over 10 km of floodplain since 1883, most of which occurred before 1950, and which represents soil removed from the catchment. His review of the literature indicates the problem is widespread.

Areas in which arable farming is practised have also been subjected to land degradation. Heathcote (1987), referring to Australia in 1888, indicates that croplands were becoming exhausted because of lack of manuring and constant cropping. Chartres (1987) has also addressed this problem and refers to the abandonment of wheat cultivation within 20 years of its initiation during the 1870s in the semiarid northern districts of South Australia. Similarly, in the Mallee woodlands of Victoria, clearance for cultivation in the early 1900s created instability in the light sandy soils, so soil exhaustion and soil erosion had become significant problems by the 1930s. They remain significant problems today, as are salinisation and waterlogging which afflict irrigated lands (Heathcote, 1994). These issues will be considered in Chapter 6.

The advent of Europeans and the rapidity with which settlement advanced had a substantial impact on Australia's forests. Today the extent of native forests is only about one-third of what it was before European settlement. Heathcote (1994) and

Williams (1988) have described the character of the forests and woodlands which vary from tropical rainforests in the coastal regions of the Northern Territory and northern Queensland through subtropical and warm temperate rainforest in southern Queensland and New South Wales to cool temperate rainforest in southern Victoria and Tasmania. Almost everywhere logging, burning and the invasion of access tracks detrimentally affected the forest cover. In addition, the location of many forest types in areas of intensive land use, often near Australia's major cities, has produced land-use conflicts. The most significant impact on Australia's forests involve clear-felling for woodchip industries and in many instances, particularly in Tasmania, such exploited forests have been replaced by plantations of exotic pines instead of native species. Much of the impact of Europeans on Australia's forest was achieved through the use of fire regimes that were very different to those of the Aboriginals. Pyne (1991) states: 'What began as a prison quickly swelled into a broad biotic invasion that utterly reconstructed the environment of the island continent. The haste of settlement was breathtaking, and everywhere accented with fire'. Along with indiscriminate felling, the European use of fire has also been called a war of destruction; see Kaleski quoted in Williams (1988). New South Wales forests declined in extent from 25 to 11 million acres between initial colonisation in 1788 and the year 1900. The cleared land was used for crop cultivation, as it was wherever the colonists advanced. The clearance of woodland and often the use of fire is recorded in pollen diagrams from a variety of sites, e.g. Terang, Western Victoria (D'Costa and Kershaw, 1995) and Lake Bolac, Western Victoria (Crowley and Kershaw, 1994).

One of the most significant introductions from Europe was the rabbit, which now occupies the arid zone south of latitude 23° S (Adamson and Fox, 1982). In particular, rabbits have influenced the establishment of the western myall (*Acacia papyrocarpa*) which requires a particular sequence of favourable rains to promote germination. Since the arrival of the rabbit, and despite at least three periods of appropriate rainfall, there has been no regeneration of this tree because of browsing and all the existing trees in a rapidly declining population became established before European settlement. Adamson and Fox also point out that about 10 per cent of the Australian flora has been introduced since European settlement with many species having been accidentally imported as weeds of cultivation (Michael, 1994). Examples (Wace, 1988) include blackberries (*Rubus* spp.), pampas grass (*Cortaderia selloana*) and privet (*Ligustrum lucidum*). Like the rabbit, some of these introduced plants also initiated environmental change and in some instances the repercussions are only now being felt. For example, Lonsdale and Braithwaite (1988) have reported on the spread of a tall prickly mimosa shrub (*Mimosa pigra*) in the wetlands of northern Australia where it is producing dense thickets that shade out native plants on which many native animals depend. When the plant was introduced into Australia is unclear, but Lonsdale and Braithwaite attribute its arrival to early colonial botanists, who were investigating the potential of new crops, and suggest that its success has at least been partly due to another introduced species, the Asian water buffalo (*Bubalus bubalis*). Since the 1880s this has escaped from captivity and by the 1970s large herds were causing considerable ecological damage due to overgrazing, especially on the floodplains of the Adelaide River. Depleted of the native plant species, this habitat rapidly became colonised by *Mimosa pigra*, which created extensive shrublands in place of green meadowlands and control measures have so far only been partially successful in halting its spread.

In addition, the advent and development of European settlement and associated farming practices have been responsible for the decline in the numbers and populations of many native plant and animal species. Saunders and Hobbs (1989) report that in Western Australia some 104 species of plant have become extinct, most of which were natives of the wheat belt centred on Perth. Out of 46 mammal species 13 have also disappeared from the region and 9 of these 13 are now extinct on the Australian mainland. Further data on the impact of European settlement on Australia's fauna and flora are given in Table 5.4. Humphries and Fisher (1994) assert that some 70 per cent of Australia's floral and faunal communities have been altered since initial colonisation.

Europeans had a similar impact on New Zealand, which was settled half a century later than Australia. Forest clearance, accelerated soil erosion, land degradation and the introduction of exotic species occurred. Some of these changes are recorded in palaeoenvironmental investigations, e.g. in the Upper Awatere catchment in the west of the Kaikoura Range, South Island (McGlone and Basher, 1995), which also provide evidence for pre-

Table 5.4 Impact of European settlement on Australia's flora and fauna

A. Impact of European settlement on Australia's flora

- 70% of the floral communities have been altered
- 65% of the original tree cover has been removed
- 75% of the rainforest has been cleared for grazing and agriculture
- 165 species of plants (out of 20 000) are now extinct
- 209 species of plants are considered endangered
- 784 species of plants are considered vulnerable
- These data indicate that 5% of the flora is extinct or under pressure

B. Impact of European settlement on Australia's fauna

- 20 species of mammals (out of 263) are now extinct
- Examples of these include the thylacine, the Alice Springs mouse, 4 species of wallaby, 4 species of hopping mice and 2 species of bandicoot
- At least 5 species of birds have become extinct (out of 522)
- The flightless birds have been most acutely affected, e.g. the Tasmanian emu

Source: based on data in Hobbs and Hopkins (1990) and Humphries and Fisher (1994)

European human impact on New Zealand's vegetation communities. In central Otago, South Island, there is evidence from more complexes in the Garvie Mountains, Old Man Range and Kawarau Gorge for burning by Maoris; further evidence for pre-European vegetation disturbance by human activity has been presented by Newnham *et al.* (1995) from Kopouatai Bog in northern New Zealand. Moreover, Blaschke *et al.* (1992) have reviewed the history of soil erosion in New Zealand and conclude that it stems mainly from the demise of the natural forest cover since European settlement.

Wilson (1993) has reviewed the evidence for forest clearance in the Catlins District of South Island for the period 1861–1991. During this period some 55 000 ha of native forest have been cleared in five distinct phases, each related to many different factors, including economic, social, legal and environmental. New Zealand has also been colonised by a wide spread of alien plants and animals, introduced by humans either deliberately or inadvertently. According to Moore (1983), nearly 59 per cent of New Zealand's flora comprises alien species, i.e. 1700 species as compared with 1200 native species. Many of the alien species are crop plants introduced to allow settlers to grow familiar foods. Atkinson and Cameron (1993) assert that, on average, 11 plant species have been introduced

annually since 1840, the year the Treaty of Waitangi was signed (annexing New Zealand by Great Britain). They also point out that many of the introduced species of plants and animals are having an adverse impact on native species; one example is the possum, initially introduced in the late 1800s to establish a fur trade, but now a pest because it has few predators and because it consumes the eggs of native birds and damages houses. Crosby (1986) also points out that when crops are introduced into a non-indigenous area so too are the weed complexes that accompany the crops. It is highly likely that the hedgehog (*Erinaceous europeus*) found its way to New Zealand in hay bales or similar. Crosby also points out that some of the crop species and weeds had their origins in the Americas, e.g. potatoes, sweet potatoes and maize, earlier adopted in Europe. The introduction of grazing animals, i.e. sheep, goats, pigs and cattle, brought new agents of environmental change; New Zealand, like Australia, had evolved without any large herbivores. Although these animals are a mainstay of the economics of Australia and New Zealand today, they have exacted and continue to exact a considerable environmental cost.

5.10   Conclusion

In the past 2 K years human beings have considerably developed their ability to manipulate natural resources, producing large-scale environmental change. The landscape of Europe was transformed by Greek and especially Roman colonists, who introduced new crops, new farming systems and new technology to their conquered lands, as well as opening up trade routes. In Europe the opulence of Roman archaeological sites attests to the significance of that civilisation.

Historical, archaeological and palaeoenvironmental studies have revealed evidence for changes in the European and British landscapes that occurred between the demise of the Roman Empire in the fifth century and the Industrial Revolution of the eighteenth century. The most significant developments which influenced landscapes in this period concern systems of land tenure, agricultural practices, including field systems and enclosures, and the economic significance of woodlands and their exploitation. It is also evident that social and economic factors were inextricably involved in landscape change throughout the historic period as social hierarchies developed, taxation systems were instituted and settlement patterns evolved. Until the early eighteenth century the population was mainly rural, but as industry came to be concentrated in the towns and developed from cottage industry to large-scale manufacture, there was a general flux of people to the towns. Industrial and mining landscapes subsequently developed.

By the time the Industrial Revolution was under way, many European nations had already become well established in other parts of the world, especially in the Americas, Africa and Australia. Just as the Romans annexed most of Europe nearly 2 K years earlier, introducing agricultural systems and exploting the wealth of their colonies, so too did Britain, France, Spain and Portugal exact their price on New and Old World peoples. Social changes brought about by exploitation of human resources were just as important as resource exploitation in transforming colonial landscapes to generate revenues for European homelands. Allthough mining of precious metals was of some significance in landscape change, the major cause was the initiation of new agricultural systems. Many of these systems involved the introduction of non-indigenous animals and crops, especially cash crops for export, many of which were unsuited to their new environments and resulted in degradation on a large scale, often with repercussions that continue today.

This review has concentrated on the evironmental changes over the past 2 K years. It is clear that social changes are just as important as the landscape itself and were frequently the motivating force underpinning environmental change. As society has developed and technology has advanced, society has become increasingly able to dominate its environment though often at great cost, as this chapter attests. This historic record provides sufficient information to teach many environmental lessons, lessons which illustrate the need for improved environmental management based on an awareness of environmental processes and systems of which human communities are an integral part.

Further reading

Boardman, J., Griffin, J. and Murray, O. (eds) (1994) *The Oxford History of the Classical World.* Oxford Unversity Press, Oxford.

Dilsaver, L.M. and Colten, C.E. (eds) (1992) *The American Environment: Interactions of Past Geographies.* Rowman and Littlefield, Lanham, Maryland.

Grove, R.H. (1995) *Green Imperialism: Colonial Expansion. Tropical Island Edens and the Origins of Environmentalism.* Cambridge University Press, Cambridge.

Heathcote, R.L. (1994) *Australia*, 2nd edn. Longman, Harlow.

Pounds, N.J.G. (1990) *An Historical Geography of Europe.* Cambridge University Press, Cambridge.

Rackham, O. (1994) *The Illustrated History of the Countryside.* Weidenfield and Nicolson, London.

CHAPTER 6

Environmental change due to post-1700 industrialisation

6.1 Introduction

Attention was drawn in Chapter 1 to fuel-powered urban-industrial systems (Table 1.2) which developed during the historic period (Chapter 5) as centres of wealth generation and resource consumption. The industry of these systems has become a major agent of environmental change, especially since the Industrial Revolution of the eighteenth century. In the ensuing centuries the atmospheric, aquatic and terrestrial environments have been affected both directly and indirectly by industrialisation and by the high concentration of human populations that characterise urban-industrial systems.

These systems have traditionally been centred on the 'heavy' industries based on the processing of minerals and agricultural products and the manufacture of consumer goods. This chapter addresses those changes in the environment that have occurred as a response. Urban-industrial systems are based on the consumption of energy, especially fossil fuel and nuclear energy. Both types of energy create environmental problems. Fossil-fuel consumption especially is now considered responsible for the onset of global warming. Both acidification, particularly important in the northern hemisphere where most industrialisation has taken place, and the enhanced greenhouse effect are products of the post-1750 industrial era. The depletion of stratospheric ozone is also a product of industrialisation insofar as damaging chemicals, such as CFCs, have been produced commercially for use as aerosol propellants and refrigerants. Moreover, urban-industrial systems generate substantial quantities of waste materials ranging from toxic chemicals to domestic rubbish; these also create environmental problems and environmental change.

All these types of environmental change could be reversed given appropriate political action. In the case of global warming, the costs would be very high, but reversal would be possible. At local and regional scales the reclamation of mine-damaged land, for example, can generate productive ecosystems. Similarly, acidification can be halted through the control of sulphurous gas emissions with subsequent ecosystem recovery.

6.2 Changes where mineral extraction occurs

The earliest human communities (Chapter 4) turned substances like stone and wood into resources through human ingenuity. Attention later turned to copper and iron as well as gold and silver ca. 8-6 K years BP. Moreover, the search for mineral wealth was a major stimulus for European expansion into Africa, the Americas and Australasia (Chapter 5). The utilisation of such resources in Europe and its colonies helped to foster the Industrial Revolution and established mineral processing *in situ*. Consequently, both developed and developing nations have experienced environmental change in areas with mineral extraction and processing. Extractive and processing plant machinery as well as waste production have created localised 'industrialised' landscapes that are often aesthetically displeasing and frequently hazardous. Extractive industries may also be responsible for altering environmental quality at a distance from the points of extraction and processing, i.e. through air and water pollution.

The precise nature of environmental change that occurs in an area of extraction depends on the type of mineral and the type of mining. Mineral resources are finite and the volumes or quantities of minerals available are known as stock resources. Stock resources are many and varied, ranging from the abundant and widespread to the scarce and localised, as illustrated in Table 6.1. This also shows that the most important methods of extraction are surface working, e.g. quarrying, opencast and strip mining, and deep mining or, in the case of oil and natural gas, deep drilling. The amount and type of waste material generated is a further factor involved in causing environmental change. For example, aggregates used for road building use up to 100 per cent of the source whereas 100 t of waste may be produced to extract 1 t of a rare metal such as tungsten.

Strip mining is the most disruptive of the methods of surface working because it affects a larger surface area than quarrying or opencast mining. There are various types of strip mining: mountain or hilltop removal produces plateaux; contour stripping involves the cutting of benches into sloping land; and area strip mining is employed in flat or gently undulating land. These techniques are used to extract 60 per cent of ca. 500×10^6 t of coal in the United States annually (ReVelle and ReVelle, 1992). The Appalachian area has experienced a great deal of mining-related environmental change. For example, contour strip mining has created 40 000 km of contour benches of which 2700 km are subject to major landslides (Toy and Hadley, 1987). In the past these landslides have destroyed neighbouring farmland, roads, reservoirs and streams. Moreover, the slopes produced by contour mining, before legislation in 1977 requiring reclamation, are subject to high erosion rates because of the lack of vegetation cover. The eroded sediment impairs the flow capacity of stream and river channels, so sedimentation increases the intensity and the frequency of the floods.

There is also evidence for considerable environmental change caused by opencast or open-pit mining in the United States as Francaviglia (1992) has discussed. He states: 'Especially since 1890, the landscape changes brought about by mining have increased dramatically in scale and impact. The topographic changes that resulted from the adoption of new technologies have been nothing short of phenomenal: certain mining districts have primary extractive features – the huge open-pit mines where early mining towns once stood or secondary and tertiary accretionary features, such as huge overburden piles and tailings ponds, covering the sites of others'. The Kennecott mine in Bingham Canyon, Utah, is the largest open-pit copper mine in the world at over 700 m deep, 3.2 km in diameter and 7 km^2 in area. It is surrounded by benches cut into bedrock; the benches carry railway lines and roads to transport the ore. Open-pit mining is also used to extract uranium ores in the United States. Toy and Hadley (1987) have discussed the impact of the dumping of overburden and waste materials from processing at the Anaconda Mine in New Mexico, the largest open-pit uranium mine in the world. It occupies 486 ha, of which 445 ha comprise 28 dump sites containing ca. 2×10^6 t of overburden or mine wastes and which are now major landscape features. The mining activities at Anaconda have affected geomorphological processes by accelerating erosion rates and disrupting local drainage networks. The tailings, which consist of sands and slurries produced from on-site ore processing, are particularly susceptible to both wind and water erosion. How these activities have influenced soil and water quality, notably level of radioactivity due to contamination with uranium wastes, is another issue about which little is publicised.

In Togo, open-pit mining is a principal method in which phosphates are extracted from a coastal sedimentary basin in the southern part of the country. According to Allaglo *et al.* (1987), phosphate is the main source of foreign exchange and extraction is centred on two sites at Hahotoé and Kpogamé in the Haho River Valley; consequently, the traditional way of life has been disrupted with many people turning to marketing foodstuffs and handicrafts. Despite the need for labour in the phosphate mines, this disruption has led to a migration of people into the towns at the expense of the rural community. Soil degradation has also occurred and the dumping of overburden has created an artificial landscape consisting of earth walls several metres high over several thousand hectares. These walls are susceptible to erosion by overland flow and are unsuitable for cultivation. Some of these problems have been overcome by introducing leguminous plants to protect the soil surface and to enhance the organic component and thus the cohesiveness of the soil.

Table 6.1 Mineral resources: their variety, distribution, output, waste to ore ratios and environmental problems

Class	Examples	Distribution and location	Ratio of ore/mineral to waste	World output measured in	Usual method of working	Possible environmental impacts	Remarks
Common rocks	Limestone, chalk, granite, sandstone, slate	Widespread and abundant	Almost all used	Billions of tonnes	Quarrying on surface of hillside	Scenic scars, loss of habitats but interesting when worked out	Except for up to 95% wastage in slate
Common rocks	Sand and gravel	Widespread and abundant	Almost all used	Billions of tonnes	Wet or dry surface pits	Voids; flooding; lowered surface levels; drainage	Sometimes creating new water habitats and recreation areas
Earths and clays	Ball clay, stoneware clay, china clay, fuller's earth	Common	All used except for overburden	Hundreds of millions of tonnes	Surface working	Lowered surface levels; drainage problems; pollution; tips from china clay	Except for china clay, where high percentage of waste material (1:10)
Common rock-forming minerals	Feldspar, dolomite, mica, quartz, fluorspar	Common but limited sources	1:1 down to 1:50	Millions of tonnes	Surface or hillside quarrying	Scenic scars	Need selective quarrying
Precious and semiprecious stones	Diamonds, rubies, sapphires, emeralds, opal, garnet, amethyst, jade	Rare	Stones are a minute percentage of waste rock	Thousands of grams (kilograms)	Open-pit sands and gravels; underground mining	Voids and scenic scars	Or found in working other minerals
Common minerals	Asbestos, vermiculite, pyrites, talc, soapstone, alum, barium, gypsum	Fairly abundant in limited locations		Millions of tonnes	Quarrying on surface or hillside	As rocks; risks of pollution to water	
Less common minerals	Graphite, sillimanite, wollastonite, Cryolite	Infrequently found		Hundreds of thousands of tonnes	Adit mining into veins and dykes	As rocks; risks of pollution to water	
Salts	Salt, rock salt, sodium salts, borax, potash, nitrate, phosphate	Common and fairly abundant when found		Millions of tonnes	Deep mining, surface quarries; alluvial mining	Waste heaps; subsidence; saline flashes	
Abrasives	Corundum, emery, pumice, commercial garnet	Rarely located but abundant when found		Millions of tonnes	Surface working of outcrop and adits into veins	As rocks	

Common metal ores (ferrous)	Magnetite, haematite, limonite	Abundant but in limited localities		Hundreds of millions of tonnes	Deep mining; pillar and stall longwall; drift mines; outcast	Waste deposits; hill and vale restoration	
Common metal ores (non-ferrous)	Bauxite, galena, nickel ores, tin ores, copper ores, zinc	Limited distribution		Millions of tonnes	Historically by deep mining, now mainly open pit: some alluvial (tin)	Voids; waste heaps; tailing dams and lakes; polluted run-off; toxic wastes	
Less common metals	Manganese, antimony, cadium, chromium, cobalt, mercury			Hundreds of thousands of tonnes	Deep mining mainly by adits; some open pit	Toxic wastes	
Rare metals	Indium, germanium, lithium, caesium, selenium, tellurium, tungsten, thorium, titanium, uranium, vanadium, zirconium	Rare to very rare	1:100 down to 1:5 000 000	Ore output varies from under one tonne to several thousand tonnes, according to scareness and demand; metals are usually measured in 100lb ounces (kilograms)	Various	Toxic wastes, radiation and similar risks	Most are only obtained as a by-product
Noble metals	Gold, silver, platinum, palladium	Very rare	Ore contains about 0.1% metal	Million troy ounces (kilograms)	Deep mines with shafts and galleries, drift mines or alluvial mining	Voids; waste heaps; scenic scars	Or as by-products
Fossil fuels	Coal	Fairly common	2:1 (deep mining) 1:15 (opencast)	Hundreds of millions of tonnes	Deep mines with shafts and galleries, or drift mines, opencast strip and auger mines	Subsidence; shale tips; scenic damage; air pollution from burning tips; liquid effluent; pollutants; temporary scenic damage by opencast mining	
	Oil	Abundant but rarely found	All used either crude, refined or in by-products	Hundreds of millions of tonnes	Land or sea walls	Oil spillage at sea or from pipelines; spoil heaps from oil shale working	
	Peat	Common	All used	Millions of tonnes	Surface working	Lowered land levels; drainage problems; may destroy or preserve bog habitats	

Source: From Blunden (1985)

Environmental problems also occur in the coastal strip, where a treatment plant was established. This has led to the removal of plantation coconut palms which, together with dust and smoke and clay dumping in the sea, means the area is much less attractive to tourists than formerly. Mining for phosphates has also damaged several islands, e.g. Nauru and Barnaba in the Pacific Ocean.

A further type of surface mining which has recently received attention is that of coral extraction. This is one of several activities causing the impairment of coral reefs; other activities include pollution via run-off from agriculture and forestry and sewage, overharvesting and overdevelopment for tourism. The depletion of coral is becoming particularly acute in many tropical islands because other materials that can be used for construction rarely occur. Thus coral is mined for lime production and as an aggregate. In Sri Lanka, for example, Rajasuriya and White (1995) report that the condition of nearshore coral is declining and even offshore reefs are being damaged by the use of explosives for fishing and bottom-set fishing nets. This is despite the existence of laws for conservation and control. At Mafia Island, Tanzania, Dulvy *et al.* (1995) have quantified the effect of mining on the live coral cover. On mined reefs, live coral was reduced by ca. 65 per cent. Moreover, fish abundance was 42 per cent lower on mined than on unmined reef and diversity was reduced by 24 per cent. Beyond the reef zone itself, shoreline erosion and loss of mangrove forest has occurred because the depleted reefs no longer dissipate wave energy effectively. Similar problems are occurring in the coral reefs of Papua New Guinea (Huber, 1994) and the Maldives (Clark and Edwards, 1994).

Deep mining or underground mining also causes environmental change. Subsidence can be a significant problem and is caused by the collapse of overlying rocks into excavated chambers. It alters drainage characteristics and may damage roads, buildings, etc., in urban and suburban areas. According to Coates (1987) subsidence has affected 25 per cent of the United States' 3250 km^2 coal mines. In such areas, sinkholes and troughs (or pits and sags) are characteristic. Sinkholes occur when overburden collapses into a chamber with steep walls and an increasing diameter with depth, whereas troughs are shallow near-circular depressions that develop as underground supporting pillars collapse, causing the overburden to sink. In the United Kingdom and the United States salt mining and coal mining have produced landscapes susceptible to subsidence. Moreover, tailings removed from deep mines over the centuries have become permanent landscape features as they have become colonised by vegetation. Where tailings derive from metal extraction their impact on the local, and distant, environment may last for centuries. For example, Merrington and Alloway (1994) have shown that a century after mining ceased in a lead–zinc mine in mid-Wales, large amounts of heavy metals are still being released into the local environment through aerial deposition and fluvial transport to soil, stream sediment and vegetation. Cadmium, copper, lead and zinc all occur in large quantities in the soils and stream sediment in particular; of these cadmium (4.2 kg y^{-1}) and zinc (1387 kg y^{-1}) occur in greatest quantities in the dissolved load of the stream. The mining of metalliferous ores may lead to the contamination of soils and plants adjacent to the mine as Jung and Thornton (1996) have demonstrated for a site in Korea. Here the mining of lead and zinc caused the contamination of nearby crops, e.g. soya beans, onions, peppers and tobacco, with cadmium and copper as well as lead and zinc. The heavy metals will thus enter food chains. In a similar study Amonooneizer *et al.* (1996) have shown that soils, plants and crops around the Obuasi gold mine in Ghana have high concentrations of mercury and arsenic.

Deep mining is also characteristic of the Kimberley region of South Africa where it is used to extract diamonds. In this part of the Karoo, diamonds were discovered accidentally in 1869 and prompted a diamond rush. Two years later diamonds were found at a site known as Colesberg Koppie, which later became Kimberley. Today, as part of its tourism industry, the original diamond-rush Kimberley is preserved as an outdoor museum. It contains houses, shops, a ballroom, printing presses, etc., and nearby is the famous 'big hole'. This was the first mine to be excavated; no longer active it is one of Kimberley's main tourist attractions. In its heyday 300 000 miners worked within its confines of 60 000 m^2. It is the largest mine in the world to have been excavated manually, initially reaching a depth of 800 m and later 1100 m when the Kimberley Mine Company took over operations. Overall, 28 × 10^6 t of earth and rock were excavated in order to extract 3 t of diamonds (based on information provided by Kimberley tourist office).

6.3 Changes distant from the source of mineral extraction

In many instances of mineral extraction, the environmental impact will occur some distance from the site of extraction. The use of fossil fuels in mineral processing contributes to the enhanced greenhouse effect, considered in Section 6.5. Among the major changes in environmental quality are the contamination of distant soils and water by heavy metals, which may have implications for wildlife and human health. Moreover, an understanding of how such contaminants circulate within the environment is essential for management and reclamation schemes.

Hart and Lake (1987) have reviewed some of the available information on the contamination of the Australian environment. The South Esk River in northeast Tasmania has been polluted by zinc and cadmium derived from the mining of tin and wolfram; the diversity and populations of macroinvertebrates as far as 80 km downstream from the source of the pollution is much lower than in upstream uncontaminated waters. Similar problems have occurred in western Tasmania's King River due to the receipt of high concentrations of copper, zinc, lead and cadmium in acidic drainage water from disused mine shafts and waste dumps. The contamination of drainage water by mining activities has also been discussed by Williams et al. (1996) in relation to Nakhon Si Thammarat Province, southern Thailand. Here tin mining from bedrock and alluvial deposits produces waste rich in arsenopyrite. Both surface water and groundwater now contain up to 500 times more arsenic than is recommended as safe by the World Health Organisation. Arsenic poisoning has been causing human health problems for more than a decade and the mining of tin in general has caused land degradation bringing it into direct conflict with Thailand's growing tourism industry.

In recent years much concern has been expressed about the threat to habitats and human health caused by the use of mercury in goldmining activities in the Amazon basin. Since 1980 a 'gold rush' has occurred in this region as hitherto inaccessible areas have been opened up. Problems associated with mercury contamination arise because the metal is used in alluvial gold mining. It is used to separate the gold from river sediments during alluvial panning, amalgamating dispersed flakes and nuggets. During this process the mercury enters the atmosphere and the aquatic systems. According to Malm et al. (1995) and Akagi et al. (1995) mercury pollution is now widespread in Amazonian goldmining areas where health problems such as neurological damage and birth effects are occurring as a result. But analysis by Malm et al. (1995) indicates that people working as gold dealers and in dealers' shops are most susceptible to mercury poisoning by inhalation or ingestion of metallic mercury, whereas people living in riverine communities are more likely to be affected by methylmercury through their consumption of fish. A study of mercury in fish in the Tucuri hydroelectric reservoir of the River Moju by Porvari (1995) has also shown there are considerable variations in concentrations between fish occupying different trophic levels. 1.3 mg kg^{-1} of mercury was found in predatory fish, 0.32 mg kg^{-1} in planktivorous/omnivorous fish and 0.11 mg kg^{-1} in herbivorous fish; Brazil's official recommendation is that mercury concentrations should not exceed 0.5 mg kg^{-1}.

The contamination of surface waters by mercury as a result of mining activities also occurs in the United States. Miller et al. have presented the example of the Comstock Lode, near Virginia City in Nevada, where mining for gold and silver began in 1859. Some 30 stamp mills were established along the Sixmile Canyon to process the ores using mercury. This resulted in the release of mercury and tailings into the Carson River. Consequently, mercury, gold and silver have accumulated in the Sixmile Canyon alluvial fan. J.R. Miller et al. (1996) estimate that 31.5×10^3 kg of mercury is lodged in the fan, much of which could be recovered to help pay for a remediation programme, along with most of the 18.2×10^3 oz of gold and 1205.8×10^3 oz of silver. Moreover, Bonzongo et al. (1996) have demonstrated that the waters of the Lahontan Reservoir of the Carson River system also have high concentrations of mercury, notably 7585 ng dm^{-3} of elemental mercury and 7.2 ng dm^{-3} of methylmercury. Other examples of mercury pollution include the Wau-Bulolo area of eastern Papua New Guinea and Slovenia. Saeki et al. (1995) have compared human hair samples from Papua New Guinea for populations upstream of goldmining activities and for gold miners that use mercury to extract the gold. In the uncontaminated group, comprising 80 individuals, mercury concentrations averaged 0.55 μg g^{-1}, in the 86 gold miners tested the mean was 1.2 μg g^{-1} and in downstream dwellers the concentrations were in between, reflecting indirect contamination. Gnamus *et al.* (1995) have analysed tissue samples from deer in the area surrounding one of the world's largest mercury mines at Idrija, Slovenia. The results show that mercury concentrations in the deer were 100 times higher than in control samples, with the highest methylmercury concentrations occurring in brain tissue.

The Carson River example of mercury pollution, referred to above, illustrates that contamination of the environment can occur over a long duration and that contaminants can accumulate in substantial amounts. Rowan *et al.* (1995) have reported a similar case in the Leadhills area of Scotland, UK. Here, the Glengonnar Water received lead-enriched drainage water from Lead mines in the area over several centuries. This has produced lead accumulation in floodplain sediments with surface concentrations of 75×10^3 mg kg^{-3}. Both floodplain sediments and former floodplain sediments that are now channel bank deposits are a continued source of lead, as it is redistributed in response to fluvial erosion and redeposition. This study illustrates how contaminants can remain concentrated and persistent in the environment unless remedial measures are taken. The problem is also highlighted by Younger's (1995) examination of water discharges from disused mines in the Durham coalfield. Concentrations of iron and aluminium are high and salts of both are deposited on stream beds; concentrations of sulphates vary in the range 130–1300 mg dm^{-3}, though they decline with distance from mine discharge points as iron sulphates are precipitated. Moreover, Gao and Bradshaw (1995) have shown that the stream remains polluted, even after a reclamation scheme on the Nant Gwydr, a tributary of the River Conwy in North Wales, operated between 1954 and 1978. They state that, despite the stabilisation programme for the tailings, the Nant Gwydr contributes ca. 1 t of zinc, 0.2 t of lead and 0.05 t of cadmium annually to the River Conwy. Although much of this derives directly from the mine itself, the tailings still generate a substantial proportion of heavy metals. In this case the remediation has not been entirely successful.

Surface-water drainage systems may also be impaired by the discharge of acid mine water. This is generated when iron sulphide minerals or pyrrhotite, often present in coal measures and metallic mineral ore, are oxidised in moist air to generate sulphuric acid. The acid then enters any water circulating within the mine and is produced as water leaches through mine tailings. According to Craig *et al.* (1996), up to 16×10^3 km of rivers and streams in the United States have been polluted in this way, causing a loss of stream flora and fauna and highly acidic soils. The problem is particularly acute in the Appalachian region, which has a long history of coal mining (Section 6.2). The mining of radioactive minerals notably uranium, poses similar problems to those mentioned above and in Section 6.2 in relation to soil and water contamination. However, radioactive waste poses an even greater hazard than the wastes generated by coal and metal mining. Fernandes *et al.* (1996) have discussed the monitoring data collected from the Pocos de Caldas uranium mine in Brazil, which they used to examine the likely environmental impact after mine closure. They show that the risk of contamination of deep groundwater would be very low but that ^{210}Pb and ^{210}Po would be released into surface water. These results show that substantial remediation will be needed after mine closure. Furthermore A.R.M. Young (1996) has drawn attention to the environmental problems associated with the uranium mine at Rum Jungle in Australia's Northern Territory, where tailings were washed into a stream and an expensive rehabilitation programme has been necessary.

6.4 Reclamation of mine-damaged land

The many programmes that have been established to rehabilitate mine-damaged land constitute environmental change. Although it is rare for such areas to be restored to their former status, artificially created ecosystems with enhanced aesthetic and economic values may result. The examples of land degradation in Sections 6.2 and 6.3 reflect the varied nature of the problems generated by extractive industries; it follows that the reclamation schemes will be equally varied. And mining companies in many parts of the world are legally obliged to undertake comprehensive surveys and impact assessments before extraction and with a view to improving aftercare.

The problems of reclaiming derelict land in Britain have been addressed in Bradshaw and Chadwick's (1980) classic work and by Finnecy and Pearce (1986). They include urban and industrial dereliction as well as mine-damaged land, though only the latter will be discussed here. These authors point out that the chief cause of dereliction is the removal of soil and one way to improve the environment rapidly is to replace topsoil. Unless this has been stored after its extraction from the site in question, it is an expensive undertaking. In consequence, many reclamation schemes involve the direct treatment of site materials, which necessitates the identification of the specific problems and deficiencies of the waste in question. In colliery spoil there are deficiencies of nitrogen and phosphorus, both essential plant nutrients; the abundance of pyrite (iron sulphide), which combines with water to produce sulphuric acid often produces a pH of between 2 and 4. The deficiencies can be overcome by dressing with nitrate and phosphate fertilisers and the low pH can be increased by adding lime. The required amounts of additives will depend on how much pyrite is present and on the quantities of iron and aluminium oxides, which can bind phosphorus and render it unavailable to plant growth. The significance of soil characteristics in relation to vegetation re-establishment on former coal mines sites has been discussed by Scullion and Malinovszky (1995). Their investigation of alder, birch and oak growth on two restored opencast coal mines in South Wales shows that soil depth and soil organic content were the most important determining factors. Soil depth was important for all species, especially alder, whereas drainage was particularly important for birch. Information like this helps to formulate reclamation strategies.

Much research has been undertaken on the reclamation of colliery waste and coal-mining areas in North America. For example, Daniels and Zipper (1995) have examined the reclamation of coal mines in the Powell River Project of the Appalachian region of the United States. They refer to the transformation of 'mine spoils' into 'mine soils' and the generation of productive land. Components of the reclamation schemes described include the replacement of topsoil and the use of topsoil substitutes. Topsoil substitutes can be produced from rock strata low in pyrites, to limit acid formation, and high in carbonates to provide cohesion. Daniels and Zipper concur with the findings on reclamation of colliery spoil in Britain insofar as adequate soil depth and soil organic matter are crucial for effective revegetation (see above). They also point out that, once an adequate topsoil has been created, the choice of species to be planted or seeded depends on the use required. In Appalachia the aim is usually to restore native forest; to achieve this, erosion

must be curtailed and soil development encouraged. This necessitates seeding with grasses and legumes, as well as the planting of tree seedlings, to stabilise soils and provide conditions for the accumulation of organic matter and nitrogen. Nitrogen can be added as sewage sludge or nitrate fertiliser whereas sawdust and tree bark add organic matter. Phosphorus fertiliser needs to be applied to overcome initial phosphorus deficiencies; in the longer term it is essential to have plant species with the potential to develop substantial mycorrhizal networks that can obtain phosphorus from diffuse sources. Given these conditions, Daniels and Zipper report that successful reclamation can be achieved. The importance of mycorrhizal activity in reclamation is also highlighted by Gould *et al.* (1996) in relation to mine sites in Kentucky. They show that mycorrhizal colonisation increases with time after reclamation and is encouraged by the availability of calcium and soil organic matter. Moreover, Harris and Zuberer (1993) have reported that the inoculation of lignite mine spoil with the bacterium *Rhizobium* (Section 9.8), a nitrogen-fixing species, initially increased the growth of subterranean clover (*Trifolium subterraneum* L.). This in turn benefited the bermuda grass crop, which was also encouraged by the addition of nitrate and phosphate fertilisers.

These examples of mine reclamation involve practices directed at returning a vegetation cover or pasture to the degraded land. Alternative land uses include the conversion of former mined sites into wetlands, which is cheaper to effect than rehabilitation. Sistani *et al.* (1995) have discussed the status of two wetlands developed on former strip mines from which coal was extracted and compared various characteristics, notably vegetation, pH and a range of extractable cations, with those of a nearby natural wetland. The results are summarised in Table 6.2, which shows there is little difference in relation to extractable cations, reflecting biogeochemical characteristics. Sistani *et al.* suggest that the conversion of mines to wetlands may compensate for lost natural wetland. Middleton (1995) has discussed comparisons between seed banks and species richness in wetlands developed in coal slurry ponds of various ages and natural ponds in the Upper Mississippi Valley, with a view to establishing management practices that accelerate the development of the vegetation. The research shows that more seeds were produced under freely drained rather than continuously flooded conditions, and more plants became established in

Table 6.2 Wetlands developed on strip-mine sites compared with nearby natural wetland in Alabama, United States

	Strip-mined wetland	Natural wetland
Dominant vegetation	Bulrush (*Scirpus* sp.) Cattail (*Typha latifolia*)	Cattail (*Typha latifolia*) Smartweed (*Polygonum* spp.)
pH of mineral and organic substance	6.7–7.4	7.7–7.8
pH of nearby soil	4.1–5.0	6.5
Extractable Ca, Mg, K, Na, Al, Fe, Mn	No consistent trends	No consistent trends

Source: based on Sistani *et al.* (1995)

both the slurry and natural ponds; biomass was also highest under free drainage. Moreover, the application of nitrate/phosphate/potassium fertiliser enhanced biomass production and seed set. Thus fertiliser application, and lime to counteract high acidity, plus the maintenance drainage characteristics could be employed to encourage the rapid establishment of a biodiverse plant community in the early stages of pond conversion.

In Australia, the reclamation of mine-damaged land has been implemented since the 1970s, as illustrated by efforts in New South Wales. The reclamation of opencast coal mines in the Hunter Valley has been discussed by Dragovich and Patterson (1995). They point out that since 1976 legislation has obliged mining companies to rehabilitate mine-damaged land. To do so, original topsoil is stored after removal then replaced after coal extraction is complete. However, the presence of topsoil itself is inadequate for rehabilitation, mainly because the topsoils in the Hunter Valley area are naturally thin, acidic and deficient in phosphorus. For successful recolonisation by vegetation, the architecture of overburden and/or mine waste must be altered to eliminate steep, unstable slopes and then covered with waste carbonaceous material. In most cases revegetation has proved to be successful despite the low pH and high aluminium concentrations whereas revegetation was very poor on untreated mine waste and overburden. Even before the 1970s, Alcoa was involved in rehabilitation; Alcoa is a major company involved in bauxite mining for aluminium extraction. According to Koch and Ward (1994), the mining of bauxite from the region between Perth and Banbury in Western Australia began in 1963. The mineral is retrieved from shallow pits of ca. 4 m depth and 1–10 ha in extent; every year ca. 450 ha are mined and restored so that little degraded or derelict land is created. Before mining commences, the topsoil is removed, beginning with the surface

5 cm containing the humic layer and followed by the rest of the A horizon. The soil is either stored to reuse after mining operations are complete or is used immediately to restore already-mined pits. After emplacement, the soil is seeded with a mixture of species including native legumes. Their association with nitrogen-fixing bacteria means that soil nitrogen is increased, along with organic matter as plants produce litter; this enhances cohesiveness. Nitrate fertiliser is also applied. All these factors increase plant cover, reducing erosion and increasing the range of species present. In the early stages, the plant community is dominated by fast-growing annual and biennial species, including many species that are not characteristic of the former jarrah (*Eucalyptus marginata*) forest. Nevertheless they stabilise the soil into which seeds of native species are dispensed. Koch and Ward have shown that by the end of the first year the rehabilitated vegetation has a similar index of ca. 25 per cent but after a decade the value rises to 50 per cent as many slow-growing and woody species of the jarrah forest become established.

Several other problems and practices associated with Alcoa's rehabilitation schemes have been reported by McChesney *et al.* (1995) and Ward *et al.* (1996). McChesney *et al.* suggested that the erratic and often low emergence of broadcast seed of forest species may be due to the alteration of microclimate resulting from mining activities. For some species, e.g. *Eucalyptus marginata* and *E. calophylla*, the absence of a canopy appears to limit early establishment from seed. These are species characteristic of the mature forest so they would normally germinate under canopy conditions. McChesney *et al.* have shown that emergence beneath a canopy was 17 per cent whereas in the open areas emergence was only ca. 6 per cent for *E. marginata*. The values for *E. calophylla* were 23 per cent and 2 per cent respectively. These differences may relate to soil moisture and diurnal temperature fluctuations. Under canopy conditions, soil moisture is higher and temperature more constant than in the open. Moreover, emergence was higher at upland rather than lowland sites, possibly because of cold-air drainage to the lowland sites. Ward *et al.* (1996) have examined the role of the timing of various rehabilitation procedures in successful re-establishment of jarrah forest. For example, ripping late in April (autumn) or scarifying in June reduces the numbers of species and the numbers of individual plants becoming established from seeds in the topsoil. The reverse was true for broadcast seeds. Moreover, scarifying before seeding in June increased species establishment from broadcast seed. Thus, to ensure high success rates of establishment from broadcast seed, pits should be ripped and sown in April. This does least to impair the establishment of species from the seed bank in the topsoil. These data help to formulate rehabilitation programmes.

6.5   Changes due to fossil-fuel use: global warming

6.5.1   Greenhouse gases: sources and sinks

Global warming is the hottest environmental issue of recent years! More has been written about the enhanced greenhouse effect and its likely impact than on any other environmental issue and in the past decade a great deal of money, time and effort have been allocated to its investigation. This reflects its environmental, social and political ramifications. Reference has already been made in Section 1.3 to the effect of atmospheric composition on global climate (Mannion, 1997b). Moreover, the long-term changes in atmospheric concentrations of both carbon dioxide and methane that occurred during the last glacial–interglacial cycle have been discussed in Section 2.4. These trends are illustrated in Figures 2.2 and 2.3 and confirm that a relationship exists between climatic change and the global carbon cycle, though the precise details of this relationship have not yet been established. For example, there are dissenting views on, and contradictory evidence for, the location of the carbon removed from the atmosphere, where it exists as carbon dioxide, during an ice age. There is also some debate as to whether the mechanism or mechanisms involved are

physical, chemical or biological, or possibly combinations of these mechanisms. An understanding of the mechanisms of carbon transfer and the role of biomass, the oceans and biogeochemistry of the past is vital to the formulation and testing of models constructed to predict future climatic change and its impact (Covey *et al.*, 1996; Melillo *et al.*, 1996). Moreover, the enhanced greenhouse effect is caused not only by carbon dioxide and methane but also by nitrous oxide, chlorofluorocarbons (CFCs), hydrofluoro-carbons (HCFCs) which are substitutes for CFCs, perfluorocarbons, ozone and water vapour (Table 6.3).

The extraction of information on atmospheric composition for the past few centuries from the polar ice cores, along with data from the past 40 years of monitoring at Mauna Loa in Hawaii (Keeling *et al.*, 1996) and the past 100 years from Antarctic snow (Battle *et al.*, 1996) allows the record of carbon dioxide and methane to be reconstructed for the past three centuries. Figure 6.1 shows that before the Industrial Revolution the concentration of carbon dioxide was ca. 270 ppm but by 1953 it had increased to ca. 312 ppm, and the Mauna Loa measurements (Keeling *et al.*, 1996) show that concentrations continued to rise to 356 ppm in 1993. The rate of increase is considered by Watson *et al.* (1992) to be 1.8 ppm per year. Although many developed nations are reducing carbon dioxide emissions as they increase energy efficiency, it is likely that global carbon dioxide concentrations will continue to increase at least in the next decade as the newly industrialising countries increase fossil-fuel consumption (Tucker, 1995). In China alone, coal production increased from 872×10^6 t in 1985 to 1080×10^6 t in 1990 but its importance as a fuel to the country as a whole decreased to 70 per cent; the deficit was made up by increases in oil and natural gas consumption (Zhao, 1994). By the year 2000, carbon dioxide concentrations will be approaching 400 ppm. A proportion of the increase will be a consequence of deforestation, especially in the tropics, which releases carbon dioxide from biomass and from soil organic matter. Other sources of carbon include cement manufacture, which generates ca. 2 per cent of total anthropogenic carbon emissions, according to Graedel and Crutzen (1993). Figure 6.1 shows the changes that have occurred in the past few centuries. The rapid increase from ca. 0.8 ppmv to ca. 1.8 ppmv since 1900, and the continued increase at a rate of

Table 6.3 Heat-trapping (greenhouse) gases and their characteristics

	Carbon dioxide (CO$_2$)	Methane (CH$_4$)	Nitrous oxide (N$_2$O)	Chlorofluorocarbons (CFCs)	Tropospheric ozone (O$_3$)	Water vapour (H$_2$O)
Greenhouse role	Heating	Heating	Heating	Heating	Heating	Heating
Sources						
Natural	Balance in nature	Wetlands	Soils, tropical forests	None	Hydrocarbons	Evapotranspiration
Anthropogenic	Fossil fuels, deforestation	Rice cultivation, cattle, fossil fuels, biomass burning	Fertilizer, land-use conversion	Refrigerants, aerosols, industrial processes	Hydrocarbons (with NO$_x$), biomass burning	Land conversion, irrigation
Atmospheric lifetime	50–200 years	10 years	150 years	60–110 years	Weeks to months	Days
Pre-industrial concentration (1750–1800) at surface (ppb)	280 000	790	288	0	10	Unknown
Present atmospheric concentration (ppbv)	360 000	1720	310	CFC-11:0.28 CFC-12:0.48	20–40 (northern hemisphere)	3000–6000 in stratosphere
Present annual rate of increase	0.5%	0.9%	0.3%	4%	0.5–2.0%	Unknown
Global warming potential (GWP)	1	11	270	3400–7100	–	–
Relative contribution to the anthropogenic greenhouse effect	60%	15%	5%	12%	8%	Unknown

Source: based on work cited in the text

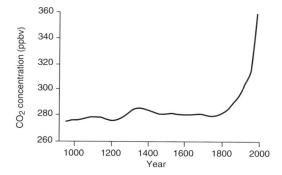

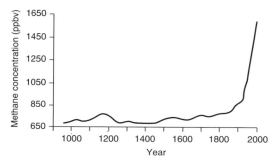

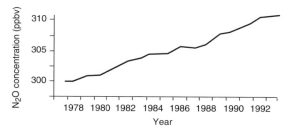

Fig. 6.1 Changes in the atmospheric concentrations of carbon dioxide (CO_2), methane (CH_4) and nitrous oxide (N_2O) (based on Houghton *et al.*, 1995).

ca. 1 per cent per annum (Watson *et al.*, 1992), is due to the extraction and consumption of natural gas, increased paddy rice production, animal wastes, biomass burning, enteric fermentation (i.e. fermentation in animal guts), landfills and sewage treatment. Overall, between $5-7 \times 10^9$ t of carbon are ejected into the atmosphere annually, approximately 50 per cent of which remains in the atmosphere. Where the remainder of the carbon is eventually stored remains as enigmatic as the location of carbon extracted from the atmosphere during an ice age (Section 2.4 and see above).

Table 6.3 shows the contribution that each heat-trapping gas emitted into the atmosphere through

direct or inadvertent human activity makes to the enhanced greenhouse effect. In addition Table 6.3 shows that the global warming potential of each gas is quite different. Global warming potential (GWP) reflects the heat-trapping capacity of individual molecules of each gas and its possible warming effect on the lower atmosphere, taking into account factors such as persistence times, i.e. atmospheric lifetime, in relation to carbon dioxide, considered to be the reference gas; see also the economic damage index proposed by Hammitt *et al.* (1996). Of particular note is the significance of chlorofluoro-carbons (CFCs), which have the highest GWP although their concentration in the atmosphere is low in comparison with other heat-trapping gases. They also have relatively long residence times in the atmosphere, which means their impact will be felt for some considerable time to come, even though their use is declining. Moreover, the hydrofluoro-carbons and perfluorocarbons now used as substitutes for CFCs are also heat-trapping gases, though they do not do as much damage to the tropospheric ozone layer as CFCs.

Any examination of global warming, while acknowledging the significance of nitrous oxide, CFCs, etc., must concentrate on carbon dioxide and methane because together these two gases are responsible for ca. 75 per cent of the current enhanced greenhouse effect. In relation to carbon dioxide there is a close relationship between world energy consumption and increasing atmospheric carbon dioxide concentrations. Figure 6.2 gives data on energy use and shows that major increases occurred after 1940, when there was a particularly large increase in oil consumption. This period has also been one of extensive and rapid deforestation (Chapter 7). Resulting emissions have altered the flux rates that operate between the atmospheric, oceanic and terrestrial reservoirs of the global biogeochemical cycle of carbon. Thus stored carbon from biomass and fossil fuels is rapidly being transferred to the atmosphere, from where a proportion is transferred to other sinks. Referring to data from a variety of sources, Flavin (1996) states that 'fossil fuel burning is now releasing about 6×10^9 tons of carbon into the air each year, adding 3 billion tons annually to the 170 billion tons that have accumulated since the Industrial Revolution'. Of the 3×10^9 t that do not remain in the atmosphere as much as 2×10^9 t are absorbed by the world's oceans (Siegenthaler and Sarmiento, 1993) and forest, and the remainder enters an as yet

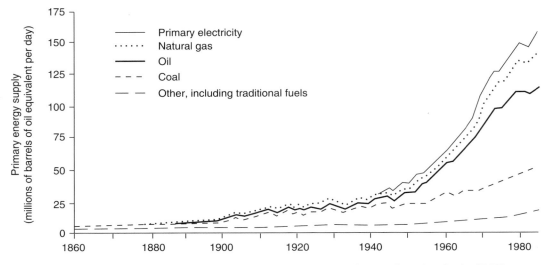

Fig. 6.2 Changes in the rate of energy consumption and sources of energy (based on Davis, 1990).

unidentified sink or sinks. Similarly about 50 per cent of the $0.5–2.5 \times 10^9$ t of carbon produced through changes in land use, especially deforestation, is retained in the atmosphere, i.e. an average of 0.75×10^9 t (Houghton, 1994). Thus, the oceans are already approaching saturation as the carbon has been absorbed (as carbon dioxide) over the years to produce carbonates and bicarbonates. A proportion of the carbon which is absorbed by phytoplankton in photosynthesis and transferred by food chains and food webs is also lost from active circulation. Organic matter produced as organisms die may return to the ocean water as nutrients but a proportion will sink to the ocean floor and become incarcerated in marine sediments, from which it enters active circulation again but only after many millennia. This is part of the so-called biological pump because it reduces the amount of carbon in the surface waters and so encourages the drawdown of carbon dioxide from the atmosphere. Houghton's review of various models indicates that 1.5– 2.54 Gt of the carbon (as carbon dioxide) produced anthropogenically enters and is stored within the oceans.

A further proportion of this carbon has a 'fertilising' effect on the Earth's biosphere, where it prompts increased primary productivity. According to Keeling *et al.* (1996) some 30 per cent of the carbon dioxide generated from fossil fuels is removed from the atmosphere to the terrestrial (biotic) sink in the northern latitudes of the northern

hemisphere, i.e. 30–60°N. This contrasts with the tropical region of the biosphere, which Keeling *et al.* consider is neither a source nor a sink of carbon dioxide; their study suggests that a further ca. 30 per cent of anthropogenic carbon dioxide is indeed absorbed in the oceans. This work supports the conclusions mentioned above in relation to the role of the global oceans as a carbon dioxide sink and suggests that the land biota of the northern latitudes comprise the 'missing sink', a possibility also suggested by Ciais *et al.* (1995) and Bender *et al.* (1996). Artificially elevated carbon dioxide levels are known to increase primary productivity (Graves and Reavey, 1996) but there is little hard evidence to corroborate the view that high-latitude forests are the repository of the 'missing sink'. LaMarche *et al.* (1984) have presented tree-ring evidence for enhanced growth in a range of subalpine conifers, including the bristlecone pine (*Pinus longaeva*) in the western United States, and it is accepted that some forests are becoming increasingly productive.

There are many published discussions on the likely mechanisms associated with the fertilising effect of elevated carbon dioxide, e.g. Woodward (1992), Leemans (1996), Schimel *et al.* (1995) and Melillo *et al.* (1993). All highlight the complexities associated with carbon dioxide fertilisation, notably the conditions affecting photosynthesis and in particular the roles of the nitrogen and hydrological cycles. For example, either limited water or limited nitrogen availability could reduce primary productivity even if carbon dioxide concentrations were doubled. However, the varied models that have been constructed to simulate the effects of carbon dioxide fertilisation on the world's biomes support the likelihood of a northern hemisphere terrestrial sink (a similar sink in the southern hemisphere is impossible because of the limited extent of landmass). Examples of such models include Smith *et al.* (1992), Woodward and Lee (1995), and Woodward *et al.* (1995) who have simulated global responses; others, e.g. Mooney *et al.* (1993), Melillo *et al.* (1995), Kittel *et al.* (1995) and Pan *et al.* (1996), have focused on North America; and yet others, e.g. Sykes and Prentice (1995), Plochl and Kramer (1995) and Kohlmaier *et al.* (1995), have concentrated on the boreal and tundra zones. The model results are variable but these, and others quoted in Gates *et al.* (1996), generally point to the northern high latitudes as a carbon sink which may be enlarged as trees migrate north in an enhanced greenhouse world. These models will be referred to again in Section 6.5.2.

The rise in atmospheric concentrations of methane is shown in Figure 6.1 and its GWP and sources are listed in Table 6.3. Ice-core data (Figs. 2.2 and 2.3) show that changes in the atmospheric concentration of methane and global temperatures are linked. For example, a doubling of methane concentrations occurred as the last ice age ended. A further doubling has occurred since ca. 1750, when its concentration was at ca. 0.8 ppmv. Today its atmospheric concentration is 1.75 ppmv, representing an average increase of approximately 1 per cent per year (Prather *et al.*, 1995). The natural sources of methane include wetlands and termite activity; in total such sources generate 160×10^6 t yr^{-1}. The largest anthropogenic sources are enteric fermentation, rice paddies, biomass burning and landfills; in total anthropogenic sources produce 3.75×10^6 t yr^{-1}. Much of this is removed from the atmosphere, mostly through its reaction in the troposphere with hydroxyl radicals (OH) but there is a 7–8 per cent difference between sources and sinks, so there is a net gain in the atmosphere. Although this does not represent a volume as large as the annual increment of carbon dioxide, the GWP of methane means that it has a significant warming effect. Moreover, the emission of methane is likely to increase in the future because of changes in land-use practices and because of feedback mechanisms caused by global warming, notably increased production from northern hemisphere wetlands and permafrost regions. Derwent (1996) has also pointed out that, as

anthropogenic emissions of trace gases increase, the volume of hydroxyl radicals in the troposphere is likely to decrease. This will allow methane concentrations to build up at an even faster rate than in recent decades as the oxidizing capability of the atmosphere diminishes. Consequently, the contribution of methane to global warming is set to increase from its present level of 15 per cent.

Apart from tropospheric ozone and water vapour (Table 6.3), whose role in global warming is not well documented, nitrous oxides and chlorofluorocarbons contribute 5 per cent and 12 per cent respectively to the enhanced greenhouse effect. According to Prather *et al.* (1995), nitrous oxide is produced in soils and water by biological activity. It has a much higher GWP than that of carbon dioxide and methane, so even small increases are significant in relation to global warming. The major natural sources are tropical and temperate soils, which emit between 6 and 12×10^6 t (N) y^{-1}. The largest anthropogenic source is cultivated soils, notably tropical pastures which have replaced forest. Other anthropogenic sources include biomass burning, agricultural systems and certain industrial processes. Overall it is estimated that 14.7×10^6 t (N) y^{-1} are produced with ca. 12.3×10^6 t (N) y^{-1} being removed, mostly through the photodissociation of N$_2$O by sunlight in the stratosphere. Data from ice cores and monitoring programmes (Prather *et al.*, 1995) indicate that before 1750 nitrous oxide concentrations were ca. 260–285 ppbv and have risen by approximately 15 per cent to 310 ppbv (Fig. 6.1). The atmospheric concentration of nitrous oxide is currently increasing by ca. 0.6 ppbv y^{-1}.

As Table 6.3 shows, chlorofluorocarbons have the highest GWP of all the heat-trapping gases. They are also responsible for stratospheric ozone depletion, discussed in Section 6.7. Stratospheric ozone depletion has a cooling effect; this offsets the warming effect that CFCs cause but it can induce other forms of environmental change due to increased ultraviolet radiation. The heat-trapping capacity of CFCs (and related compounds) is of considerable concern. Houghton (1994) states: 'It is estimated that a warming of about 0.2 °C on average (or about 20 per cent of the total amount of warming by all greenhouse gases) will have occurred in the tropics – at higher latitudes there is a compensating effect due to ozone reduction ... – due to the levels of CFCs present in the atmosphere by the end of the 1990s'. CFCs were developed in the 1930s and widely adopted as aerosol propellants, foam-blowing agents and refrigerants because of their non-flammability and low toxicity. However, they do vaporize readily, so they have been released into the atmosphere inadvertently. International agreement, the Montreal Protocol (Section 6.7) prompted by fears of ozone depletion, has reduced the volume of CFCs reaching the atmosphere (Simmonds *et al.*, 1993), though the use of CFCs in refrigerants in particular will continue to cause their emission for some time to come. However, several of the alternative compounds developed to replace CFCs are demonstrably less efficient at trapping heat. Imasu *et al.* (1995) have calculated the radiative forcing for a range of CFC alternatives, notably fluoroalcohols, fluoroethers and fluoroamines, and have found their warming potential is considerably less than for CFCs because of their shorter lifetimes in the atmosphere.

In view of the data presented above, it is surprising that evidence for global warming is still not overwhelmingly convincing (Santer *et al.*, 1996). However, there is now a consensus that the world is warming. Nicholls *et al.* (1996) have presented monitored data for the past 140 years, which show that 1986–1995 was the warmest decade of this period, with a temperature of 0.25 °C above the 1960–1990 average, and that 1994 and 1995 were among the warmest for the entire period. Overall, during the past century, the warming trend has amounted to 0.4 °C. The recorded trends are given in Figure 6.3. In the first International Panel on Climate Change (IPCC) report (Houghton *et al.*, 1990) it was suggested that a ca. 0.5 °C increase in temperature per century could reflect natural climatic variability. Now, however, it is acknowledged that the 0.4 °C increase in temperature represents a net rather than gross increase because of the counteracting cooling effect of aerosols such as sulphur dioxide and nitrous oxides, the components of 'acid rain' (Section 6.6). This was first highlighted by Wigley (1991) and has since been confirmed (e.g. Santer *et al.*, 1995; Hegerl and Cubasch, 1996; Hegerl *et al.*, 1996 and papers in Charlson and Heintzenberg, 1995). That one form of pollution may be mitigated by other forms of pollution should generate at least as much concern as global warming, stratospheric ozone depletion and 'acid rain' because this disruption of global biogeochemical cycles, i.e. Gaia in action (Section 1.3), will prove difficult to combat. This 0.4 °C warming has not, however, been globally uniform, as Parker *et al.* (1994) and Horton (1995) have demonstrated.

Much effort has also been invested in predicting the

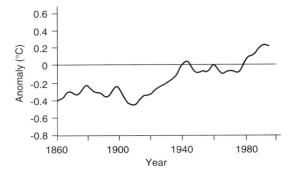

Fig. 6.3 Combined annual land-surface and sea-surface temperature anomalies for 1861–1994 relative to the mean for 1961–1990 (based on Nicholls *et al.*, 1996).

extent and rate of global warming for the ensuing decades. It is, however, generally acknowledged that available models have major limitations (Houghton, 1994; Houghton *et al.*, 1996), mainly because of the many variables involved and their complex interrelationships. These models are known as general circulation models (GCMs), the components of which have been described by Maunder (1994); examples are given in Table 6.4. Although there is much agreement between the results from different models there are also many discrepancies. To reconcile these differences the Intergovernmental Panel on Climate Change (Houghton *et al.*, 1990, 1992, 1995, 1996), established by the World Meteorological Organization (WMO) and the United Nations Environment Programme (UNEP), has presented a 'best assessment' of current understanding. There is an international group to examine the science of climatic change, another concerned with impacts and a third concerned with policy. The IPCC have used a range of scenarios, based on economic growth, energy consumption and CFC production, to estimate the total volume of emissions of greenhouse gases. These values are then used in GCMs to predict likely global warming in the early part of the next millennium. Overall IPPC suggest a rise of ca. 1 °C above that of 1994 by 2025, with further increases of ca. 0.3 °C per century. Further work has been undertaken to refine such estimates to take account of aerosol forcing. Assuming a slow reduction in the rate of economic growth and hence a slow reduction in greenhouse gas emissions, Mitchell *et al.* (1995a, b) have shown that global warming between 1990 and 2100 would be in the order of 2.7 °C, which is 0.8 °C lower than for

Table 6.4 Organisations that produce general circulation models (GCMs)

Model producers	Acronym
Geophysical Fluid Dynamics Laboratory, Princeton, USA	GFDL
Goddard Institute for Space Physics, Washington, USA	GISS
National Centre for Atmospheric Research, Boulder, USA	NCAR
Meteorological Office, Bracknell, UK	UKMO
Canadian Climate Centre in Toronto, Canada	CCC
Max Planck Institute for Meteorology, Germany	MPIM
Oregon State University, Oregon, USA	OSU

greenhouse-gas forcing alone. An additional implication of these results is that global warming is likely to accelerate if aerosol production is reduced, e.g. through measures to curb 'acid rain' (Section 6.6) such as the removal of sulphur dioxide and nitrous oxides within power-generation stations, factories, etc. Predictions such as these have been used to estimate the possible impact of global warming on sea-level ecosystems, agricultural systems, etc., as discussed below.

6.5.2 The potential impact of global warming

Global warming will have a significant impact on the Earth's natural and cultural characteristics. Increases in temperature such as those considered at the close of Section 6.5.1 will affect all aspects of the physical environment, e.g. the hydrological cycle, sea level, ecosystem structure and function, as well as those systems which are human dominated, e.g. agricultural systems. Adjustments of political and economic structures will be necessary to plan for and accommodate such changes.

The predictions of GCMs have been used to investigate the likely impact of increased temperatures on various aspects of the global environment. In relation to the hydrological cycle, the most obvious impacts will be on global sea level and the polar ice caps. How much melting will increased temperatures cause and what effect will this have on sea level? There is evidence for a reduction in ice volume from polar regions (e.g. Wadhams, 1995); Houghton (1994) states that the extent of glacier melting could have prompted a 5 cm rise in sea level i.e. 25–50 per cent of the total observed (10–20 cm) sea-level rise that has occurred in the past century. The remaining rise in sea level was most likely due to thermal expansion of ocean waters. For the period 1990–2030 Wigley and Raper (1992) predict that sea level will rise on average by

8 cm due to thermal expansion with an additional rise of ca. 8.4 cm due to polar ice and Alpine glacier melting. This represents a change in the rate of rise from $1 \, \text{mm y}^{-1}$ to ca. 3–$6 \, \text{mm y}^{-1}$. Like global warming itself, sea-level rise will not be globally uniform but its impact will be considerable. Moreover, the predicted increases may be considerable underestimates due to the difficulties of establishing the dynamics of the West Antarctic ice sheet.

Even the lowest estimated rise in sea level will have an impact. Clayton (1995), for example, has drawn attention to the existing problems caused by coastal erosion and the costs associated with coastal protection. Moreover, he points out that for the United States alone the cost of a 30–40 cm sea-level rise, predicted to occur by 2100, would cost between $5–10 billion, equivalent to 0.2 per cent of the gross domestic product (Yohe *et al.*, 1996). Although the United States could find this money and the expertise to make the necessary adjustments without significant loss of life, there are many other countries which do not enjoy such resources. Low-lying and island nations are particularly at risk and those in many tropical regions have the added hazard of the increased occurrence of tropical cyclones. As Varley (1994) has pointed out, natural disasters tend to have a much greater impact in less developed countries than in developed nations; less developed nations have a high vulnerability due to their inadequate infrastructure.

Among the nations most at risk from rising sea levels are Egypt and Bangladesh. Not only are they low-lying countries, both of them are densely populated. The possible impacts of sea-level rise have been discussed by Broadus (1993), who points out that these countries have extensive deltaic regions susceptible to subsidence. This would exaggerate sea-level rise. Both also rely heavily on agricultural production from their deltas and both have high population growth rates of 2.2 per cent

Table 6.5 Predicted precipitation and run-off for nine major rivers in high latitudes under a CO_2-doubled climate and the percentage changes from model-generated values for the present climate

	Predicted precipitation	Predicted run-off ($km^3 y^{-1}$)	Change (%) Precipitation	Run-off
Yenesei	1 770	660	29	41
Lena	1 712	730	15	27
Ob	1 564	620	47	37
Amur	1 438	308	3	8
Severnay Dvina	322	173	34	26
Kolyma	692	508	37	45
Indigirka	275	169	21	28
Mackenzie[a]	1 504	711	21	21
Yukon[a]	824	638	15	26

Source: based on van Blarcum *et al.* (1995)
[a]Alaskan, the others are in northern Asia

per annum according to the World Resources Institute (1996). The plight of Bangladesh is made worse by its vulnerability to storm surges caused by cyclones and flooding caused by heavy rainfall events in the Himalayas. In Europe the Netherlands is most vulnerable to sea-level rise as ca. 50 per cent comprises coastal lowlands, many of which are below sea level. Although it is also densely populated, the Netherlands at least has the financial wherewithal and engineering expertise to make the necessary adjustments. According to de Ronde (1993), mitigation measures will need to include the raising and strengthening of dykes and additional pumping facilities, the cost of which will be in the order of $10 billion. It is interesting to speculate on the funds the EU will need to make available to help its member states in dealing with the widespread problems of global warming. However, it is also likely that some island nations will disappear entirely, e.g. the Marshall Islands, Tuvalu and Kiribati. Others, e.g. the Maldives, will be severely affected. As well as losing land area much of their groundwater will be lost and/or adversely affected by the intrusion of saline water, and revenues will be restricted as tourism diminishes. Moreover, such nations have few resources to combat these problems.

The impact of global warming on the hydrological cycle is not confined to sea-level change. For example, Chiew *et al.* (1995) have simulated run-off and soil moisture availability in 28 Australian catchments under conditions suggested by several GCMs for the years 2030 and 2070. In wet, tropical catchments such as those of northeast Australia, the GCMs predict an increase in run-off of up to 25 per

cent. For drier southeast Australia, however, the GCMs do not agree but show a variance of ±20 per cent for run-off in 2030. For catchments in the South Australian Gulf, a decrease of 35 per cent in annual run-off is predicted with an increase of 10 per cent for Tasmanian catchments; for the west coast, simulations are highly variable at ± 50 per cent. Thus, the most erratic predictions are for the drier catchments, not the wetter catchments, and soil moisture changes in the drier catchments can be equally erratic. Overall the results point to the necessity of planning and anticipation of altered drainage patterns and all that such changes imply for domestic water supplies. In contrast, Wang and Allard (1995) have examined the impact of climatic change in a permafrost area of northern Quebec, Canada, where cooling rather than warming has occurred over the past 40 years. They suggest that cooling is likely to continue, decreasing the depth of the active layer by 20 cm and 30 cm over till and gneiss respectively, thus reducing the rate of permafrost creep. This is likely to occur elsewhere given the mosaic of global warming and cooling that is likely to characterise an enhanced greenhouse world (Section 6.5.1).

Van Blarcum *et al.* (1995) have used a GCM to estimate the monthly river flow for nine major rivers in high latitudes under doubled atmospheric carbon dioxide concentrations. Their results are summarised in Table 6.5, which shows that precipitation and run-off increase in all cases, often by more than 20 per cent. The analysis of Van Blarcum *et al.* also shows that the start of spring run-off and the timing of maximum run-off occur earlier in the year, in many cases in late March rather than in April.

Table 6.6   GCM predictions of global temperature, precipitation and vegetation change

A.   Predictions of changes in temperature and precipitation according to four GCMs

GCM[a]	Change in mean global temperature (°C)	Change in precipitation (%)
OSU	2.84	7.8
GFDL	4.00	8.7
GISS	4.20	11.0
UKMO	5.20	15.0

B.   GCM predictions of changes in the extent of five major biomes

GCM[a]	Land area (10^3 km^2)					
	Tundra	Hot and cold desert	Grassland	Dry forest	Mesic forest	Fraction of terrestrial land changing cover (%)
CURRENT	939	3 699	1 923	1 816	5 172	–
OSU	−302	−619	380	4	561	39.4
GFDL	−515	−630	969	608	−402	48.0
GISS	−314	−962	694	487	120	44.3
UKMO	−573	−980	810	1 296	−519	55.0

Source: based on Smith *et al.* (1992, 1995)
[a]Table 6.4 explains the GCM acronyms

There are complications for river-catchment management, including HEP production, the diversion of water for irrigation and future dam construction. The results agree with those of Ingram *et al.* (1996) for the southeastern Hudson Bay area, for which predictions include increases in precipitation and run-off and an earlier than usual break up of sea ice. As well as influencing precipitation patterns and run-off, global warming will affect the biota of both freshwater and saline water bodies. In relation to saline water bodies, Southward *et al.* (1995) have examined records for the past 70 years of marine communities in part of the English Channel and off the coast of southwest Britain. During warm periods, warm-water species have increased in abundance with some reversal during cool periods. Latitudinal shifts of up to 200 km in the range of species have occurred and Southward *et al.* suggest that, if temperatures increase by 2 °C in the next 50 years, latitudinal shifts of 320–640 km will occur and the composition of the marine biota will alter substantially. It is probable that species now common in the Bay of Biscay will colonise the English Channel, and those common in the western English Channel will migrate into the Irish Sea.

Changes in the hydrological cycle caused by global warming, as well as temperature increases, affect all the components and processes within ecosystems. Prediction of the impact of global warming on the world's ecosystems is at least as difficult as the prediction of global climatic change itself. This reflects the integrity of environmental systems and the use of imperfect GCM data on which to base forecasting. Nevertheless, it is important for conservation and planning to obtain information on likely future ecosystems and agricultural systems. Information on agricultural systems is particularly important because of the need to predict the quantity and location of future food supplies. Although there are many differences between the results of various GCMs (Table 6.4), all agree that high latitudes are likely to experience a greater degree of warming than middle and low latitudes. The data on river flow and precipitation in high latitudes given in Table 6.5 reflect such predictions. Moreover, all ecosystems are likely to be affected by the fertilising effect of increased atmospheric concentrations of carbon dioxide (Section 6.5.1). In consequence, many attempts have been made to model the future distribution of vegetation communities. Some of these models have been mentioned in Section 6.5.1.

Smith *et al.* (1992, 1995) have calculated likely changes in the extent of five major biome types (cf. Holdridge, 1967) using four GCMs. The results are given in Table 6.6, which shows that the overall trends are similar for all the models, except

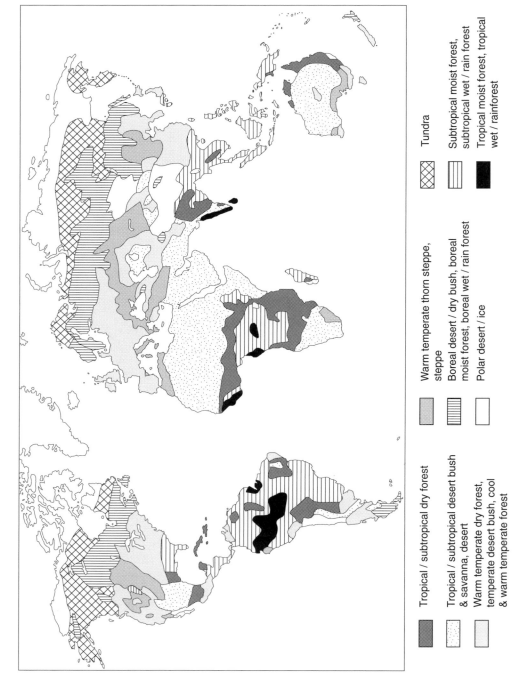

Fig. 6.4 (A) World vegetation zones today (based on Holdridge life zones).

Tropical / subtropical dry forest

Tropical / subtropical desert bush & savanna, desert

Warm temperate dry forest, temperate desert bush, cool & warm temperate forest

Warm temperate thorn steppe, steppe

Boreal desert / dry bush, boreal moist forest, boreal wet / rain forest

Polar desert / ice

Tundra

Subtropical moist forest, subtropical wet / rain forest

Tropical moist forest, tropical wet / rainforest

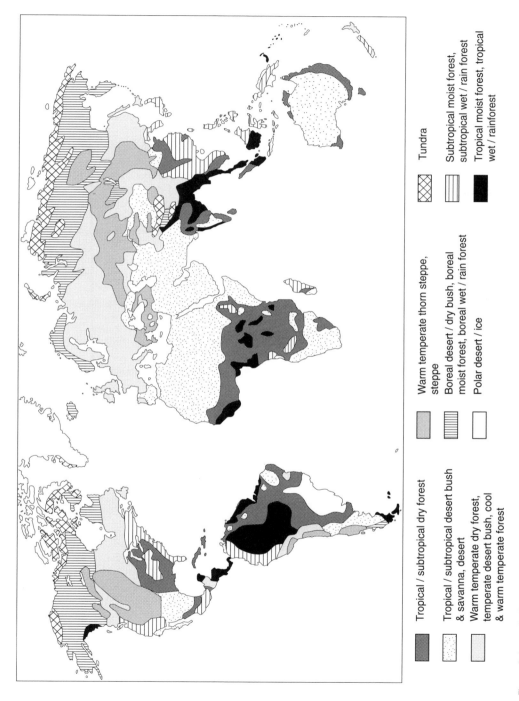

Legend:

Tropical / subtropical dry forest

Tropical / subtropical desert bush & savanna, desert

Warm temperate dry forest, temperate desert bush, cool & warm temperate forest

Warm temperate thorn steppe, steppe

Boreal desert / dry bush, boreal moist forest, boreal wet / rain forest

Polar desert / ice

Tundra

Subtropical moist forest, subtropical wet / rain forest

Tropical moist forest, tropical wet / rainforest

Figure 6.4 contd. (B) World vegetation zones as predicted by the general circulation model of the UK Metereological Office (based on Smith *et al.*, 1995).

predictions on the extent of mesic forests, but that variations occur in the magnitude of change. In particular, the area of tundra will diminish considerably, as will the extent of cold and hot deserts, whereas grasslands and forests will expand; tree lines will extend further north and to higher altitudes than at present. It is likely that carbon fixation (Section 6.5.1) will increase by ca. 10 per cent as forest increases but it is unclear what impact the invasion of tundra by boreal forest will have on the tundra's stored carbon. If, as suspected, these extensive wetlands become sources of carbon dioxide rather than sinks, the additional carbon dioxide will reinforce global warming (see Section 6.5.1). Woodward *et al.* (1995) have also devised a predictive model of global primary productivity and vegetation community distribution based on a range of observed parameters relating to controls on photosynthesis, e.g. temperature and soil moisture as well as nitrogen, carbon and hydrological budgets. Success with simulating modern vegetation and productivity patterns implies the model can be used to predict future patterns under elevated carbon dioxide concentrations. Figure 6.4 is an example of one attempt at predicting the distribution of the world's major ecosystems in a warmer world.

Many simulations of likely future vegetation patterns have been undertaken at regional scales. For example, Lenihan and Neilson (1993, 1995) have used the Canadian Climate Vegetation Model (CCVM) to determine the response of Canadian vegetation communities to doubled concentrations of atmospheric carbon dioxide, i.e. a doubling of pre-industrial carbon dioxide concentrations, which is predicted by IPCC (Houghton *et al.*, 1990) to occur between 2030 and 2050. Their model incorporated parameters that are known to be fundamental influences on plant growth and reproduction, notably degree days, absolute minimum temperature, depth of snowpack, actual evapotranspiration and soil moisture deficit. One of the main reasons for this study is the significance of boreal forests to Canada's wood and woodpulp industry and the need for sustainable management policies. Lenihan and Neilson used the GISS and GFDL (Table 6.5) GCM predictions of future climate to determine the effect of CCVM parameters on vegetation formations, forest type and species dominance. Their results are summarised in Figure 6.5, which shows that the GISS and GFDL models give different results in terms of the extent

of change but generally concur in relation to direction of change. In common with other models of high-latitude vegetation change (e.g. Smith *et al.*, 1992), the boreal and temperate formations expand and migrate northwards as Arctic and subarctic tundra formations decline in extent. Figure 6.5 also shows that individual tree species react in different ways to predicted global warming with species of spruce (*Picea mariana* and *P. glauca*) being reduced in extent by 20–30 per cent. The GISS and GFDL models give very different results for *Picea banksiana*. This, Lenihan and Neilson believe is due to the different ways the models take into account absolute minimum temperatures, snowpack depth and summer actual evapotranspiration, to which the species is particularly sensitive. The changes in the distribution of individual species will alter the forest types; and there is the emergence of a new category for which there is no analogue among present-day forest types, and which is most prevalent under the GFDL scenario.

However, these and many other models fail to take into account the ability of plant species to disperse themselves and whether or not such mechanisms will ensure survival against the backdrop of a warming process that is likely to occur rapidly. This has been discussed by Dyer (1995), who also points out that dispersal capabilities will be impaired as vegetation communities become fragmented. A comparison of wind-dispersed and bird-dispersed species, using computer models for landscapes with multifaceted land use and land cover, showed that bird-dispersed species were particularly sensitive to habitat fragmentation. Consequently, Dyer suggests that many plant species will be unable to migrate. Numerous other studies have been directed at examining the physiological response of individual species to climate warming. Repo *et al.* (1996) report the delaying effect of elevated temperatures on the frost hardening of 20–25 year old Scots pine in Finland, whereas Kellomaki *et al.* (1995) suggest that frost damage may occur if growth begins early during warm periods in late winter then the warm periods are followed by cooling. Kramer (1995) has also noted that temperatures just after leaf unfolding are critical for many European tree species. This and the study by Kellomaki *et al.* (1995) indicate that many tree populations could be adversely affected if greenhouse climatic regimes are characterised by erratic temperatures in late winter and early spring.

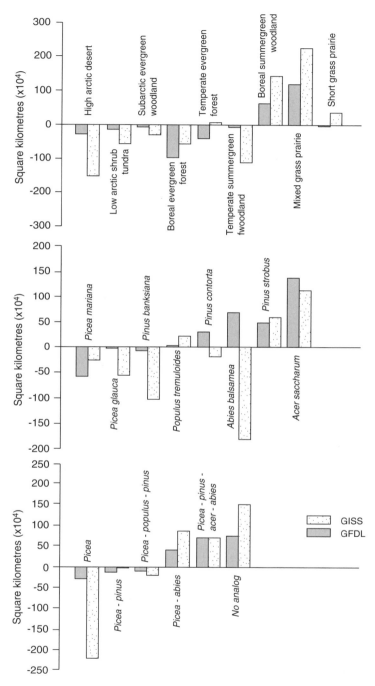

Fig. 6.5 Effects of climatic change on Canada's forest: predicted by the GISS and GFDL general circulation models for doubled pre-industrial carbon dioxide concentrations (based on Lenihan and Neilson, 1995).

Some studies have also been directed at modelling the effects of climatic warming on animal populations. Ries and Perry (1995) have examined the impact of elevated temperatures on the growth rates and food consumption of brook trout (*Salvelinus fontinalis*) at high-altitude Appalachian streams in West Virginia, United States. Their results show that if temperatures increase by 2 °C or less, fish growth would increase, but a 20 per cent increase in food consumption would be necessary, possibly increasing to 40 per cent with a temperature increase of 4 °C. Moreover, Keleher and Rahel (1996) have reported that any increase in temperature would result in a reduction in the habitat of salmonid fishes in Wyoming, United States. A 1 °C rise in temperature would reduce habitat by 16.2 per cent, and an increase of 4 °C would cause a 53 per cent reduction. The impact would be even more severe in the Rocky Mountain region with a 2 °C and 4 °C rise in temperature reducing the habitat by 35.6 per cent and 62 per cent respectively. This is because salmonid fish require mean July air temperatures of 22 °C or less. Moreover, as fish were forced to move to higher elevations their populations would become fragmented and gene pools would be reduced. In a large-scale study of Australia, Brereton *et al.* (1995) have drawn attention to the possibility of a reduction in the bioclimatic range of 41 species of fauna with some species likely to experience a disappearance of their bioclimatic zones. These studies all highlight the relatively high risk of extinction that is associated with the impact of global warming on the species being studied. However, such species were selected because of their likely susceptibility to climatic change and it must be remembered there have been periods of rapid climatic change in the past, when extinctions have occurred under natural circumstances. The late-glacial world is a case in point (Section 3.5). Much depends on the ecological tolerances and thresholds of individual species. What is different in the greenhouse as compared with the late-glacial world is the fragmented and vulnerable condition of so many communities and habitats of both plant and animal species due to human activity.

Simulating the impact of global warming on agriculture is at least as difficult as simulating its impact on hydrological systems and on plant and animal communities. Nevertheless, it is desirable that governments, etc., should have some indication of what to expect since political security and wealth generation depend on an adequate food supply. It is also sobering to consider that the lives of some 6.2 billion people depend on agriculture and this number increases daily. Apart from the direct impact of global warming on crop productivity, i.e. through its influence on soil moisture, nutrient availability, etc., there is the physiological response of crop types to consider (as in the case of European trees mentioned above), including the effect of carbon dioxide fertilisation (see above), and the impact of pests and diseases, which may themselves have been influenced by climatic change, e.g. altered ranges. The effect of global warming on crop and pasture production will in turn influence patterns of livestock production. There will also be economic and scientific responses to the impact of global warming on the world's agricultural systems which introduces a different set of variables that are difficult to quantify and have to be considered alongside the environmental and physiological factors.

Many experiments in crop physiology have been conducted to evaluate the impact of doubled carbon dioxide concentrations. The results generally indicate that crop productivity would increase. Yields in grain crops increase by as much as 36 per cent and those of cotton increase by 100 per cent (Nilsson, 1992; Wolfe and Erikson, 1993). However, such increases are rarely observed in field crops, some of whose yields have decreased (Bazzaz and Fajer, 1992). For those crops which show increased yields it is likely their use of water and available nitrogen is enhanced under elevated carbon dioxide concentrations. Such characteristics may facilitate the introduction of crops to areas and soils hitherto unsuitable. Furthermore the rise in temperatures predicted for high latitudes will probably encourage the introduction of crops, notably grain crops, into areas that currently experience too short a growing season. Some of the problems of predicting the impact of global warming on agricultural productivity are illustrated by the recent work of Ohta *et al.* (1995), who have examined the cultivation of paddy rice in China. Despite having the largest population in the world, China manages to feed its people but continued population growth means it could be susceptible to food shortages if environmental conditions became unfavourable. Conversely, China's role as a world power would be considerably enhanced if it could increase food production substantially. Work by Ohta *et al.* shows that conditions of doubled pre-industrial carbon dioxide would increase temperatures sufficiently to allow the transplanting of rice seedlings from nurseries 20–30 days earlier than occurs now; the warming would also extend the period available for

Table 6.7 Predicted regional impact of global warming on crop yields

	Temperature change	Soil moisture	Impact on yield
North America			
USA	Warming	Decrease	Decrease in maize, wheat, soya bean
Canada	Warming	Decrease	General decrease though northern limit of crop will move poleward
Central America	Limited warming	Decrease	May need more irrigation
South America			
Brazil	Limited warming	Varied	Increases overall
Andes	Limited warming	Increase	Some gains; cultivation possible at higher altitudes
Europe	Warming, especially in the north	Varied	Scandinavian wheat increases of 10–20% yield; losses in south Europe due to drought; increases in Alpine zone
Former USSR	Warming	Varied	Aridity increases in high latitudes, decreases in mid-latitudes
Middle East	Limited warming	Decrease	Up to 40% reduction in wheat yields
Africa	Limited warming	Varied	Decrease in arid and semiarid regions; varied elsewhere
Eastern Asia	Limited warming	Increase; also flooding	Increase, e.g. rice; if monsoon in southeast decreases, crop yields will fall
Japan	Warming	Increase	Increases of rice yield by 2–5%
Pacific Islands[a] Australasia	Limited warming	Varied	Some gains, e.g. New Zealand; some losses, e.g. Australia

Source: based on work cited in the text

[a]Pacific Islands will lose productive land to the sea as sea levels rise; this is likely to have a greater impact than global warming on crop yields

rice cultivation by 30–50 days, thus allowing a northward extension of cultivation, which could also be extended into semiarid areas. This would increase China's rice production considerably. However, shallow paddy fields expose considerable volumes of water, which under increased temperatures would be susceptible to raised evaporation rates. Under these circumstances water could be a limiting factor to rice production, which could decrease as a result of less available water. Hubbard and Floresmendoza (1995) have also addressed the issue of potential crop expansion in the United States for soya bean, wheat, sorghum and maize. Their deliberations, like those of Ohta *et al.* for China, reflect the different responses of the various crops to global warming and the difficulties of assessing the effect of temperature increase on water availability. Rotmans *et al.* (1994) have predicted that with a temperature increase of 1.85 °C by 2050 the cultivation zones in Europe will shift northward by 100–200 km.

Regional and/or national assessment of the impact of global warming on agriculture (e.g. Parry,

1990a, b, 1992; Tegart *et al.*, 1990) have shown that some regions will experience increased crop productivity whereas others will experience decreases, as shown in Table 6.7. It is unlikely there will be any major global change in agricultural productivity; although its distribution will alter to some degree, those countries which experience food deficits are likely to continue to experience them. Of particular concern are the arid and semiarid regions of Africa and Asia; their marginality for crop production may increase as weather patterns become increasingly erratic and they may experience rapid population growth. Rosenzweig and Parry (1994) and Rosenzweig *et al.* (1995) have also attempted to estimate food production but have used information on world food trade as well as crop growth models and GCM predictions of climatic change. Their results show that although global food production will not alter very much the distribution of crop production will change with most benefits in the temperate zone. In a study of future agriculture in England and Wales, Parry *et al.*

(1996) have included the predicted effect of climatic change on food prices. They demonstrate that the amount of land under agriculture would be reduced from present levels due to improved productivity caused by technological innovation. The results show there is little overall change in the total area under cereals but its location will change with a move into the fens and Lincolnshire from East Anglia. However, there have been many criticisms of this type of work. Reilly (1994) has pointed out that, especially in a global context, there is too much emphasis on the main cereal crops which are not necessarily the dominant crops in developing countries.

6.6 Changes due to fossil-fuel use: acidification

Acidification emerged as a major environmental problem in the late 1960s, when Scandinavian ecologists drew attention to declining fish stocks and the European atmospheric chemistry network provided data to show that precipitation was becoming progressively more acidic. The problem is now considered to be acute, especially in industrialised countries and has already prompted the initiation of preventative measures at an international level.

Although the term 'acid rain' is often used loosely to describe air pollution that involves the formation of acids in the atmosphere, it is acid deposition that creates environmental problems. As shown in Figure 6.6, the main gases involved in the production of acid deposition are sulphur and nitrogen oxides, especially sulphur dioxide (SO_2), nitrogen dioxide (NO_2) and nitrous oxide (NO) (together these latter two are known as NO_x). Although small amounts of these gases are produced by natural processes such as volcanic eruptions, the major input into the atmosphere is due to the burning of fossil fuels and industrial processes. As Figure 6.2 shows, energy consumption has increased during the past two decades, and though this has exacerbated the greenhouse effect (Section 6.5), it has also increased emissions of sulphur and nitrogen oxides. Once in the atmosphere, these gases combine with hydroxyl radicals or monatomic oxygen to produce acids, as shown schematically in Figure 6.7, in reactions that are enhanced by sunlight. These processes have been described by Wellburn (1994)

and occur in the troposphere, where the acids become incorporated into clouds (because they have high water solubility). This can create pH values as low as 2.6 in clouds. This can have important implications for high-altitude vegetation communities, especially forests, that are directly exposed to the cloud base. Ultimately, as depicted in Figure 6.6, the acid rain may be deposited hundreds of kilometres from the source of pollution as precipitation occurs, a process that Mohnen (1988) describes as 'a direct consequence of the atmosphere's self-cleaning nature'. This is the process of wet deposition. Alternatively, dry deposition of sulphur and nitrogen oxide as gases, aerosols or dry particles can occur, usually close to the emission source (Fig. 6.6). Both types of deposition can create environmental problems and it is generally considered that some two-thirds of this acidic pollution is due to sulphurous emissions while one-third is due to nitrous emissions.

The occurrence of acid rain varies spatially and temporally in relation to emission sources and prevailing meteorological conditions, especially wind direction. Several regions which have been severely affected include Scandinavia, northern Europe, eastern Canada and the northeast United States, all of which are located in the industrialised region of the temperate zone in the northern hemisphere and which experience relatively high precipitation rates on areas of acid bedrock. Such areas are particularly vulnerable to acid deposition damage not only because they are in receipt of pollution but because acid bedrock, with its attendant poor soils, offers little buffering capacity.

Until recently the developing world was generally considered to be free of environmental problems caused by acid deposition. However, rapid increases in industrialisation have occurred since the mid-1970s and there is now evidence to show that acidification is occurring in some countries, notably China and India. This is reflected in Table 6.8, which shows that emissions of both sulphur dioxide and nitrous oxides have increased by more than 50 per cent in the last decade. Foell et al. (1995) report that a project entitled Acid Rain and Emissions Reduction in Asia was established in 1989 to monitor the changes in air quality and acid deposition in some 23 countries. Their data reflect not only substantial declines in air quality and widespread increases in acid deposition but also the occurrence of transnational pollution. This has been further highlighted by Chung et al. (1996), who have

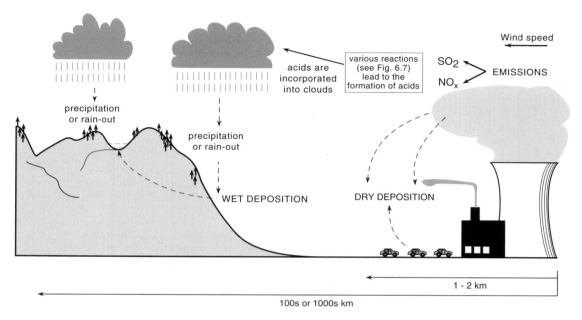

Fig. 6.6 Formation and deposition of acid pollution.

expressed concern that acidification is now widespread in the Korean peninsular with much of the pollution deriving from China. Moreover, Matsuura (1995) has anticipated that the problem will spread to Japan if China continues to industrialise so rapidly without desulphurisation technology to mitigate the problem. In China itself there are many areas that have been adversely affected by deposition. For example, damage to trees, stone and concrete structures in the cities of Chongquing and Guiyang is attributed to acid deposition (Zhao, 1994) and many urban centres have poor air quality with a high incidence of respiratory disease. There are also many areas in South America, notably Brazil and Venezuela, where acid deposition is having a damaging impact (Rodhe and Herrera, 1988).

The release of sulphur dioxide and nitrous oxide (Fig. 6.7) from coal, oil and natural gas represents a disruption of the biogeochemical cycles of sulphur and nitrogen. Essentially, human intervention is accelerating the flux of these substances from the lithosphere into the atmosphere and back into the lithosphere, where they not only affect terrestrial and aquatic ecosystems but also the fabric of the fuel-powered urban-industrialised systems. The Scandinavians were among the first to recognise the acidification problem. Their efforts prompted

widespread research throughout the industrialised nations, leading eventually to two major research initiatives directed at establishing pollution histories. In the United Kingdom, Norway and Sweden such research was undertaken as part of the Surface Waters Acidification Programme (SWAP), and a similar project, the Palaeoecological Investigation of Recent Lake Acidification (PIRLA) was established in North America. Not only have these projects demonstrated that acidification is widespread, they have also shown that the timing and degree of acidification are not uniform. The majority of the research undertaken under SWAP and PIRLA relies on the sensitivity to pH of diatoms, unicellular green algae that are widespread in aquatic habitats. Thus the changes in the composition of diatom assemblages can be determined by extracting the frustules (skeletal remains) of diatoms from lake-sediment cores and reconstructing the pH histories. The lake sediments may also contain the remains of other organisms such as cladocera or chrysophytes, soot particles and a geochemical record of heavy-metal deposition. These can augment diatom-derived pH histories. Battarbee (1994) has reviewed the results of the palaeoecological subproject of SWAP. This involved the establishment of a calibration data set of surface-sediment diatoms and chemistry from no less than 170 lakes. In

A Sulphurous and sulphuric acids

SO_2 is emitted from natural and anthropogenic sources and dissolves in cloud water to produce sulphurous acid:

$$SO_2 + H_2O \longrightarrow H_2SO_3 \rightleftharpoons H^+ + HSO_3^-$$

Sulphurous acid can be oxidized in the gaseous or aqueous phase by various oxidants

$$SO_2 \xrightarrow{\text{oxidant}} SO_3$$

Aqueous sulphur trioxide forms sulphuric acid:

$$SO_3 + H_2O \longrightarrow H_2SO_4 \rightleftharpoons H^+ + HSO_4^- \rightleftharpoons 2H^+ + SO_4^{2-}$$

B Nitrous and nitric acids

N_2O is emitted by the process of denitrification and although relatively inert it is a greenhouse gas. NO and NO_2 (collectively designated as NO_x) are produced by combustion processes and lightning. They are involved in many chemical processes, some of which damage the ozone layer in the stratosphere:

$$O_3 + NO \longrightarrow NO_2 + O_2$$

Other chemical processes may generate ozone in the troposphere causing photochemical smogs:

$$NO_2 \xrightarrow{\text{light}} NO + O$$
$$O + O_2 \longrightarrow O_3$$

In addition, nitric and nitrous acids may be produced:

$$2NO_2 + H_2O \longrightarrow HNO_3 + HNO_2$$

These acids are components of acid rain along with sulphurous and sulphuric acids

Fig. 6.7 Major components of acid rain: a simplified scheme to show how they form in the troposphere.

combination with data on dissolved organic content (DOC) and aluminium, various transfer functions were derived and used to reconstruct pollution histories. Battarbee states that the results indicate 'the overwhelming importance of acid deposition as the cause of acidification'. Many lakes in southwest Scotland have been adversely affected with pH changes of up to 1.2 units, e.g. Loch Grannoch. Some of the SWAP results are given in Table 6.8 along with a selection of PIRLA results (Charles, 1990; Charles and Smol, 1994). The PIRLA results reflect the sensitivity of lakes in New England and the Great Lakes Region to acid pollution (see also Hall and Smol, 1996).

All the lakes cited in Table 6.9 have experienced considerable changes in pH as a result of industrialisation with the earliest declines of pH occurring around 1840–1850. Many other studies outside the scope of either PIRLA or SWAP have identified similar trends in susceptible regions. For example, Cumming *et al.* (1994) have reconstructed the pH histories of 20 Adirondack Park Lakes (United States) using chrysophyte remains. Chrysophytes are particularly sensitive to pH and the analyses of Cumming *et al.* show that most of these lakes have become acidified as sulphate deposition accelerated in line with the intensification of industrialisation in the late nineteenth and early twentieth centuries. Moreover, in a study of four lakes in the west of Ireland, at considerable distance from pollution sources, Flower *et al.* (1994) have used diatom analysis to show that two have become acidified, a process which began in the 1860s. The concept of critical load has been formulated to extend the value

Table 6.8 Examples of changes in the pH of lake waters since ca. 1840

Site[a]	Location	pH change	Approximate date of initial pH change
Loch Enoch	SW Scotland	0.9	1840
Loch Grannoch	SW Scotland	1.2	1925
Loch Dee	SW Scotland	0.5	1890
Round Loch Glenhead	SW Scotland	1.0	1950
Loch Laidon	SW Scotland	0.7	1850
Grosser Arbersee	West Germany	0.8	1965
Kleiner Arbersee	West Germany	0.8	1950
Gårdsjön	Sweden	1.5	1960
Hovvatn	Norway	0.75	1918
Holmvatn	Norway	0.5	1927
Malalajärvi	Finland	1.0	1950
Hirvilampi	Finland	0.7	1950
Big Moose Lake	USA	1.0	1950–1960
Woods Lake	USA	0.4	1930s
Ledge Pond	USA	0.6	1880s
Beaver Lake	Canada	0.6	1950s
Lake B	Canada	1.6	1955
Lake CS	Canada	1.0	1955

Source: based on work cited in the text
[a]This is just a selection of the many lakes that are known to have become acidified since ca. 1840

Table 6.9 Changes in the emissions of sulphur dioxide and nitrous oxides in specific countries 1980–1993

	1980	1990	1993
Sulphur dioxide			
Canada	4 643	3 326	3 030
USA	23 780	21 060	20 622
Poland	4 100	3 210	2 725
Russia	12 123	10 166	NA
UK	4 903	3 754	3 188
China	13 370	19 990	NA
Japan	1 263	876	NA
India	2 010	3 070	NA
Nitrogen oxides			
Canada	1 959	1 999	1 939
USA	21 469	21 373	21 001
Russia	2 578	3 050	NA
UK	2 395	2 731	2 347
Japan	1 622	1 476	NA
China	4 910	7 370	NA
India	1 670	2 560	NA

Source: based on data in United Nations Environment Programme (1993) and World Resources Institute (1996)
NA = not available

of such data to management. Critical load is the amount of acid deposition that can be absorbed by an ecosystem or environment without causing positive feedback, i.e. enviromental change, including ecological change. The critical load for any given lake or ecosystem will depend on many variables, including the buffering capacity of soils and bedrock. Moreover, critical load is the central

concept of a protocol signed at Oslo in 1994, a few years after the Montreal Protocol of 1987. Signatories to the Oslo document committed themselves to the reduction of acidifying emissions. Essentially, emissions produced or reaching an area must be below the damage level. Consequently, methods for calculating critical loads are important. One approach is suggested by Battarbee *et al.* (1996)

using a diatom-based model. The premise is that diatom sensitivity to pH change can be used as a reliable indicator of an ecological threshold.

In lakes that have undergone accelerated acidification, the ecological impact has been marked. Increased concentrations of hydrogen ions can cause the potentially fatal release of sodium cations and chloride anions from freshwater organisms. In addition, the deposition of sulphate is often accompanied by heavy metals, which can cause deleterious effects. There may also be disruptions to phosphorus and nitrogen biogeochemical cycles. All of these changes affect plant and animal growth and their reproductive potential. Overall, acidification alters the structure and function of aquatic ecosystems. As Muniz (1991) has discussed, there are changes in all types of aquatic biota, altering food chains and food webs and generally reducing biodiversity as only the most acid-tolerant species survive. Low pH slows down the breakdown of organic detritus because it inhibits microbial activity; this means that minerals are not recirculated as rapidly as they would be in non-acidified ecosystems. There is also some evidence for the adverse impact of acidification on marine ecosystems though this occurs indirectly through the leaching from soils of trace metals, which eventually accumulate in coastal waters to inhibit phytoplankton growth (Granéli and Haraldson, 1993). The acidification of the breeding ponds of many North American amphibians is now considered to be a threat to their survival, as Tattersall and Wright (1996) have demonstrated. Their experiments on the toad, *Bufo americanus*, have shown that exposure to water with pH 6 resulted in an increase of between 20 per cent and 80 per cent in the amount of ammonia excreted by embryos and tadpoles. This loss of nitrogen may account for high tadpole mortality rates. In the United Kingdom the natterjack toad is adversely affected by acidification.

There is also the possibility that lakes and terrestrial ecosystems undergoing acidification are simultaneously being affected by global warming, as Wright and Schindler (1995) have discussed. For example, the increase in temperatures predicted for the boreal zone could cause increased mineralisation of nitrogen and oxidation of organic compounds, including those containing sulphur. The acids so produced would mobilise aluminium; both would reach lakes to reinforce the acidification process and contribute to high concentrations of aluminium.

Moreover, a decline in precipitation caused by global warming would cause a decline in lake water levels, so deepwater habitats for organisms requiring a narrow annual temperature range would be reduced and ultraviolet radiation would penetrate to depths currently not reached. On the basis of the artificial addition of sulphuric acid to lakes in the Experimental Lakes Area of northwestern Ontario, Canada, Schindler *et al.* (1996) have demonstrated that increasing acidity and warming have had complex but interactive impacts. In combination they have produced a decline in dissolved organic carbon content and since such substances absorb solar radiation their decline has facilitated the penetration of solar radiation, including harmful ultraviolet-B (UV-B), to greater than usual depths. Schindler *et al.* suggest this had a more intense effect than exposure to UV-B caused by stratospheric ozone depletion (Section 6.7). Moreover, they argue it is possible that some of the ecological damage, e.g. the loss of species ascribed to acid deposition, may be due to increased UV-B radiation. Research on Swan Lake, near Sudbury in Canada, which has been subject to considerable acid deposition generated by nickel and copper smelting, has led Yan *et al.* (1996) to draw similar conclusions. The impact of UV-B radiation (Section 6.7) will intensify as ozone depletion, global warming and acidification continue.

However, from an economic perspective most interest has centred on fish stocks, especially salmonids, which are particularly sensitive to declining pH. Fish can be adversely affected by several aspects of acid deposition, including the gradual accumulation of hydrogen ions over long periods, short-lived pulses of acid water following rapid snowmelt in the spring, or flushing of acidified soils by heavy rain and the increased concentration of heavy metals, notably aluminium, caused by their removal from catchment soils. Fish are particularly vulnerable when pH declines to below 5.5; stocks decline not because of high mortality rates, which tend to occur only when pulses of acid water occur, but through recruitment failure. A combination of pH 5.5 or less plus high concentration of aluminium in calcium-deficient waters may promote physiological changes in fish, especially in immature fry, that cause death by interfering with the gills. The first evidence for a decline in fish stocks came from Scandinavia in the early 1970s. Historical records of fish catches reflect a progressive decline since the 1880s. For example, Hesthagen *et al.*, quoted in

Mannion (1992b), have shown that fish populations are almost extinct in some Norwegian countries. They also show that catches of Atlantic salmon in 10 rivers of Aust-Adger in southern Norway and Vest-Agder, Rogaland, have declined from ca. 53 t in the 1880s to less than 1 t in the 1980s. Many lakes and streams in Scotland and Wales have also suffered declines in fish populations, especially brown trout.

The impact of acid rain deposition on terrestrial ecosystems occurs in two ways. Firstly it has a direct effect as wet or dry deposition occurs on vegetation, and secondly it has an indirect effect through its influence on soil properties and soil flora and fauna. However, deteriorating ecosystem health may not be entirely due to acidification, as ozone accumulation in the troposphere and global warming may also be contributing to this phenomenon. Acid deposition increases soil acidity, decreases nutrient content and mobilises heavy metals. As hydrogen ions accumulate, cations such as calcium, sodium and magnesium are displaced and may be leached as soluble salts from the system. Thus the soil becomes nutrient impoverished. Nutrient cycling is further influenced by microbial activity which is also sensitive to pH; as pH declines, bacteria become less important than fungi as decomposers. The loss of earthworms and other soil fauna reduces soil mixing and aeration. All of these processes cause soil impoverishment. Although such processes can also occur where afforestation programmes, involving extensive conifer plantations, have been carried out on acidic soils and peats, it is now acknowledged that a major cause of soil degradation is acid deposition. This is exemplified by Likens *et al.* (1996), who report change in soil and stream chemistry in the Hubbard Brook Experimental Forest, New Hampshire, United States for the past 50 years. They have shown that substantial declines in total soil calcium and magnesium have occurred; these substances have been removed from the forest ecosystem in drainage water. Although the amount of calcium and magnesium in the drainage water has declined since the late 1960s, the store in the catchment soils has been so seriously depleted that Likens *et al.* suggest it will delay ecosystem recovery as acid deposition declines. This decline is compounded by a reduction in the amount of calcium deposited from precipitation. Another study on the effect of acidic deposition on soils has been undertaken on the podzols of the Kola Peninsula, Russia (Koptsik and

Mukhina, 1995). Here pH decreased, and calcium and magnesium cations declined as exchangeable aluminium increased under simulated rainfall of pH 4.5. These results occur with observed soil characteristics in the vicinity of a nickel smelter, the source of acid deposition. Koptsik and Mukhina state that the loss of magnesium and calcium could be the cause of forest damage in the area. In addition, Graveland and Vanderwal (1996) have shown that eggshell defects in forest passerine birds in the Netherlands are due to acid rain. This occurs because calcium is removed from the soil by rain of low pH and thus causes a decline in snails. In turn, lower snail densities concur with high incidence of eggshell defects.

Acid deposition has been identified as a cause of forest damage in Europe and North America, though it may also promote forest growth. In Germany's Black Forest, for example, die-back involving defoliation and crown thinning was first reported in the 1960s. Since then a systematic monitoring programme of Europe's forests, which began in 1986, has shown that the problem is widespread (Innes, 1992; Elvingson, 1995). Not all of this can be attributed to acid deposition; other causes are disease, drought and pests. Nevertheless, it is also possible that, where critical loads of acid deposition are exceeded, tree growth is sufficiently impaired to render species more than usually susceptible to drought, frost, etc. The enhanced leaching of nutrients by acidified drainage may contribute to this, as might the adverse effects of low pH on soil micro-organisms. Mazurski (1990) has mapped the distribution of forest damage in Poland in relation to atmospheric sulphur dioxide concentrations and shown there is a high degree of correspondence between the two. Forest damage is greatest around Katowice, where up to 100 μ m^{-3} of sulphur dioxide occurs. This derives not only from within Poland itself but also from its industrialised neighbours, reflecting the transnational nature of acid pollution. The overall damage to Poland's forests amounts to an annual timber loss of 3×10^6 m^3 and there are attendant losses in game; together these amount to $200 million. Forest damage may also occur through occult deposition, i.e. deposition from clouds directly onto tree canopies, and by the deposition on soils of heavy metals, which can impair tree growth. Widespread die-back has also been noted in the United States, where red spruce forests are particularly vulnerable, especially in New England. Conversely, the negative impact of acid deposition can be offset by the fertilising effect of enhanced sulphur and nitrogen deposition.

Acid deposition not only affects aquatic and terrestrial ecosystems, it also affects the fabric of urban areas and human health. The discolouration and solution of urban buildings, monuments and statues, etc., is well known in many of the world's major cities. These conditions are caused by industrial and vehicular emissions of sulphur dioxide and nitrous oxides and the severity of the impact relates to the type of stone used; these emissions can also corrode metals, especially iron and iron alloys (e.g. Vertes *et al.*, 1995; Knotkova *et al.*, 1995). Experimental work on Portland and Monks Park limestone and various marbles has shown that recession rates, based on the amount of released calcium, was highest for the Monks Park limestone and lowest for the marble (Johnson *et al.*, 1996). Many of London's limestone buildings have been adversely affected as a result of the susceptibility of these building materials to acid emissions. The problem is also acute in Manchester, and in Athens many ancient monuments are under threat. Sandstones are also susceptible to damage by acid emission, as exemplified by the study of Nord and Tronner (1995) on the Royal Palace and the Riddarholm church in Stockholm. These are constructed mainly of calcitic Gotland sandstone and were built in the eighteenth century. Both show evidence of serious decay, notably a loose surface, gypsum formation, discolouration, signs of exfoliation and efflorescence caused by salt formation. In relation to human health, acid emissions adversely affect respiratory and cardiovascular systems. During the past decade there has been a major increase in respiratory disorders, including asthma, which may be related to high pollution levels caused by traffic. According to Friends of the Earth (1995), some 1200 deaths are caused annually by sulphur dioxide pollution, incurring annual health costs of £2.6 billion. The mobilisation of heavy metals from soils could indirectly contaminate drinking water to cause other health problems.

These examples illustrate the widespread occurrence of acid rain and its role in environmental change, and because it is now acknowledged as a major pollution problem inroads have been made into developing ameliorative strategies. Some of these strategies have produced further environmental change. Short-term solutions have been sought to

alleviate immediate problems and liming has been successfully employed in some lake ecosystems to restore productivity. In Lake Gårdsjön in Sweden the increased availability of phosphorus after liming, due to complex interactions between the lime, dissolved oxygen concentrations, organic carbon and nitrogen, have led to enhanced phytoplankton productivity (Broberg, 1988; Fleischer *et al.*, 1993). Results from this study also suggest that liming reduces the amount of aluminium in solution, which is advantageous to fish populations; its reduction may also contribute to enhanced primary productivity because aluminium is thought to depress the rate of carbon dioxide uptake in aquatic photosynthesis. Liming has also been carried out at Loch Fleet, a small upland acid lake in southwest Scotland, and Howells (1990) reports that pH improved rapidly, even within 5 days in some parts of the lake, with simultaneous declines in aluminium concentrations. Experiments involving the liming of catchment soils have also been undertaken and, as Blette and Newton (1996) have shown for Woods Lake in the Adirondack Mountains of the United States, rapid improvements in soil and lake pH can occur (Newton *et al.*, 1996). Activities such as liming do not provide a long-term solution to acidification since they do not address its causes. The most significant measures that can be taken relate to curbing the sources of the pollutants. This is an obvious solution and its efficiency is borne out by Wright and Hautis (1991), who have monitored the effects of shielding the acid-sensitive Lake Risdalsheia in southern Norway from acid precipitation. After construction of a roof and the application of 'clean precipitation', changes caused by acidification were reversed. In particular, the flux of strong-acid anions in the run-off, mainly sulphate and nitrate, decreased as acid deposition decreased. Wright and Hautis report that nitrate concentrations decreased by 60 per cent within 2 weeks of commencement and the sulphate concentrations began to decline with 4 months.

How much of a reduction in pollutant emissions is necessary appears to vary within and between regions, and depends on critical loads (see above). Schindler (1988) has suggested that annual sulphate deposition must be limited to 9–14 kg ha^{-1} in order to protect the most sensitive aquatic ecosystems. Since annual values of 20–50 kg ha^{-1} are currently recorded in eastern North America, significant reductions must occur before the problem is eliminated. Although it is unlikely that such a goal will be achieved, it is already apparent that attempts to curb emissions in Europe that began in the early 1980s have had an effect. The Geneva Convention, signed in 1979 and effective from 1983, as well as the so-called '30 per cent club', a protocol to this convention signed in 1985 by many acid-producing countries, constitute examples of international collaboration aimed at solving environmental problems. The '30 per cent club' was signed by 20 European countries, including the former Soviet Union, but not by the United States, Britain and Poland; its aim was to reduce sulphur dioxide emissions by some 30 per cent of 1980 levels by 1993. The recognition that different areas react in different ways to acid deposition prompted the need for a discriminating instead of a blanket approach. Consequently, a new protocol was signed in 1994 to replace the 30 per cent club (see above). This new protocol takes into account critical loads and has produced specific target reductions in emissions for each country. There are many ways to achieve reductions; desulphurisation is used in the power stations of west Germany and at some power stations in Britain; other methods include the use of less sulphur-rich fossil fuels, the development of alternative energy sources and more efficient technologies for energy provision from fossil fuels. Table 6.8 gives data on changes in sulphur dioxide emissions for several countries between 1980 and 1993. It shows that the various protocols have achieved some success, especially for the industrialised countries. However, it is notable that sulphur dioxide emission control remains poor in newly industrialising countries, where levels have increased (see above). Moreover, emissions of nitrous oxides have increased overall with only a few nations showing reductions. Protocols to curb these emissions are still needed.

There are already numerous examples for the recovery of lake ecosystems where pollutant emissions have been reduced. Battarbee (1994) has reported that diatom communities in the most recent sediments of several Scottish lochs are responding to pH declines due to post-1970 reductions in British acid emissions. Similarly, Henriksen et al. (1988) have suggested that lower sulphate loadings in lakes of southern and eastern Norway, although still somewhat elevated, may be due to a general decrease in the western European consumption of heavy fuel oils since 1979. There are further examples of lake recovery in eastern Canada, where smelter closures and controls on sulphur dioxide emissions in the past 15 years have reduced emissions to about one-third of their early 1970s values (Gunn and Keller, 1990). Many of these lakes now have lower concentrations of aluminium and toxic metals as well as lower sulphate concentrations and increased pH, and in some cases trout have become re-established. Moreover, it is to be hoped that developing countries will take note of the lessons to be learned from the developed nations and avoid, as far as possible, the ecological consequences of acid deposition. However, there is also the possibility that reducing sulphurous emissions will remove one of the factors that may be holding in check enhanced greenhouse warming (Section 6.5.2). This suggests that mitigation of one environmental problem, by reducing sulphur dioxide emissions, may actually accelerate the rate at which another environmental problem, enhanced greenhouse warming, becomes manifest.

6.7 Ozone and lead problems

In addition to increases in the concentrations of greenhouse gases and acidification, which have initiated mechanisms of environmental change, human activities have caused changes in the ozone layer in the stratosphere and increasing concentrations of atmospheric lead. These have significant implications for human health as well as for the Earth's ecosystems.

The issue of the stratospheric ozone layer is quite separate from the increasing concentrations of tropospheric ozone that contributes to enhanced greenhouse warming (Section 6.5) and relates to its disturbance by long-lived pollutants (see review in Crutzen, 1996). Figure 6.8 shows the position in the stratosphere of the ozone layer; this is important because it filters out incoming ultraviolet (UV) radiation and thus acts as a screen against UV-B radiation, which can increase the occurrence of some forms of skin cancer and cataracts. Ozone is formed by a reaction between oxygen and ultraviolet radiation, and is depleted naturally in three possible ways, as shown in Figure 6.8. Concern was first expressed about the possibility of accelerated stratospheric ozone depletion in the early 1970s due to the advent of supersonic aircraft which fly in the lower stratosphere and emit nitrogen oxides. Although these gases are potential catalysts for the destruction of ozone, they are now acknowledged as

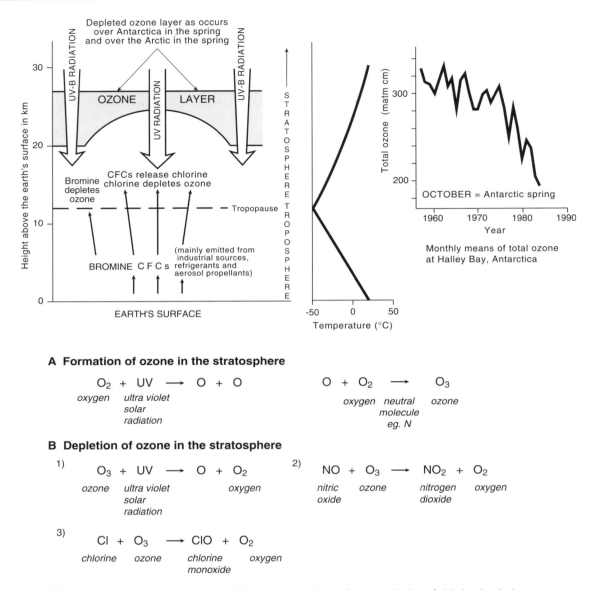

A Formation of ozone in the stratosphere

$$O_2 + UV \longrightarrow O + O$$

oxygen   ultra violet
         solar
         radiation

$$O + O_2 \longrightarrow O_3$$

oxygen neutral ozone
       molecule
       eg. N

B Depletion of ozone in the stratosphere

1)
$$O_3 + UV \longrightarrow O + O_2$$

ozone   ultra violet   oxygen
      solar
      radiation

2)
$$NO + O_3 \longrightarrow NO_2 + O_2$$

nitric   ozone   nitrogen   oxygen
oxide   dioxide

3)
$$Cl + O_3 \longrightarrow ClO + O_2$$

chlorine ozone   chlorine   oxygen
             monoxide

All these processes occur naturally but 3) is accelerated through the production of chlorine (and other halogens) from CFCs.

Fig. 6.8 Stratospheric ozone depletion and its relation to UV-B radiation (data on graph are generalised from Farman *et al.*, 1985).

a minor factor in comparison to the effect of another group of pollutants known collectively as Freons or chlorofluorocarbons (CFCs). These compounds are non-toxic, non-flammable and chemically inert liquids or gases. These properties which make them useful for a wide range of applications, including aerosol propellants, refrigerants, cleansers for electronic components and in the production of foam plastics. According to Clarke (1986), global production of the two most common CFC gases, Freon 11 ($CFCl_3$) and Freon 12 (CF_2Cl_2) rose rapidly from less than 50×10^3 t per annum in 1950 to 725×10^3 t per annum in 1976, of which some 90 per cent is released directly into the atmosphere. The remaining 10 per cent, mostly used for refrigerants, is eventually released when the refrigerators are discarded. Although CFCs are released into the troposphere, which underlies the stratosphere (Figure 6.8) and wherein they act as greenhouse gases (Section 6.5), they do not rapidly degrade and eventually enter the stratosphere. There they are subject to intense UV radiation – the same band of radiation as absorbed by ozone. This can lead to photodissociation and the release of chlorine atoms (Figure 6.8) which destroy ozone; other halogens have a similar effect, especially bromine, which is used as the fungicide methyl bromide (Danilin et al., 1996). According to Maunder (1994), a single chlorine atom released from a CFC can catalyse ozone destruction to re-emerge unchanged as many as 100 000 times before it combines with other chemicals.

The Nobel laureates Rowland (1996) and Molina (1996) have reviewed developments in stratospheric chemistry. They show that even before there was clear evidence for destruction of the ozone shield, theoretical work suggested that increasing emissions of CFCs would lead to a build-up of chlorine of sufficient magnitude to impair its protective function. In addition, the inert nature of the CFCs means they have long residence times in the atmosphere, ranging from 60 to 110 years (Table 6.3). This ensures that, even if no further emissions were to occur, the destructive process would continue for some considerable time. The first significant data were presented by Farman et al. (1985) of the British Antarctic Survey, who established that an ozone 'hole' had occurred in the stratospheric ozone layer over Antarctica each spring since 1977. For the period 1977–1984 the concentration of ozone decreased during the spring by ca. 40 per cent. Overall the Earth has lost 4–8 per cent of its stratospheric ozone. The greatest thinning occurs between $10°$ S and $20°$ S but thinning has also been recorded between $1°$ N and $60°$ N (Abbatt and Molina, 1993). The impact on humans is likely to be considerable, especially in the southern hemisphere (see below).

Additional work, e.g. Herman et al. (1995), Jones and Shanklin (1995) and Jiang et al. (1996), has confirmed that stratospheric ozone over Antarctica is regularly depleted in the spring, i.e. October. It has also been established that the trend of ozone depletion was greatest in the early 1980s as anthropogenic emissions of chlorine increased. Moreover, there have been reports of stratospheric ozone depletion elsewhere. For example, Kirchhoff et al. (1996) recorded a substantial decrease of ozone over southern Brazil in October 1993 which was probably related to ozone depletion over Antarctica as the hole extended north in a narrow belt. The seasonal depletion of ozone over the Arctic is also now well established (Santee et al., 1995) though thinning of the ozone layer is much less extensive than over the Antarctic; so-called mini-holes are thought to occur over Europe (Peters et al., 1995). The hole over the Antarctic in spring has certainly been evident from remotely sensed images. When it was initially reported, it was considered the Antarctic ozone hole might be a short-lived phenomenon. Now its regular reappearances, sometimes with increased intensity, reflect its persistence; see Maunder (1994) for details of the extent of the hole in 1992 and 1993, and Downey et al. (1996) for details of the hole in 1994.

Although it is not yet clear what are the effects of stratospheric ozone depletion, or what they will be in the future, there is concern about possible increases in skin cancer due to the enhanced receipt by the Earth's surface of UV-B radiation, 99 per cent of which is usually absorbed by stratospheric ozone. Indeed medical statistics show a rapid increase in the incidence of human skin cancers, especially in Australia and New Zealand. How much of this is due to stratospheric ozone depletion is difficult to ascertain because recreational and fashion habits have also increased exposure to solar radiation. At low exposure rates, ultraviolet radiation is beneficial because it promotes the production of vitamin D in the skin of higher organisms, but at high exposures it becomes harmful because it damages proteins and DNA molecules and it suppresses the natural immune system. Nebel and Wright (1996) state: 'If the full amount of ultraviolet radiation falling on the stratosphere reached Earth's surface, it is doubtful that any life could survive; plants and animals alike would simply be 'cooked'. Even the small amount (less than 1%) that does reach us is responsible for all the sunburns and more than 700 000 cases of skin cancer and precancerous ailments per year in North America, as well as for untold damage to plant crops and other life-forms'. Moreover, Ramaswamy et al. (1996) report that stratospheric ozone depletion is likely to be responsible for a cooling trend in the lower stratosphere that has occurred over the past 11 years, adding yet another complexity to the study of global climatic change.

As stated in Section 6.6, some of the damage to aquatic ecosystems attributed to acidification may in fact be due to increased UV-B either because of ozone depletion or because of reduced dissolved organic content. Moreover, if enhanced UV-B radiation has an adverse impact on phytoplankton populations in lakes and oceans, there will be implications for secondary productivity, including fish production, with implications for food supplies. Hader *et al.* (1995) have reviewed some of the literature on this issue and suggest that 7×10^6 t of fish per year could be lost if there was a 16 per cent depletion of stratospheric ozone which would cause a 5 per cent loss in phytoplankton. There is also evidence for a damaging effect on macroalgae, seagrasses and the developmental stages of fish, shrimp, crab and amphibians, some of which is caused by the formation of the ozone hole in the spring and so increasing UV-B radiation at a crucial time in their life histories. Schofield *et al.* (1995) have shown that increased UV-B inhibited carbon fixation rates in communities of Antarctic ice algae by ca. 4 per cent during the morning hours to as much as 23 per cent at the end of the day. In addition Wynn-Williams (1996) has shown that succession and community development of algal communities in Antarctic soils are likely to be altered as UV-B increases. Such increases may also have an impact on global biogeochemical cycles which, as Zepp *et al.* (1995) discuss, could alter the sources and sinks of greenhouse gases. This once again reflects the interplay between various forms of pollution. Increased UV-B in terrestrial ecosystems is likely to influence the production and decomposition of plant matter, which are important fluxes between the atmosphere and biosphere in the biogeochemical cycle of carbon. For example, increased UV-B could accelerate litter decomposition or, conversely, it could retard decomposition if the radiation prompts changes in the chemical composition of living biomass that reduces its biodegradability on death. Such changes would also impinge on microbial activity and the decomposer component of the ecosystem. Gehrke *et al.* (1995) have provided evidence that changes in the litter quality, notably an increase in tannins and a decrease in alpha-cellulose, occur under enhanced UV-B conditions in subarctic regions and that some decomposer fungi decline in numbers.

Even before the availability of the data of Farman *et al.* (1985), mitigating measures at international level were being proposed by politicians in the mid-1970s (see Rowlands, 1995, for an account of the politics of ozone depletion). The first formal agreement was the Vienna Convention for the Protection of the Ozone Layer, established in 1985. This led to a first protocol on the control of CFCs, the so-called Montreal Protocol 1987, which came into force in January 1989. It was concerned with limitations on the production and emissions of CFCs, notably a freeze in consumption of CFCs at 1986 levels by mid-1990, a 20 per cent reduction by mid-1994 and a further 30 per cent reduction by mid-1999. Between 1990 and 1994 several amendments were made to the original protocol. As O'Riordan (1995) has discussed, the effects of what he calls environmental diplomacy since 1987 have been considerable. Other substances, e.g. methyl chloroform, halons and carbon tetrachloride, have been included in the protocol, and in November 1992 the fourth meeting of the Montreal Protocol agreed to phase out CFCs by 1996 with a 75 per cent reduction by 1994. Alternatives to CFCs have also been developed, particularly as refrigerants: hydrochlorofluorocarbons (HCFCs) are not as effective at destroying ozone but they still have a detrimental effect; they are also heat-trapping gases and therefore contribute to the enhanced greenhouse effect (Section 6.5). Consequently, controls on HCFCs have been agreed as part of the Montreal Protocol. These measures will limit damage to the stratospheric ozone layer but will not eliminate its continued destruction because of the long lifetimes of CFCs and the effects of HCFCs, nor lead to its replenishment, the engineering of which has been discussed by Avnir *et al.* (1995). Moreover, Holmes and Ellis (1996) believe that all ozone-depleting substances must be eliminated in order to allow the complete replenishment of the stratospheric ozone layer.

The enhanced greenhouse effect, acid deposition and ozone depletion are all examples of atmospheric pollution that have occurred as a consequence of industrialisation. Increasing concentrations of heavy metals in active circulation also derive from industrialisation and fossil-fuel combustion and they represent perturbations to biogeochemical cycles (Figure 6.9 shows the biogeochemical cycle of lead). Besides lead, heavy metals include mercury, zinc, cadmium, copper and

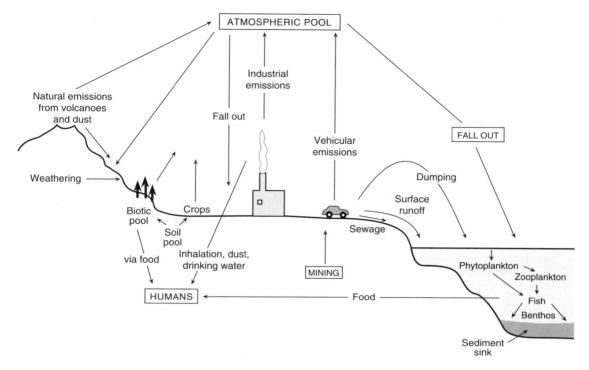

Fig. 6.9 The biogeochemical cycle of lead and its pathway to humans.

arsenic, high concentrations of which can be harmful to human health as the following examples show. Evidence from the palaeoenvironmental record for enhanced lead concentrations in the biosphere has been reviewed by Livett (1988) and Mannion (1995b). For example, many of the lake sediments discussed in Section 6.6, which contain palaeoecological evidence from acidification, also contain a record of increasing lead concentrations. In addition, analyses of heavy-metal concentrations from Arctic and Antarctic cores (e.g. Hong *et al.*, 1994; Boutron *et al.*, 1995) allows the magnitude of post-industrial changes in their atmospheric concentrations to be assessed against baseline data from pre-industrial times. For example, Boutron *et al.* have shown that the concentrations of lead, cadmium, zinc and copper show marked increases in Greenland snow since ca. 1800 to the mid-1960s. Of these metals, lead increased most markedly due to its use as an antiknock additive to petrol which began in the 1940s. Since the 1960s Boutron *et al.* record a distinct decrease following the phased elimination of lead in petrol. This is also reflected in air quality data from the United States. Pirrone

et al. (1996a,b) have shown that lead concentrations in Detroit declined by nearly 10 per cent in the year 1980. Similar but not synchronous patterns of lead deposition have been recorded in Antarctic ice. They reflect the introduction of unleaded petrol to Australia in 1986 (Simpson and Xu, 1994). Wolff and Suttie (1994) have presented data for the past 70 years and show that between 1920 and 1950 mean values were $2.5\,ng\,kg^{-1}$; between 1950 and 1980 average values increased to $6\,ng\,kg^{-1}$ with a reduction thereafter. These studies indicate that up to 99 per cent of lead in polar ice is derived from anthropogenic sources rather than natural sources such as soil-derived dust and volcanic activity. The introduction of lead to petrol in the late 1940s is also reflected in other environmental archives, as Satake *et al.* (1996) have shown in relation to the bark pockets of tree trunks on the remote Yakushima Island and at Nikko, 100 km north of Tokyo, Japan. Lead concentrations of $0.1-0.22\,\mu g\,g^{-1}$ accumulated more than 200 years ago in bark pockets of the conifer *Cryptomeria japonica*, but lead concentrations in the outer bark of $1.4\,\mu g\,g^{-1}$ occur on Yakushima

Island rising to $150\,\mu g\,g^{-1}$ in Nikko. Despite the introduction of unleaded petrol to many developed countries, lead concentrations continue to remain high in many urban areas. This reflects vastly increased volumes of vehicular traffic, as car ownership continues to increase, increased emissions from industries and the degree of control exerted through legislation on vehicle emissions. In Valencia, Spain, the lead content of some 583 samples of airborne particulate matter collected over a year showed that the highest concentrations of lead occurred during periods of peak traffic flow (Molto et al., 1995). Moreover, Orlando et al. (1994) have shown that in Liguria, Italy, blood levels of shopkeepers in low-traffic streets were lower ($7.06\,\mu g\,dl^{-1}$) than for shopkeepers in high-traffic streets ($8.30\,\mu g\,dl^{-1}$), though even $8.30\,\mu g\,dl^{-1}$ is below the level recommended by the EU. In countries where emission control is limited, vehicular traffic is the main source of atmospheric lead, e.g. Varanasi in India (Tripathi, 1994), Harare in Zimbabwe (Sithole et al., 1993), Ibadan in Nigeria (Onasanya et al., 1993) and Riyadh in Saudi Arabia (Alsaleh and Taylor, 1994).

Elevated lead concentrations, and those of other heavy metals, can derive from sources other than petrol. For example, Anglinbrown et al. (1995) have reported on high concentrations of lead, cadmium and zinc in allotment soils from the Hope Flat district of Jamaica, a residential district on a heavily mineralised soil. They suggest that concentrations in soil and water from mine adits may be a potential health hazard, especially to children, who are particularly susceptible to neurological damage (see below). Reference was made to the possibility of heavy-metal contamination of the wider environment by mining activities in Sections 5.2 and 5.3. A further example of this is reported by Sanchez et al. (1994), whose analysis of river biota, sediments and water draining an old lead–zinc mine in Spain reflects contamination by nine heavy metals. Samanta et al. (1995) have highlighted the health risk of several hundred thousand people living in the neighbourhoods of some 40 factories producing lead ingots and lead alloys in Calcutta, India. There is little emission control and, since factory chimneys are only 15–25 m high, the local atmosphere is heavily polluted not only with lead but also with arsenic, cadmium and mercury, as reflected by analysis of soils, plant matter, etc. The hazards of smelting have also been highlighted by the study of Lent et al. (1992), who have extracted cores from estuarine sediments close to a lead smelter at Port Pirie, South Australia. They show that concentrations of lead, zinc and cadmium increased markedly in the 1950s and 1960s with highest concentrations occurring downwind of the smelter. In a similar study based on birch leaves in the area surrounding a nickel–copper smelter at Monchegorsk, northwestern Russia, Kozlov et al. (1995) have demonstrated that concentrations of nickel, copper and iron were between 6 and 12 times greater near the smelter when compared with background regional measurements. In particular, the concentrations of nickel and copper have increased by factors of between 3 and 5 in the past two decades.

What is the role of these elevated heavy-metal concentrations in generating environmental change? The anthropogenic perturbation of such biogeochemical cycles has altered from the natural distribution of heavy metals; the elevated concentrations can cause damage to ecosystems (Section 6.6) and impair human health. Acute lead poisoning can cause liver, kidney and neurological damage; exposure to elevated concentrations of lead in air and water is considered to cause anaemia and mental retardation in children. Mental retardation has been reviewed by Morgan (1996) and Harrison (1993). Even in countries where there are emission controls, some groups of children remain susceptible to the effects of lead pollution, as Brody et al. (1994) have reported for the United States. They estimate that 8–9 per cent of children, i.e. 1.7 million, have blood lead concentrations of at least $0.48\,\mu mol\,dm^{-3}$ and are therefore at risk. Most are from urban areas and low-income minority families, which reflects the role of socioeconomic factors in the exacerbation of environmental hazards. Proximity to a concentrated source of pollution is also cause for concern, as Galvin et al. (1993) have shown for residents adjacent to the Pasminco lead and zinc smelter, Lake Macquarie, New South Wales, Australia. Moreover, despite the increasing use of unleaded petrol, Foner (1993) has determined that some Israeli children are at risk of lead blood levels rising above $10\,\mu g\,dl^{-1}$, now considered hazardous. Although there is no conclusive evidence that high concentrations cause neurological damage in children, a study in Denmark has shown a relationship between neonatal jaundice and neurological deficits caused by enhanced exposure to lead (Damm et al., 1993). This suggests that neonatal jaundice predisposes children to damage by lead pollution.

6.8 Changes caused by disposal of waste materials

Waste materials are an inevitable consequence of fuel-powered urban-industrial ecosystems because they are the foci of industrial activities, energy consumption and high population densities. High population densities produce waste materials that can be crudely divided into sewage and domestic rubbish, and their disposal creates environmental change both inside and outside the urban area. In addition, the concentration of industrial activities within the urban complex produces a great variety of waste materials that require treatment and/or disposal. Emphasis here will be placed on wastes produced by processing and manufacturing industries and their environmental impact. Urban systems are also fuelled by both agricultural systems and energy sources. The environmental consequences of agriculture will be examined in Chapter 7, and in Section 6.5 the impact of fossil fuels has been discussed, so this section will consider problems associated with nuclear power, an alternative energy source promoted in the developed world since the 1950s.

High population densities in urban areas give rise to vast quantities of waste water and sewage as well as domestic rubbish. The removal of sewage from urban areas only began in the 1840s; it often used to contaminate water supplies, causing epidemics of diseases like typhoid and cholera. Such problems still occur in many developing countries but even in developed countries, where sewage treatment is widely employed, contamination of fresh waters and the nearshore marine environment frequently results from sewage disposal. Although it mainly consists of water, sewage is rich in nitrate and phosphate, of which no more that 50 per cent is removed by treatment. The remainder is released into drainage systems and is a major contributory factor to cultural eutrophication, which can profoundly alter the characteristics of aquatic ecosystems (Section 7.5). Since agricultural run-off also contributes to this process, it will be examined in Chapter 7. Sewage also causes pollution by promoting bacterial growth in coastal waters. This can cause human health problems; if the sewage is untreated it will contain pathogens for a variety of diseases, e.g. polio, dysentery, typhoid and cholera. The offshore disposal of sewage via outfall pipes causes the formation of beds of sewage residues. A.R.M. Young (1996) reports that severe pollution of

Sydney's beaches prompted the water board to install deep-ocean outfall pipes in the early 1990s. Despite this the sewage does not disperse far from the coast.

Disposal of domestic rubbish presents a different set of problems. The nature of waste material varies worldwide according to level of development and per capita income. The World Resources Institute (1996) states that 'with increased wealth, the composition of wastes changes from primarily biodegradable organic materials to plastics and other synthetic materials, which take much longer to decompose'. In Britain most municipal waste, which includes industrial or commercial rubbish along with domestic rubbish, is dispatched to some 4000 landfill sites. London alone produces some 15×10^6 t of waste every year, of which almost half is disposed of in London itself, mostly in disused gravel pits. Much of the remaining waste is transported to landfill sites in neighbouring counties. These sites can create environmental problems which depend on the nature of the waste. Noxious odours may constitute a local nuisance, especially during the active life of the landfill. Gases such as hydrogen sulphide can be produced by the action of sulphate-reducing bacteria if the landfill contains sulphate waste and rotting vegetable material. Methane is also generated where rotting organic matter ferments under anaerobic conditions. According to Meadows (1996), 2×10^6 t of methane are generated annually from Britain's landfill sites, which represents ca. 46 per cent of the nation's total generation of methane (Aitchison, 1996). This can continue for a considerable time after the landfill site has been closed, which can be particularly hazardous because at sufficient concentrations methane can form an explosive mixture with air. Numerous problems have occurred in Britain as a result of building over landfill sites in which methane has accumulated to dangerous levels. In some cases explosions have damaged buildings and in one case landfill gas diffused into a school playground. Nitrous oxide is also produced by landfill and, as Tsujmoto *et al.* (1994) have shown for such sites in Osaka, Japan, emissions are twice as high from active sites than from closed sites, a pattern also exhibited by methane emissions. Both methane and nitrous oxide are greenhouse gases (Section 6.8).

As well as problems associated with methane, landfill sites are subject to subsidence that can cause structural damage to buildings. Moreover, groundwater may become contaminated by leachate

Table 6.10 Potential resource savings through recycling

	Potential saving (%)			
	Aluminium	Steel	Paper	Glass
Energy consumed	90–97	47–74	23–74	4–32
Water consumed	NA	40	58	50
Air pollution emissions	95	85	74	20
Water pollution discharges	97	76	35	NA
Mining wastes	NA	97	NA	80

Source: based on United Nations Environment Programme (1991a)
NA = not available

from the landfill and this can only be prevented by using heavy-duty impermeable linings. This is particularly important where hazardous waste is disposed in landfill sites. As Yakowitz (1993) has discussed, some 75 per cent of the 24×10^6 t of hazardous wastes produced annually in countries of the Organisation for Economic Cooperation and Development (OECD) are so disposed. The absence of adequate containment in landfills can lead to pollution of the wider environment. For example, the decline in water quality in wells near Raipur, India, during the wet season is attributed to the use of an abandoned limestone quarry as a waste-disposal site from which leachate contaminates the aquifer (Bodhankar and Chatterjee, 1994). The problem was sufficiently acute to result in the hospitalization of people who were using the wells for drinking water. Similar problems occur elsewhere in the developing world where only between 30 and 50 per cent of solid waste is collected for legal disposal. The remainder is burned or dumped in unauthorised landfills, which are health hazards and cause pollution. The World Resources Institute (1996) cites the example of Balut in Manila, the Philippines, where 650 t of solid waste are dumped daily. This accumulation has caused 34 ha of Manila Bay to be reclaimed. Problems with landfill sites similar to those in Britain also occur in the United States where ca. 75 per cent of the solid waste it produces is disposed of in landfill sites. However, increasingly stringent regulations in the developed world and the lack of suitable sites for additional landfills means there is a pressing need for alternative methods of disposal. Apart from the use of deep-well injection and surface improvements – deep wells are used for hazardous waste (see below in relation to nuclear waste) and surface improvements are used for highly contaminated hazardous liquid waste – other possibilities include recycling and incineration.

Recycling is becoming increasingly commonplace. This is not realistic for all types of waste but is feasible for glass, paper, aluminium, plastics, steel and textiles. Recycling is being increasingly organised in many developed countries and facilitated by local authorities through the provision of collection points for recyclable materials or a collection service. Public participation is an important prerequisite for the effectiveness of these programmes (Lober, 1996). Recycling is commonplace in many developing countries; municipal waste is considered to be an economic resource and many people earn a living as wastepickers by collecting glass, cans, etc., for resale. The Chinese have a long history of recycling, especially in agriculture; the ethic also pervades municipal waste recovery and recycling (Yang and Furedy, 1993). Recycling may be either primary or secondary. In primary recycling the waste material is reconstructed into its original form, i.e. newspapers are recycled into newsprint; in secondary recycling the waste materials are made into products, e.g. paper from rags. The amount and variety of materials recycled vary enormously from country to country. Within the European Union, the Netherlands recycles 62 per cent of its glass in comparison with the United Kingdom's 15 per cent. Even within a single country there may be considerable variations from one region to another: New York recycles only 5 per cent of its solid municipal waste whereas Los Angeles and Seattle recycle 21 per cent and 48 per cent respectively (Nebel and Wright, 1996). There are many advantages to recycling, as Morris (1996) has discussed. Firstly, raw materials are conserved, which means there is less pressure on primary minerals and landfills sites. Secondly, other resources, notably energy and wastes are conserved (Table 6.10), and pollution is curtailed.

Incineration is increasingly being considered as an

alternative means of waste disposal to landfills, although it is more expensive. There are some 50 incinerators in the United Kingdom treating ca. 10 per cent of waste, and 130 in the United States treating 16 per cent of waste. Incineration has many advantages and many disadvantages. The high organic content means the volume of solid municipal waste can be reduced by up to 90 per cent, the implication being that the residual material occupies a comparatively small landfill space though target management is required (Hjelmar, 1996). Toxic or hazardous substances are concentrated in the resulting fly ash, which is produced in the gaseous component and can be trapped by emission controls. The residual bottom ash can also be treated for metal recovery and used as an aggregate in road construction. For example, thermal processes such as sintering or vitrification (the binding of the metals in powder or silica) can be used to 'fix' heavy metals in fly ash so they are not susceptible to leaching when the material is placed in landfills (Wunsch et al., 1996; Venkatesh et al., 1996). Most important, many incinerators have facilities to generate electricity, facilities which can be used to power the plant itself and for industrial and domestic consumption. An ideal incineration plant should produce electricity and should have emission controls to negate the production of 'acid rain' along with facilities to recycle useful substances from the various ash products (Pickens, 1996). In relation to the generation of heat-trapping gases, Aumonier (1996) has shown that incineration is the most efficient means of waste disposal. Incinerators can also be manipulated to prevent or reduce the formation of other potentially damaging substances such as polychlorinated dibenzo-p-dioxins (PCDDs) and polychlorinated dibenzofurans (PCDFs); emissions of these substances are considerably reduced if coal is used as a source of energy, because it contains sulphur (or if sulphur is added to natural gas) with which the chlorine combines (Raghunathan and Gullett, 1996).

Among the disadvantages of incineration is the production and subsequent deposition of mercury (Lindquist, 1995), the main source of which is batteries. According to Pirrone et al. (1996b) anthropogenic emissions of mercury increased by ca. 4 per cent during the 1980s and are now decreasing at a rate of 1.3 per cent annually. These data, however, mask the fact that in the developed world emissions increased by between 4.5 and 5.5 per cent up to 1989 and since then have stabilised; in

developing countries the increase in emissions continues at a rate of between 2.7 and 4.5 per cent annually. Moreover, incinerators are the major source in the Americas, western Europe and Africa; coal combustion is the principal source in Asia, the former Soviet Union and eastern Europe whereas in Oceania it is the mining and smelting of lead and zinc. Concerns in the United States about such emissions is leading to a tightening of controls for both municipal waste and medical waste incinerators (Krivanek, 1996). In New Jersey, for example, an emissions standard of $28\,\mu g/dscm$ (dry standard metre) is expected to be in place by the year 2000 with current interim emissions at about $65\,\mu g/dscm$ (Aucott and Winka, 1996). The PDDDs and PCDFs mentioned above are often known as the dioxins and furans respectively, and both are hazardous to human health. The dioxins may adversely affect hormone mechanisms and damage the immune system and are implicated in some forms of cancer. This is one reason why there is often public alarm when a new incinerator site is proposed. Moreover, epidemiological studies have shown that the incidence of cancer is higher in populations close to incinerators. Elliott et al. (1996) have analysed cancer incidence in some 14 million people living near 72 incinerators in the United Kingdom and shown there is a significant decline in risk of all cancers with distance from the incinerators. In particular, they found a relatively high risk of liver cancer, a risk which requires further investigation.

All of the problems associated with the disposal of domestic waste apply equally to the disposal of solid industrial waste. However, some industries produce very different types of waste, which have a varied range of environmental impacts. Aquatic and soil environments are most at risk from industrial wastes, especially those which contain heavy metals and synthetic chemical compounds, and this can have serious repercussions for human health. Detergents can also cause considerable water pollution. These are synthetic compounds, whose residues enter drainage systems from sewage and industrial waste water. Many are not biodegradable, especially tetrapropylene benzene sulphonate (TBS), although there is now legislation in many countries that requires detergents to be formulated with a considerable proportion of biodegradable substances. These compounds can reach relatively high concentrations in nearshore and offshore waters where untreated sewage is emitted into the sea. In addition, the use of these detergents to disperse oil slicks can have more severe ecological effects than the oil itself.

Heavy-metal pollution may occur in aquatic ecosystems as a result of industrial emissions and in extreme cases it may be a threat to human health. Methyl mercury poisoning can impair sensory, visual and auditory functions, especially in foetuses. There are several examples of methyl mercury poisoning that have proved fatal on a relatively large scale. One of the best-known examples occurred in Japan in the period 1953–1975 around Minamata Bay and the Agaro River (Smith and Smith, 1975). The discharge of methyl mercury came from a local chemical plant engaged in the production of acetaldehyde; it entered marine food chains and mercury-rich residues became concentrated in higher trophic levels comprised of shellfish and fish. These were harvested and consumed by local fishermen, resulting in approximately 100 deaths and disablement for up to 1000 people. (Section 6.3 considers mercury pollution of waterways due to its use in gold mining.)

Groundwater and soils may also be contaminated with industrial waste products. Waste products are many and varied but one important and widespread group comprises synthetic organic compounds, some of which are non-biodegradable. They are used in many manufacturing processes, e.g. plastics, synthetic fibres and rubber, solvents, pesticides and paints. Included in this group are the polychlorinated biphenyls (PCBs). These were first produced commercially in 1929, mainly for electrical appliances. Since the 1970s fears about their environmental and human impact has led to a rapid decline in PCB use, and production ceased by 1979. These compounds are stable as well as toxic; if incinerated at low temperature, i.e. below 1200 °C, dioxins may form (see above). Moreover, vast volumes of PCBs have been buried in landfill sites and may prove to be hazardous in the future. There were two major human health incidents in the 1970s, in Japan and Korea, due to the consumption of rice oil contaminated with PCBs. The oil not only caused minor skin irritations but also birth defects and high rates of miscarriage. PCBs in the environment can undergo biological magnification as they pass along food chains. This is exemplified by Bobovnikova et al. (1993), whose work on PCB levels in vegetation and commercial vegetable and berry production around the Serpukhov capacitor plant in Russia has shown that high concentrations

of PCBs in breast milk occur along with a higher than average incidence of disease in babies. Another example is provided by Abramovitz (1996), who states 'A person would need to drink Great Lakes' water for more than 1000 years to ingest as much PCBs as eating a two-pound trout'. This reflects pollution problems experienced in the Great Lakes of North America, where substances such as mercury, dioxins, chlordane and DDT contribute to the impairment of water quality. Populations of birds of prey and marine mammals also have high concentrations of PCBs in their tissues, which may render them increasingly susceptible to disease.

The disposal of nuclear waste poses many problems. About 5 per cent of the world's energy requirements are currently supplied by nuclear power. How much this will increase in the future is debatable because of increasing concerns about the safety of reactors and the disposal of radioactive wastes. These concerns have heightened since 1986 in the wake of the Chernobyl disaster in the Ukraine, in which a reactor exploded causing death and long-term health problems (see below). Currently there are 428 nuclear reactors in 30 countries worldwide but mostly in the developed world (see the anonymous review in *ATW International Zeitschrift für Kernenergie*, 1996, **41**(3), 209–21). For example, there are 54 in France (Beuneche, 1996) and 37 in Britain, respectively supplying 75 per cent and 20 per cent of the electricity generated in each country; in the United States there are 109 operational plants producing 24 per cent of the country's electricity. Although there are plans for expansion in France, Japan, India and Korea, many nations, e.g. Australia, have abandoned plans for nuclear facilities because of public approbation based on human and environmental health concerns. However, all existing nuclear plants produce waste materials which are radioactive to a greater or lesser degree, as shown in Table 6.11. There are also low-level radioactive wastes produced by industry and medicine; their disposal is governed by strict regulations and may take place through incineration, landfill tipping (e.g. Meck, 1996) and discharge of liquid waste into the sewerage system. However, these methods are inadequate for the disposal of intermediate- and high-level wastes, though there are no generally accepted alternatives.

Problems arise because of the presence of radioactive isotopes that generate large quantities of heat and which will remain active for centuries or even millennia. Consequently, they will need to be isolated from the biosphere for between 10^4 and 10^5

Table 6.11 Types of nuclear waste

Designation	Examples	Disposal
Low-level wastes	Discarded working clothes Floor sweepings and refuse Laboratory equipment Milling wastes	Sealed landfill repositories
Intermediate-level wastes	Chemical sludge Metal fragments	Sealed landfill repositories? Underground geological sites?
High-level wastes	Reprocessing solvents Spent fuel rods	Land-based geological sites? Deep-sea geological sites? Burial in ice caps?
Transuranic wastes	Americum-243 Curium-242 Plutonium-239	As for high-level wastes

years to prevent contamination of soils, water, etc., and health problems. For this reason there have been numerous proposals to use geological formations that would provide secure repositories. However, the uncertainty associated with such long timescales and public opposition have been major stumbling-blocks to the development of these sites. For example, Gascoyne (1996) has examined Canadian proposals for geological disposal involving an examination of the geochemical, hydrogeological and microbiological factors at depths of 500–1000 m in the shield region of ancient igneous rocks. This and other studies, e.g. Pedersen (1996), have highlighted the hitherto neglected factor of microbial activity on containment canisters and the radionuclides themselves. Preliminary research on the use of micro-organisms to degrade nuclear waste (a form of biotechnology), is considered in Section 9.7. In addition, Horseman and McEwen (1996) have drawn attention to the likely impact of the heat that is generated by such waste and its effect on containment rocks of an argillaceous nature. In the United Kingdom and Spain it has been suggested that disused salt mines and geological salt formations would provide suitable repositories (e.g. Delascuevas and Pueyo, 1995), though none have so far been used. Hydrogeological factors in relation to a proposed disposal site near the Sellafield nuclear plant, Cumbria, UK, have been presented by Hazeldine and McKeown (1995), where the overall upward movement of water, as shown by a model simulation, could render the site unsuitable. Alternative possibilities to land-based disposal include deep-ocean repositories, though again there is much opposition to this, especially since there are

major concerns about the discharge of low-level waste into the Irish Sea from Sellafield. The issue is generally so sensitive that the majority of intermediate- and high-level nuclear waste is being stored until the technology for disposal is improved (Robertson, 1996).

The accidental release of waste, even low-level waste, generates considerable controversy. Such leakages have occurred from Sellafield, UK and from the Cap de la Hague, France. Their long-term effects on ecology and human health remain to be seen. Over the years there has certainly been much speculation about the role of nuclear facilities in the creation of clusters of childhood leukaemia and other cancers. The occurrence of the two together suggests a link but its nature has so far eluded researchers. Conversely, the effects of major nuclear explosions have become well documented because of the atomic bombs dropped on Hiroshima and Nagasaki, Japan, in 1945, which caused nearly 200 000 immediate deaths and health problems for many more thousands and their children. The more recent tragedy in 1986 at Chernobyl, the most serious of all the accidents to have occurred at a nuclear plant, has focused the world's attention on the potentially devastating impact of nuclear fallout. In an ecological and agricultural context, the release and subsequent deposition of radionuclides had major effects, especially in areas of the northern hemisphere that received rainfall originating over central Europe. The substances deposited include caesium-134, caesium-137, iodine-131, strontium-90 and plutonium-239. In the environment caesium behaves like potassium and enters food chains. Its uptake by vegetation and eventually by grazing animals led many European governments to impose bans on the marketing of livestock for human consumption. The countries most badly affected were in Scandinavia but sales of lamb were banned even in Cumbria, UK. According to Brynildsen, *et al.* (1996), caesium contamination of sheep and reindeer is still a significant problem in Norway and elsewhere.

The implications for human health arising out of the Chernobyl disaster will become increasingly apparent as time progresses. The number of actual deaths at the time of the explosion was comparatively low; there were less than 100 deaths within a month. However, there are many health problems associated with contamination by radioactive fallout, as Rytomaa (1996) has discussed. The only possible increase in congenital abnormality so far detected is that of Down's syndrome but in the future others are likely to occur. Evidence for genetic change in rodents (Baker *et al.*, 1996) and increased mutilation frequencies in germline mutations in children (Dubrova *et al.*, 1996) have been detected in the Chernobyl region, the implication being that future problems will occur. In reality, the major problem is a high incidence of childhood thyroid cancers. Rytomaa (1996) reports that in some parts of the Ukraine and Belarus, and in the Bryansk region of Russia the increase has been more than 100 per cent. Moreover, homes, farms and whole communities have been destroyed. The legacy of Chernobyl in its many contexts will live on well into the next millennium.

6.9   Conclusion

There is no doubt that industrialisation, especially during the past 200 years, has exacted a heavy toll on global environments. The exploitation of mineral resources has scarred, and will continue to scar, those landscapes which have developed over mineral-bearing rocks. As illustrated by the example in this chapter, the technology and expertise are available to limit and ameliorate the environmental change that such activities create *in situ*. Stricter controls and effective reclamation schemes, facilitated by legislation, could also go a long way towards curtailing the wider detrimental consequences of mineral extraction. Although the impact of such activities is readily apparent, in contrast to the atmospheric pollution and its subsequent effects on terrestrial and aquatic

environments, remediation is not simple or cheap. The enhanced greenhouse effect, accelerated acid deposition, high heavy-metal concentrations and stratospheric ozone depletion are all products of the industrial age but have only come to light in the past three decades. They are all contributing insidiously to environmental change, which occurs sufficiently slowly that ecological thresholds are only now being recognised. All are related to the interplay between the various terrestrial and atmospheric processes, involving global atmospheric, geological, geomorphological, hydrological, pedological and biological systems. The systems are united by energy transfer and biogeochemical fluxes, and it is these fluxes that have been so perturbed by human activities. Drastic measures are now needed to counteract many of their potentially major effects on civilisation. Such threats to human well-being will require international cooperation between politicians and much forethought in the drafting of energy policies. Energy policies will need to consider the advantages and disadvantages of nuclear and renewable sources of energy and compare them with fossil fuels. Moreover, problems of waste disposal are reaching massive proportions worldwide with implications for primary resource exploitation, energy supplies and hazards to human health. Recycling is set to become a major feature beyond the year 2000.

Further reading

Craig, J.R., Vaughan, D.J. and Skinner, B.J. (1996) *Resources of the Earth: Origin, Use and Environmental Impact*, 2nd edn. Prentice Hall, Upper Saddle River NJ.

Erisman, J.W. and Draaijers, G.P.J. (1995) *Atmospheric Deposition*. Elsevier, Amsterdam.

Graedel, T.E. and Crutzen, P.J. (1995) *Atmosphere, Climate and Change*. W.H. Freeman, New York.

Harris, J., Birch, P. and Palmer, J. (1996) *Land Restoration and Reclamation*. Longman, Harlow.

Hill, R., O'Keefe, P. and Snape, C. (1995) *The Future of Energy Use*. Earthscan, London.

Houghton, J.M., Meiro Filho, L.G., Callender, B.A., Kattenburg, A. and Maskell, K. (eds) (1996) *Climate Change 1995: the science of climate change*. Cambridge University Press, Cambridge.

O'Riordan, T. (ed.) (1995) *Environmental Science for Environmental Management*. Longman, Harlow.

Strzepek, K.M. and Smith, J.B. (eds) 1995. *As Climate Changes: International Impacts and Implications*. Cambridge University Press, Cambridge.

CHAPTER 7

The environmental impact of agriculture in the developed world

7.1 Introduction

The development, expansion and significance of agricultural systems since the initial domestication of plants and animals 10 K years ago was discussed in Chapters 4 and 5. Since 10 K years BP, agriculture has had profound effects on the global environment. Habitats and landscapes have been altered and fragmented, and species extinction has accelerated as ecosystems have been transposed into agroecosystems to support fuel-powered urban-industrial systems. The application of science to agriculture has contributed to these changes, mainly as a result of manipulating trophic energy flows, biogeochemical cycles and the hydrological cycle i.e. through the use of artificial fertilisers, crop-protection chemicals and irrigation. Artificial fertilisers augment naturally occurring nitrates and phosphates which are essential to growth, and crop-protection chemicals simplify trophic relationships by eliminating so-called pests that use energy, nutrients and light in competition with the crop. Although landscape change may be rapid and observable, other significant impacts have occurred gradually as inadvertent consequences of high-technology agriculture, e.g. cultural eutrophication. Soil erosion has also become a significant problem in developed countries as a direct consequence of agricultural methods and salinisation is fast becoming a problem in semiarid regions due to irrigation. Consequently, 'industrialised agriculture', itself a response to politically based agricultural policies and economic pressures, is a major cause of ecological and environmental change.

7.2 Landscape change: loss of habitats and biodiversity

The increasing reliance of agriculture on technology since 1750 has produced extensive landscape change in the developed world (Sections 5.6, 5.8 and 5.9). As agriculture has intensified since the Second World War, older agricultural landscapes have been destroyed and replaced by those that characterise the 1990s. Much of this intensification has been achieved by political manipulation and in many developed nations agriculture is controlled by interventionist policies rather than being a free-market enterprise. Although such policies were established, e.g. in Britain during the 1940s, political control of agriculture became highly significant in Europe due to the formation in 1957 of the European Economic Community (EEC), now the European Union (EU), which Britain joined in 1973. The operation of a Common Agricultural Policy (CAP) has protected agriculture and stimulated the development of agribusiness, especially via intervention price agencies. In effect, target prices for specified commodities are guaranteed and grants are available for agricultural improvements. Coupled with scientific advances, these policies have transformed agricultural practices and in so doing have created widespread landscape change by directly and indirectly promoting intensified food production. Surpluses of many animal products and crops are today produced to excess and necessitate extensive storage within the EU or cheap disposal to external countries. Among the most notable environmental changes that have occurred are the loss of biodiversity through extinction associated with the

Table 7.1 Data on land-use change since 1700

A. Changes in land cover

	Forests/woodlands (%)	Grassland/pasture (%)	Croplands (%)
Global data 1700–1980	−18.7	−1.0	466.4[a]

B. Land area characteristic

	Undisturbed (%)	Partially disturbed (%)	Heavily disturbed (%)
Europe	15.6	19.6	64.9
Asia	43.5	27.0	29.5
Africa	48.9	35.8	15.4
South America	62.5	22.5	15.1
North America	56.3	18.8	24.9
Australasia	62.3	25.8	12.0
World: all land	51.9	24.2	23.9
World: excluding rock, ice, etc.	27.0	36.7	36.3

C. Changes in extent of forest

	Total forest area in 1990 (10^6 ha)	Percent change in total forest area 1981–1990[b]	Percent change in extent of natural forest 1981–1990
Tropical Africa	529.82	−7.0	−7.2
Tropical Asia and Oceania	338.03	−6.7	−11.1
Tropical Latin America and Caribbean	924.19	−7.2	−7.5
All tropical countries	1 792.03	−7.1	−8.1
Temperate Africa	15.27	−4.2	−9.4
Temperate Asia and Oceania	159.33	5.2	−3.4
Temperate Latin America	43.28	−4.7	−6.2
All temperate developing countries	217.88	2.4	−4.5

D. Rate of increase in croplands

	Temperate regions (10^6 ha y^{-1})	Tropical regions (10^6 ha y^{-1})	Global (10^6 ha y^{-1})
1850	1.6	1.5	2.4
1900	1.8	3.0	5.8
1950	9.0	4.0	14.0
1970	4.0	7.0	11.0

Source: panel A, Richards (1990); panel B, Hannah *et al.* (1994); panel C based on FAO (1995), quoted in World Resources Institute, 1996; panel D, Houghton (1994)
[a] The increase in croplands includes the following increases
  6 700% in North America
  1 930% in Latin America
  1 280% in southeast Asia
  1 060% in Australasia
[b] Includes plantation forests, which are increasing in extent
[c] Based on 1995 data from FAO quoted in World Resources Institute (1991)

loss and fragmentation of natural and seminatural habitats.

Several recent assessments of land-cover and land-use change reflect the significance of agriculture as a major agent of environmental change globally. Some of these data are given in Table 7.1, which shows that of all the continents Europe is the most disturbed, with only 15.6 per cent remaining undisturbed (Hannah *et al*, 1994). It has thus become what Hannah *et al.* characterise as a human-dominated environment. Much of this change is due to the spread and intensification of agriculture since its inception in the Near East 10 K years ago. Most ecosystems of Europe have been transformed for either arable or pastoral agriculture. The temperate forests have been particularly severely affected along with temperate grasslands. Although a proportion of this change occurred over several millennia, the past two centuries have witnessed accelerated change because of the Industrial Revolution, which fuelled population growth. For example, between 1850 and 1980 the world's forests and woodlands decreased by 603×10^6 ha (equivalent to 11 per cent of the total), of which 15 per cent, i.e. 9.5×10^6 ha, occurred in the temperate zone (R.A. Houghton 1996a). However, it is not possible to determine the extent of change in temperate grasslands as no distinction is made between grasslands and pasture, though Houghton states that much of the land converted to agricultural land during this period was temperate grassland; a further complication involves the conversion of woodland/forest to pasture, the values for which are included in the grassland/pasture category of Table 7.1. Nevertheless, the data of Table 7.1 indicate that the expansion of cultivated land varied considerably from region to region. Large increases occurred in North America and Australasia, over 6000 and 1060 per cent respectively (Richards, 1990). These increments reflect the expansion of Europe into the New World and the Antipodes. Alayev *et al.* (1990) also point out that the largest expansion of arable agriculture in Russia occurred between 1850 and 1920, mainly in the Russian plain.

The world's wetlands have also been adversely affected by agriculture. As far back as the Roman period, wetlands have been drained for agriculture (Section 5.2), a case in point being the fenlands of East Anglia, UK. In the Mediterranean basin there is ample evidence for wetland demise caused by agricultural expansion. Many deltaic regions and river floodplains have been adversely affected, e.g. the Camargue in France and the Po in Italy. In the former Tamisier and Grillas (1994) report that over the last 50 years some 30×10^3 ha of the Camargue's wetlands have been lost, mainly for agriculture but also for salt extraction and industry. Moreover, the remaining wetlands are intensively managed with the provision of dykes to increase the influx of freshwater and decrease the influx of salt water. The resulting decline in salinity and flooding frequency have altered the habitat's biota and caused a loss of biodiversity.

Many of the Mediterranean's coastal lagoons are adversely influenced by high inputs of nutrients from artificial fertilisers, especially phosphates and nitrates. According to Mesnage and Picot (1995), this causes cultural eutrophication (Section 7.5) and, in particular, the creation of a reservoir of phosphate that is fixed in lagoon sediments and which can be released under certain conditions to promote algal blooms. Even in the United States, where there is little apparent need for additional agricultural land, the extent of wetlands has declined by ca. 47 per cent, with the largest losses occurring in California and the Midwest (Dahl, 1990; Mitsch and Gosselink 1993). Mnatsakanian (1992) has reported that many hundreds of thousands of hectares of wetland in Belarus were drained during the 1970s to develop croplands. Despite the high cost and technical input, the lack of infrastructure caused people to leave the land, so vast areas of degraded bog are now subject to wind and water erosion. In some parts of the world, wetlands can be altered or destroyed by agricultural activities in their catchments. One such example is the wetland of Australia's lower Murray River basin, which has been substantially altered by the receipt of saline irrigation waters (Finlayson, 1991). Other examples of wetland demise are given in Dugan (1993).

These data on land-cover change for the developed world reflect only a proportion of the global land-cover change that has occurred in the past few centuries; it is also occurring apace in the developing world, as discussed in Chapter 8. Apart from reflecting the magnitude of biospheric change which has been promoted by agriculture the demise of forests, woodlands and wetlands signifies substantial alterations in the global biogeochemical cycling of carbon, with all the implications it has for global climatic change (Section 6.5). All of these vegetation types are important stores of carbon; their biomass represents carbon taken out of active circulation. Deforestation and wetland drainage return the carbon to the atmosphere, thus contributing to the enhanced greenhouse effect. According to R.A. Houghton (1995), the reduction of 603×10^6 ha of forest and woodland alone between 1850 and 1980 (see above) has released 100 Gt of carbon into the atmosphere. In addition, the conversion of mires and bogs (excluding coastal wetlands) to other land uses is estimated to generate between 63 and 85×10^6 t of carbon per year (Maltby and Immirzi, 1993). Although considerably less than emissions produced by fossil-fuel burning (Section 6.5), this alteration of the biosphere is significant because it has not only generated carbon but also represents the loss of important carbon sinks. However, carbon emissions from similar habitats in the tropics are set to increase, so they will aggravate the enhanced greenhouse effect. Thus, while the destruction or alteration of a habitat occurs on a local or regional scale its impact is ultimately global.

The expansion of agriculture has caused a diminution in the global extent of forests, etc., at regional levels, but alongside this, many habitats have become increasingly fragmented. Both clearance and fragmentation have caused, and will continue to cause, species extinction. Just how fragmented each habitat and how small each fragment must become before there is a loss of biodiversity and a threat to its viability, are matters of much debate (Dobson, 1996; Poethke *et al.*, 1996; Henle *et al.*, 1996). Fragmentation causes a reduction in the genetic resource base of the biota. As fragments decrease in size, and unless there are linking corridors to facilitate interbreeding, i.e. genetic exchange, the habitat becomes increasingly non-viable as population numbers decline; extinction will eventually occur. Fragmentation is a serious problem in Europe because of its long history of agriculture and because of the intensive use of crop protection chemicals (Sections 7.7 to 7.9) and artificial fertilisers (Section 7.5). The losses of European wetlands mentioned above are examples of habitat fragmentation. In East Anglia, UK, the long history of drainage and intensification of agriculture have caused the disappearance of its fens, leaving only a few isolated examples. Remaining fens, as Fojt and Harding (1995) have demonstrated, have been altered substantially because of changes in water regime, increased nutrient availability (from fertilisers) and lack of traditional management. The demise of lowland heathlands in the United Kingdom provides another example of habitat fragmentation. Such areas experience conflicts of interest from recreation provision, aggregate extraction and urban spread as well as agriculture. Jarvis (1993) has shown that the fragmentation of the Dorset heaths has had an adverse impact on several bird populations, notably the Dartford warbler. Not only has the extent of these heathlands declined from nearly 40×10^3 ha since ca. 1780 to a value of 7900 ha today, but fragmentation has become acute. This has had a detrimental effect on invertebrate populations, as Webb and Thomas (1994) have discussed. They show there is a negative relationship between community diversity, area of heath and degree of

isolation, i.e. small and isolated fragments are more biodiverse than larger non-isolated fragments. Heathlands and other natural habitats in the Llŷn Peninsula of north Wales, UK, have been subject to similar pressures as Dorset. Blackstock *et al.* (1995) have reported that a comparison of documented habitat extent in 1920–22 and 1987–88 reflects an overall reduction in 3629 ha, or 44 per cent, of seminatural habitats. Dry and wet heathland have been particularly adversely affected, with declines of 51 per cent and 95 per cent respectively, involving losses of 1500 ha and 1738 ha. Most of this heathland has been converted to agricultural grassland.

Loss of biodiversity is a major problem elsewhere, as is illustrated by the fynbos of the Cape Peninsula of South Africa. Richardson *et al.* (1996) report that 65 per cent of the original area of natural vegetation has been altered significantly since European colonisation of the region in 1652, mainly through agriculture, urbanisation and the spread of alien species (Cowling *et al.*, 1996). There are many invertebrate species and 2285 plant species in the area, making it one of the most biodiverse regions in the world but effective conservation measures are essential. Similar problems in Canada have been addressed by McLaughlin and Mineau (1995), and Lamberti and Berg (1995) have provided an example of the indirect effects of changing land use on the macroinvertebrate community of a stream in Indiana, United States. An increase in sedimentation caused a major shift in the composition of the population involving an 86 per cent decrease in filter feeders and a 292 per cent increase in collector-gatherer feeders. There may also be indirect effects of clearance for agriculture, as Lyons *et al.* (1996) have demonstrated for southwestern Australia. Here the replacement of natural vegetation with winter wheat has reduced the flux of sensible heat to the atmosphere during winter and spring; this may be responsible for an observed decrease in winter rainfall in the region, which will affect remaining natural vegetation communities in the medium to long term.

Why does the preservation of biodiversity matter? Biodiversity is important in an environmental and heritage context insofar as it has been an important component of global biogeochemical cycles throughout geological time and thus a regulator of global climates (Mannion, 1997b); it has also provided many of the resources on which humanity depends (Mannion, 1997a). However, there has been some debate as to whether or not some species are 'ecologically redundant', i.e. they duplicate the role of other species and contribute little that is novel and/or their demise would not be significant (Gitay *et al.*, 1996). The emergence of biotechnology has contributed to a reconsideration of biodiversity as a resource, especially as a repository of useful substances for medicine, crop protection and bioremediation (Chapter 9). The loss of biodiversity through extinction (see below) means that opportunities to discover such substances and other goods are lost. In addition, there have been two parallel, and coincidentally complementary, lines of research which strongly imply that biodiverse ecosystems are more productive than less biodiverse ecosystems under identical environmental conditions. This is not a new proposal; it was initially suggested by Charles Darwin but without substantiation. Now Naeem *et al.* (1994, 1995) have provided corroborating experimental evidence through a series of experiments in an Ecotron, a series of chambers in which conditions can be controlled, at Silwood Park, Berkshire, UK. Three different ecosystems were constructed with 9, 15 and 31 species, each one reflecting changes in the decomposers, herbivores and parasites as well as the primary producers. During a 6 month period, monitoring reflected distinct patterns of productivity: it increased by 200 per cent and 300 per cent as biodiversity increased from 9 to 15 to 31 species. It has not yet been shown why this should occur; it could relate to light interception, which influences primary productivity, and/or increased efficiency of nutrient use.

Complementary real-world research is provided by Tilman and Downing (1994) on native grassland ecosystems of Minnesota, United States. They have monitored plots over 11 years and have shown how the most biodiverse plots are more resistant to drought and capable of recovering more rapidly than their less biodiverse counterparts. In this case the maintenance and recovery of stability was assessed using measurements of biomass. Additional work by Tilman *et al.* (1996) confirms these results. This involved the planting of 147 plots with seeds of 1, 2, 4, 6, 8, 12 and 24 species, with at least 20 plots per category. Productivity, as reflected in percentage plant cover, increased with increasing biodiversity. Moreover, the nitrate content below the root zone, i.e. inaccessible to the plants, was much reduced in the most biodiverse plots. This suggests at least one nutrient was being used most effectively, with

reduced losses from the root zone, in plots of high biodiversity. These results parallel those of Naeem *et al.* (1994, 1995) and indicate that biodiversity, carbon capture and possibly stability are interrelated.

These data should intensify concerns about the high rates of extinction that are associated with modern society. The determination of extinction rates is itself controversial (World Conservation and Monitoring Centre, 1992), not least because a large proportion of the world's biota remains undescribed and because there is little historical baseline data against which to compare modern inventories. Moreover, some groups, notably micro-organisms and insects, have been poorly documented when compared with plants and animals. Some estimates of extinction rates are given in Table 7.2. During the past it was the spread of agriculture that caused most extinctions; this accelerated with the expansion of Europe into the New World and Australasia (Smith *et al.* 1993). Now urbanisation and industrialisation, with attendant pollution, have begun to cause additional extinctions. Whichever estimate of the rate of extinction is nearest the truth, the problem of extinction is of global importance.

7.3 Soil degradation, erosion and conservation

Although the demise and fragmentation of natural and seminatural habitats represent a readily observable change to the environment, there is increasing concern about environmental change due to the more insidious soil degradation and erosion. These problems tend to be associated with developing rather than developed countries, but this is a fallacy. Developed countries are just as seriously affected as developing countries but the repercussions in terms of agricultural productivity, food shortages and even loss of life are not so severe or newsworthy. Over the past 20 years, soil degradation and erosion have been acknowledged as serious problems for the developed world, even in Europe, so remediation and conservation programmes have therefore been established.

Soil erosion involves the removal of soil particles from one place and their deposition somewhere else. It is influenced by many factors involving all the components of Earth-surface systems, nature of land use and whether or not conservation measures are

Table 7.2 Estimating rates of extinction and species under threat of extinction

A. Estimates of extinction and their authors[a]

0.2–0.3% of all species per year	Wilson (1988)
5–15% of forest species by 2020	Reid and Miller (1989)
2–8% loss 1990–2015	Reid (1992)
2–3% of rain forest species per decade	Ehrlich and Wilson (1991)
50–100 species per day	Myers (1993a)
10 000–20 000 species per annum	Stork (1993)
0.04% of animals and 0.2% of plants since 1600	Smith *et al.* (1993)

B. Birds and higher plants[b]

	Total no. of birds	Number threatened	Total no. of higher plants	Number threatened
Greece	398	9	4 900	539
Portugal	441	7	2 500	240
Spain	506	10	NA	896
UK	590	2	1 550	28
USA	768	46	16 320	1 845
Canada	578	5	2 920	649
Australia	751	51	15 000	1 597
New Zealand	287	45	2 160	236

C. Reptiles and amphibians

	Total no. of birds	Number threatened	Total no. of amphibians	Number threatened
France	32	2	32	2
Italy	40	3	34	9
Japan	66	10	52	11
Bulgaria	33	1	17	0
South Africa	299	36	95	16
UK	8	0	7	0
USA	280	23	233	16
Australia	748	42	205	20

[a] These estimates represent a 1–11% loss of species per decade
[b] Panels B and C are based on data in World Resources Institute (1996)
NA = not available

undertaken. The rate of this natural process of terrain deflation has been altered by the initiation and spread of agricultural systems during the last 10 K years BP (Section 4.5). Not only does soil erosion affect the land from where the soil is removed, it also affects the areas where soil is deposited, because irrigation systems, reservoirs, etc., become infilled. Conversely, soil degradation occurs *in situ* and may not necessarily be due to agriculture. Indeed there are so many forms of soil degradation, which is a catch-all phrase for a multitude of soil quality problems. The forms of soil degradation that are not due to agriculture, and hence will not be discussed here, are acidification due to fossil-fuel use, industrial pollution, pollution by mining (Section 6.2) and nutrient impoverishment due to deforestation. Soil degradation caused by agriculture includes salinisation (Section 7.4), contamination with pesticides, loss of nutrients and

loss of organic matter. These are forms of chemical degradation, to which must be added physical degradation. Physical degradation involves compaction, sealing and crusting. Chemical and physical soil degradation and soil erosion may all occur together. Table 7.3 shows the global extent of water and wind erosion and chemical and physical degradation.

Physical degradation is least extensive and accounts for only 4 per cent of the total degraded land area. Of the 83×10^6 ha affected by compaction, etc., 33×10^6 ha are in Europe, and as Oldeman (1994) points out, most of this is due to the use of heavy machinery which characterises industrialised agriculture. The processes cause soil to be poorly aerated and they restrict water movement, adversely affecting soil biota by reducing biodiversity. According to Stefanovits (1994) compaction is the

Table 7.3 Global extent of soil erosion and soil degradation

	Extent (10^6 ha)			
	Water erosion	Wind erosion	Chemical degradation	Physical degradation
Africa	227	186	62	19
Asia	441	222	74	12
South America	123	42	70	8
Central America	46	5	} 7	5
North America	60	35		1
Europe	114	42	26	36
Oceania	83	16	1	2
World	1 094	548	240	83

Source: based on 1991 data from UNEP quoted in Oldeman (1994)

major cause of physical soil degradation in Hungary, especially on chernozem soils. The topsoil loses its structure through injudicious ploughing regimes, and the base of the ploughed layer compacts. This may be sufficiently severe to cause stagnant water pools to accumulate. Turnock (1993) has reported that similar problems occur widely in the sub-Carpathians, Romania, where 7.6×10^6 ha also suffer from loss of humus, another characteristic of physically degraded soils. Such developments seriously limit agricultural productivity, as McGarry (1993) has reported for cotton-growing areas of Queensland, Australia. Here compaction causes the rooting systems of the cotton plants to spread horizontally rather than vertically.

Chemical degradation of soils is more widespread than physical degradation. As Table 7.3 shows, it is particularly prevalent in the developing world (Chapter 8) but it is also significant in Europe, where 3×10^6 ha are adversely affected by loss of nutrients. The loss of humus is a major underpinning cause of nutrient loss because organic compounds can complex valuable nutrients. Loss of humus can occur for a variety of reasons, the most important being soil erosion and lack of replenishment of the humic layer if no crop residues are allowed to remain in the soil. Mnatsakanian (1992) has reported that humus loss due mainly to soil erosion is particularly significant in the Ukraine: 'When in the end of the XIX century the famous Russian soil scientist ... Vassily Dokuchaev, observed Ukrainian chernozems, the richest of them had humus contents of 11–12%. Now the richest soils have 5–6% of humus, and average humus content in 1989 was about 3.2%'. Such problems occur throughout eastern Europe (Carter and Turnock, 1993). In common with many other parts of Europe, Stefanovits (1994) reports that the use of artificial fertilisers has obscured productivity losses

due to chemical degradation of many of Hungary's soils. He also points out that, as well as overall losses of humic materials, there has been a change in its structure to organic compounds of lower rather than higher, and more stable, molecular weight. Factors such as the use of sewage sludge, animal sewage and pesticides also influence soil chemistry, mostly detrimentally. Pankhurst's (1994) review of pollution in agricultural soils highlights the impact that pesticide residues and heavy metals (derived from sewage sludge) can have on soil biota, notably loss of biodiversity. In view of the role of richly as opposed to poorly biodiverse ecosystems in primary productivity and nutrient capture (Section 7.2), the impact of physical and chemical soil degradation could prove to be highly significant for the future productivity of agricultural systems, especially against the backdrop of global warming (Section 6.5).

Figure 7.1 summarises the factors involved in soil erosion (Lal, 1994). There are three key environmental variables: the erosivity of rainfall, i.e. the capacity of rainfall to erode which is determined by its extent and frequency; the erosivity of wind, i.e. its strength and frequency; and the erodibility of the soil, i.e. how easily soil particles can be dislodged and removed. The key cultural variables concern land use, management and population pressures. Many combinations of these factors can occur giving rise to various intensities of soil erosion that range from minimal to severe. Moreover, as Oldeman (1994) has demonstrated (Table 7.3), soil erosion affects all the continents, with water erosion occurring on some 1094×10^6 ha and wind erosion occurring on some 548×10^6 ha. Although the largest proportion occurs in the developing world, i.e. 76 per cent of the land affected by water erosion and 83 per cent of the land affected by wind erosion, soil erosion is a major

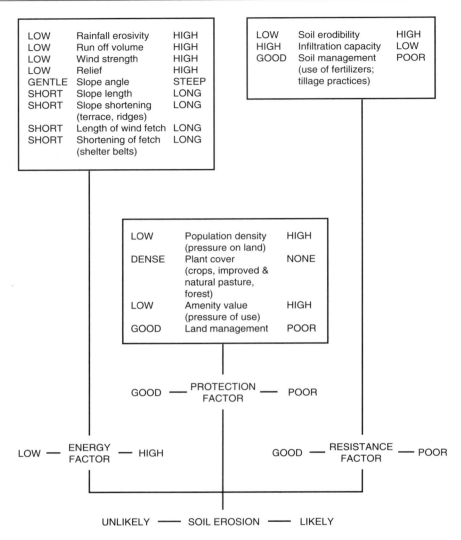

Fig. 7.1 Factors affecting soil erosion (from Morgan, 1986)

environmental problem in the developed world, where much of it is due to overgrazing and injudicious agricultural practices. In the past two decades there has accrued a considerable volume of information on soil erosion and conservation (see for example Morgan, 1995; Agassi, 1996).

According to Evans (1990), approximately 37 per cent of the arable land of England and Wales is at risk from soil erosion, though there are many difficulties associated with the measurement of soil erosion (Morgan, 1995). Methods generally focus on retrospective assessments based on the deposition of eroded material or the monitoring of erosion over long periods of time. The distribution in some profiles of caesium-137, a radionuclide produced from the testing of thermonuclear weapons between the 1950s and 1970s, has been used to determine past rates of soil erosion. This element is deposited from the atmosphere by rainfall and is absorbed by clays in the soil. Its horizontal distribution is the result of this deposition plus or minus any subsequent deposition or erosion of soil particles. Consequently, the comparison of a caesium-137 profile from an undisturbed site, e.g. nearby pasture

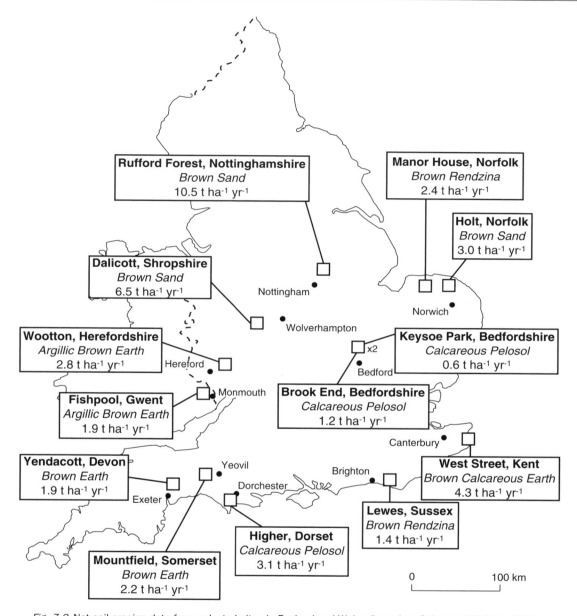

Fig. 7.2 Net soil erosion data from selected sites in England and Wales (based on Quine and Walling, 1991).

or woodland, provides a reference site against which the profile from a disturbed site of the same soil type can be compared to determine the extent of soil deposition or erosion. Such work is exemplified by Quine and Walling (1991) for a range of sites in England and Wales. Their results, which have been compared with other estimates of soil erosion to confirm the veracity of the methodology, reflect considerable variation in the erosion rates, as shown in Figure 7.2. The lowest rates of ca. $0.6\,t\,ha^{-1}\,y^{-1}$ occurred on clay soils which exhibit a high degree of coherence whereas the highest rates of ca. $10.5\,t\,ha^{-1}\,y^{-1}$ occurred on sandy and sand-loam soils. Using many of the sites illustrated in

Figure 7.2, Evans (1993) has shown that erosion occurs frequently, often more than once in three years, even on soil types not particularly susceptible to physical erosion. He has also shown that, in areas liable to erosion, at least 5 per cent of the arable land is affected with a worst case of almost 14 per cent in Nottinghamshire. Other examples of soil erosion in the United Kingdom are given in Boardman and Evans (1994). However, there have been few studies aimed at determining the impact of soil erosion on crop yield, as Biot and Lu (1995) have pointed out. Their experimental work in East Anglia has shown that where 15 cm and 25 cm of topsoil was removed from deep sandy soils the yield of barley declined by ca. 15 and 45 per cent when compared with non-desurfaced soils. This was caused by a decrease in water, organic matter and available nitrogen. The experiment shows that a rapid and severe erosion episode can alter productivity substantially, and although it represents an extreme situation it is likely that less dramatic events would cause pro rata losses.

Europe's loam soils, because of their friability, are also susceptible to erosion with reported annual rates of erosion of $3–100 \, t \, ha^{-1}$. Most of this is considered to be the result of industrialised agriculture, which has replaced traditional intercropping, and the replacement of pasture with arable land. Vandaele and Poesen (1995) have monitored soil erosion in the Belgian loam belt where, despite low slope gradients, soil erosion values for two catchments are in the range $5.4 \, m^3 \, ha^{-1} \, y^{-1}$ to $8.2 \, m^3 \, ha^{-1} \, y^{-1}$. They found that 60–70 per cent of the soil loss was due to rill and ephemeral gully erosion in late spring and early summer, probably due to an increased frequency of intense storms. Not only was soil erosion itself a problem, but Vandaele and Poesen report that most of the soil is removed from the catchment to be deposited as flood-derived sediments in downstream villages. The conversion of pasture to arable land in the former Soviet Union has similarly given rise to serious problems of soil erosion. According to Libert (1995), more than $40 \times 10^6 \, ha$ of land in Kazakhstan, the Volga, Ural, Siberian and Far East regions were converted to arable land between 1954 and 1960 as part of Khrushchev's attempt to improve Soviet food supplies. Agriculture was unsuccessful on at least 25 per cent of this land due to its initial unsuitability and much of it is today liable to erosion, notably wind erosion. Libert reports that ca. 20 per cent of all of the former

Soviet Union's agricultural land and 28 per cent of arable land is subject to soil erosion, as shown in Figure 7.3. In relation to water erosion the Caucasus region has been most severely affected because of cultivation of steep slopes, the cultivation of row crops between which gullies form, the high volume of snowmelt and the frequency of rainfall. In an attempt to combat this problem some $94 \times 10^3 \, ha$ in Georgia have been converted to pasture. Figure 7.3 reflects the significance of wind erosion but since this is associated in many regions with desertification, it is examined in Section 7.4.

The world's most productive croplands in North America are as susceptible to erosion as the croplands of the former Soviet Union, though conservation measures in North America are more widespread. Despite this about one-third of US croplands experience soil erosion and in North America as a whole $60 \times 10^6 \, ha$ are affected by water erosion and $35 \times 10^6 \, ha$ are affected by wind erosion (Table 7.3). Reference has already been made in Section 5.8 to the impact of Europeans and their agriculture during the colonial era, when many soil erosion regimes were established. The case of the Piedmont region was cited, where the cultivation of sloping land generated sufficient sediment to form so-called bottomlands of the Alcovy River swamps (Trimble, 1992). According to the review in Johnson and Lewis (1995), agricultural practices resulted in soil erosion on 95 per cent of the Piedmont's area with 10 per cent being so severely affected as to lose not only its topsoil but also its subsoil. Despite this, the region today is agriculturally productive due to natural and engineered rehabilitation. This involved the abandonment of cropping, beginning in the 1930s, and allowing the land to undergo secondary vegetation succession back to forest. This process was enhanced by soil conservation measures, tree planting and the introduction of wheat and soya beans as crops to replace cotton and maize (as legumes, the soya beans improved the nitrogen content of the soil).

Dregne (1995) has recently reviewed soil erosion and its relationship with productivity in Australia and New Zealand. Sheet, rill and gully erosion are prevalent in eastern Australia whereas wind erosion is a serious problem in the dry cropland areas of southern and western Australia. Nevertheless, the major losses of productivity are a consequence of impaired soil structure (see above). In contrast, New Zealand's major problem concerns the widespread mass movement in the so-called steeplands. In both

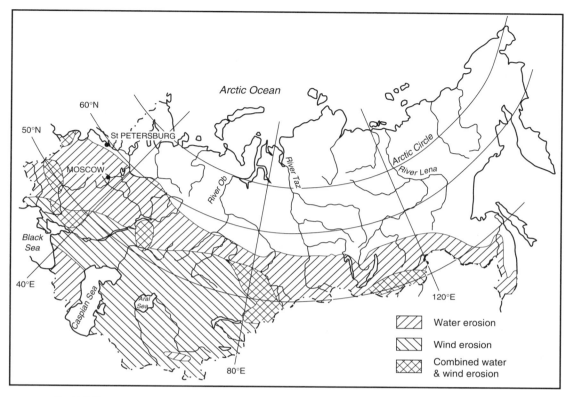

Fig. 7.3 Water and wind erosion: affected areas in the former Soviet Union (based on Libert, 1995).

countries the problem has been exacerbated by European colonisation and the removal of natural vegetation cover (Section 5.9), though as A.R.M. Young (1996) points out, soil erosion and associated landscapes are natural features, just as they are in the semiarid region of the United States. According to Edwards (1993), some $2.7 \times 10^6 \, km^2$ or 51 per cent of the area used for agriculture (including pasture) in Australia are degraded in some way. Approximately 44 per cent of this can continue to be used if appropriate management practices are employed, e.g. contour cultivation, and crop–pasture rotation, etc. The other 56 per cent requires conservation measures involving some form of engineering, e.g. contour banks and soil dams.

In Australia the cereal-producing region of New South Wales and the sugarcane region of Queensland are particularly badly affected. The sugarcane region has been examined by Prove *et al.* (1995), who report that the average loss of soil of $148 \, t \, ha^{-1} \, y^{-1}$ occurs on lands that are cultivated conventionally, i.e. with ploughing. With no tillage,

i.e. direct drilling, erosion rates can be reduced to less than $15 \, t \, ha^{-1} \, y^{-1}$. Such methods can also increase yields and the biodiversity of soil fauna (Radford *et al.*, 1995). Moreover, Edwards (1993) has shown that crop prices influence rates of soil erosion in New South Wales; when grain prices are high, cropping will extend into marginal areas, hence accelerating erosion. Blaschke *et al.* (1992) have reviewed soil erosion problems in New Zealand, where land above 1000 m is particularly susceptible. This is partly cultural and is due to pastoral and arable activities, and partly natural because the land is tectonically active. Blaschke *et al.* show that landslip occurrence is highest at $7.3 \, km^{-2}$ in pasturelands but only $2.5 \, km^{-2}$ in the Taranaki district of southwest North Island.

Much recent work on soil erosion in North America has involved an assessment of the effectiveness of conservation measures. This is illustrated by the work of Richardson and King (1995), who have shown that, on a clay soil, zero tillage (no ploughing) reduced the loss of sediments,

Table 7.4 Soil conversion methods to combat soil erosion by water

A. Reducing run-off

Conservation tillage: requires coverage of at least 30% of the soil surface with crop residues
    (a) *Mulch tillage*: the soil is tilled so that crop residues are retained on the soil surface (also trash farming, stubble mulch tillage)
    (b) *Reduced tillage*: less tillage is needed if weeds are controlled by alternative means, e.g. herbicides
    (c) *No tillage*: crops are planted without seedbed preparation
Mulches: apart from the crop residues mentioned above, which are a type of mulch, other materials can be used as mulches, e.g. stones, gravel, paper, coal and bitumen, though costs are relatively high
Cover crops: these are close-growing crops, e.g. grasses, legumes and small-grain crops, and are grown specifically for protection of the soil
Chemical additives: these may be used to prevent surface sealing of the soil and so encourage water infiltration; they include phosphogypsum and various polymers

B. Retaining run-off

Contour tillage: ploughing, planting and cultivating occurs across the slope of the land, i.e. along the contour; this may give rise to ridge–furrow or lister tillage
Furrow dyking: is the formation of small earthen dykes or dams across the furrows of a ridge–furrow system; rainfall is thus captured and soil infiltration is encouraged
Soil pitting or chain dyking: involves the formation of small depressions (pits) at close intervals to retain water and encourage infiltration; chain dyking involves similar pitting but with special equipment
Level terraces: the creation of terraces is particularly suitable for steeply sloping land; it prevents run-off
Land levelling: is a variation on terracing; it involves levelling of the land to create a series of steps; each step may have a raised outer lip to trap water

C. Controlling run-off

Land smoothing: involves moving soil within fields to create a smooth surface; it gives an even distribution of water
Stripcropping: protective crops are grown in alternating strips along the contour
Graded furrows: are designed to carry excess water from fields at low flow rate; this minimises sediment transport
Graded terraces: the objective is the same as for graded furrows
Variations of bench terraces: may be inward or outward sloping and reduce run-off and sediment transport
Intermittent terraces: are often used for orchards, especially where land is steep
Discontinuous parallel terraces: are valuable on heavily dissected landscapes
Land imprinting: a series of furrows across the slope of the land are interconnected to assist water distribution; this is useful for degraded rangelands
Tillage per se: tillage is modified to suit local conditions
Other practices: vertical mulching, slot mulching, deep tillage, profile modifications, contour-strip rainfall harvesting, etc.

Source: based on Unger (1996)

nitrogen and phosphorus when compared with conventional tillage. In another conservation-based approach, Evans *et al.* (1995) have demonstrated that if crop residues are put into irrigation furrows, the increased roughness causes eroded soil particles to be deposited, including soil particles with fertilisers attached. Moreover, Edwards *et al.* (1995), using field and laboratory data from Prince Edward Island, Canada, have investigated the effect of mulching on cool season erosion. Their results imply that the optimum rate of mulch application for effective erosion control is $4\,t\,ha^{-1}$ of straw. Another Canadian study (Larney *et al.*, 1995) on prairie soils of Alberta has established that the addition of irrigation water cannot compensate for the loss of soil moisture due to topsoil erosion and that crop yields decreased in step with depth of soil removal from non-irrigated plots. As Table 7.4 shows, there are many ways of manipulating mulches, types of tillage and topography to effect the conservation of soils subject to water erosion. Such measures can curtail soil erosion, along with other environmental problems such as cultural eutrophication (Section 7.5) by reducing the loss of nutrients from the agricultural system. Gaynor and Findlay (1995) have shown that conservation rather than conventional tillage in southwestern Ontario increased the amount of phosphorus, as orthophosphate, in drainage water destined for Lake Ontario. Conservation measures to protect soil from wind erosion include shelterbelts; an example of their effect has been presented by Dejong and Kowalchuk (1995) from Saskatchewan, Canada. Their data show that shelterbelts ca. 6 m tall and 200 m apart substantially reduced soil losses from medium- and coarse-textured soils with little difference between sheltered and unsheltered fine-textured soils. Fryrear (1995) has also shown that the maintenance of only a 4 per cent ground cover of crop residues reduced soil losses through wind erosion by 15 per cent whereas 50–70 mm ridges reduced soil erosion by 98 per cent on a sandy loam soil in Texas.

These issues will be reconsidered in Section 8.3 in relation to the developing world.

7.4   Desertification

Although desertification is considered to be a major environmental issue, affecting a large proportion of humanity (e.g. Myers, 1993a), there is much confusion as to what it actually comprises. As Thomas and Middleton (1994) state, 'the term desertification has been used to refer both to processes of change and states of the environment'. It is a form of land degradation specifically associated with arid and semiarid environments and although the UNEP's definition (Mendoza, 1990) is now widely accepted, it does have shortcomings, as Thomas and Middleton (1994) have discussed. UNEP define desertification as 'a comprehensive expression of economic and social processes as well as those natural and induced ones which destroy the equilibrium of soil, vegetation, air and water, in the areas subject to edaphic and/or climatic aridity'. Continued deterioration 'leads to a decrease in, or destruction of the biological potential of the land, deterioration of living conditions and an increase of desert landscape'. There has also been much debate as to the cause of desertification, notably about the roles of natural change versus culturally induced change and, because of media coverage, it is often considered to be a feature of the developing rather than the developed world. It is now accepted that both natural and cultural factors are involved and that desertification also occurs in the developed world.

Figure 7.4 shows the areas of the world at risk from desertification; outside the developing world the major areas at risk are the southwest United States, parts of southern Europe, much of Australia, parts of South Africa and the central Asian republics (formerly part of the Soviet Union). However, there is much debate as to how much of the world's land is desertified and at risk from desertification. This is because of different definitions being applied, difficulties of monitoring and the use of different criteria (Thomas and Middleton, 1994). Two estimates for the extent of desertification in the areas cited above are given in Table 7.5. Although the estimates do not agree, it can be concluded from these data that desertification is indeed extensive in these regions. Not only is it a cause of environmental degradation, it is also responsible for economic losses as crop and animal productivity decline; yet in all these nations desertification is primarily caused by poor agricultural practices. Moreover, remedial measures can often limit the damage and bring about rehabilitation.

A classic case of desertification occurred in the United States during the 1930s. It became known as the Dust Bowl because large and destructive dust storms were generated and affected extensive areas of the Great Plains, notably Kansas, Colorado, Oklahoma, Texas and New Mexico. The problem arose because of a prolonged period of drought and increased pressure on the land from migrant farmers. Crop cultivation began in the region during the 1870s as waves of migrants moved westward; the early 1900s enjoyed higher than average rainfall, so agriculture was successful. The prairie was ploughed up in ever increasing amounts to grow wheat, leaving the soil over vast tracts of flat land vulnerable to wind erosion. High prices for the wheat and increasing mechanisation also encouraged the spread of crop cultivation. However, in 1931 drought brought circumstances hitherto never experienced by Great Plains farmers. Nearly 3×10^6 ha were severely affected by wind erosion and, according to Worster (1979), about 3.5 million people abandoned their homes and most moved west. Despite major advances in soil conservation in the Great Plains, and other parts of the southwestern United States, such problems still recur. The most positive outcome of these events was the heightened awareness of the need for conservation measures. As a result, a distinguished research programme into soil erosion and its impact was initiated. Moreover, as Kassas (1995) points out, drought occurred again in the Great Plains in the 1950s. This time its impact was much less conspicuous because a range of measures were by then adopted to safeguard people and the environment. Such measures included the establishment of the Soil Conservation Service in 1935 to develop soil conservation procedures, federal financial support to undertake ecological restoration, carrying-capacity policies for rangelands, improved transport via railways and

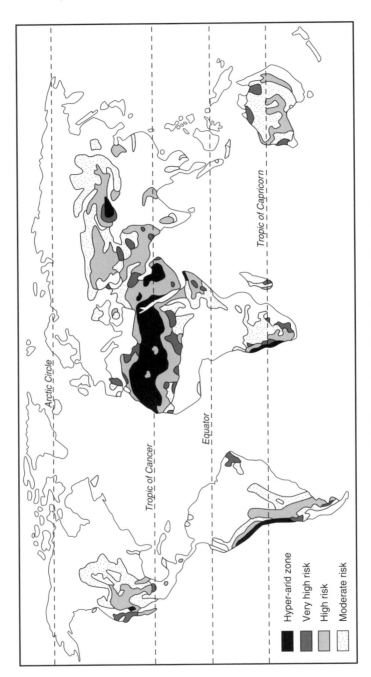

Fig. 7.4 Worldwide risk of desertification (based on Myers, 1993a).

Table 7.5 Two ways to estimate the extent of desertification

	Rangelands		Rain-fed croplands	
	Extent (ha)	Percent of total	Extent (ha)	Percent of total
North America[a]	475×10^6	75	74×10^6	16
North America[b]	300×10^6	42	85×10^6	39
Australia[a]	656×10^6	55	42×10^6	34
Australia[b]	450×10^6	22	39×10^6	30
Former USSR[a]	not available		not available	
Former USSR[b]	250×10^6	60	40×10^6	30

[a]United Nations Environment Programme (1991c)
[b]World Resources Institute (1989)

roads, improved cattle breeds, fencing and improved water availability. However, overgrazing continues to be a cause of current desertification in the United States. It depletes the grass cover, which is so heavily grazed that it cannot recover, leaving the soil bare and subject to both wind and water erosion. This has been a problem since the 1880s, when buffalo were eradicated as ranchers became established. Overgrazing with cattle not only led to soil removal but also the encroachment of unpalatable shrubs such as sagebrush and mesquite.

Overgrazing is cited as an underpinning cause of desertification in the subtropical zone of the eastern Cape and the Karoo of South Africa. Kerley *et al.* (1995) assert that the subtropical thicket of the eastern Cape, comprising shrubs, lianas and herbs has been subject to intensive grazing by sheep and cattle to such an extent that biodiversity has declined substantially since 1800. Biomass and primary productivity have also declined. The situation in the Karoo has been reviewed by Hoffman *et al.* (1995) and Dean *et al.* (1995), who present conflicting evidence from a variety of sources. They suggest that the problem is not widespread nor is it irreversible despite extensive post-colonial cattle-rearing activities. The most significant change that has occurred is the decline in grasslands which have been invaded by shrubs due to grazing pressure (Bond *et al.*, 1994). Similar processes have occurred in Australia; again this has been due to the 'expansion of Europe' and the introduction of European agricultural practices. Although both arable and pastoral agriculture can cause desertification, pastoral agriculture is the biggest cause in Australia due to the introduction of hard-hooved animals in large numbers. According to Ludwig and Tongway (1995), some 42 per cent of the arid and semiarid lands of Australia have become desertified, a process involving the loss of perennial grasses from grasslands and savannas and their replacement by unpalatable

shrubs and soil movement. The process whereby such changes occur has been detailed by A.R.M. Young (1996) who indicates that overgrazing leads to soil erosion and reduced soil moisture. This does not favour grass growth; the increased penetration of water into insect holes and dehydration cracks creates a store of moisture at depth, moisture that grasses cannot utilise but which can be retrieved by deeper-rooted shrubs. Young states 'In the mulga country of south-west Queensland woody shrubs that stock will not eat (for example, *Eremophila* spp.), or thickets of mulga with no grass understorey, are still spreading into pastures that were once grassy and lightly timbered'.

The mismanagement of irrigation systems can also lead to desertification. If salts are allowed to concentrate in the soil due to the overapplication of water, leakage from canals and poor drainage, crop growth will be poor and agricultural land may be abandoned. The regrowth of natural vegetation will be slow, or may not take place if salt concentrations are very high. This leaves the soil vulnerable to soil erosion and loss of soil biota, leading to desertification. This has occurred in many parts of the central Asian republics of Tadzhikistan, Turkmenistan, Uzbekistan, Kyrgyzstan and Kazakhstan. Together they have 9.4×10^6 ha of irrigated land, representing nearly 45 per cent of the irrigated land of the former Soviet Union, (Goskomstat quoted in Libert, 1995, though Libert asserts that the true figures may be increased by 30 per cent). Much of this land was brought under irrigation in the 1970s and 1980s, when hectarage doubled using techniques that were not designed for specific areas but which were centrally planned. Coupled with poor water management some of this vast area of irrigated land has been abandoned. According to Roskomzem, quoted in Libert (1995), some 400×10^3 ha of irrigated land in the former

Soviet Union were abandoned in 1991–92. In the Asian republics the problem is exacerbated by overgrazing on the grasslands. Libert (1995) indicates that in the plains of Turkmenistan, Uzbekistan and southern Kazakhstan 65×10^6 ha have been damaged by overgrazing. Desertification is also associated with the problems that beset the Aral Sea, as will be discussed in Section 7.6.

However, desertification is most acute in the steppelands of Kalmykia, west of the Caspian Sea. The problem began after the Second World War as 3.6×10^6 ha of grasslands were used to overwinter sheep from nearby states (Libert, 1995). Year by year the number of sheep increased; arable cropping was introduced during the 1960s to produce animal feed, and desertification set in. Although nearby states ceased sending their sheep for winter pasture because its quality had declined, state collective farms were established. They intensified activities, so by 1986 a total of 1.5 million sheep were grazing 3.6×10^6 ha, i.e. double the carrying capacity. The extent of severely damaged pasture increased from 32 per cent in 1959 to 76 per cent in 1986. By the 1980s, 40×10^3 to 50×10^3 ha of pasture were being lost annually. There are currently 650×10^3 ha of drifting sand dunes and salinised soils and more than 2×10^6 ha of pasture are degraded, representing 20–30 per cent loss of production capacity. Moreover, Zonn (1995) states that in 1986 the General Scheme of Desertification Control was implemented, including dune reclamation by tree planting, pasture rotation, planting of fodder plants, etc. The scheme has been successful in its early stages but once the pastures revert to management by the original farms there are fears that the problems will begin again.

There is some debate as to whether or not desertification is actually occurring in Europe, a problem that is due to the difficulty of precisely defining desertification (see above) and the fact that it comprises a range of processes rather than a single process. In Mediterranean Europe it is generally considered to represent the final stage of soil degradation (Chisci, 1990), involving a substantial loss in biological productivity. Inevitably soil erosion plays a significant role. There are examples of desertification in Spain and Italy. It is possible that fire contributes to desertification in Spain, as Carreira and Niell (1995) have discussed. Their work on a semiarid gorse scrubland in southeast Spain has shown that large amounts of nutrients are removed in material eroded from burnt

plots, which is disadvantageous for primary productivity in the long term. Land management in the central part of the Ebro Valley is also considered to be the cause of desertification (Quine et al., 1994). In a comparison of soil erosion from cultivated and uncultivated plots, the cultivated experience a rate of soil removal five times higher (1.6–$2.5 \, \mathrm{kg \, m^2 \, y^{-1}}$) than the uncultivated (0.2–$0.4 \, \mathrm{kg \, m^2 \, y^{-1}}$), though Quine et al. suggest the high rates could be reduced by developing a bush or shrub cover to protect the soil. The situation in Greece has been discussed by Yassaglou (1990) who asserts that the problem occurs on steeply sloping land, especially on limestone, where soils are naturally thin and any soil loss leads to a decline in primary productivity. Yassaglou suggests that the salinisation of irrigated soil in Greece may also give rise to desertification. Boix et al. (1995) have asserted that intensive grazing, controlled firing and abandoned agricultural terraces may all contribute to desertification where climatic conditions allow.

7.5   Water quality: cultural eutrophication

Since the first agricultural systems were established 10 K years BP (Section 4.5), farmers have sought to influence soil fertility because the presence of available nutrients in the soil is essential for crop production. The early agriculturalists in the Middle East, especially in the valleys of the Nile and Tigris, recognised the importance of fertile silt deposited by floods and in many cases instituted measures to control the floods so that silt deposition was ensured in crop-growing areas. Similarly, the use of animal manure and human sewage has been recognised throughout history as a means of increasing soil fertility and of improving soil structure. The most important nutrient elements that are essential to plant growth are nitrogen and phosphorus; if they are in short supply within the soil, plant productivity will be impaired. Until the Industrial Revolution of the 1700s, enhancement of soil nutrients was achieved by manuring in traditional farming systems (Section 5.3 and 5.4) that involved both crops and livestock; the livestock were often allowed to roam on fields left fallow, so organic matter and nutrients were replenished until a further crop was planted. In addition, many crop rotations involved leguminous crops, which enhanced the nitrogen content of the soil by the activity of nitrifying bacteria in root nodules, and the ploughing in of plant residues after the crop was harvested.

Today, however, developed nations have largely replaced these practices by high-technology agriculture, wherein nutrients are replenished by the application of artificial fertilisers. Artificial fertilisers were first produced in the early 1900s, when the Haber–Bosch process was developed. This facilitated the fixation of nitrogen from the atmosphere, a process that was hitherto only possible via the action of bacteria. Nitrate, phosphate and potash fertilisers have been widely available since the Second World War and they have been applied on a large scale in developed countries. Together with mechanisation, their application has been responsible for increased food production in these areas; and since the 1960s artificial fertilisers have been increasingly used in developing countries. Although the increase in food production is a beneficial effect of their use, research over the past two decades has revealed there are also detrimental effects of artificial fertilisers. The most significant disadvantage is the cultural eutrophication of freshwater aquatic ecosystems in receipt of drainage from agricultural land. Consequently, many wetland habitats have suffered a detrimental change in their flora and fauna, often exacerbated by sewage inflow that is also nutrient rich. In addition, some groundwater and aquifers are now known to contain nitrate levels that are considered hazardous to human health and which are creating problems for domestic water-supply industries. Cultural eutrophication has also occurred and is continuing to occur in estuaries and coastal environments to the detriment of their ecology.

Figure 7.5 illustrates the growth of fertiliser use, which has been especially marked since 1960. Moreover, the use of nitrate fertiliser has outstripped the use of phosphate and potash. The production of nitrate fertilisers represents an important component of modern agribusiness, along with crop-protection chemicals. Both have a significant environmental impact, though both have contributed substantially to the increase in global food production. The major benefit has, however, been in the developed world, where increasing fertiliser use has been paralleled by increasing mechanisation and the 'industrialisation' of agriculture. The world's major food producers, including the United States and the European Union, have achieved their status by employing

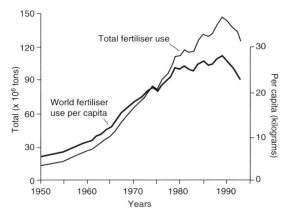

Fig. 7.5 Trends in world fertiliser use (based on FAO data quoted in Brown, 1994).

such technologies, all of which are underpinned by fossil fuels. The increasing use of fertilisers accompanied the intensification of agricultural production on existing cropland, the expansion of which was curtailed through planning policies and lack of suitable land. Fertiliser use has also increased substantially in some developing countries. In China, for example, the lack of land into which agriculture can expand and a large population of 1.115 billion has led to a huge increase in fertiliser use. The significance of nitrate fertiliser production is reflected in the fact that the Haber–Bosch process now uses 1.3 per cent of world energy production (Gilland, 1993). This contributes to the production of heat-trapping gases and global climatic change (Section 6.5) as well as acidification (Section 6.6). Phosphate fertilisers are produced from phosphate-rich sediments which are mined. Nowadays the United Kingdom alone imports 1.5×10^6 t of phosphate annually from Senegal, Morocco and Tunisia (Body, 1987).

The use of nitrate and phosphate fertilisers represents a manipulation of the nitrogen and phosphorus biogeochemical cycles. In both cases there is an acceleration in the flux of each nutrient from the atmospheric pool (as nitrogen gas) and from the lithospheric pool in the case of phosphate, to the soil pool. Not all of the fertiliser is retained in the soil or taken up by crops; a proportion is removed from its place of application and enters drainage water or groundwater; nutrient-rich drainage may also reach coastal regions. Such a process of nutrient enhancement occurs naturally under conditions of non-cultivation. The erosion and leaching of nutrients enriches water bodies, thereby providing nutrients for algal growth and the general maintenance of trophic structures. This process is known as eutrophication. Cultural eutrophication is the acceleration of these processes by direct or inadvertent human activity. Although the use of nitrate and phosphate fertilisers is one cause of cultural eutrophication, other contributory factors include the emission of untreated or partially treated sewage, urban run-off and the disruption or removal of vegetation cover, which leaves soils vulnerable to additional leaching and erosion. Generally, however, non-point fertiliser use is the main source of nutrients; nitrate, because of its high solubility, is readily flushed out of soils into drainage water and/ or groundwater whereas phosphate, which is not very soluble, is removed from its site of application by the removal of soil particles to which it is attached; phosphate is therefore associated with soil erosion (Section 7.3). Other agricultural sources of these nutrients are animal sewage, especially where animals are housed indoors, which may also be used as a fertiliser, and crop residues. Crop residues are often ploughed back into the soil, or retained as a protective cover (Section 7.3), and are then subject to microbial decomposition; the resulting nitrates may be leached.

The increased input of nutrients into freshwater environments promotes the growth of algae, the reproduction of which would otherwise be constrained by nutrient scarcity. This increase in primary productivity is initially beneficial, but as cultural eutrophication proceeds, the water quality declines and the biotic community is adversely affected, including the diminution of biodiversity. Algal blooms develop as algal productivity increases; and as their density increases, the oxygen in the water is consumed and the floating algal mat inhibits the replenishment of oxygen from the atmosphere. Consequently, other organisms in the aquatic system die as anaerobic conditions develop. It is also likely that the lack of oxygen will reduce ferric salts to ferrous salts, causing water discolouration. Coupled with the production of smelly hydrogen sulphide, such changes are aesthetically displeasing. In addition, the seepage of nitrates into groundwater is problematic because they are associated with human health problems.

Eutrophication is widespread in the United Kingdom (Moss, 1996), as exemplified by a recent survey of 102 wetland Sites of Special Scientific Interest (SSSIs) in England (Carvalho and Moss, 1995). Eighty-five showed evidence of eutrophication

and 78 had already become eutrophic or even hypereutrophic, though in about half of these sites the major cause was sewage. One area that has been particularly badly affected is East Anglia, especially the Norfolk Broads, where water quality and biodiversity have declined as a result. In the early 1960s Broadland waterways had phosphate concentrations of ca. $80 \, mg \, dm^{-3}$ whereas values as high as $300 \, mg \, dm^{-3}$ occur today. Turbid water allows little light penetration and is characteristic of these eutrophic regimes, along with a few species of algae in abundance. The rapid growth of algae plus the decomposition of dead algae inhibits the diffusion of oxygen into the water, causing a loss of fish and invertebrate species whereas the decline of plant communities along the banks has caused erosion to accelerate. According to Blunden and Curry (1996), only four of the 41 broads had the traditional range of aquatic plants by the early 1980s. Cultural eutrophication is also a problem in rivers, as Muscutt and Withers (1996) have discussed in relation to phosphate loads in the rivers of England and Wales. They found that orthophosphate concentrations in 98 rivers ranged from less than 0.01 to $7.85 \, mg \, dm^{-3}$. The highest concentrations occurred in urbanised catchments, where they are attributed mainly to sewage, and in rural catchments with low flow rates, where agricultural practices were more important. Inputs of animal wastes sometimes provide point sources for nutrients.

Phosphate pollution is a significant problem in Europe, especially where intensive animal farming takes place. Behrendt and Boekhold (1993) have suggested that a surplus of phosphorus has occurred in the soils of European countries because it is often applied in excess either as inorganic fertiliser or manure. This surplus may be as much as $45 \, kg \, ha^{-1} \, y^{-1}$ and much of it has been stored in soils which have become saturated; continued application of fertilisers and/or manure means that leaching to groundwater and removal to rivers and lakes will take place. Behrendt and Boekhold highlight the 300×10^3 ha of Dutch sandy soils that are considered to be saturated with phosphorus and which will contribute to cultural eutrophication in the future. Moreover, wetlands of the Netherlands are at risk from eutrophication (Wolff, 1993). Many European rivers are also characterised by high nitrate loadings. For example, Tolba et al. (1992) report that average values for nitrate in European rivers are $4500 \, mg \, dm^{-3}$; this compares with $100 \, mg \, dm^{-3}$ outside Europe. Further examples of riverine eutrophication include those reported by Ekholm (1994) and Kauppi *et al.* (1993) for watercourses in Finland. Kaupi *et al* point out that the rivers of the intensively cultivated region of south and west Finland experience algal blooms and turbidity.

Outflows from rivers and streams with high nutrient concentrations from northern Europe and Scandinavia are contributing to the cultural eutrophication of the North Sea and the Baltic Sea respectively. The situation in the North Sea has been examined by Riegman (1995), who indicates that one symptom of eutrophication along the Dutch coast has been a doubling of annual algal biomass over the last 30 years. Here the input of phosphate is so high it is no longer a limiting factor to algal growth, which is more limited by nitrate availability. One result of this change is a shift to dominance of large algae, e.g. *Phaeocystis*, which cannot be grazed by microzooplankton. This alters the structures of food webs and food chains, which can also be affected by changes in the oxygen content of the water. The impact of cultural eutrophication on the Baltic Sea has been discussed by Enell and Fejes (1995), who suggest it is one of the world's most polluted seas. It receives drainage from many countries, mostly involved in 'industrialised' agriculture and it has a limited outflow to exchange water with the Atlantic Ocean. Currently $1.409 \times 10^6 \, \mathrm{t \, y^{-1}}$ of nitrogen reach the Baltic Sea, of which ca. 70 per cent derives from rivers and non-point sources. This alone is 50 per cent more than Enell and Fejes suggest is acceptable to maintain the ecosystems of the Baltic Sea. In a specific study of diatom communities in the Gulf of Riga, Sakson and Miller (1993) have shown that areas of high eutrophication, notably river estuaries, are characterised by diatom communities of varied composition; diatom communities were more or less constant in areas with moderate eutrophication. The communities become less diverse with increasing nutrient content. Concerns about the adverse impact of cultural eutrophication on the Baltic Sea have prompted policies to reduce nutrient input considerably as the millennium approaches.

The Mediterranean Basin, which like the Baltic Sea is enclosed with only a small outlet, has been affected by cultural eutrophication along extensive areas of its coastline. For example, the Adriatic Sea receives drainage from agricultural land and sewage effluent. According to Barmawidjaja *et al.* (1995), the nutrient load of the Po Delta has increased steadily since 1900, with intensification around 1930. Oxygen deficiency

developed a decade ago, and these anoxic conditions caused faunal changes in the ecosystem. Problems in the Aegean Sea have also been reported (e.g. Balopoulos and Friligos, 1994). The regular growth of algal blooms, consisting of diatoms and dinoflagellates, can deplete local oxygen levels and lead to fish kills. Dinoflagellate reproduction can sometimes occur so rapidly that so-called red tides develop over tens of kilometres. Coupled with the production of noxious smells, red tides reduce amenity values and deter tourists; health hazards from eating shellfish may also occur.

Increasing concentrations of nitrates in many of Europe's aquifers are creating additional problems, not least because the World Health Organisations (WHO) and the EU have established 'safe limits' for nitrates in drinking water. The directive is that nitrates should not exceed $50 \, \mathrm{mg \, dm^{-3}}$. Some water companies in the United Kingdom, notably Severn Trent and Anglian, are experiencing difficulties meeting this standard (Dudley, 1990); the standard originated from the alleged but unproven link between nitrate and stomach cancer, and its proven link with methaemoglobinaemia ('blue baby' syndrome). The problem of high nitrate concentrations in groundwater may intensify because the nitrates emerging today may be the result of fertiliser use several decades ago; the potential time lag between application to soils and emergence in aquifers means there is no short-term solution. Similar problems are occurring throughout Europe and are especially manifest in areas of intense agriculture. There are also problems with phosphate; for example, De Smet *et al.* (1996) describe the impact of the use of pig manure as a fertiliser in west Flanders, Belgium. Here, sandy loam soils have a limited capacity for the absorption of phosphate and since the water table is only 1 m below the surface, contamination is likely.

Beyond Europe, cultural eutrophication is a problem in North America and, to a lesser extent, in Australia. In Section 5.8 reference was made to the impact of European colonists on North American landscapes and the onset of eutrophication in many lakes as deforestation occurred and agriculture was initiated. The sediments derived from these activities and the remains of aquatic organisms (e.g. diatoms) affected by them have been analysed in a range of lakes extending from the East Coast to the West Coast. However, as in Europe, the advent of artificial fertilisers in the early 1900s and the subsequent intensification of use has brought about substantial environmental change. For example, Boggess *et al.*

(1995) and Havens *et al.* (1996) have discussed the case of Lake Okeechobee, a subtropical lake in Florida. Alterations began in the early 1900s and accelerated eutrophication has recently occurred, following a doubling in phosphorus concentrations from animal wastes from dairies (Havens *et al.* 1996); nitrogen has now become the main limiting factor. As well as an accumulation of phosphorus in the lake sediments, change in the biota include the replacement of diatoms by cyanobacteria and a dominance of the fauna by oligochaetes (bristle worms). Over many years, concerns have been expressed about the condition of the Great Lakes, which have experienced cultural eutrophication since initial settlement in the early 1800s (Section 5.8). Schelske and Hodell (1995) have detailed the history of Lake Erie since this time and have shown that phosphorus loading increased gradually until the late 1940s; thereafter it increased substantially until the 1970s. Consequent on management practices to curtail phosphorus inputs, its loading has declined since the mid-1970s. On the basis of carbon isotope data, the peak of eutrophication was followed by a peak in productivity in the mid-1970s, which has since declined in line with phosphorus.

A major example of river and coastal eutrophication caused mainly by agricultural activities is Chesapeake Bay and its tributaries. Chesapeake Bay has an area of $162 \times 10^3 \, km^2$ with a hinterland of intensively farmed cropland and a small outlet to the Atlantic Ocean; one of its main tributaries is the Susquehanna River, which contributes 50 per cent of the land-derived water, 40 per cent of the nitrate and 21 per cent of the phosphorus. Other tributaries include the Potomac, Patrixent and Choptank rivers. According to Boynton *et al.* (1995), the nutrients entering Chesapeake Bay, including those from atmospheric sources, are converted into organic and particulate forms soon after they enter the estuary. The data presented by Boynton *et al.* for total nitrogen and total phosphorus reflect 600 per cent increases in nitrogen loading, and 1300–2400 per cent increases in phosphate since pre-colonial times. In the Susquehanna River alone, annual loads of phosphorus and nitrogen in the past 11 years have varied substantially, with a fourfold increase in phosphorus and a doubling of nitrogen. For the other tributaries, diffuse sources of nutrients contributed 60–70 per cent of the total, though some of the nitrogen loading is due to the deposition of nitrous oxide generated from other sources (Section 6.5). The restricted outflow of Chesapeake Bay also means that

its ecology has been significantly altered in the past 50 years. Problems of cultural eutrophication in the United States have also been reported by Moore *et al.* (1995) and Mozaffari and Sims (1994), who highlight problems associated with the use of poultry waste as manure; McNeal *et al.* (1995) point out that the use of nitrate fertilisers in Florida for citrus fruit production is causing elevated nitrate concentrations in Lake Manatee, a reservoir used for drinking water, through the contamination of groundwater. The survey by Spalding and Exner (1993) on nitrate in US groundwater has shown that high nitrate concentrations occur in groundwater in regions where cropland soils are well drained and irrigated, mainly west of the Missouri River. In addition, Peckol *et al.* (1994) have suggested that changes in the dominant species within macroalgal communities in Waquoit Bay, Massachusetts, are due to nitrogen loadings from groundwater.

Examples of the occurrence of cultural eutrophication in Australia are given by Young *et al.* (1996) and McComb and Davis (1993). They have shown that wetlands and coastal bays are particularly susceptible because nutrients are readily flushed out of the sandy soils of the catchment, especially as streamflow is seasonal. The nutrients derive from a variety of sources but the agricultural use of fertilisers is a major contributor. The nutrient enhancement in coastal bays causes algal blooms, mainly of cyanobacteria and a decline in seagrass abundance. Invertebrate and fish populations tend to be adversely affected by reduced oxygen concentrations and waterbirds have been killed by ingesting algal toxins and outbreaks of botulism. There is also evidence for the large-scale eutrophication of the Great Barrier Reef lagoon in the vicinity of Low Isles. Bell and Elmetri (1995) have shown that phytoplankton concentrations have risen over the past 65 years with an increase in dinoflagellates. This is attributed to enhanced nutrient inputs due to agricultural practices. Such alterations to algal communities may be responsible for the increase in the crown of thorns starfish, which in turn is overgrazing coral, reducing the biodiversity of the reef and its value as a tourist attraction.

There are many ways in which the problem of cultural eutrophication can be mitigated. These can be divided into two groups: measures to combat symptoms and measures to combat the underlying causes. Treatment processes combat symptoms; they can reduce nitrate levels in aquifers used for domestic water supplies. Two methods are the establishment of denitrification plants and the mixing of nitrate-rich water with nitrate-poor water;

both are costly. The use of buffer zones between agricultural land and watercourses has also been proposed (e.g. Haycock *et. al.*, 1993) because riparian vegetation communities sequester nitrogen in the biomass, and bacteria within the community denitrify nitrates to emit gaseous nitrogen to the atmosphere. Such communities would also trap sedimentary phosphorus before it reached watercourses. Jansson *et al.* (1994) have discussed the use of a similar buffer zone to reduce cultural eutrophication off the southern coast of Sweden, though in this case they recommend a series of ponds rather than a single wetland unit. Furthermore, Gumbricht (1993) has highlighted the significance of submerged macrophytes for the sequestration of both phosphate and nitrate.

However, the measures required to tackle the source of the problem concern agricultural practices; farmers need to be persuaded to reduce applications of fertilisers. A pilot scheme to reduce nitrate pollution was established in the United Kingdom during 1989 (Archer, 1994). Ten nitrate sensitive areas (NSAs) and nine nitrate advisory areas (NAAs) were established by the Ministry of Agriculture Fisheries and Food. The NSAs are governed by voluntary restrictions on the use of artificial fertilisers, slurry, etc., with financial compensation being paid to farmers who participate by entering into a five-year contract, during which time they will alter their farming practices, and possibly their land use, to curtail nitrate losses. Farmers in the NAAs receive advice but no compensation. It remains to be seen how effective these measures prove and whether or not they have any impact on the loss of phosphates. Similar measures have been introduced in Europe and North America where set-aside, the removal of land from agricultural productivity, is equally important.

7.6 Soil and water quality: salinisation

Just as significant as cultural eutrophication is the problem of salinisation in watercourses and water bodies. This is a problem in arid and semiarid regions, notably where irrigation is practised. It is most acute and widespread in the developing world rather than the developed world, so the processes involved are considered in Section 8.6. The problem arises because of the relationships illustrated in Figure 7.6, i.e. irrigation systems expose a large surface area of water

to evaporation. High evaporation rates, which characterise arid and semiarid areas, increase the salt concentrations in the irrigation water and in the irrigated soils. The irrigated soils may also be subject to waterlogging, and together with salinisation, waterlogging decreases crop productivity as well as creating problems of water supply. According to Postel (1996), agriculture uses 65 per cent of all the water extracted from rivers, lakes and aquifers for human uses. This compares with 25 per cent used in industry and 10 per cent for domestic use.

Figure 7.7 gives details of global water usage and emphasises the significance of agriculture as the chief consumer of water. According to the World Resources Institute (1996), between 225×10^6 ha and 250×10^6 ha of the Earth's surface is irrigated, mainly for crops; some 60 per cent of this is in Asia, especially in India, China and Pakistan (Section 8.6) with smaller but significant areas in Europe, the United States and the former Soviet Union, notably the Asian republics. Irrigation is the major cause of irretrievable water loss, though it remains a vital component of the world's food-production systems. Along with fertilisers (Section 7.5), improved crop varieties and crop-protection chemicals, irrigation development underpinned the so-called Green-Revolution of the 1970s, when it expanded at the huge annual rate of 2.2 per cent, falling to 1.9 per cent in the 1980s (World Resources Institute, 1996); it is now at 0.8 per cent. This reduction in the rate of increase is because of the adverse environmental impact of irrigation as well as the high costs now associated with construction and maintenance. Of the environmental degradation due to irrigation that has occurred in the developed world, the most severely affected areas are in the southwest of the United States, the Murray–Darling Basin of Australia and the region around the Aral Sea in the republics of Turkmenistan, Uzbekistan and Kazakhstan (the republics also suffer from desertification, as discussed in Section 7.4).

In the United States the area most severely affected by soil and water salinisation is the Colorado River basin and the parts of California that use its water for irrigation, as shown in Figure 7.8. Since 1923, when the Colorado River Compact was signed by six of the seven states through which it flowed, eight large dams have been constructed along with extensive aqueducts to provide water for irrigation. Even before the seventh basin state, Arizona, joined the compact in 1944, when the Mexican Water Treaty was also signed, the Hoover Dam had been constructed (1930s) along with the All-American Canal to bring water

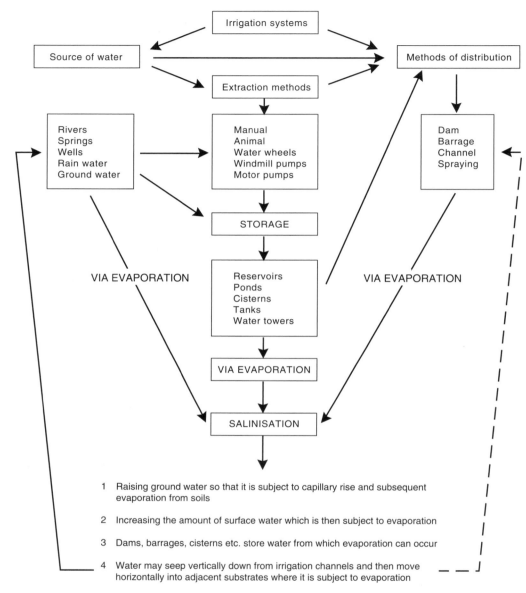

Fig. 7.6 Relationships between irrigation systems and salinisation.

into California's Imperial Valley for irrigation. All of the signatories, plus Mexico, were pledged specified volumes of water which, according to Craig *et al.*, (1996), amounted to 16.5×10^6 acre feet (2.04×10^{10} m^3). These allocations were, unfortunately, based on flow values during a series of wet years and have since proved to be unrealistic as average flows are ca. 13.1×10^6 acre feet (1.62×10^{10} m^3).

Problems have also arisen because of high salinity levels, as much as 1500 ppm, in the water reaching Mexico and because a Navajo Indian reservation is also pressing for irrigation water. Not all these states have demanded their full allocation, and neither have the Navajo Indians, but demands are increasing. The latest project to be completed, the Central Arizona Project, has meant Arizona using its full allocation

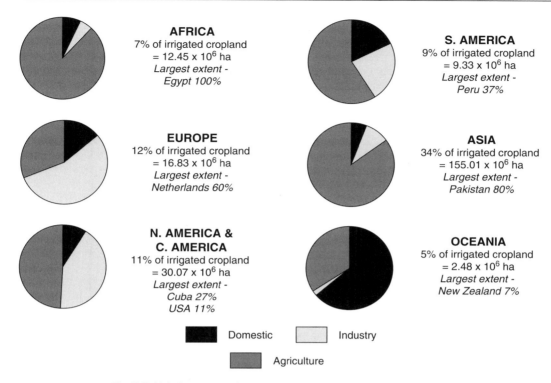

AFRICA
7% of irrigated cropland
= 12.45 x 10⁶ ha
Largest extent -
Egypt 100%

S. AMERICA
9% of irrigated cropland
= 9.33 x 10⁶ ha
Largest extent -
Peru 37%

EUROPE
12% of irrigated cropland
= 16.83 x 10⁶ ha
Largest extent -
Netherlands 60%

ASIA
34% of irrigated cropland
= 155.01 x 10⁶ ha
Largest extent -
Pakistan 80%

N. AMERICA &
C. AMERICA
11% of irrigated cropland
= 30.07 x 10⁶ ha
Largest extent -
Cuba 27%
USA 11%

OCEANIA
5% of irrigated cropland
= 2.48 x 10⁶ ha
Largest extent -
New Zealand 7%

Domestic   Industry   Agriculture

Fig. 7.7 Global water use (based on World Resources Institute, 1996).

and California having to relinquish 1×10^6 acre feet $(1.2 \times 10^9 \, m^3)$ of its allocation, which it has used to supply San Diego and Los Angeles. The exposure of a substantially increased surface area of water through reservoirs, etc., has led to increased salinity. This has caused problems for Mexico's use of the water. Consequently a desalinisation plant has been constructed at Yuma, Arizona at a cost of $350 million, some 30 times the initial cost of providing irrigation water to California. According to Craig *et al.* (1996), this will become operational in the late 1990s to provide Mexico with 1.5×10^6 acre feet $(1.9 \times 10^9 \, m^3)$ of water with salt content reduced to ca. 800 ppm, for irrigation. The saline water produced by the plant, with a salt concentration of ca. 8200 ppm will be emitted via canal into the Gulf of California.

Problems of water availability and water quality are likely to increase in the future. Charbonneau and Kondolf (1993) have pointed out that between 1980 and 1990 California's population has increased by 25 per cent, prompting urban spread, though water demands have decreased due to conservation measures (Diaz and Anderson, 1995). Much of this urban spread has occupied some 13 per cent of prime agricultural land, with a further 48 per cent on natural habitats and marginal farmlands. At the same time, the extent of irrigated farmland has increased but 66 per cent of this increase has occurred on land of poorer quality than used for urban spread. The implication is that these poorer croplands will need increased volumes of fertilisers and will be subject to soil erosion. Soil and water salinisation and cultural eutrophication, as well as soil erosion, are thus set to increase if such trends continue or accelerate. Other ramifications of intensive water use have been discussed by Lemly *et al.* (1993) in relation to arid wetlands throughout the western United States. Firstly, water flows have been reduced because of withdrawals for irrigation and urban consumption, and secondly, contaminants have caused the poisoning of migratory birds on at least six wildlife refuges. A case in point is the Kesterson Refuge in the San Joaquin Valley. Here high selenium concentrations built up to reach $4.2 \, mg \, dm^{-3}$ in 1982–83, which is considerably in excess of the $0.01 \, mg \, dm^{-3}$

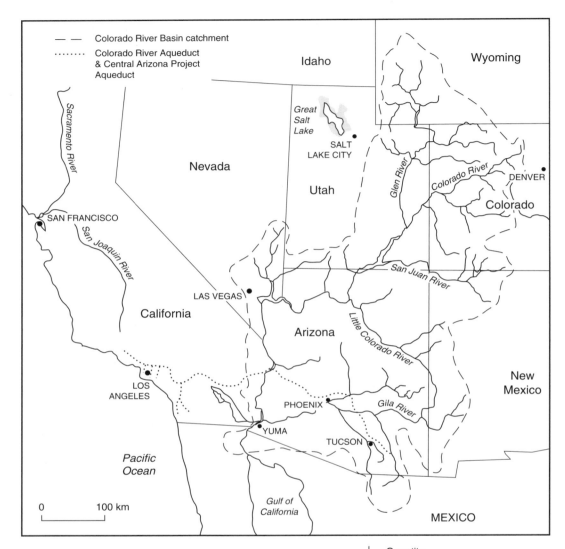

Water demand	Quantity (x 10^{12} litres)
In-basin consumptive uses (75% agricultural)	14.26
Central Arizona Project (rising to 3.5 x 10^{12}l)	1.24
Mexican allotment (1944 Treaty)	1.86
Evaporation from reservoirs	1.86
Bank storage at Lake Powell	0.62
Phreatophytic losses (water-demanding plants)	0.62
Budgeted total demand	20.46
(1930 - 1980 average flow of the river	16.12)

Fig. 7.8 Water resources in the Colorado Basin of the United States (based on Christopherson, 1994).

recommended by the United Statess Environmental Protection Agency (EPA) for safe drinking. The high concentrations result from the import of selenium salts from the western side of the San Joaquin Valley via the Sacramento–San Joaquin river system. By the time the drainage water reaches the Kesterson Refuge, salt concentrations are very high. Much is absorbed by plants and animals; and concentrations in their tissues may be as much as 3000 ppm. When grazed by waterbirds these concentrations can cause deformities in chicks and death in both young and adults. High selenium concentrations have also been a problem in the Cibola National Wildlife Refuge in the lower Colorado River valley (Welsh and Maughan, 1994).

The area most severely affected by salinisation in Australia is the Murray–Darling basin of Victoria and New South Wales (reviewed in A.R.M. Young, 1996). The basin is a closed groundwater and sedimentary basin with soils that are naturally high in salts but fertile; crops grown there represent 45 per cent of the value of Australia's agricultural production and it contains some 70 per cent of Australia's irrigated land. The high dependence on irrigation means it experiences problems that are similar in nature and magnitude to those discussed earlier for the southwest United States. An additional problem is the relatively high natural salinity of groundwater. All of these factors, plus the enhancement of salinity by evaporation from irrigation waters and the unwise choice of crop in some areas, e.g. paddy rice growing on 110×10^3 ha, have created major environmental and water-supply problems. For example, in the lower reaches of the Murray River near Adelaide, salinity concentrations are 10 times higher than in the upper reaches of the river, i.e. 500 ppm as compared with 50 ppm. Lawrence and Vanclay (1992) report that crop losses, which may be as much as 33 per cent in the Kerang Irrigation District, amount to an annual cost of ca. $100 million. In order to reduce the salinity of soils and water, conservation measures have been introduced since the 1970s. Examples include the use of tiled and lined drains along with piped water supplies to reduce water losses through seepage and evaporation. To reduce the demand for irrigation, water rice cultivation is restricted, notably to areas with heavy clay soils, which have a low permeability; and evaporation basins have been constructed into which highly saline water is pumped to remove the salt for industrial uses.

Figure 7.9 illustrates the distribution of irrigated land in the former Soviet Union, where at least 9 per cent, ca. 21×10^6 ha, of arable land is irrigated (Libert, 1995). Since 1970 the extent of irrigated land has doubled, with annual increases in the 1970s of 600×10^3 to 700×10^3 ha; by 1990 this rate of increase had been halved. As Figure 7.9 shows, most of the former Soviet Union's large rivers have been exploited. The situation is crystallised by Libert's statement that 'the large rivers of the ex-USSR are no longer rivers in the real sense of the word, but instead strings of water reservoirs separated by giant dams'. Dams were often constructed to provide hydroelectric power as well as, or instead of, irrigation water. Libert is also of the opinion that the environmental degradation created by irrigation is more acute in the former Soviet Union than elsewhere in the world. Salinised soils and water resources are certainly extensive, probably because of the decision-making process in a centrally planned economy; irrigation policies and water management were not directed to local or regional conditions or needs.

One region of the former Soviet Union which has been particularly acutely affected is the Aral Sea basin (Fig. 7.9). Here, salinisation and desiccation are an indirect effect of irrigation, rather than a direct effect. Micklin (1988, 1992) has pointed out that between 1960 and 1987 the level of the Aral Sea has declined by 13 m and its area has decreased by 40 per cent. Located in the desert region of the south-central Soviet Union, the Aral Sea is a large, shallow, saline water body; its level is determined by the balance between loss from evaporation and inflow from rivers, groundwater and precipitation. Although the lake has a long history of flooding and desiccation associated with the climatic changes of the Quaternary period, its current problems are mainly the result of human activity, though periods of low precipitation in 1974–75 and 1982–86 are also contributory factors. The major reason for creation of a negative water balance has been the removal of irrigation water from the Amu Dar'ya and Syr Dar'ya rivers that normally contribute most of the Aral Sea's water. This is not a recent innovation in the Aral Basin; even in 1900 some 3×10^6 ha of land were irrigated, increasing to 5×10^6 ha by 1960.

By 1985 the irrigated area had increased to nearly 6.5×10^6 ha; it reached 7.6×10^6 ha by 1987, requiring some 104 km^3 of water. This increase in water use has not been counterbalanced by reductions in natural losses and has been aggravated by expansion of the irrigated region into dry steppe

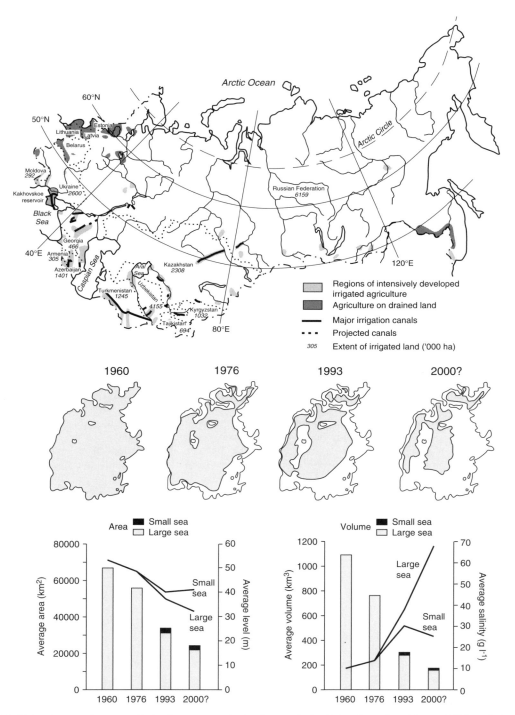

Fig. 7.9 Major rivers and extent of irrigation systems in the former Soviet Union (based on Libert, 1995) and changes in the Aral Sea during the period 1960–2000 (based on Micklin's update in Glantz *et al.*, 1993).

(e.g. Golodnaya in Hungary), which has required vast quantities of water to mitigate soil-moisture deficit, and the construction of huge reservoirs, which expose water to the atmosphere and thus promote evaporation. The Karakum Canal also diverts at least $14 \, km^3$ of water annually from the Aral Sea. These changes have created what many scientists have described as an ecological disaster, giving rise to dust and salt storms, groundwater depression as well as concerns about water supplies and economic loss in agriculture, fisheries and hunting and trapping. According to Micklin (1994) 35 million people depend on the Aral Sea in some way; apart from causing economic losses, its desiccation is causing health problems. There are indications that such large-scale desiccation has prompted climatic change, involving a trend towards continentality with a decline in relative humidity and a reduction in the length of the growing season. The exposure of ca. $27 \times 10^3 \, km^2$ of bottom sediments between 1960 and 1987 not only produced a massive salt accumulation that inhibits revegetation but also a loose mobile surface that is susceptible to deflation. Some ca. $43 \times 10^6 \, t$ of salt are removed from the Aral basin annually and are deposited as aerosols in rain and dew over 1.5×10^5 to $2.0 \times 10^5 \, km^2$. Although calcium sulphate is the most abundant salt, environmental degradation is created by sodium chloride and sodium sulphate. Particularly toxic to plants, these two sodium salts have contributed to crop losses on an economically significant scale. Salinisation of the lake itself has adversely affected aquatic productivity: only four of 24 native fish now remain and commercial fish catches have virtually disappeared with associated employment consequences. Declining levels of surface water have also had adverse effects on ecologically significant deltaic regions, initiating desertification and habitat destruction so that commercial hunting and trapping have declined. Depressed groundwater levels have caused the desiccation of wells and springs and created problems for the supply of drinking water.

Since 1989 the decline in water content has divided the Aral Sea into a northern small sea and a southern large sea (Fig. 7.9). The water levels in the small sea have risen since 1990 and there is some evidence of recovery (Aladin *et al.*, 1995). Despite many attempts to produce a recovery plan, Saiko (1995) asserts that the issue has been compounded by the disintegration of the Soviet Union. The problem is now shared by a number of separate states, all with economic ills and all of them blaming Russia and its desire for cotton. As a result, a plan devised in the 1980s to divert waters from the Ob and Irtysh rivers of Siberia has been revived. The environmental consequences of such a scheme are likely to be just as dire as the problem it aspires to address.

7.7 The impact of crop-protection agents

Crop protection chemicals are compounds produced artificially and on a large scale for use in agriculture, horticulture and silviculture. The rationale is that such chemicals improve crop-production efficiency by providing protection from disease, reducing competition and controlling growth. In effect, sunlight, water and nutrients are channelled toward desired crops to increase yields. Biopesticides are another group of crop-protection agents; they comprise populations of organisms such as bacteria, fungi and viruses which are pathogenic to specific pests. Biopesticides are discussed again in Section 9.5 since they constitute a form of biotechnology. Table 7.6 gives the general categories of pesticides that are available. According to Wood Mackenzie (1996), the global agrochemical market is worth ca. $\$30 \times 10^9$ and amounts to $4 \times 10^6 \, t$ of pesticides (data are for 1995). Many pesticides are active against a wide range of organisms, e.g. paraquat and glyphosate are described as broad-spectrum herbicides because they kill a wide range of plant species. Other pesticides are selective as they are designed to kill specific pests, e.g. pirimicarb, a carbamate insecticide which controls aphids. Animal health products are also considered under the heading of crop-protection agents. These are aimed at the improvement of animal productivity, i.e. secondary productivity, as they prevent or cure disease. Similarly, plant-growth regulators are included here because they enhance productivity by controlling crop architecture to maximise the use of available light, water and nutrients and to improve harvesting efficiency.

Crop-protection chemicals underpin most modern agricultural systems, especially in the developed world and increasingly so in the developing world (Dinham, 1995), notably Brazil, China and India. Although new developments are occurring through biotechnology (Section 9.5), including crops with

Table 7.6 Pesticide categories

	Herbicides	Fungicides	Insecticides	Acaricides	Nematicides	Molluscicides	Rodenticides	Plant growth regulators
Target organisms	Weeds	Fungi	Insects	Mites	Nematode worms	Snails, slugs	Rodents	Crops
Examples of pesticide types and typical products	*Phenoxyacids* 2,4-D MCPA	*Pyrimidines* Ethirimol Fenarimol	*Organochlorines* DDT Dieldrin	*Organochlorines* Dicofol	*Fumigants* Methylbromide Dibromo-chloropropane	*Aldehydes* Metaldehyde	*Coumarins* Warfarin Coumatetryl Brodifacoum	*Hormones* Gibberellic acid Ethylene
	Bipyrilliums Paraquat Diquat	*Alanines* Furalaxyl Ofurace	*Organo-phosphates* Terbuphos Dimethoate	*Organotins* Cyhexatin	*Carbamates* Aldicarb Oxamyl	*Carbamates* Methiocarb Thiocarboxime	*Inorganic* Zinc phosphide	*Triazoles* Paclobutrazol
	Phosphonates Glyphosate	*Triazoles* Triadimefon Flutriafol	*Carbamates* Aldicarb Carbofuran	*Tetrazines* Clofentezine	*Organo-phosphates* Fenamiphos			
	Pyrazoliums Difenzoquat		*Pyrethroids* Permethrin Cypermethrin Fenvalerate Tefluthrin Cyhalothrin					
	Aryloxy-phenoxyacetic acids Diclofop-methyl fluazifop-butyl		*Avermectins*					
	Nitro-diphenylethers Fomesafen							
	Acetanilides Alachlor							

inbred engineered insect resistance, the pesticide market is likely to remain significant well beyond 2000. Crop protection used to be achieved through crop rotation (Sections 5.3 and 5.4) to promote the biological control of insect pests and through hand weeding to remove plant pests. Crop-protection chemicals are a relatively recent development; they stem from the period just after the Second World War when the political climate stimulated agricultural production. By this time artificial fertilisers were already being used widely (Section 7.5); this stimulated the development of other artificially produced chemicals that could magnify plant and animal productivity with little concomitant increase in labour. Moreover, wartime research into nerve gases and defoliation agents generated a chemical database from which many modern crop-protection chemicals were initially derived. Crop-protection agents inevitably promote environmental change by deflecting energy flows and nutrient cycles; they do this by simplifying the biodiversity of agricultural systems. Alongside the successes there have also been failures, both ecological and environmental; the misuse of crop-protection agents can also present a threat to human health, and because they are produced mainly in the developed world, their prices are controlled by agrochemical companies, making much of the technology exclusive to industrialised countries.

Van Emden and Peakall (1996) have discussed the wide range of chemicals that are available for crop protection, etc. In the United Kingdom alone there are some 1400 approved pesticide products available for use in agriculture and horticulture (Ivens, 1993). Numerous others are used in many developing countries, although they are no longer sanctioned in developed nations; add to these all the new compounds currently under development and the sum total represents a powerful array of chemical tools with which to manipulate agricultural systems. Some of the early pesticides, notably the organochlorine compounds (see below), were far-reaching, affecting areas to which they had never been directly applied, e.g. Antarctica. The ecological and human health implications of intensive pesticide use were highlighted by Rachel Carson's classic book *Silent Spring*, published in 1962. One of the political consequences of *Silent Spring* and similar outcries has been the establishment of regulatory authorities, which have stringent criteria for product licensing and registration. In the United Kingdom

the regulator is the Ministry of Agriculture, Fisheries and Food (MAFF) and the United States regulator is the Environmental Protection Agency (EPA). Figure 7.10 gives details of toxicity and environmental tests required by regulatory bodies such as those for the registration of new crop-protection compounds. The World Health Organisation (WHO) and the Food and Agriculture Organisation (FAO) have additional requirements for determining the toxicity of pesticide residues in food, as has been discussed by Kuiper (1996) and Cleveland (1996). The tests have become increasingly stringent in order to safeguard both the public and the environment, and rightly so. Nevertheless, there are periodic scares about the safety of pesticides and residues, as illustrated below with reference to Alar®, a plant-growth regulator withdrawn from the US market in 1989.

These adverse impacts of the early postwar synthetic pesticides have been widely publicised. The early chemicals dominated pesticide markets for 20–30 years; they were developed to combat pests as agriculture became industrialised and oriented towards monoculture. Produced in the laboratory and without natural counterparts, the compounds generally comprised stable and persistent structures containing a toxophore, i.e. a toxin-carrying component. Examples are the organochlorines, and the most well known of them is dichlorodiphenyltrichloroethane (DDT). When it was first discovered to have insecticidal properties, DDT was heralded as a major breakthrough in pest control, initially for public health purposes and later for treating crops. According to Mellanby (1992), DDT was first used as a crop-protection chemical in 1941 to control an outbreak of Colorado potato beetle in Switzerland. It was also used on a large scale during the Second World War to prevent disease outbreaks among civilians and army personnel; an epidemic of typhus in Italy during 1943 was arrested by DDT sprays to control body lice. For several years after, it was a major adjunct of the World Health Organisation's malarial control programme. DDT was considered to be a major step forward in pest control for several reasons. It was thought to have a low acute toxicity to mammals, which gave it a big advantage over the thiocyanates, and did not cause skin irritation, so handling was relatively safe. DDT is a stable compound that does not readily degrade, which provides advantages for storage and transport. It was found to be active against a wide variety of insect pests, requiring only cuticle contact,

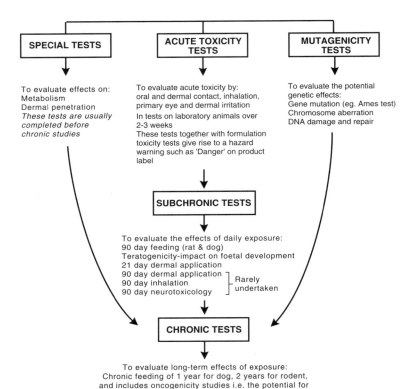

1   **Metabolism studies in crops and animals**

   A   Identification of crop metabolites

   B   Identification of rotated crop metabolites

T   C   Feeding radiolabelled pesticide and/or plant metabolites to ruminant (eg. goat) and hen, followed by identification of metabolites in tissues, milk and eggs

T   D   Determination of pesticide accumulation in fish and identification of metabolites

3   **Residue Studies**

   A   Analysis of pesticide and metabolite residues in:
     i   Crops and processed crops
     ii   Rotated crops
   T iii Animal tissues, milk, eggs
     iv Soil
   T v   Water (for aquatic pesticides)

T   B   Run-off studies (for pesticides toxic to fish)

T   C   Groundwater studies and analysis for pesticides which leach

T : Denotes studies that have been triggered by laboratory studies which suggest adverse effects

2   **Environmental studies in the laboratory**

   A   Identification of products of hydrolysis and photolysis in water, soil and air

   B   Identification of metabolites formed in soil under aerobic and anaerobic conditions

   C   Soil leaching studies on the pesticide and its soil metabolites

4   **Ecology tests**

   A   Laboratory tests on non-target organisms:
     i   Fish toxicity (acute and partial life cycle)
   T ii   Fish full lifecycle
     iii Toxicity to *Daphnia* (acute)
   T iv Toxicity to *Daphnia* (lifecycle)
     v   Toxicity to birds
     vi   Effects on bird reproduction
     vii Effects on bees

T   B   Mesocosm studies involving the identification of pesticide effects on aquatic organisms and fish in synthetic ponds

T   C   Other field studies determined by laboratory tests eg. monitoring pesticide residue levels in avian diet if pesticide is toxic to birds

Fig. 7.10 Environmental tests required for the registration of a new pesticide by the United States EPA (based on a personal communication from B.D. Cavell).

not ingestion via the stomach. Moreover, DDT manufacture is relatively easy; it can be undertaken in rapidly constructed plant. The stability of DDT is now known to be environmentally disadvantageous; not only is it chemically stable to the conditions it meets after spraying (e.g. sunlight, water), but it largely withstands the metabolic processes used by non-target organisms to detoxify exogenous chemicals. Nevertheless, its success as a combatant against widespread disease is unquestionable, especially in view of the crisis period during which it was developed.

In terms of the environmental impact, the use of DDT has had some disastrous consequences. Although it is not very soluble in water, a characteristic originally believed advantageous, DDT and its metabolites, DDD and DDE, are readily transferred through the trophic structure of ecosystems, i.e. the food chains and food webs. This occurs because DDT and its metabolites are stored in oil and fats that constitute part of the body tissues of animals, and they are passed on from one trophic level to the next, often undergoing a process known as biological magnification. This means that organisms higher up the trophic structure will contain ever increasing amounts of DDT residues. This is considered to be the reason why populations of birds of prey have been so adversely affected by DDT. Potts (1986) cites eggshell thinning due to DDT residues as a major reason why there has been a decline in populations of the peregrine falcon and sparrowhawk. In the longer term, broad-spectrum insecticides like DDT can have adverse effects on populations of insect pests because they alter the natural relationship between the pests and their predators, often other insects. Thus they can adversely influence natural biological control. Horn (1988) cites the example of the European red spider mite, which did not become a pest of apple crops until after the use of DDT and organophosphorus compounds effectively eliminated the mite's natural predators. During the 1960s Dempster described a similar case of disruption to biological control in a series of papers that examined the impact of DDT on cabbage-white butterfly (*Pieris rapae*) populations in a crop of Brussels sprouts (Mellanby, 1992). The initial impact of spraying brought about a significant reduction in the pest, but because of the chemical's non-target-specific activity, the natural predators of the pest were also eliminated. The cabbage-white ultimately resumed its significance as a pest because of this and because the persistence of DDT in the soil

prevented ground-living predators from regaining pre-application population levels.

However, DDT is not the only organochlorine insecticides to promote unexpected and unwanted ecological consequences. Aldrin, dieldrin and endrin (cyclodienes) all derive from the same chemical family and dieldrin, in particular, has had a considerable environmental impact. Its use and effects have been reviewed by Lehman (1993) and Perkins and Holochuk (1993). The use of dieldrin as both a sheep-dip and seed dressing, and the fact that it is more toxic to mammals than DDT, has caused declines in populations of birds of prey and otter in Britain. Earth Report (1988) also cites the example of declines in sandwich tern populations in the Wadden Sea of the Netherlands, where the number of breeding pairs fell from more than 40 000 in the 1940s to only 1500 in 1964. As in the case of DDT, biological magnification occurs as concentrations of fat-soluble residues of these compounds increase with trophic prominence.

Organophosphate insecticides, developed at the same time as the organochlorines, have also had wide-reaching effects. In these compounds the toxophore is a phosphorus-containing group which can be attached to a variety of organic molecules to attack insect nervous systems; they are similar in mode of action to the carbamates (van Emden and Peakall, 1996). They are not so persistent in the environment as the organochlorines but are highly toxic to mammals and birds (Edwards, 1993). Dempster (1987), for example, has reported that the use of carbophenothion to treat cereal seeds resulted in several cases of poisoning of greylag geese; in 1971 some 500 geese died after feeding on recently sown fields and in 1974–75 there were deaths of both greylag and pinkfooted geese. Furthermore, spraying of oilseed rape with triazophos, another organophosphate pesticide, has caused the death of honey-bees. Dinham (1993) reports that concerns about pollution of the Mediterranean Sea with organophosphates has led to Mediterranean countries, except Albania, agreeing to phase out their use by 2005.

The environmental problems created by these early synthetic pesticides and the increasing stringency of registration requirements (Fig. 7.10) stimulated agrochemical companies in the 1970s to begin the search for less hazardous pesticides. Attention turned to natural products such as the so-called insect powder, which was used as an insecticide in the nineteenth century. This is derived from the dried flower heads of pyrethrum (*Chrysanthemum cinerariaefolium*), a plant native to the Caucasus–Iran region of Asia and to the Adriatic coast of Yugoslavia (Davies, 1985). Pyrethrum was later grown commercially in east Africa as a source of pyrethrins, the active insecticidal compounds. This was the basis for developing a new range of insecticides called the pyrethroids. Once chemical structures were ascertained for the pyrethrins, they provided a template for the development of pyrethroids, their synthetic analogues.

A range of pyrethroids (Table 7.6) are now available that represent the culmination of some 60 years of research, much of which was undertaken at Rothamsted Experimental Station in Britain and subsequently continued by several chemical companies such as Sumitomo, Roussel Uclaf and Zeneca. Zeneca, then a part of ICI, marketed permethrin in 1977. Together with cypermethrin, deltamethrin and fenvalerate, permethrin is a broad-spectrum contact insecticide which is active against lepidopterous larvae; it is therefore suitable for use in cotton crops to combat cotton bollworm and tobacco budworm. The advantages of pyrethroid insecticides relate to their non-susceptibility to biological magnification. There is no evidence to show this occurs in food chains, as it does with DDT. Moreover, the toxicological effects of the pyrethroids on mammals appear to be relatively slight (Litchfield, 1985), as deduced from tests required by the EPA (Fig. 7.10), and then only at higher dosages than would normally be applied in crop spraying. This is another advantage of the pyrethroids; they are so potent as insecticides that only low dosages are required to bring about a substantial decrease in insect populations. Green *et al.* (1987) advise that an application rate of only ca. $50\,g\,ha^{-1}$ is necessary and that new products may have an application rate of as little as $5\,g\,ha^{-1}$ in some crops. Thus, although pyrethroids are difficult and expensive to produce, they are still cost-effective for the farmer to use.

Leahy (1985) has reviewed existing data on the fate of pyrethroids in the environment. The use of radioactively labelled pyrethroids (i.e. molecules containing a radioactive carbon) in field experiments has shown they are easily degraded, especially under aerobic soil conditions. This is because their chemical structure retains some of the biodegradable features of the pyrethrins. Even under anaerobic conditions, breakdown occurs via similar pathways to produce carbon dioxide. The rate of degradation varies; cypermethrin has a half-life of 1–10 weeks and permethrin has a half-life of 5–55 days. Moreover, leaching experiments in soils suggest that, once in the soil, pyrethroids are virtually immobile, largely because they are hydrophobic, which means they do not readily dissolve in soil water; instead they bind with soil organic components. Thus, because of rapid degradation and immobility, there is less chance that pyrethroids will contaminate the environment beyond the site of application, and to date there have been no deleterious effects noted on soil microflora and microfauna. There is, however, evidence to show that pyrethroids are toxic to bees (Ivens, 1993) and fish. Hill (1985) reports that fish are five times more sensitive to pyrethroids than mammals and that pyrethroids can deplete populations of aquatic arthropods, though some arthropods can recover rapidly. Fish toxicity is a disadvantage for treating rice crops, because fish are often present in paddy fields and may be harvested for human consumption. As a result, the pesticide industry has developed pyrethroids that are now less toxic to fish. This is an important market because rice is grown on a widespread basis as a staple crop.

A recent addition to the pyrethroid market is tefluthrin. This is the first pyrethroid designed specifically as a soil insecticide for the treatment of soil pests mainly in corn (maize) crops. Such insects, like rootworm and cutworm, are usually treated by carbamates and organophosphates, the disadvantage of which include their toxicity to non-target organisms and their susceptibility to leaching into groundwater. Tefluthrin is more potent than these older insecticides but it retains acceptable toxicity levels on non-target species, including mammals. Preregistration trials conducted in the United States indicate that tefluthrin is not toxic to seeds and will not give rise to contamination of groundwater via leaching (Zeneca promotional literature).

Although the pyrethroid insecticides have enjoyed considerable success with relatively few adverse impacts, some groups of pests are developing resistance; sometimes this is caused by overuse of pesticides or, conversely, it can result in hugely increased applications. Either way, once resistance develops, new pesticides are required. An emerging generation of insecticides are the avermectins, natural compounds produced by the actinomycete *Streptomycetes avermitilis*, and the milbemycins, a related group of compounds produced by another soil micro-organism. In order to reach the market-place, these compounds will have to have minimal environmental impact.

In terms of market share, herbicides are the most important crop-protection chemicals. They may be either systemic herbicides, i.e. they enter the plant via the roots, or contact herbicides, i.e. death of the target plant occurs when the herbicide reaches the foliage. Like the organochlorines, the phenoxyacid herbicides (also known as hormone herbicides) were developed during the late 1940s; examples include 2,4-dichlorophenoxyacetic acid, (2,4-D) and 4-chloro-2-methylphenoxyacetic acid (MCPA), both widely used for weed management in cereal crops. These herbicides mimic the plant hormone, auxin, which controls cell elongation and root development. Death eventually occurs through excessive growth. According to Edwards (1993), these herbicides do not cause serious environmental problems, mainly because they are not persistent and show low toxicity to humans and other mammals. There have, however, been some inadvertent effects insofar as they have caused a decline in broad-leaved species in hedgerows due to imprecise foliar spraying. Moreover, the increase in the monocotyledon wild oats (*Avena sativa*), is also likely to be a result of using hormone herbicides; this is because wild oats are unaffected by hormone herbicides while their dicotyledon competitors are eliminated. Potts (1986) has suggested that widespread use of hormone herbicides in the United Kingdom and the rest of Europe has contributed to the decline in partridges; this is because they eliminate the food plants of insects and larvae on which the partridge chicks depend. The problem of wild oats as a pest of arable agricultural systems prompted a search for other herbicides to combat both monocotyledons and dicotyledons.

These new herbicides were the bipyridyliums. Examples are paraquat and diquat. These are contact herbicides that work by inhibiting photosynthesis. They are widely used to clear weeds from arable fields before seeding and thus to facilitate 'no tillage' agriculture. This is considered to be environmentally beneficial because it eliminates ploughing and reduces soil erosion. These contact herbicides are used where farmers employ direct drilling, the simultaneous application of the herbicide, crop seed and fertiliser. Few adverse environmental impacts of these herbicides have been reported, though they are highly toxic to

mammals. Their immobility in soils reduces the risk of contamination of the wider environment, though there may be some seepage in soils containing low levels of organic matter and clay. Although it is an organophosphate, glyphosphate is another broad-spectrum herbicide which has the same effects as the bipyridyliums. Moreover, it has a longer-lasting effect than paraquat. Glyphosate acts by inhibiting the production of essential amino acids in plants. It is inactivated in the soil through its absorption by soil minerals (Bewick, 1994). However, glyphosate is toxic to fish and should not be used if there is the possibility of contaminating watercourses.

One of the first fungicides to be used was Bordeaux mixture, comprising copper sulphate and quicklime. This had adverse environmental impacts, especially where it entered rivers and streams. Bordeaux mixture is no longer widely used, nor are the organomercury compounds, the first carbon-based synthetic fungicides; this is because of their toxicity to wildlife. Fungicides are usually applied as dusts or as seed dressings and are only really effective when they are used in a preventative context. Compounds developed more recently include the triazoles, which work by inhibiting the biosynthesis of steroid hormones in fungi. There is, however, increasing fungal resistance to the triazoles, so they are no longer as effective in their control of crop diseases such as mildews, blights and scabs. The phenylamides comprise a recently invented group of fungicides which act by inhibiting the synthesis of ribonucleic acid. An emerging group of fungicides are the strobilurins; modelled on natural products derived from a specific fungus, they were first registered for use in 1996.

Despite the stringent regulatory requirements of MAFF and the EPA, problems will occasionally arise with pesticides after many years of use, causing them to be withdrawn from the market. This occurred with Alar®, a plant-growth regulator designed to manipulate firmness and colour in apples. According to Hathaway (1993), Alar was first registered for use in food in 1968 and was withdrawn in 1989 because of fears that its breakdown product could cause cancer in children. The EPA has legislated that Alar residues are no longer acceptable even in minute quantities. Some pesticides have aroused concerns about the contamination of water bodies and groundwater, with problems for the provision of drinking water. In view of their widespread use, it is surprising that the adverse effect of pesticides on the environment has not been more widespread.

7.8 Conclusion

Agricultural practices have undoubtedly transformed the landscape of developed nations and affected environmental quality. The loss of wildlife habitats has changed the character of landscapes and caused habitat fragmentation as well as loss of biodiversity. Indeed habitat fragmentation is a serious threat to conservation in most developed nations. High-technology agriculture has also produced depletion of the soil resource, so that soil erosion is now recognised as a significant problem in Britain and Europe as well as Australia and North America, and it derives from the replacement of traditional crop rotations by industrialised monoculture and the consequent loss of organic residues that are so important for promoting soil cohesiveness. In semiarid regions, notably the United States, Australia and the South Asian republics, the intensive use of irrigation has caused degradation of the soil resource by excessive salinisation.

Since the Second World War there has been an escalation of chemical usage in agriculture. Increased applications of artificial fertilisers, while improving harvests, have had deleterious effects on freshwater and coastal ecosystems. The enhanced nutrient inputs into these ecosystems have promoted cultural eutrophication that has altered trophic structures and, in severe cases, adversely affected commercial fisheries and tourism. In terms of environmental quality, there are few developed nations that do not have problems with groundwater quality. High nitrate levels in groundwater used for domestic water supplies may become potential health hazards which require expensive long-term mitigation. The use of crop-protection chemicals has also caused environmental change; organochlorine compounds such as DDT have had especially profound effects on wildlife. Herbicides to control broad-leaved weeds have altered the balance of weed populations so that, in solving one problem, others are created which

require a further battery of chemical tools. In the wake of tighter legislative controls, there has been the development of more environmentally benign compounds suich as the pyrethroids. This at least signifies a more responsible attitude on the part of governments and agrochemical concerns. However, the relatively short history of these compounds may mean that their long-term effects have yet to emerge; the rapid increase in fertiliser use after the Second World War was clearly never intended to produce the contaminated domestic water supplies that are now manifest. Obviously it is essential to promote efficient food production and ridiculous to expect to return to prewar traditions, but whatever new developments science brings to agriculture, there is no room for complacency. Society must also bring to bear the considerable powers of science in order to solve some of the problems it has already created. Moreover, if scientists should use their potential with vision, then politicians could wield their rhetoric more effectively by improving the balance between the interests of agriculture and the interests of conservation.

Further reading

Gardner, B. (1996) *Farming for the Future*. Routledge, London.

Gaston, K.J. (ed.) (1996) *Biodiversity: A Biology of Numbers and Difference*. Blackwell, Oxford.

Heywood, V.H. (ed.) (1995) *Global Biodiversity Assessment*. Cambridge University Press, Cambridge, and UNEP, Nairobi.

Mannion, A.M. (1995) *Agriculture and Environmental Change: Temporal and Spatial Dimensions*. J. Wiley, Chichester.

Mason, C.F. (1996) *Biology of Freshwater Pollution*, 3rd edn. Longman, Harlow.

Pimental, D. (ed.) (1993) *World Soil Erosion and Conservation*. Cambridge University Press, Cambridge.

Settele, J., Margules, C., Poschlod, P. and Henle, K. (eds.) (1996) *Species Survival in Fragmented Landscapes*. Kluwer Academic, Dordrecht.

Thomas, D.S.G. and Middleton, N.J. (1994) *Desertification: Exploding the Myth*. J. Wiley, Chichester.

van Emden, H.F. and Peakall, D.B. (1996) *Beyond Silent Spring*. Chapman and Hall, London.

CHAPTER 8

The environmental impact of agriculture in the developing world

8.1 Introduction

Population growth, foreign aid and the desire to earn foreign currency have, in the past 50 years, stimulated the expansion of agriculture in developing nations and have thus prompted environmental change. Although there are many success stories in terms of increased food supply (China can now feed its population of 1.115 billion), the expansion and intensification of agriculture have also brought about environmental degradation. As in the case of the developed world (Chapter 7), the impact of agriculture is manifest in the loss of natural habitats and associated loss of biodiversity. This is affecting all types of natural ecosystems but the loss of tropical forests is of greatest concern because of the impact it is likely to have on global climate and because of the loss of species which might generate useful substances, e.g. pharmaceuticals and agricultural chemicals, in the future. The removal of the natural vegetation cover gives rise to soil erosion and degradation; this may reach such proportions that agricultural systems become unproductive. *In situ* soil degradation, which causes loss of production, can also occur. However, there are examples of successful soil conservation which demonstrate that land degradation can be reversible and productivity restored. This is also true of desertification, which is more extensive in the developing world than in the developed world. Nevertheless, it is a particularly serious problem in Africa and Asia. Agricultural practices have also had a major impact on water quality in the developing world. For example, cultural eutrophication is becoming increasingly significant as developing countries turn to artificial fertilisers to boost agricultural production. Moreover, the problem of soil and water salinisation is becoming widespread as ill-conceived and mismanaged irrigation systems exact their toll.

8.2 Landscape change: loss of natural habitats and biodiversity

The loss of natural habitats, especially the loss of tropical forests, must rank along with global warming as one of the most important environmental issues of the 1990s. The loss of tropical forests in particular is not only a loss of biodiversity and hence of genetic resources but also has implications for climatic change because tropical forests are major sinks of atmospheric carbon dioxide. The loss of trees diminishes the flux of carbon dioxide from the soil via photosynthesis and reduces the storage capacity of the biosphere for carbon. Indeed the removal of trees (i.e. biomass), especially through burning which often precedes the cultivation of land, releases their stored carbon into the atmosphere along with carbon from the litter and soil, thereby contributing to the enhanced greenhouse effect (Section 6.5). There is usually an increase in carbon flux to the atmosphere when other vegetation types are removed for agriculture, i.e. savannas, wetlands, mangroves.

Determining the rate and extent of deforestation is as difficult as determining the rate and extent of soil erosion and desertification. This is because the term 'forest' can be defined in different ways and because quoted rates of deforestation are often estimates rather than precise determinations (Downton, 1995). Even satellite imagery can be interpreted in different ways (Whitmore and Sayer, 1992). Figure 8.1 shows the past and current extent of tropical forests which occupy some 40 per cent of tropical regions (Vanclay, 1993). The remaining 60 per cent is too dry, comprising arid and semiarid lands. Approximately half of the tropical forest is rain forest; the rest comprises seasonal forests, savanna woodland and several types of open forest. According to the World Resources Institute (1996), the FAO defines forest in developing (mainly tropical or subtropical) countries as land with a minimum tree crown cover of 10 per cent. Forest and other wooded land within this definition declined by 2 per cent globally, i.e. 100×10^6 ha between 1980 and 1990. Much of this change occurred in tropical regions; here the extent of forest and other wooded land declined by 3.6 per cent during 1980–1990 with an 8 per cent decline in forest, i.e. 163×10^6 ha. The hectarages involved in these changes are given in Table 8.1, which also gives data on the changes in forest and woodland cover that have occurred in selected developing countries. These data show that the greatest extent of forest clearance has occurred in South America, notably in Brazil (Myers, 1993b). This rather overshadows the problems experienced on some tropical islands such as Hainan Island, China, though here Zhou (1995) has shown that the mean rate of deforestation is 2.02 per cent, nearly double the global rate. Despite the limitations of these and other data (see reviews in Mannion, 1995a; Adger and Brown, 1994) it is clear that deforestation is occurring extensively and rapidly.

What does this mean for loss of biodiversity and species extinction? Again there are no irrefutable data available on which to examine these issues with any certainty, as is also the case for the developed world (Section 7.2). There is the additional complicating factor that inventories of biota are even less comprehensive than they are for the developed world. Nevertheless, it is generally considered, because of high biodiversity in the tropics and the accelerating rate of habitat loss, that the loss of species is sufficiently great to warrant international action to halt habitat loss. Some estimates of extinction rates are given in Table 7.2, which shows the variation that occurs. However, even the loss of 1 per cent of the Earth's species per decade represents a considerable loss and, as Table 7.2 shows, losses for tropical regions are likely to be in excess of this. Such losses not only represent a loss of opportunities for the future, i.e. the

provision of goods, but may also lead to the impairment of tropical ecosystems as mediators of biogeochemical fluxes, especially those of carbon; see the discussion in Section 7.2 on the work of Naeem et al. (1994, 1995). Preliminary work on this aspect of tropical ecosystem function has been reported by Silver et al. (1996) who show that the maintenance of biodiversity preserves the integrity of nutrient and energy fluxes.

Table 8.2 gives one estimate of the likely decline in numbers of tropical forest species between 1990 and 2015; the values range between 3 and 18 per cent, depending on the rate of deforestation and the region. Many of the species lost will not even have been recorded or classified. Dobson (1996) states that 'on a global scale we will commit 4 to 8 per cent of the world's tropical forest species to extinction in the next 25 years and as many as 9 to 19 per cent over the next 50 years'. If there are roughly 5 million species on Earth, and between 50 and 90 per cent live in tropical forests, this would amount to approximately 4000 and 14 000 species per year, or 10–38 per day. Not only is this a potential loss of wealth through the generation of useful products, it is an irreversible loss. It would, for instance, be possible to halt global warming and, as discussed in Sections 6.6 and 7.5, it is possible to reverse acidification and eutrophication, and to reclaim derelict land (Section 6.4). It is not possible to recover an extinct species. In this respect, biodiversity loss is a more crucial and pressing problem that most others discussed in this book. Moreover, pressures created through global warming (Section 6.5) may accelerate extinction rates.

Although it is possible to generalise about the loss of opportunities caused by extinction, it is impossible to quantify them. Moreover, what obtains at present in relation to the use of biota to produce goods that generate wealth is unlikely to obtain in the future. As science and technology progress, components of the biosphere which now provide only services (these are intrinsically valuable) may become wealth generators through their provision of food, fibre, fuel, building materials or fine chemicals, etc., as biotechnology expands (Mannion, 1995c, 1997a). However, the pharmaceutical industry illustrates the value and potential value of biota. The medicinal properties of species have been utilised by many of the world's population; one such example is Chinese traditional medicine. In addition the beneficial properties of many natural substances have been recognised by

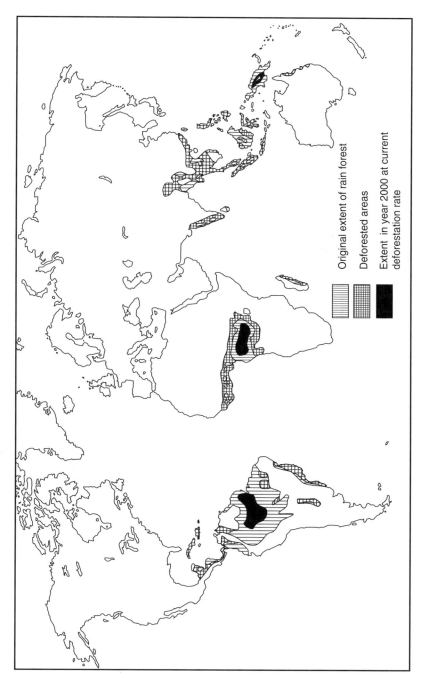

Fig. 8.1 Distribution of tropical rain forests, the extent of present deforestation and an estimate for the year 2000 (based on *Earth Report*, 1988).

Table 8.1 Changes in the extent of forest and woodland during the period 1981–83 to 1991–93

	Extent in 1991–93 (10^6 ha)	Percent change since 1981–83
Africa		
Ivory Coast	7.080	−24.4
Liberia	1.707	−13.2
Mauritius	0.044	−24.1
Nigeria	11.400	−20.3
Tanzania	33.500	−14.4
Ghana	7.943	−8.0
Central America		
Cuba	2.403	−9.2
El Salvador	0.104	−18.8
Nicaragua	3.223	−24.3
Panama	3.260	−18.0
Asia		
Thailand	13.557	−13.8
Bangladesh	1.896	−12.2
Cambodia	11.667	−11.3
Vietnam	9.639	−8.8
Indonesia	111.258	−3.7
Malaysia	20.347	−1.8
South America		
Paraguay	12.983	−32.5
Venezuela	29.828	−8.2
Brazil	488.833	−4.8
Colombia	49.633	−5.8
Global	4 168.956	−3.6
All tropical countries	1 761.228	−8.1
Tropical Africa	527.697	−7.2
Tropical Asia and Oceania	315.391	−11.1
Tropical Latin America and the Caribbean	918.140	−7.5

Source: based on World Resources Institute (1996)

Table 8.2 Percentage loss in tropical forest species estimated between 1990 and 2015

Region	Loss of species	
	Deforestation at 10×10^6 ha y^{-1}	Deforestation at 15×10^6 ha y^{-1}
Africa	3–6	4–9
Asia	5–11	8–18
Latin America	4–8	6–13
All tropics	4–8	6–14

Source: based on Dobson (1996)

pharmaceutical companies and turned into widely used drugs. The majority of these substances are plant based, being derived from secondary metabolites, though as the world's animals and micro-organisms are explored chemically, the number of natural-product pharmaceuticals will increase substantially.

The status of plant-based drugs has been reviewed by Kinghorn and Seo (1996), who point out that some 80 per cent of the world's population rely on traditional medicine and that compounds derived from higher plants are components of ca. 25 per cent of prescriptions dispensed in the United States. Their annual worth is 15.5×10^9 at 1990 prices (Principe, 1991). Examples include codeine (from the opium poppy), a well-tried and tested substance, digitalin (from the foxglove), also well tried and tested, and taxol (from the Pacific yew), recently used to treat ovarian cancer.

Although the chances of finding a suitable substance lie in the range one-in-10,000 to one-in-100,000 new, robotised, methods of screening for pharamaceuticals and agricultural chemicals (note that the pyrethroid insecticides discussed in Section 7.7 are natural products as are the avermectins) render such prospecting feasible. To ensure that the hosts of the resource, i.e. plant or animal, benefit from it the Biodiversity Treaty established at the Earth Summit meeting in Rio de Janeiro in 1992 embodied the idea that profits should be shared between the host and industry. This is exemplified by the relationship between Merck, the United States' largest pharmaceutical company and INbio, Costa Rica's National Biodiversity Institute (Reid *et al.*, 1993) which Reid (1996) refers to as a 'gene co-op'. The objective is that 50 per cent of the royalties paid by Merck to INbio from sales of successful drugs derived from Costa Rican forests will be used to fund the work of the Costa Rican government's national park service. Thus biodiversity helps maintain itself (Mannion, 1995b). Many other components of the Earth's biodiversity are increasingly being used to provide valuable services for environmental remediation and the production of valuable substances through biotechnology as will be discussed in Section 9.4.

The significance of tropical deforestation to the global carbon budget has been examined by Brown (1993) and Adger and Brown (1994). Adger and Brown have reviewed some of the estimates, which range from 0.4 to 2.9×10^9 t of carbon for 1980. Sedjo (1993) has calculated that 1.7×10^9 t of carbon have been released annually due to tropical deforestation as estimated by the FAO. In a specific case study of Jambi Province, Sumatra, Murdiyarso and Wasrin (1995) have shown that the conversion of 60×10^3 ha of primary forest to agricultural lands resulted in the release of up to 0.03×10^9 t of carbon annually. Moreover, Veldkamp (1994) has shown that even when pasture is established on previously forested land in Costa Rica, the pattern of carbon release varies with age; decomposition of tree roots means that higher amounts of carbon are released in the early stages of pasture establishment than in the later stages. Studies on two soil types showed that the loss of soil organic carbon over a

25 year period of pasture following deforestation ranged from 21.8×10^6 g ha^{-1} for a Eutric Hapludand to 1.5×10^6 g ha^{-1} for an Oxic Humitropept. Esser (1995) has also suggested that approximately 25 per cent of the global releases of carbon dioxide during the period 1860 to 1978 derived from land-use changes, a major component of which was deforestation in Monsoon Asia. Esser also suggests that the magnitude of the contribution of this region into the next millennium will depend on how agricultural systems and policies respond to the need to increase food production, i.e. whether agricultural systems will expand at the expense of tropical forest or whether production methods will intensify on existing agricultural land. This issue is relevant to other parts of the tropics and reflects the social and political factors that underpin deforestation. These will be discussed below.

Emissions of other greenhouse gases have also been enhanced as a result of tropical deforestation. For example, Granli and Bockman (1995) have reported that emissions of nitrous oxide (N_2O) from tropical soils have increased following deforestation; Hao and Ward (1993) report that biomass burning, often associated with deforestation, releases methane. Hao and Ward assert that some 84 per cent of global emissions of methane derive from the tropics, though not all of this is due to deforestation as paddy fields are significant producers. Nevertheless, the proportion due to biomass burning may have increased by 9 per cent in the past 10 years as deforestation has accelerated. As discussed in Section 6.5, emissions of these gases contribute to global warming. However, this is only one aspect of the impact of deforestation on climate. Eltahir (1996) has examined the role of tropical rain forests in large-scale atmospheric circulations: large-scale deforestation alters heat and energy balances, reducing boundary layer entropy; overall this could weaken large-scale circulation. McGuffie *et al.* (1995) reach similar conclusions and add that the disruption of circulation patterns over tropical regions initiates climatic change in middle and high latitudes. Consequently, large- and small-scale deforestation are likely to alter climate through a variety of mechanisms which involve species loss and/or community loss.

Agriculture is a principal cause of deforestation, but it is not the only cause; logging and mineral extraction have a substantial effect. Forest regions are being opened up by logging and mining companies, and this provides access for shifting cultivators. Additional

Table 8.3 Factors associated with tropical deforestation

1970s	1980s
A. Deforestation in Asia	
Rural population growth: increase in shifting cultivation	Value of wood products: growth in timber industry
	Rural population growth: increase in shifting cultivation
	High rural population density
Overall: low number of variable and not much regional variation	
B. Deforestation in Africa	
High population density	High population density
Gross national product	Debt in a few cases
Debt	Coastal location of forest
Size of rainforest	High rural population growth
Presence or absence of roads	
Coastal location of forest	
Rural population growth in some countries	
Overall: increased number of variables compared with Asia and considerable regional and national variation	
C. Deforestation in Latin America	
High rural population growth	Gross national product
Gross national product	Presence or absence of roads
Presence or absence of roads	Debt
High rural population densities	High rural population densities
Size of forests	Value of wood products: growth in timber industry
Debt	especially in Central America
Overall: many variables and considerable regional variation, including marked difference between Central and South America	

Source: based on Rudel and Roper (1996)

threats come from construction, mainly for hydroelectric provision in the Amazon basin, fragmentation, and deforestation to grow coca (the source of cocaine) in Peru (K.R. Young, 1996), a rather abstruse form of agriculture. Fragmentation has been discussed by McCloskey (1993), who reports that only 33 per cent of primary rain forests occur as large wilderness blocks whereas the other 66 per cent are fragmented by roads, rivers, etc. The fragmented rain forests are particularly susceptible to colonisation by shifting farmers. However, Turner and Corlett (1996) have demonstrated that even relatively small fragments of less than 100 ha have important conservation value, especially for small vertebrates, invertebrates, plants and micro-organisms, notably in areas like Singapore, the coastal forests of Brazil, the islands of the Philippines and elsewhere in southeast Asia where there are no large areas to conserve.

Agriculture can take many forms (Mannion, 1995a) and in relation to tropical deforestation the two most important causes are the spread of cattle ranching and shifting cultivation. These in turn are underpinned by diverse driving forces. Cattle ranching is a major factor in Latin America but is

not so important elsewhere. Almost everywhere there is a relationship, albeit crude, between deforestation and population growth. Flint's (1994) analysis of land-cover change and population growth in southeast Asia between 1880 and 1980 shows that population increased by 262 per cent as the extent of cultivated land increased by 86 per cent and forest cover decreased by 29 per cent. Flint suggests that deforestation was also linked to local and external economic conditions. In a study concerned with all tropical forests, Rudel and Roper (1996) have highlighted the diversity of causal factors and the interplay between population growth and other variables. The other variables are given in Table 8.3 in relation to Asia, Africa and Latin America. Rudel and Roper conclude that the issue of conservation must be tackled regionally because of differences in the underlying causes of deforestation between regions. Moreover, the situation is likely to be quite different on small islands when compared with continental regions. This has been highlighted by Paulson's (1994) study of Western Samoa. Here much of the tropical forest has been converted to agriculture since 1950 despite the operation of a controlled, traditional tenure system. Population growth has been an important stimulus, along with cash cropping. Involvement in cash cropping has been voluntary but has provided rewards that have been accommodated by a traditionally flexible land-tenure system and local competition between families for status and power. As Paulson points out, 'forest conservation will not take place unless local leaders and land managers perceive the forest to have more value when kept intact than when converted. The caveat for Western Samoa is that its forests are not particularly rich in timber or non-timber products'. Whether this is true or not is incidental to the environmental services provided and the potential goods that might be developed in the future. Paulson's commentary reflects a short-term view.

For many other people the situation is not one of improving standards but of survival. Myers (1993c), for example, believes the demise of more than 50 per cent of tropical moist forest is caused by 'shifted' cultivators. These are people with no tradition or heritage of shifting cultivation but who are forced into it because of rapid population growth in urban areas and poor living conditions. These landless people enter the forests in search of a 'better life' though they have little knowledge about raising crops. According to Fearnside (1993), some 30 per cent of the deforestation in Brazil's Amazonia region is caused by small farmers clearing plots of less than 100 ha. The other 70 per cent is due to medium or large ranches. Ranching is also a major reason for deforestation in Central America. Both tax incentives and ready markets for beef encourage the spread of ranching, some of which is undertaken by foreign interests. Tax incentives are given in order to encourage foreign investment and in some cases land has actually been purchased from shifting cultivators, who then move further into the forest. Many of these ranches which are several thousand hectares to allow for low carrying capacity, are abandoned as they become unprofitable and pasture quality declines. Besides weed infestation, pasture quality declines through nutrient depletion following the loss of their major source, the trees. This does not always deter ranchers, who may just move on to create another ranch. Shifting cultivators and ranchers together cleared $19 \times 10^3 \, \text{km}^2 \, \text{y}^{-1}$ for 1988–1989 and $11 \times 10^3 \, \text{km}^2 \, \text{y}^{-1}$ for 1990–1991 (Fearnside, 1993).

Further stimuli to deforestation are resettlement programmes operated by governments; Brazil has opened up Rondonia in Amazonia, and through its Repelita programme, Indonesia has encouraged settlement of its less well-populated islands. The total number of resettled people in Indonesia is more than 5 million and the overall aim is to resettle 65 million; this will have a major impact on the forests of Indonesia's outer islands. Land tenure is another encouragement to deforestation. On the one hand, landowning elites and multinational corporations are influential politically and will often oppose rural and agrarian reforms. On the other hand, most of the best agricultural land is owned by these elites, who engage in plantation agriculture and ranching to produce exports. This is especially true of Latin America. Here about 50 per cent of all rural families are either landless or farming small plots of less than a hectare (Utting, 1993). These people may also enter the forests in search of a living. This problem will escalate as population continues to grow rapidly.

Mangrove forests have also been affected by deforestation. Examples include the Gambia Estuary, traditionally used for rice cultivation, and the mangrove ecosystems of Cameroon and Nigeria, where vegetables are produced. According to Jagtap et al. (1993) there are 3150 m² of mangroves along the coasts of India, mainly along the east coast and the Andaman Islands. This figure represents only 70 per cent of the original mangrove forest as 30 per cent has been lost to agriculture or urban development. The mangroves of the Sundarbans, situated where the Ganges, Brahmaputra and Megha rivers enter the Bay of Bengal, have also diminished in extent due to the spread of agriculture. Foote et al. (1996) have discussed the overall plight of mangrove forests in India and suggest that the construction of open-water shrimp ponds is another major factor in deforestation as it is in the Philippines. According to Primavera (1991), since 1975 some 70 per cent of the Philippine mangroves have been replaced by shrimp ponds for what is essentially shrimp farming. This conversion of mangrove forest to shrimp ponds has repercussions for water quality in remaining mangroves.

8.3 Soil degradation, erosion and conservation

Soil degradation is a serious global problem, as discussed in Section 7.3. However, Oldeman's (1994) analysis of the results of the UNEP project entitled

Global Assessment of Soil Degradation (GLASOD) shows that the extent of chemical soil degradation, i.e. soils with depleted nutrients or which are salinised (Section 8.5) are most extensive in Africa, Asia and South America. The greatest extent of soil suffering from nutrient depletion is in South America, where it amounts to 68×10^6 ha, and in Africa where it comprises 45×10^6 ha. Although Asia has only 15×10^6 ha of nutrient-depleted soil, it has by far the largest extent of salinised soil, i.e. 53×10^6 ha compared with 15×10^6 ha in Africa and 2×10^6 ha in South America. Oldeman asserts that globally some 56 per cent of chemical soil degradation is caused by unsuitable agricultural activities or practices and that a further 28 per cent is caused by deforestation. Both factors are widespread in the developing world. Much of this chemical degradation is caused by a loss of organic matter from the soil, hence the loss of organic compounds that bind metals such as calcium and magnesium. These metals are required for plant growth. As discussed in Section 8.2, deforestation and the initiation of pasture on tropical soils leads to a net loss of carbon, reflecting loss of organic matter, for many years after deforestation. Moreover, the rapid loss of nutrients from pasture soils, mainly through leaching as well as the decline in organic content, can lead to their abandonment (Section 8.2). Heavy-metal pollution may also occur in agricultural soils. Zhao (1994) has reported heavy-metal pollution on 13×10^6 ha of soil in China through the use of untreated wastewater from industrialised areas. Problems include mercury and cadmium pollution.

Physical soil degradation, i.e. compaction, sealing and crusting, is most extensive in Europe (Section 7.3) though an equivalent extent of physically degraded soils can be found in Africa (18×10^6 ha), Asia (10×10^6 ha) and South America (4×10^6 ha) combined. In West Africa, for example, Lal (1993) reports that upland soils, which are naturally nutrient-poor, rapidly lose their organic matter with cultivation and erosion, causing soil compaction. Moreover, the soil compaction reduces aeration and water infiltration, producing an overall loss of productivity. These changes have been recorded by Allegre and Cassel (1996) when slash and burn agriculture is replaced with continuous cropping systems in the Yurimaguas area of Peru (the eastern Andean foothills). Their research shows that mechanical clearing to prepare the land had a significant impact on bulk density and water

infiltration rate whereas subsequent cropping accelerated soil erosion to rates as high as $53 \, \text{t ha}^{-1} \, \text{y}^{-1}$. Soil compaction was most severe where grazing was practised. The problems were reduced where agroforestry was practised, e.g. alley cropping involving hedgerows with crops, including crops for grazing in between. Similar changes in soil characteristics have been recorded by Reiners *et al.* (1994) as lowland tropical rainforest in Costa Rica has been converted to pasture and by Woodward (1996) in Amazonian Ecuador. Both natural soil compaction and compaction due to machinery are now evident in the Santa Cruz lowlands of tropical eastern Bolivia. According to Barber (1995), some 50 per cent of the soils in this annually cropped area have experienced moderate to severe compaction. This imposes restrictions on root growth and water penetration.

Soil erosion is brought about by wind and/or water. It can involve many different processes (Table 8.4), as discussed by several contributors in Agassi (1996). The global extent of soil erosion is shown in Figure 7.2; this shows that it is particularly serious in mountainous tropical regions and dry tropical regions. Thus it is a major problem in many developing countries in which it is often associated with desertification (Section 8.4). As in the rest of the world, soil erosion is an integral component of natural denudation processes, though its rate has been greatly accelerated by human activity, especially injudicious agricultural practices. Soil erosion affects millions of people, not only in the areas from which soil is removed but also in areas where it is deposited. If deforestation continues at the rates estimated by the FAO (Section 8.2) then rates of soil erosion will accelerate as well. According to Oldeman (1994), erosion by water is the most widespread form of soil erosion, affecting 1100×10^6 ha globally. Wind erosion affects ca. 550×10^6 ha and occurs mostly in arid and semiarid areas. The continental distribution of water and wind erosion is given in Table 8.5, which shows that water and wind erosion are particularly extensive in Asia and Africa. However, the data on severely eroded land are particularly important because much of this land is no longer suitable for agriculture. In the developing world, where there is rapid population growth and efforts are being made to improve standards of nutrition, the consequent loss of agricultural productivity is a travesty and an incentive for further deforestation. Moreover, Döös (1994) estimates that between 5 and 10×10^6 ha y^{-1}

Table 8.4 Processes involved in soil erosion

A. Erosion by water

Rainsplash: may consolidate the soil surface by encouraging the formation of a crust that reduces the infiltration capacity and thus promotes greater surface run-off, another erosive agent; may disperse the soil by detaching particles which may then move downslope

Overland flow: occurs when storm water or prolonged rainfall produces a saturated soil, causing sheet flow or a series of braiding channels in which soil particles are carried downslope

Subsurface flow: may occur in subsurface layers of the soil or in pipes transporting fine and colloidal particles downslope

Rill erosion: occurs where overland flow becomes concentrated in channels and, although ephemeral, rills are very significant in the downslope movement of soil particles

Gully erosion: these are permanent features which enhance erosion by headwall and bank erosion

Mass movements: slides, rockfalls and mudflows can cause large amounts of material to be moved downslope

B. Erosions by wind[a]

Suspension: fine particles of small diameter are transported high in the air and often over long distances

Surface creep: coarse grains are rolled along the ground

Saltation: grains are moved along the ground surface in a series of jumps

[a]The significance of wind as an agent of erosion depends on the velocity of the moving air and the amount of protection afforded by the land

Table 8.5 Distribution of water and wind erosion

	Land area (10^6 ha)	Fraction strongly eroded (%)	Fraction of global total (%)	Major cause
		Extent of water erosion		
Africa	227	44.90	20.75	grazing/overcropping
Asia	441	16.55	40.31	grazing/overcropping
South America	123	9.75	11.24	grazing/overcropping
Central America	46	50.00	4.20	grazing/overcropping
Global total[a]	1 094			
		Extent of wind erosion		
Africa	186	4.84	33.94	grazing/overcropping
Asia	222	6.76	40.51	grazing/overcropping
South America	42	0.00	7.66	grazing/overcropping
Central America	5	20.00	0.91	grazing/overcropping
Global total[a]	548			

Source: based on 1992 data from UNEP quoted in Oldeman (1994)
[a]The global total is greater than the column total because the table does not include every continent

of croplands are lost due to soil erosion; even taking the lower value, this could represent an annual loss of 5×10^6 t of grain. The magnitude of the problem is also illustrated by Pimentel's (1993) claim that in Asia, Africa and South Africa soil erosion rates on cropland are between 20 and $40\,t\,ha^{-1}\,y^{-1}$ but the renewal rate is at best around $1\,t\,ha^{-1}\,y^{-1}$.

Examples of soil erosion in Latin America include the work of Garciaoliva *et al.* (1995) on Mexican pasture soils and Michelena and Irurtia (1995) on wind erosion in La Pampa Province, Argentina.

Garciaoliva *et al.* have shown that on formerly forested areas in the Chamela region the top 5 cm, containing the majority of nutrients, etc., could be lost in as little as 25 years, with the greatest erosion occurring immediately after forest clearance. On average, erosion rates amounted to $13.2\,t\,ha^{-1}\,y^{-1}$ with soil removal being caused by water erosion. In contrast, Michelena and Irurtia focused on a naturally semiarid region where measured soil erosion rates were between 9.4 and $54\,t\,ha^{-1}\,y^{-1}$. The highest rates occurred on cultivated areas in the southeastern part of La Pampa Province whereas the lowest rates occurred on pasture in the eastern part of the province. Enhanced soil erosion occurs below 2000 m in the Los Santos Forest Reserve of Costa Rica, caused by deforestation and the use of pesticides which is widespread and poorly controlled (Kappelle and Juarez, 1995).

Agricultural systems in the Andean countries of South America are also accelerating soil erosion, though the increasing use of conservation measures is curtailing this to some extent. Dehn (1995) has reported on the various options available for soil conservation in the Ecuadorian Andes, where the effect of bench terraces, infiltration ditches and contour bunds have been determined. All three techniques gave the same reduction in soil erosion, but contour bunds and infiltration ditches proved more cost-effective after totalling the labour costs, the loss of arable land due to construction and the monetary costs. This was for a region characterised by smallholdings and rural depopulation. This study reflects the social dimensions of implementing soil conservation as well as the practical considerations. Grazing and fuelwood collection can cause soil degradation, and sloping ground will accentuate their effects. This has been demonstrated by Oyarzun (1995), who examined infiltration rates and soil erosion rates in the Bio Bio River basin of the Cordillera de La Costa in central Chile. Infiltration rates varied from 280 mm h^{-1} for native forest to as little as 0.9 mm h^{-1} for degraded thicket; most of the difference is due to the loss of organic matter from the degraded thicket. Erodibility rates were also considerably higher for degraded thicket than for native forest. However, it is possible to reduce erosion rates as Allegre and Rao (1996) have shown for Yurimaguas, Peru (see above). Here contour hedgerows with intercropping conserved not only 73 t ha^{-1}y^{-1} of soil but also 287 mm of rainfall.

As Table 8.5 shows, there are 663×10^6 ha of eroded land in Asia, of which 66 per cent is due to water erosion and 33 per cent is due to wind erosion. Much of this eroded land is in China, as discussed by Zhao (1994) and Dazhong (1993). Indeed Zhao (1994) goes so far as to state: 'Soil erosion is probably the most serious land degradation problem in China. The annual soil loss caused by water erosion (not including wind erosion) totals about 5 billion (10^9) tons, of which about two-fifths empty into the sea, and about three-fifths are deposited on the neighbouring lowlands'. Soil erosion in some regions is

contributing to desertification, as discussed in Section 8.4. The regions of China most severely affected by soil erosion are given in Figure 8.2. According to Dazhong, 42×10^6 ha of the land under cultivation are characterised by serious water and wind erosion, this is ca. 33 per cent of the 130×10^6 ha of land currently being cultivated. For a country with a population of 1.12×10^9 and in which the land available for cultivation is only ca. 33 per cent of the average for the world, soil erosion is a major problem and has prompted widespread soil conservation.

Figure 8.2 shows that the most serious soil erosion occurs in the Loess Plateau and the southern region. According to Zhao (1994), erosion is so severe on the Loess Plateau that the often extensive systems of deep gullies can be described as 'badlands'. Moreover, much of the soil eroded enters the Yellow River, whose name derives from the silt it carries. Of the 2200×10^6 t of soil eroded annually, approximately 1600×10^6 t enter the river and the rest is trapped by dams and accumulates in reservoirs. Such severe problems arise because the yellow loess (a wind-blown sediment derived from the deserts of the continental interior during the Quaternary period; see Section 2.2) is friable with a high erodibility, especially once its natural cover of steppe grassland is removed for cultivation, as occurs over 25 per cent of the plateau. Average erosion rates are 60 t ha^{-1} y^{-1} from cultivated land, though they can be as high as 100 t ha^{-1} (Dazhong, 1993). Most of the erosion is due to run-off following intense rainfall (Zhang *et al.*, 1994), which gives rise to sheet, splash and rill or gully erosion (Table 8.4). Collected from some 35 tributaries, the 1600×10^6 t of sediment carried by the Yellow River is now 25 per cent more than it was 40 years ago. The sediments deposited in the river channels themselves, especially in the Yellow River, produce a raised river bed at certain places, e.g. Kaifeng City, Hunan Province, where the river bed is 8 m higher than the city. This 'suspended' river increases susceptibility to flooding, a serious problem in China. Siltation in irrigation channels, reservoirs, etc., also reduces the water-storage capacity. Soil conservation schemes have been established in an effort to counteract some of these problems. One such attempt is the middle Yellow River Valley, where tens of thousands of earth dams have been constructed on gullies in the Loess Plateau. According to Leung (1996), preliminary results indicate high success rates.

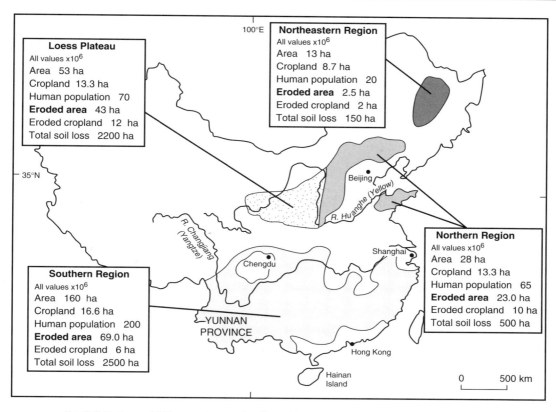

Fig. 8.2 Regions of China most severely affected by soil erosion (based on Dazhong, 1993).

In the southern region (Fig. 8.2) erosion rates $(40\,t\,ha^{-1}\,y^{-1})$ are not as high as in the Loess Plateau. This region has an area of 160×10^6 ha, 10 per cent of which is cultivated, mainly for paddy rice. Water erosion is most prevalent and the major drainage system, that of the Yangtze River, carries a much increased sediment load compared with a few decades ago. According to Dazhong (1993), this has impaired the transportation capacity of the river and necessitated regular dredging in places. It is not clear how this problem will be affected by the construction of the Three Gorges Dam, which despite the controversy surrounding its impact on the environment, is now under construction. In Yunnan Province (Fig. 8.2), in the west of this region, Whitmore *et al.* (1994) have used palaeoenvironmental methods on lake sediments to determine the history of soil erosion. They show that human activity has increased erosion by a factor of 15 in relation to natural erosion rates. Equally important is their conclusion that the

export of phosphorus has increased by a factor of 19; phosphates are vital for plant growth and often a limiting factor in ecosystem and agricultural productivity. To compensate, increased amounts of phosphate fertilisers are used. Despite efforts to conserve soil via government policies, there is still a problem with the acquisition of basic data (Higgitt and Rowan, 1996).

Elsewhere in Asia, soil erosion is serious in India (Khoshoo and Tejwani, 1993). Here 113.3×10^6 ha are subject to water erosion and 38.7×10^6 ha are subject to wind erosion; overall 5.33×10^9 t of soil are eroded annually, half of which enters India's rivers. A further 10 per cent is deposited in reservoirs. Equally important is the fact that this erosion results in the loss of 5.37×10^6 t of plant nutrients. According to Abrol and Sehgal (1994) loss of topsoil has occurred on 150×10^6 ha, mainly through water erosion. The areas most severely affected are the hill regions of the Outer Himalayas, the Punjab, northeast and southeast India. In their survey of erosion rates, Singh *et al.* (1992) have shown that they vary considerably from less than $5\,t\,ha^{-1}\,y^{-1}$ to more than $80\,t\,ha^{-1}\,y^{-1}$. Moreover, the average soil erosion rate is ca. $40\,t\,ha^{-1}\,y^{-1}$ in the northeastern states, where shifting cultivation is practised, though rates of $170\,t\,ha\,y^{-1}$ can occur in the early years after clearance on sloping land. In Rajasthan (northwest India) water erosion and salinisation have affected 33 per cent of the area (Raina *et al.*, 1993). In the Himalayan foothills of northern India, Grewal *et al.* (1994) have shown that agroforestry, using *Leucaena* spp., is effective in conserving soil, water and nutrients. Vetiver grass (*Vetiveria zizanoides*) has also been used for the construction of hedgerows to combat soil erosion in India and Nepal. Parts of northeastern Pakistan, also in the Outer Himalayas, experience erosion rates of between 30 and $150\,t\,ha^{-1}\,y^{-1}$ (Ellis *et al.*, 1993).

As Table 8.5 shows, soil erosion is a significant problem in Africa, where it affects 413×10^6 ha, 14 per cent of the total land area. It occurs in all of Africa's climatic zones, and in the arid and semiarid regions it is associated with desertification (Section 8.4). In Ethiopia, for example, 3.8 per cent of the land is totally degraded and experiences a rate of soil loss of $70\,t\,ha^{-1}\,y^{-1}$; the next highest soil loss is $42\,t\,ha^{-1}\,y^{-1}$ from cropland (Hurni, 1993). This may reduce productivity by 1–2 per cent per year and it is a crucial issue in a country that has a rapidly growing population and where food shortages occur regularly. According to Gade (1996), deforestation in Madagascar's highland zone has caused large-scale soil erosion as well as floods. Erosion is exacerbated by regular burning to manage pasture; consequently, cropping and pastoral agricultural systems have substantially declined in productivity. However, the effectiveness of conservation measures is reflected in the report of Banda *et al.* (1994). This details changes in rates of soil loss from a 44 per cent slope used for maize cultivation in Malawi. Soil loss was $80\,t\,ha^{-1}\,y^{-1}$ before planting with *Leucaena leucocephala*; after hedgerow planting it declined to $2\,t\,ha^{-1}\,y^{-1}$ and maize productivity was maintained at $1.5–2\,t\,ha^{-1}$ compared with $0.8\,t\,ha^{-1}$ or less on an unhedged control site. Another success story has been recorded by Michels *et al.* (1995), whose project involved the use of millet residue to reduce wind erosion in Niger's Sahel zone. They found that $2000\,kg\,ha^{-1}$ of residue was necessary to reduce erosion but with $500\,kg\,ha^{-1}$ there was little effect. Success has not been repeated in southern

Zimbabwe, where Hagmann (1996) has shown that water from contour ridges, constructed to control rill erosion, can actually contribute to it. Water from roads and waterways also contributes to rill erosion. Lewis and Nyamulinda (1996) have highlighted soil erosion problems in Rwanda, where agriculture on steep slopes is commonplace and gives rise to erosion rates of $68.2\,t\,ha^{-1}\,y^{-1}$ on average. Much of this is caused by human activity such as hoeing rather than via fluvial action, which is curtailed by the planting of grass and/or shrub strips.

Although research into soil erosion and soil conservation has a long history in Kenya, it has not always met with success (Pretty *et al.*, 1995); but the adoption of a more holistic catchment approach is now generating rewards. In terms of practicalities, Gicheru (1994) has experimented with various soil management practices, e.g. conventional tillage, tied ridges and crop-residue mulching, and their effect on soil moisture, a crucial factor for crop growth in semiarid Kenya. The results showed that the application of crop residues improved soil moisture profiles more than the other types of management, probably because of the water-retaining capacity of organic matter. This is corroborated by Kiome and Stocking (1995), who report that trash lines (i.e. crop residues) were particularly effective means of conservation and were recognised as such by farmers. Erosion can be reduced in grazing lands by the maintenance of at least a 55 per cent plant cover, as Snelder and Bryan (1995) have determined from rainfall simulation experiments on the degraded rangelands of Kenya's Baringo District. At a cover of 55 per cent or above, soil loss varied between 0 and $7.3\,g\,m^{-2}$, but at 25 per cent cover or less it increased to $80\,g\,m^{-2}$ for 60 min storms. One area of Kenya in which soil conservation measures have been effective is the Machakos District, where interrill, rill and gully erosion were widespread (rill and gully erosion combined to produce sheet erosion). Tiffen *et al.* (1994) have reviewed the historical evidence for soil erosion and show that since the 1940s erosion has been considerably reduced. Average rates in the range $5–15\,t\,ha^{-1}\,y^{-1}$ currently occur, though there is much spatial variation. Most erosion occurs on grazing lands whereas erosion on cultivated land has been reduced through improvements in terracing rather than alterations in cropping systems. Erosion has even been reduced on grazing lands, apparently as a result of land registration, etc.; consequently, there are now few communal grazing areas in Machakos.

Numerous soil erosion studies have been undertaken in west Africa, in both the savanna and forest zones, as Lal (1993) has reviewed. Rates of erosion vary from $5 \, t \, ha^{-1} \, y^{-1}$ on cropland in the forest zone to as much as $600 \, t \, ha^{-1} \, y^{-1}$ on bare-ploughed soil. Lal asserts that declines in crop productivity are a direct consequence of soil erosion and that rehabilitation of degraded soils is essential. In a series of experiments on Alfisol degradation following deforestation and initial planting using various methods, Lal (1996a, b, c) has shown that bulk density and penetration resistance greatly increased. This was caused by machinery, a reduction in soil faunal activity and exposure of soil to rain and sun. Over six years, and under a variety of cropping regimes which included a ley phase and grazing, soil properties continued to degrade, mainly due to the loss of clay and organic matter from near-surface horizons. Moreover, the infiltration rate of water was substantially reduced following deforestation and cultivation, and it was associated with the increase in bulk density and penetration resistance. Even when manual rather than mechanised clearance was practised, the soil quality deteriorated. Fallowing, however, proved beneficial as it led to the restoration of the soil's structural properties. However, the soil's chemical properties deteriorated as cultivation continued, despite applications of artificial fertiliser to suit the maize–cowpea rotation and the use of various, supposedly advantageous, cropping systems, including some forms of agroforestry. This loss of nutrients is attributed to leaching, erosion and harvest removal. Lal's data also show that soil erosion was at its highest in the first year after land clearance and then decreased; an erosion rate of $7.6 \, t \, ha^{-1} \, y^{-1}$ was recorded immediately after deforestation, declining to $0.2 \, t \, ha^{-1} \, y^{-1}$ in the second year after deforestation. Moreover, erosion was most severe when mechanised forest clearance was used and when ploughing was employed instead of no-tillage seedbed preparation. Alley cropping successfully reduced soil erosion.

8.4 Desertification

Definitions of desertification and the controversies associated with it were addressed in Section 7.4 in relation to the developed world. Moreover, the processes of soil degradation and erosion, as discussed in Section 8.3, may also contribute to desertification in arid and semiarid regions. Although it is a significant problem in the developed world, desertification is most extensive in the developing world, where it is a factor often associated with drought, famine and loss of life, especially in the sub-Saharan (Sahel) zone of Africa. However, desertification is also a serious problem in China with some relatively minor occurrences in South America (Fig. 7.4). Table 8.6 shows the extent of desertified land in relation to the total extent of drylands in Asia, Africa and South America and highlights the severity of the problem in Africa and Asia (desertification in the central Asian republics is discussed in Section 7.4). Moreover, Dregne et al. (1991) have estimated the economic cost of the problem, i.e. income lost, which amounts to 9.296×10^9 for Africa, 8.313×10^9 for Asia and 2.084×10^9 for South America; in all cases the greatest economic losses derive from degraded rangelands. Over the years there has been much debate as to whether or not desertification is caused by human activity or natural geomorphic processes that are reacting to climatic change. As reflected by the economic data of Dregne et al., it certainly has an impact on people and their livelihoods. Moreover, Thomas and Middleton (1994) confidently state 'People cause desertification'. However, they add the caveat that it may not necessarily be a response to specific activities, such as grazing or rain-fed cropping, but a response to their intensity, which is influenced by political policies and population growth. Indeed most of the social variables that influence tropical deforestation (Table 8.3) also affect desertification, especially in Africa.

The African Sahel has received most of the media coverage. This is because in the 1970s several factors conspired to cause loss of life: a period of drought followed a period of higher than average rainfall, and economic problems followed independence for many Sahelian nations as populations continued to grow (Thomas and Middleton, 1994). An environmental vicious circle was created: desertification contributed to economic problems and agricultural systems responded in such a way as to exacerbate desertification. The occurrence of drought at a moment when other factors were unfavourable tipped the balance and brought about much human hardship. Thomas and Middleton (1994) point out that in the early 1970s estimates of human deaths range from 50 000 to 250 000. The FAO have also estimated that 3.5 million head of

Table 8.6 Data on drylands and extent of desertification

Land area (10^6 ha)	Africa	Asia[a]	South America
Total extent of drylands	1 959.00	1 949.00	543.00
Total agriculturally used drylands	1 432.60	1 881.40	420.70
Total degraded irrigated land	1.90 (18%)	31.81 (35%)	1.42 (17%)
Total degraded rain-fed cropland	48.86 (61%)	122.28 (56%)	6.64 (31%)
Total degraded rangeland	995.08 (74%)	1 187.61 (76%)	297.75 (76%)
Total degraded agriculturally used drylands	1 045.84 (73%)	1 311.70 (69.7%)	305.81 (72.7%)

Source: based on Dregne *et al.* (1991)
[a]Includes central Asian republics discussed in Section 7.4

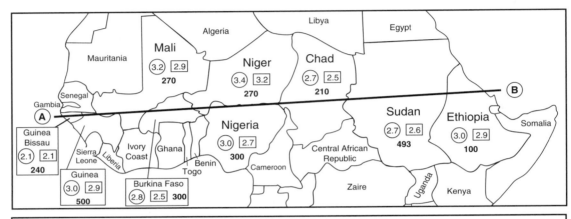

Fig. 8.3 The countries of sub-Saharan Africa (the Sahel) with data on food production (from Morgan and Solarz, 1994) plus data on GNP and population growth (based on World Resources Institute, 1996).

cattle died in the Sahel in 1972–1973. Although these tragic events brought the world's attention to desertification, it remains a serious problem in the Sahel, where population continues to increase rapidly (Fig. 8.3). There is also much debate as to whether the extent of desertification is increasing. Tucker *et al.* (1991) have used remotely sensed images to show that the 200 mm y^{-1} precipitation isoline, an accepted definition for desert conditions, has shifted 130 km south of its 1980 position. Moreover, Morgan and Solarz (1994) have shown that recently there has been only a low rate of growth in food production, and even a decline for some nations (Fig. 8.3). However, in some areas

there is little evidence for environmental degradation, as Schlesinger and Gramenopoulos (1996) have reported for the Sudan. Their analysis of archive satellite images and air photographs for 1943–1994 shows that tree density has remained more or less constant, despite persistent drought over the last 30 years or so. This shows that vegetation is not necessarily a reliable indicator of climatic conditions in these circumstances, probably because no threshold is being crossed. In contrast, Hulme (1996) has analysed precipitation and temperature trends in nine dryland regions for the period 1990–1994 and has found evidence for a drying trend only in the Sahel. This is despite

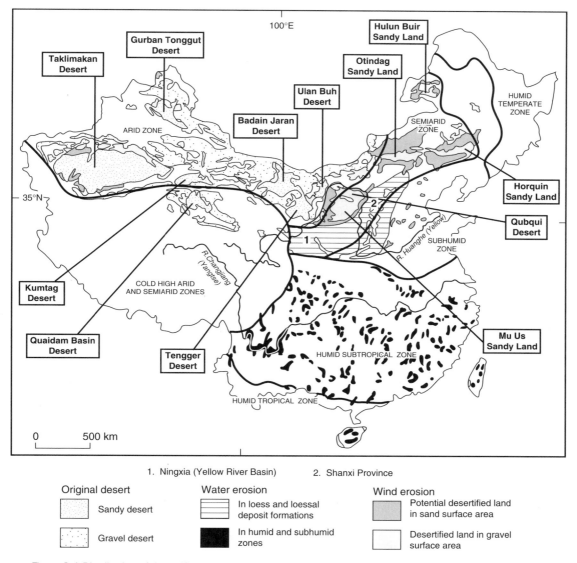

1. Ningxia (Yellow River Basin)   2. Shanxi Province

Original desert

☐ Sandy desert

☐ Gravel desert

Water erosion

In loess and loessal deposit formations

In humid and subhumid zones

Wind erosion

Potential desertified land in sand surface area

Desertified land in gravel surface area

Figure 8.4 Distribution of desertified lands in China (based on United Nations Environment Programme, 1992).

evidence for warming everywhere except the Sahel; Hulme suggests that the production of dust due to desiccation may be reradiating solar radiation and so preventing a warming trend.

There is, however, some localised evidence for land degradation. At three sites near Mopti in Mali, Nickling and Wolfe (1994) have noted dune formation on deltaic soils that were formerly cultivated. Dunes have formed because of cultivation and grazing during periods of drought.

Lindquist and Tengberg (1993) have shown that environmental degradation occurred from the late 1960s to 1990 in northern Burkina Faso. Satellite images, etc., show that droughts began in the late 1960s and caused intense degradation, including the formation of large areas of bare ground. Although rainfall has increased since 1985, the degraded areas have not recovered, partly because of repeated droughts and partly because of continued human pressure. Tengberg (1995) has also documented the formation of nebkha dunes in the region; these are considered to be an indicator of land degradation. Land-use changes have also occurred in response to economic factors. Irrigated rice production for domestic consumption has been introduced in the Senegal River basin (Venema and Schiller, 1995) to reduce the foreign debts of Senegal, Mauritania and Mali, who control the river basin. However, the scheme has been a failure from the social, economic and environmental points of view. Widespread abandonment of plots has increased the amount of desertification and is keeping productivity low; this situation is likely to intensify if drought conditions persist. As Ebohon (1996) has discussed, projects such as these do little to assist Africa in achieving sustainable development. Not only do problems ensue in areas which become decertified but problems may also occur for neighbouring regions. Westing (1994) has discussed the issues surrounding the growing number of displaced people in Africa, where 3 per cent of the population is displaced and where numbers are increasing at a rate of 1.5 million annually. Many of these people become displaced because of warfare but also because of poor subsistence conditions, often associated with land degradation. International agencies and other countries have to feed and clothe these 'environmental refugees'.

Asia's largest areas of desertified land occur in China (Fig. 8.4), where desertification is occurring at a rate of $150 \times 10^3 \, ha^{-1} \, y^{-1}$; at least 25 per cent of it is caused by inappropriate arable practices (Smil, 1993a). According to Zhao (1994), the problem is most severe in the northwest, especially in semiarid regions, where rain-fed agriculture is practised under erratic rainfall conditions. The Mu Us sandy land (Fig. 8.4) has been particularly badly affected by expanding desert, an encroachment which began ca. AD 828 (middle Tang Dynasty) when the suburbs of the city of Tongfan (now known as the White City ruins) received wind-blown sand. After the Great Wall was constructed in 1473, along the southern edge of the Mu Us area, it effectively separated pastoral land from cultivated land. Nevertheless, blown sand began to appear in the arable fields. Zhao also reports that between 1644 and 1911 the Qing government encouraged the cultivation of a large proportion of the southeastern part of the Ordos Plateau (where the Mu Us sandy land occurs). The ploughing of this extensive grassland exacerbated the process of desertification and contributed to the formation of a belt of mobile

sand about 60 km wide along the Great Wall. Despite successful efforts to rehabilitate the area in the 1950s and 1960s, the Cultural Revolution prompted a return to injudicious cultivation and renewed desertification. Efforts have been under way since 1978 to establish sustainable agricultural practices in this region. Fullen and Mitchell (1994) have also observed the desertification process in the Mu Us region, which they report occurs initially in localised areas of rangelands to produce 'blisters'. The blisters enlarge and coalesce.

Although China has experienced a great deal of desertification, it has also undertaken considerable research on reclamation. With a large and growing population China can ill afford to lose agricultural land in this way. Measures to reclaim degraded land include the use of windbreaks, irrigation, checkerboards of straw or clay to increase surface roughness, land enclosures with fences to manage stocking rates, and chemical treatment of sodic soils to improve soil structures. In order to reduce desertification and curtail soil erosion, the Project of Protective Forest System was established in 1978. By 1985, according to Kebin and Kaiguo (1989) some 5.3×10^6 ha had been planted with trees, shrubs, etc.; this included 700×10^3 ha of forest as a sand fixation belt, 700×10^3 ha of shelterbelts to protect arable land and 170×10^3 ha of shelterbelts to protect pasturelands.

Desertification is also a problem in India and Pakistan (Fig. 7.4), where it is attributed to poor management of irrigated and/or agricultural land. As Figure 7.4 shows, the area of India most at risk is in the northwest, the part of Rajasthan that borders the Thar Desert. The Indira Gandhi Canal Project introduced water to the region from the Punjab for irrigation; in some cases this has created salinisation and waterlogging (Section 8.5) and even the abandonment of land. Along with a high rate of population increase, concomitant intensification of agriculture outside irrigated areas and pressures on woodland for fuel, the water from the canal project has contributed to desertification. Problems of land degradation in Rajasthan are exemplified by those in the Aravalli Hills, north of Udaipur (Mishra et al., 1994), and in the Churu and Jhunjhunu districts of northwest Rajasthan. Rain-fed agriculture in northwest Rajasthan produces poor returns; pastures are overgrazed and stocking rates (of goats, sheep and cattle) vary from 1.63 to 12.08 animals per hectare but may be as high as 35 animals per hectare during the dry season. These factors all contribute to land degradation. According to Dhir (1993), the degradation of pastures is particularly severe. Desertification is widespread in Pakistan (Ali and Mirza, 1993) and has caused considerable degradation of tropical thornforest in the Punjab (Khan, 1994). According to Grainger (1990), the growth of rain-fed cropping, even in areas where rainfall is particularly precarious, has caused a 75 per cent increase in livestock densities in the remaining pasturelands. This overstocking has increased erosion rates. Similar problems have arisen in the Gulbarga region of Karnataka State (Ghosh, 1994).

Desertification is much less of a problem in South America than in the rest of the developing world (Table 8.6) though there are some problems in Chile and Argentina. Gutierrez (1993) has suggested that the dry farming of wheat in the Andean foothills of north-central Chile, with overgrazing and the use of woody biomass for domestic fuel and mining, have all contributed to land degradation. The removal of native plant species has also allowed the invasion of introduced annuals; this reduces biomass and nutrient availability in the long term.

8.5 Soils: salinisation, alkalinisation and waterlogging

Salinisation, alkalinisation and waterlogging in soils are components of desertification (Fig. 7.6); although they are problems in some parts of the developed world (Section 7.6), they are particularly serious in the developing world. Salinisation is a process whereby the concentration of salts, notably sulphates, chlorides and carbonates, increases in soils, groundwater and surface water that are exposed to the atmosphere. Alkalinisation occurs where the salts comprise sodium compounds, mainly carbonates and bicarbonates. Soil salinisation and alkalinisation may occur naturally in arid and semiarid regions, where high evaporation rates bring salts to the surface by capillary action; landscape features such as salt flats develop as a result. Such salt accumulations can deter vegetation growth, leaving the regolith susceptible to wind and water erosion. Human activity, via agricultural systems which depend on irrigation, can aggravate these difficulties by increasing the atmospheric exposure of water in the soil and water in irrigation channels themselves. In extreme cases the salt concentration

may become so high that land and/or irrigation systems are abandoned for agricultural purposes. Salt accumulation in soils may also lead to deterioration of soil structure and the development of a hard crust; both reduce productivity. In addition, surface waters and groundwaters may become salinised where they are in receipt of drainage water from irrigated areas; the problem is compounded when these waters are used to supply additional irrigation systems or to replenish systems from which the saline water was initially derived (Fig. 7.6). Permanent waterlogging may occur in areas where continuous irrigation is practised, so the land is never allowed a fallow period. The resulting rise in the water table promotes salinisation, causes waterlogging and adversely affects soil biotic activity and aeration. Thus, improperly managed irrigation systems can actually impair crop productivity, the very factor they were designed to increase.

Figure 7.7 illustrates the distribution of irrigated land in relation to total cropland by continent; Asia has the largest extent of irrigated cropland, 34 per cent; in Africa it is 7 per cent; and in South America it is 9 per cent. Figure 8.5 shows the extent of irrigated land in the developing countries that use most irrigation, along with the proportion of irrigated croplands. Pakistan irrigates 80 per cent of its croplands, while Egypt irrigates 100 per cent. Irrigation played a significant role in the Green Revolution of the 1970s because it generated a major boost in food production, especially in the developing world. As discussed in Section 7.6, irrigation increased substantially during this period and into the late 1980s. Its growth has now slowed, partly because of environmental costs but also because of international concerns about large dam projects. According to Alexandratos (1995), about half of the projected increase in crop production between 1995 and 2010 will derive from the expansion of irrigation, despite its reduced growth rate (this is an FAO projection).

According to the World Bank Report (Umali, 1993) $20\text{--}30 \times 10^6$ ha of the world's irrigated land, i.e. 8 per cent, is considered to be seriously affected by salinisation and a further $60\text{--}80 \times 10^6$ ha are likely to be moderately affected (Postel, 1994). Table 8.7 gives the extent of this land for each of the countries mentioned in the following paragraphs. The problem of salinisation is so acute in Pakistan that crop yields have declined by 30 per cent (Postel, 1993). In a country with a high rate of population growth and in view of the considerable

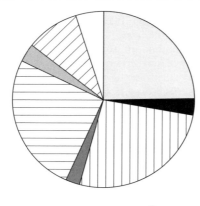

Total irrigated land (ha^6)

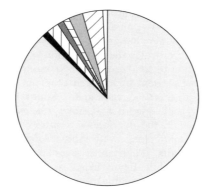

Irrigated land as a % of
total cropland

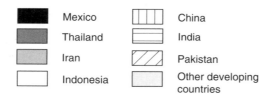

Fig. 8.5 Extent of irrigated land (based on Postel, 1993) and extent of irrigated land as a percentage of the total cropland (based on World Resources Institute, 1996).

salinised; in the Punjab 1.396×10^6 ha are salinised, 21.4 per cent of which are strongly salinised. The irrigation systems in these regions and in Baluchistan employ a variety of constructions. These include canals and tubewells, as well as several types of traditional lift system (the use of a bucket on a scaffold), karez irrigation involving the construction of hand-dug horizontal wells that collect water from the base of hills and channel it into fields as far as 10 km away (e.g. Kahlown and Hamilton, 1994), and sailaba or run-off farming (Kahlown and Hamilton, 1996). Moreover, the Indus Basin is characterised by an extensive irrigation system based on barrages, link canals and relatively large dams (e.g. the Mangle and Tarbela dams, which also produce hydroelectric power). This is a product of the Indus Water Treaty, established between India and Pakistan in 1947 when the two countries became separate states. Many subsequent disputes were resolved by the World Bank, the architect of a new treaty in 1960 (McCaffrey, 1993). However, waterlogging and salinisation had developed in the area by the 1850s, mainly because of canal irrigation and overuse of water.

Developments in irrigation and mismanagement have been discussed by Khan (1991) and Umali (1993). Both point out the problems associated with perennial irrigation and constant seepage from unlined canals. Even in a semiarid region such as this, water reaches the soils at a rate considerably in excess of evapotranspiration. This causes the water table to rise; when it reaches 5 m below the surface, capillary action begins to move water upwards, carrying salts to the surface where they are deposited. This continued addition of salts to the soil accelerates as the water table rises. According to Khan (1991), crop productivity is adversely affected when the water table reaches 3 m below the surface, and crop production may have to be abandoned when it reaches 1.5 m below the surface. Eventually, the water table approaches the surface to cause permanent waterlogging. Other factors have also contributed to this problem, notably the need for increased food productivity because of a rapidly growing population; this has led to a reduction or abandonment of fallow periods. Moreover, the original design of the irrigation systems did not give adequate consideration to flushing and drainage requirements. Despite the establishment of the Salinity Control Reclamation Project (SCARP), Pakistan's crop yields are now among the lowest in

costs associated with the rehabilitation of salinised land, the human costs and the environmental costs are high. According to a government survey conducted in the early 1980s (Umali, 1993), the Sind and Punjab are the worst-affected areas of Pakistan in relation to salinisation. In the Sind 2.597×10^6 ha are affected, 37.5 per cent of which are strongly

Table 8.7 Extent of salinised land in selected countries

	Land area (10^3 ha)		
	Total cultivated	Total irrigated	Salt affected
East Asia			
Australia	768 680	1 664	85
Bangladesh	9 133	1 441	2 500
Burma	9 971	971	600
China	95 674	47 822	
Indonesia			13 200
Kampuchea			1 300
Malaysia	4 150	304	4 600
Philippines	9 900	1 180	400
Thailand	16 363	2 378	1 500
Vietnam	5 770	980	1 000
South Asia			
India	169 080	43 050	7 000
Pakistan	20 500	15 440	
Latin America			
Colombia	113 890	350	64
Mexico	24 750	5 000	
Peru	3 400	1 190	250
Near East			
Egypt	100 140	2 830	226
Iran	164 880	3 500	23 500
Iraq	43 490	3 670	1 500
Israel	2 070	173	28
Morocco	9 327	1 570	250
Syria	18 480	500	110
Turkey	78 060	1 813	3 333

Source: based on data in Umali (1993); some data were collected in the 1970s

the world, though reclamation schemes are annually returning some 80×10^3 ha to productivity (Umali 1993). Improvements to irrigation systems in Pakistan, and elsewhere, need not necessarily involve vast costs. This is highlighted by Ahmad *et al.* (1996), who have shown that earthen ponds provide good water storage but their efficiency is increased with a physically or biologically based lining, e.g. manure or vegetation. Also a problem in Bangladesh, salinisation is a significant factor in limiting crop production during the winter season. Rahman *et al.* (1995) have shown that the productivity of salt-tolerant cultivars of wheat, barley and pulses can improve productivity significantly. The combined use of irrigation and fertiliser increased wheat yields by 80 per cent and the water-use efficiency of wheat increased with fertiliser application.

In India nearly 5×10^6 ha of land are salinised as a result of irrigation; a further ca. 1.15×10^6 ha of land are waterlogged, according to Smedema,

quoted in Umali (1993). Figure 8.6 gives the location of sites on which these data are based and which reflect the regions in which the adverse effects of irrigation have been significant. Section 8.4 has already mentioned desertification in Rajasthan, where S. Singh *et al.* (1994) report that, besides extensive soil erosion and deposition, many croplands have been adversely affected by soil salinisation and waterlogging. Moreover, Jaglan and Quereeshi (1996) have reviewed the overall impact of irrigation based on the Indira Gandhi Canal. Although they point out many of the advantages of the canal, they also emphasise that a substantial part of its command area will probably be waterlogged by 2000, and that within only five years of the canal's construction, strong soil alkalinisation has developed in some areas, including the formation of calcareous pans. Tyagi *et al.* (1993) have highlighted problems of salinisation in the Lower Ghaggar Basin, supplied with water by the Bhakra Canal System (Haryana State), where

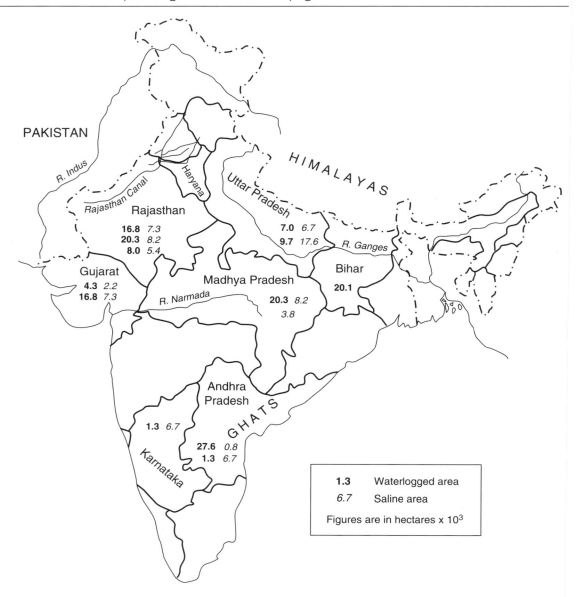

Fig. 8.6 Salinisation and waterlogging: Indian sites where assessments have been made (based on data in Ulima, 1993).

waterlogging will occur in more than 50 per cent of the area by 2003 if improvements are not made to irrigation systems. Problems in this region of irrigated rice and wheat production have also been discussed by Singh and Singh (1995). They highlight the employment and economic difficulties that ensue when the productivity of agricultural land begins to decline; farm income and employment opportunities are reduced and standards of living fall. Consequently, the economic system becomes as unsustainable as the agricultural system.

This land degradation does not bode well for India's agricultural future, especially since it is experiencing a population growth rate of 1.9 per cent (World Resources Institute, 1996). Nevertheless, it is likely that India's continued development will require an increasing dependence on food production from irrigated areas, as Pike (1995) has discussed. At least 55 per cent of agricultural output currently derives from irrigated lands, though productivity is declining in some areas because of salinisation, etc. Consequently, there is a pressing need for a reconsideration of irrigation management strategies, rehabilitation of irrigation-damaged land and possibly an expansion of irrigation systems. Expansion of irrigation systems may become the sole means of increasing food production if, as a result of global warming, drought becomes increasingly persistent in India's semiarid areas, where rain-fed agriculture is currently important. This is a possibility throughout the world's semiarid zones. Careful management will be required to improve groundwater irrigation in India (Minhas, 1996), often either saline or alkaline, especially as high installation costs have so far prevented the use of sprinklers and drip irrigation. Moreover, under experimental conditions, D.P. Sharma *et al.* (1994) have devised irrigation programmes that combine saline groundwater with fresh canal water to grow crops in the dry season without reductions in productivity. However, it is essential that the regime of water application to field crops should be carefully determined in relation to crop type, soil type and crop water requirements throughout its life cycle; indiscriminate or permanent application of water is unnecessary and environmentally damaging. Advice on such matters, based on experimental work, needs to be passed on to farmers. Besides conserving water, Singh *et al.* (1996) have experimentally determined that on a sandy-loam soil the date of seedling transplantation and the type of irrigation regime in northern India can substantially improve crop productivity. Of three transplant dates, 31 May proved to be the best because it avoided attack by the shootborer, which reduced the yield of the crop transplanted on 16 June. It also used 15.4 per cent less irrigation water. Moreover, Singh *et al.* found that submergence for two weeks after transplanting with subsequent irrigation every two days saved ca. 73 per cent irrigation water when compared with continuous shallow submergence, the traditional method. Oswal (1994) has examined many possibilities for water conservation and management in rain-fed agricultural systems; this work is important because improved productivity from rain-fed ecosystems will limit the need for the expansion of irrigation.

One possible solution to problems in salinised croplands is to find crops which have developed a tolerance to salinity. Flowers and Yeo (1995) have indicated this will be essential given further global food requirements, and crop breeding programmes may well use genetic engineering (Section 9.5) to develop such cultivars. However, this tackles the symptoms rather than the causes of salinisation and could lead to an increase in salinisation. Nor is it a solution for the rehabilitation of lands that are already salinised. G. Singh *et al.* (1994) have discussed various possibilities that are available. For example, soil amendments have been used to rehabilitate salt-affected soils in the Indo-Gangetic Plain; the addition of gypsum (magnesium sulphate) to irrigated soils with high quantities of sodium carbonate flushes out sodium sulphate leaving magnesium carbonate. Overall such treatment improves infiltration rates, crop emergence rates and grain yield (Joshi and Dhir, 1994). G. Singh *et al.* (1994) have also discussed a range of successful reclamation projects that have been undertaken using various agroforestry techniques. In some cases it has been possible to restore land for the resumption of normal crop production. Various species of trees, identified as salt tolerant, can be manipulated under various irrigated and topographical situations to effect reclamation. Examples include *Acacia nilotica* (*Acacia* spp.). Agroforestry also reduces soil erosion and provides fuelwood as well as improving soil chemistry and structure. Salinisation is also a problem in Bangladesh, and is a significant factor in limiting crop production during the winter season. Rahman *et al.* (1995) have shown that the productivity of salt-tolerant cultivars of wheat, barley and pulses can significantly improve productivity. The combined use of irrigation and fertiliser increased wheat yields by 80 per cent and the water-use efficiency of wheat increased with fertiliser application.

Irrigation-damaged lands in China comprise 11.1×10^6 ha which are salinised (Umali, 1993), though Zhao (1994) suggests that some 20 per cent of China's 100×10^6 ha of farmland are affected to some extent by salinisation, i.e. 20×10^6 ha. Moreover, Smil (1993a) has estimated that over the period 1989–2000 China is likely to lose between 0.25×10^6 ha and 0.50×10^6 ha of farmland because of irrigation damage. Water efficiency is low in many of China's irrigated lands; loss through seepage via unlined channels and canals and evaporation can be as high as 80 per cent. According to Xiong *et al.* (1996), high soil salinity has been a problem in the irrigated lands of the Yellow River in Ningxia Hui (Fig. 8.4) since 214 BC, when the first canal was built. They state that by 1985 some 120×10^3 ha of the region's 300×10^3 ha of irrigated land had become affected. Long-term irrigation and overirrigation have caused a rise in the water table hence high salinities in the soil, as have poor management and unlined canals etc.; similar situations occur in India and Pakistan, as discussed above. Additional factors in Ningxia Hui include the high level of the Yellow River during the flood season, emphasis on paddy rice, which means that water covers large areas continuously and is thus subject to evaporation, introduction of fish ponds to improve protein consumption in rural areas and the flat terrain with many lakes.

Remotely sensed data (Qiao, 1995) indicate that in Shanxi Province (Fig. 8.4) 33 per cent of the irrigated land is salinised. As Zhao (1994) discusses, inland basins like Xinding are particularly susceptible to salinisation, mostly due to overirrigation and poor drainage. Zhao also refers to the Black Elbow Plain, in the Yellow River basin, where 50 per cent of the 200×10^6 ha of irrigated farmlands are salinised. Between 0.8 and 1.6×10^6 t of salts are deposited annually. Nevertheless, China has achieved remediation of salinised land, exemplified by the North China Plain, where salinisation considerably reduced the productivity of ca. 4×10^6 ha of cropland during the 1950s and the 1960s. Improved irrigation systems and practices, including deep wells, have lowered the groundwater level and reduced the amount of salinised cropland by 1×10^6 ha.

Many countries in the Middle East rely heavily on irrigation to support their agricultural systems. Not only does this create environmental problems, it also generates considerable political problems, i.e. hydropolitics, as has the use of Indus waters by India and Pakistan (see above). A survey of hydropolitics has been provided by McCaffrey (1993), and Kliot (1994) has given a detailed review of the situation in the Middle East. Egypt has been heavily dependent on the Nile throughout its history and all its 2.8×10^6 ha of cropland are irrigated. Despite this dependency, irrigation efficiency is only 44–58 per cent, rising to 65 per cent if the reuse of drainage water is taken into account. Water losses are high; at least 20 per cent of the 40.9×10^9 m^3 of water used for irrigation is lost and losses may be as

high as 50 per cent. This is due to losses from the 30 000 km of canals; they are in a poor state of repair and give inadequate control over water flows. Kliot asserts that Egypt's plans to reclaim additional sectors of the Western Desert through irrigation are overambitious in view of water availability and the high wastage rates. Moreover, Pearce (1992) contends that salinisation and waterlogging will also thwart these plans. Approximately 1 t of salt is currently deposited annually per hectare; this requires an annual expenditure of $2 billion for mitigation measures.

In Iraq 2.7×10^6 ha are irrigated, i.e. about 25 per cent of its total cropland (Kliot, 1994). The water is extracted from the Tigris and the Euphrates. The irrigation is inefficient with up to 40 per cent losses due to high evaporation rates, unlined canals and poor control. Salinisation is a serious problem that is being addressed through the introduction of modern drainage systems, but Kliot indicates there is no available data by which to measure success. Irrigation efficiency in Jordan is also low, at 46 per cent, though there is a scheme in hand to replace open concrete canals with pressure pipes. Reducing the exposure of water will reduce evaporation and salt concentrations whcih will curtail soil salination.

Salinisation may also affect groundwater. This may be caused directly through poor irrigation management or indirectly through the intrusion of salt water into aquifers that have been depleted of fresh water. According to Khair *et al.* (1992), the removal of water from near-coastal aquifers in Lebanon has caused just such a problem. Salinity increased from 340 mg dm^{-3} in the early 1970s to 22 000 mg dm^{-3} in 1985. Agriculture in Libya is highly dependent on groundwater for irrigation; the water for this is obtained from the mining of aquifers. Elassward (1995) asserts that the water mined in some parts of the country exceeds the natural replenishment by 500 per cent. The decline of water levels in coastal aquifers has allowed the intrusion of sea water. Moreover, it is likely that Libya and its neighbours will increase aquifer mining to support agriculture, as Nour (1996) has discussed in relation to the Nubia Sandstone aquifer in North Africa. The use of groundwater for irrigation can influence groundwater properties in other ways, limiting its use for purposes other than irrigation. This is illustrated by Alsulaimi *et al.* (1996), who have reported on a rise in the volume of dissolved solids in brackish groundwater lenses in northern Kuwait. Total dissolved solids (TDS) have increased as water has been withdrawn for irrigation then returned to the aquifer (ca. 66 per cent), taking with it dissolved solids. They estimate that if irrigation continues at the 1989 rate or above, only 10–25 per cent of wells will have TDS concentrations of less than 7500 ppm. This has implications for the provision of drinking water as groundwater lenses with TDS concentrations of less than 1000 ppm are usually the only wells used for this purpose. Problems may ensue where sewage effluent or industrial wastewater are used for irrigation, ranging from bacterial contamination to high concentrations of heavy metals. Some aspects of this have been discussed by Farid *et al.* (1993) in relation to Egypt (Section 8.6 considers the impact of agriculture on water quality).

8.6 The impact of agriculture on water quality

Agriculture influences water quality in many ways. The most significant impacts include the addition of nutrients either directly as a consequence of artificial fertiliser use or indirectly through the release of animal sewage into drainage systems and wetlands. In many cases this causes cultural eutrophication (Section 7.5). Other contaminants that derive from agriculture include pesticides and heavy metals. Heavy metals are washed out of the soil by irrigation water or occur because of the use of industrial wastewater for irrigation. Groundwater as well as surface water may be affected by these factors and cultural eutrophication of coastal waters may occur.

According to Rast and Thornton (1996), cultural eutrophication is becoming increasingly significant in the developing world, especially in countries where the use of artificial fertilisers has increased (Lijklema, 1995), as shown in Table 8.8. Rice production in Asia has increased by an average of 2.7 per cent per year due to fertiliser use, irrigation, etc. (Dedatta, 1995). However, the uptake of nitrogen in tropical rice systems is low, usually between 30 and 40 per cent; the rest enters the wider environment. The analysis by Justic *et al.* (1995) of 10 of the world's largest rivers, including the Yangtze, Yellow, Zaire and Amazon, shows that nitrogen and phosphorus loadings have increased substantially in the recent past. Although increased loadings have caused coastal eutrophication, the

problem is not as severe as in the Adriatic (Section 7.5) or the Gulf of Mexico. According to Sommaruga *et al.* (1995), similar problems are occurring in the Rio de la Plata, though some of the phosphorus loading derives from detergent. Moreover, three areas of Uruguay have been identified with lakes, etc. becoming eutrophic because of inputs of phosphorus from fertilisers. Carney *et al.* (1993) have provided evidence for the increased influx of nutrients and sediment into Lake Titacaca (Bolivia, Peru) which has occurred as traditional raised fields have been abandoned in favour of flat pastures and fields.

As Table 8.8 shows, the use of artificial fertilisers has increased considerably in Asia during the past two decades. And the increase in fish farms in both freshwater and coastal waters has led to locally high concentrations of nutrients which are altering the chemistry and biology of these waters. In the Philippines this has occurred in Laguna-de-Bay, a lake south of Manila. Reyes and Martens (1994) have shown how the introduction of milkfish in pens caused a major decrease in primary productivity, and there is evidence for increased nitrate and phosphate loadings since 1973, reflecting cultural eutrophication. Eutrophication caused by agricultural run-off (and other wastes) is also impairing coral reefs in Philippine waters (Food and Agriculture Organisation, 1993). Nixon (1995) has identified Korea, India and Pakistan as primary areas for eutrophication, including coastal eutrophication. Hoedong Lake in Korea is experiencing cultural eutrophication from both point and non-point sources (Choi and Koo, 1993). The non-point sources are probably due to agriculture. Similar problems in India are exemplified by Lonar Lake, Maharashtra, where blue-green algal blooms of *Spirulina, Arthrospira* and *Oscillatoria* occur (Badve *et al.*, 1993). Indeed Badve *et al.* have suggested that because *Spirulina* is a protein-rich alga it could be produced commercially as single-cell protein food. Eutrophication has also been reported as a problem in Gujarat in northwest India (Rana and Kumar, 1993) and off the Kerala coast of south India, where high phosphate concentrations have been recorded in marine sediments (Nair and Balchand, 1993).

There is also some evidence for eutrophication in Africa, especially from Lake Victoria, the focus of many research programmes. Gophen *et al.* (1995) have reviewed the changes in Lake Victoria since 1954, when the Nile perch was introduced as a source of protein for local people and for possible

Table 8.8 Annual fertilizer usage (kg) per hectare of cropland in selected developing countries

	1983	1993		1983	1993
El Salvador	113	106	Algeria	22	17
Jamaica	58	107	Burkina Faso	5	6
Mexico	60	71	Cameroon	7	3
Trinidad and Tobago	66	51	Egypt	363	357
			Ethiopia	1	6
Ecuador	30	31	Gambia, The	15	4
Guyana	21	24	Kenya	20	27
Paraguay	5	14	Libya	43	49
Peru	22	44	Madagascar	5	3
Surinam	162	49	Malawi	30	51
Uruguay	31	72	Mali	5	10
Venezuela	41	65	Morocco	29	29
			Mozambique	5	1
China	184	261	Namibia	0	0
India	46	73	Nigeria	9	16
Indonesia	58	85	Rwanda	1	2
Iran Islamic Republic	68	52	Senegal	11	11
Iraq	17	52	Sierra Leone	2	6
Jordan	47	34	Sudan	3	5
Korea Democratic People's Republic	409	315	Uganda	0	0
Korea Republic	331	474	Zaire	1	1
Malaysia	116	212	Zambia	13	16
Nepal	16	31	Zimbabwe	55	55
Pakistan	59	101			
Philippines	40	61			
Sri Lanka	90	111			
Thailand	25	54			
Vietnam	57	136			

Source: based on World Resources Institute (1996)

commercial exploitation. This caused the extinction of many endemic (haplochromine) fish species which altered the trophic structures within the lake. The Nile perch is itself a piscivore and many of the species it caused to become extinct were phytoplankton/zooplankton/detritus consumers. This led to a reduction in the consumption of phytoplankton and detritus. Two species increased: the endemic species *Rastrineobola argentea* and the prawn *Caridina niloticus*, wreaking significant changes in the fishery status and fish industry of the lake. Moreover, eutrophication has been occurring since 1960 with restricted vertical mixing and anoxia at depth alongside increased phytoplankton productivity and a change from diatoms to blue-green algae. Both the internal changes in the lake and the external changes in the catchment, i.e. agriculture, deforestation and urban development, contributed to this eutrophication. External changes in particular caused an influx of nutrients, with increases in nitrogen from the 1920s and increases in phosphorus from the 1950s. Gophen *et al.* point out that these alterations have increased pressure on the remaining endemic fish species. The deoxygenation

of deep waters in the lake has been investigated by Hecky *et al.* (1994), who compared the vertical variations recorded in 1960–61 with those recorded in 1990–91. Most significantly, oxygen concentrations in hypolimnetic (deep) waters are now lower than in 1960–61 and remain low for a longer period. Moreover, Mwebazandawula (1994) has reported on the changes in the zooplankton of Lake Victoria since 1931; the significant alterations in the community reflect the changes in predation and eutrophication mentioned above.

Eutrophication can also occur in reservoirs, especially those that trap silt derived from the soil erosion of agricultural land. This may be accelerated if the land is fertilised with animal or mineral fertiliser. Alaouimhamdi and Aleya (1995), for example, have shown there is a net accumulation of phosphorus in the Al-Massira Reservoir in Morocco. Many small ponds in China have also experienced eutrophication because of intensive fish farming and the high use of mineral fertilisers that produce high inputs of nutrients. As Table 8.8 shows, the use of fertiliser in China has increased from 184 kg ha^{-1} in 1983 to 261 kg ha^{-1} in 1993

(World Resources Institute, 1996). Beaumont (1989) notes that a major cause of eutrophication in the River Jordan is the receipt of drainage of mineral fertilisers as well as inputs of sewage and organic matter from the erosion of peat from the former Lake Huleh. Cultural eutrophication in the Nile has also increased significantly since the construction of the Aswan High Dam, which became operational in the early 1960s. Although this allowed Egypt to expand its area of irrigated land, doubled the production of electricity and provided flood control for the lower reaches of the Nile, it created several disadvantages. In particular, it caused silt to be trapped, reducing the input of nutrients that were formerly deposited by annual floods. Although the dam provided much-needed water, it reduced nutrient inputs to such an extent that mineral fertilisers became essential to maintain Egypt's agriculture, especially as it now has the lowest amount of arable land per capita in Africa and has a high rate of population growth (Biswas, 1993). Today, the Nile's water and canals have a major problem with the water hyacinth (*Eichhornia crassipes*), a species introduced from South America. The water hyacinth reproduces rapidly and thrives in nutrient-rich waters where it has no predators; eventually it congests waterways and irrigation canals. The dense growth can also inhibit oxygen diffusion from the atmosphere and cause fish kills. In addition the kariba weed (*Salvinia molesta*) occurs widely and causes similar problems. There is also evidence that the use of fertilisers is influencing groundwater quality in some of Egypt's recent reclamation areas (Awad *et al.*, 1995).

The intensive exploitation of Egypt's water resources can lead to other problems of contaminated drinking or irrigation water, as Siegel *et al.* (1994) have discussed for the Manzalah Lagoon, Nile Delta, where high concentrations of mercury, lead, zinc, copper, silver and selenium have been recorded from recent sediments. These metals have accumulated following the industrial development stimulated by the cheap electricity from the Aswan High Dam. The wastewaters from the enterprises, and solid wastes, are dumped in the delta area, including the Barhr El-Baquar drain. From here the metals are deposited in the Manzalah Lagoon. Two concerns arise: firstly, waters from this drain may be used for irrigation, opening up the possibility that heavy metals will enter the food chain; secondly, some lagoons are being reclaimed for agriculture, so

food crops are cultivated on metal-contaminated sediments. Heavy-metal pollution is also reported in Nile sediment and beach samples (Awadallah *et al.*, 1996). Cai *et al.* (1995) have reported a parallel situation in China. River water used to irrigate cropland in Dayu County, Jiangxi Province, contains high concentrations of cadmium because it drains tailings from tungsten ore dressing plants. As a result, local residents have apparently been exposed to high levels of cadmium in crops for the past 25 years. Cai *et al.* report that 99.5 per cent of orally ingested cadmium derives from rice and vegetables; smokers absorb even more cadmium because they smoke locally grown tobacco. Such exposure over the long term may lead to kidney problems. In view of the scarcity of water in many parts of the developing world, and a growing trend to overcome this by reusing wastewater, it is essential for the water to undergo treatment before its use in irrigation.

Alexandratos (1995) maintains that the use of pesticides in developing countries increased markedly in the 1960s and 1970s. He states 'In general, the demand for chemical pesticides increases with increasing land scarcity and market access'. In terms of active ingredients, 530×10^3 t of pesticides were used throughout the developing world during 1985; this was equivalent to ca. 20 per cent of global consumption. The highest use was in east Asia, including China, which used 38 per cent of the total for the developing world; Latin America accounted for 30 per cent with the Near East and north Africa together using 15 per cent, south Asia used 13 per cent and sub-Saharan Africa used only 4 per cent; the 1985 data are from Alexandratos (1995). Consumption increased by ca. 1 per cent between 1985 and 1990; it has fallen slightly since then. Alexandratos states 'Overall ... a reasonable estimate seems to be that pesticide consumption in developing countries will continue to increase but at a slower pace than in the past. Most of the growth is likely to be in South and East Asia and in Latin America'. Some 50 per cent of pesticides used in the developing world are insecticides. It must also be borne in mind that many developing countries allow the use of pesticides which are banned in developing countries, including DDT (Section 7.7), though its use will diminish in the future. (Some countries use DDT as a public health control instead of an agricultural pest control.)

Pesticides are widely used in China in order to meet the country's food requirements. Along with increased irrigation and fertiliser use, pesticides have helped China to avoid the famines that occurred up to 1965. However, there is evidence that surface waters and groundwaters have been contaminated with pesticides. Dudgeon (1995) has reviewed some of the evidence for pesticide pollution in China's rivers, including the Yangtze and the Pearl. Hong et al. (1995) have also demonstrated that the sediments of Xiamen Harbour (and Victoria Harbour, Hong Kong) are contaminated with organochlorine pesticides, including DDT. They also suggest that these and other organic pollutants have altered the structure of benthic communities. In the long term, China and other developing countries will intensify their use of integrated pest management (IPM) strategies. By using a range of techniques, IPM reduces the dependence on chemicals and decreases the rate at which organisms become pesticide resistant, especially insects (Section 7.7). Moreover, China and India have active research programmes in biotechnology and genetic engineering, including the production of insect-resistant and herbicide-resistant crop plants (reviewed in Mannion, 1995a, d). This is discussed further in Section 9.6.

At Narova in the northern state of Uttar Pradesh, the River Ganges is contaminated with organochlorine pesticides, including DDT, and organophosphate pesticides (Rehana et al., 1996). DDT is also found in mothers and their infants, even in Delhi (Nair et al., 1995) and in soils at high altitudes (Singh and Agarwal, 1995). According to Misra and Bakre (1994), DDT, its derivatives and other organochlorine compounds are present in the tissues of migratory birds found around Mahala Reservoir, near Jaipur. They suggest these residues are the result of biomagnification as the substances have become concentrated through food chains which include fish from the reservoirs. Coastal waters have also been contaminated with organochlorine pesticides; Tanabe et al. (1993) have recorded high concentrations in the blubber of netted dolphins from the waters of India's southeast coast. Moreover, Shailaja and Singbal (1994) have reported high concentrations of DDT, its derivatives and dieldrin in a range of zooplankton and bottom-feeding fish in the same area, the Bay of Bengal. These results are supported by Iwata et al. (1994), whose survey of organochlorine residues in river water and sediment in tropical regions shows that higher concentrations are present in developing nations, e.g. India, Thailand, Vietnam and

Malaysia, than in developed nations, e.g. Japan and Australia.

Similar contamination occurs within and around the coast of Latin America. Reyes et al. (1996) have suggested that organochlorine pesticides may be responsible for the decline in the shrimp fishery in the state of Sinaloa, Mexico, where DDT is still used, mainly for the control of malaria. Dolphins, sediments and mussels also provide evidence for organochlorine contamination in the Gulf of Mexico (e.g. Sericano et al., 1993; Salata et al., 1995), some of which derives from pesticide use in North America during the 1950s and the 1960s. Organochlorine compounds, but not DDT, have been found in the Uruguay River in South America (Janiot et al., 1994).

Examples of groundwater contamination by pesticides include the case of community water supplies of Ibadan, Nigeria (Sangodoyin, 1993) and tubewell contamination by ricefield pesticides in the Philippines (Castaneda and Bhuiyan, 1996). The Philippine tubewells were used for domestic water supply but had pesticide residue concentrations (e.g. endosulfan) in excess of international standards. It was also found that pesticide residues increased under irrigated agriculture, light soils and wet-season cultivation. Thus, in this case, the management of agriculture contributed to the problem. High concentrations of organochlorine pesticides are present in groundwater in Egypt and in the Nile. In the Kafr El Zayat Governorate the problem is compounded by the proximity of a pesticide manufacturing plant (Dogheim et al., 1996).

8.7 Conclusion

Agriculture has undoubtedly transformed the environments of all developing countries. And although agriculture is essential, not only for the support through food supply for indigenous populations but also to provide a sound base on which economic progress can be made, it is unfortunate that injudicious land-use practices have so often negated the very factor they sought to improve: increased productivity. The reasons for this are manifold and often rooted in the historic past, though colonialism (see Sections 5.7 and 5.9) cannot be used as a perpetual excuse. This is especially true in the light of the many more recent and post-independence projects that have failed, despite the often sizeable investments in the form of aid. This is not to exonerate outside influences *in toto* because even the most altruistic schemes have regularly failed to accommodate the specific requirements of local groups and their inherent land-management skills. Moreover, the fundamental social structures and traditional land-management practices of many indigenous populations have often been dismissed as more or less irrelevant to modern-day needs.

The very inception of agriculture (Section 4.3) was based on the manipulation of natural biotic resources, emphasising the close relationship between environment and agriculture in all its forms. The harnessing of fossil-fuel energy and other technological developments, including their applications to agricultural systems, have opened up a new range of possibilities for agriculture in the developed world, but all this has happened over a relatively long time and is not without environmental implications. Such expertise is not easily extrapolated to regions where the environment is markedly different. All agricultural systems are underpinned by energy flows and biogeochemical cycles, and when they are disrupted they bring about environmental degradation. In the developing nations, mostly located in the tropical and subtropical regions, the workings of these fundamental flows and cycles are only now becoming clear. In many instances this is too late to prevent desertification and soil erosion. However, degradation can still be halted by careful consideration of environmental processes and social requirements, as this chapter has shown.

Land use has to be balanced to achieve sustainability. Land, productivity and conservation should not be inimical or mutually exclusive. The fact they have frequently become so is a fundamental error which will exact a substantial toll in the longer term.

Further reading

Agassi, M. (ed.) (1996) *Soil Erosion, Conservation and Rehabilitation*. Marcel Dekker, New York.

Brown, L.R. (ed.) (1996) *State of the World 1996*. Earthscan, London.

Chadwick, D.J. and Marsh, J. (eds.) (1993) *Crop Protection and Sustainable Agriculture*. J. Wiley, Chichester.

Gleick, P.M. (ed.) (1993) *Water in Crisis: A Guide to the World's Freshwater Resources*. Oxford University Press, Oxford.

Greenland, D.J. and Szabolcs, I. (eds.) (1994) *Soil Resilience and Sustainable Land Use*. CAB International, Wallingford.

Greping, Q. and Jinchang, L. (1994) *Population and the Environment in China*. Lynne Reinner, Boulder Co, and Paul Chapman, London.

Grigg, D. (1995) *An Introduction to Agricultural Geography*, 2nd edn. Routledge, London.

Mannion, A.M. (1995) *Agriculture and Environmental Change: Temporal and Spatial Dimensions*. J. Wiley, Chichester.

CHAPTER 9

Other agents of change: forestry, recreation and tourism, biotechnology

9.1 Introduction

Not all of the environmental changes that have occurred in the recent past are due to the impact of industry and agriculture. Throughout history, both have involved either the manipulation and/or the destruction of the world's forests. This continues today as the natural forest resource is managed to provide wood and wood-based products in both the developed and the developing world. Afforestation programmes are under way in many nations to augment the rapidly diminishing natural forest resource and in some cases to halt environmental degradation. Both forest management and afforestation are responsible for direct and indirect environmental change. Moreover, since the Second World War, increasing automation and especially the rise in living standards of the developed world have increased the amount of leisure time, which in turn has brought increased pressures on the environment via recreation, including sport and tourism. The development of these industries has also been facilitated by improved communications and transport, as well as the need to generate income in areas that have been adversely affected by economic recessions or which are unsuitable for other kinds of economic activity such as agriculture. Similarly, the need to generate foreign currency has stimulated tourism in regions that would have been almost inaccessible 20 years ago. However, these activities often have adverse and environmentally degrading consequences, especially where wildlife habitats and wildscapes are concerned. In some instances, they are leading to the destruction of the specific resource that attracts visitors. At the very least, the growth of tourism has affected the environment by prompting the construction of hotel and resort complexes, nature trails and safari parks etc.

Another recent innovation is biotechnology. Biotechnology involves the use of organisms in industrial and technological processes and it is being applied to a variety of environmental management problems such as waste treatment. Associated with biotechnology is the recent development of genetic engineering. Although its origins lie in the manipulation of plant and animal species that began ca. 10 K years BP (Section 4.5), genetic engineering had been confined to plant and animal breeding programmes. Now, however, science has provided the means to manipulate the basic genetic material of plants and animals i.e. deoxyribonucleic acid (DNA), and already this has generated exciting progress in forensic science, pharmacology and disease control. Genetic engineering is also being applied to the production of new crop strains and agents, such as viruses, that can protect existing crops from pests and frost. Much international debate currently relates to the circumstances in which engineered species should be released into the environment and whether or not their impact will be entirely beneficial. Genetic engineering thus constitutes an important potential agent of environmental change.

9.2 Forestry and afforestation in the developed world

Chapters 4 and 5 have shown that forests have featured significantly in the historic development of

Table 9.1 Wood and paper production in the major wood-producing countries of the developed world

	Roundwood production 1991–93 (10^3 m^3)	Percent change since 1981–83	Processed wood production 1991–93 (10^3 m^3)	Paper production (10^3 m^3)	Percent change since 1981–83
Europe total	**319 100**	**5[a]**	**111 093**	**68 233**	**36**
Finland	37 663	6[a]	8 155	9 304	52
France	43 617	15	13 695	7 652	48
Germany	36 245	NA	22 362	12 998	NA
Russian Federation[b]	244 488	NA	53 594	5 115	NA
Sweden	59 907	17	13 022	8 503	39
UK	6 197	50	4 046	5 115	56
North America total	**663 703**	**NA**	**198 887**	**91 945**	**NA**
Canada	172 703	20	62 582	16 900	28
USA	491 000	19	136 305	75 045	31
Oceania total	**44 140**	**25**	**7 593**	**2 803**	**31**
Australia	19 860	17	3 929	2 007	39
New Zealand	15 110	51	3 329	796	13

Source: based on FAO statistics quoted in World Resources Institute (1996)
[a] Denotes exports
[b] Not included in regional totals
NA = not available

the world's industrialised countries (Schama, 1995). This is reflected in the fact there is little natural forest left in western Europe as a whole and that forest resources were much prized by colonial powers in the 1800s and 1900s (Section 5.8 considers this in relation to North America). Forestry is a significant business sector throughout the developed world though many developed countries rely on developing nations for wood and wood products that cause forest loss and forest degradation; this will be examined in Section 9.3. Thus, the management of forests on a global basis constitutes an agent of environmental change. Furthermore in response to dwindling forest resources, many nations have afforestation programmes on both small and large scales. In some cases, as in the United Kingdom, afforestation is widespread and like agriculture the rationale behind it is based on economics, not ecology. Like forest manipulation, afforestation modifies the environment and must therefore be regarded as a significant agent of environmental change.

Table 9.1 gives wood and paper production figures for selected developed countries. These are the major wood producers in the developed world with Canada and the United States being particularly significant. After a precarious beginning in the colonial period (Section 5.8), US and Canadian forestry is now practised on a near sustainable basis involving a wide range of silvicultural techniques. The transition from

unsustainable to sustainable resource use involves a change from exploitation to careful management. This is essential if renewable resources such as forests are to be maintained in a state that is conducive to future use. Although exploitation and management are both agents of environmental change, management is conservational because the objective is optimum return on a long-term basis. In contrast, exploitation tends to involve the rapid diminution of resources, often to such an extent that the capacity for renewal is destroyed and optimum returns are short-lived. Good management is more or less synonymous with sustainable development. However, the transition from exploitation to management is underpinned by a change in economic and social attitudes, from those based on short-term high returns for a few to those that recognise the need for sustainability and equitable resource exploitation in order to achieve long-term worthwhile returns (Toman and Ashton, 1996).

These transitions occurred in North American forestry in the period between early colonial occupation in the seventeenth century and 1920 (Williams, 1989). As Figure 9.1 shows, the forest regions of North America are both extensive and diverse, which makes it difficult to offer a generalised historical perspective. What is clear, however, is that the early European colonists entered a land in which Native Americans had already influenced forest character (Section 5.8). In

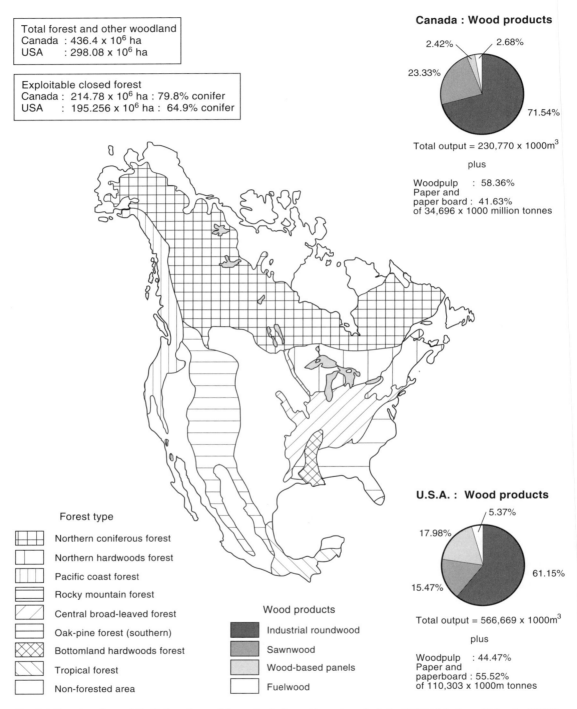

Fig. 9.1 Forest regions of North America and the output of wood-based products in 1985 (data from Richards, 1987).

addition, the temporal basis of North American forest use has varied in relation to the extension of early colonial settlers across the continent, but as Table 9.2 illustrates, the turning-point in US forest history came in 1891 with the General Revision Act. At this time, there were fears of a timber famine because earlier laws had failed to redress the impact of lumber companies and agricultural development; and it was becoming increasingly apparent that public lands were being sacked of their natural timber resources. This Act, however, allowed the establishment of forest reserves and marked a change in attitude to the nation's forests, hitherto considered inexhaustible. This change from exploitation to conservation was undoubtedly influenced by growing industrialisation and urbanisation and by the recognition of the need to manage catchments, water supplies, etc. Nevertheless, the prime objective of forest management continues to be timber production for a variety of purposes.

As Mather (1990) observes, the forest area of the United States has remained roughly the same since ca. 1920. This is a response to sustainable management policies which encourage growth and entail the maintenance of forests, not at optimum timber volume but at optimum growth rates. The adoption of such policies at a critical time in forest resource use averted what politicians like President Theodore Roosevelt described as a potential timber famine. Forest policies, securely ensconced in an effective and enforceable legislative infrastructure, inaugurated in 1891 and subsequently reinforced by later statutes (Table 9.2), were essential controls on the transposition of natural forests into managed forests. As in any stable, mature ecosystem, losses via the death of organisms, in this case trees, will be balanced by new growth. Tree removal in such a system will inevitably lead to forest demise, as occurred in American history during the period 1607–1900, but if the ecological system is carefully managed there are large-scale opportunities to reap a timber harvest without jeopardising sustainability. This depends on the relationship between annual growth and annual harvest; maximum tree growth is attained before old age. Thus, if a forest is managed to achieve maximum growth rather than maximum timber volume, a characteristic of mature climax forest, it can both thrive and produce a substantial harvest in the long term (provided no exogenous effects such as climatic change occur and as long as soil nutrient status is maintained).

Today the US national forest system comprises 95×10^6 ha of forest land, approximately 25 per cent of the national total; the remaining area is privately owned. Hurt (1994), reviews Forest Service management since 1945. Canada, on the other hand, has 94 per cent of its forest area in public ownership. As Repetto (1988) has pointed out, much of the US national forest consists of old-growth timber and is often in less accessible areas; this contrasts with many privately owned forests, where productivity is higher and there is greater accessibility. This is partly a result of private acquisition of forest lands before the General Revision Act of 1891, when it made more economic sense to annex high-quality, easily accessible forests. Nevertheless, the wilderness value of the national forests and the adoption of multiple-use policies by the US national forest system mean that conflicts of interest continue to arise because actual timber production is often uneconomic (Table 9.2). Consequently, many conservationists have argued against the commercial production of timber, proposing instead that forest areas should be maintained as wilderness for conservation and recreation purposes. Most of the nationally owned forest is in the western United States and is dominated by softwood. In contrast, almost 90 per cent of private commercial forests are in the east, where hardwoods dominate (Fig. 9.1). Today the United States is one of the largest producers of wood and wood products (Table 9.1). Production is centred on the Pacific northwest and the south; the various products are given in Figure 9.1. The significance of forestry in the United States is reflected in the amount spent on research and development for forestry and forest products. The total current expenditure is 1.3×10^9, of which ca. 885×10^6 derives from the private sector, ca. 820×10^6 for forest products (Ellefson and Ek, 1996). This investment is likely to increase as the range of uses for forest products and services expands, including environmental protection.

Timber production is maintained by a variety of practices which, since colonial times, have transformed the character of North American forests. Although extensive tracts of virgin forest remain, from which sawnwood and plywood are derived, there has been a growing trend since the Second World War towards fibre-based products from plantations and second- and third-growth forests. Silvicultural practices are designed to achieve sustained yield involving optimum growth

Table 9.2 Development of forest policy in the United States

Pre-1607 North Americans	Species selection in some areas. Hunting, gathering in all forest regions. Clearance for cultivation in some forest regions.
1607–1783 Colonial settlers	Exploitation of eastern US forests for timber, especially for local fuel and building (Section 5.8) as well as colonial government's (UK) requirements for shipbuilding. Overall control of policy lay with local administrators and the dictats received from colonial masters.
1783–1830 Post-independence	The Treaty of Paris in 1783 recognised the independence of the United States and the control of forest lands, generally considered to be inexhaustible, passed to the Confederate government but hostilities against the British continued. The war of 1812, for example, put new pressures on the forest resources due to army requirements and especially the need to form a strong navy. By 1817, due to depletion of ships, Congress instituted a law to reserve public-domain lands supporting oak and cedar, but enforcement was difficult and illegal cutting continued. Nevertheless, a precedent was set giving Congress the right to control the use of public lands.
1830–1891	The peopling of the continent continued and the demand for land, as setlement proceeded westward, resulted in the massive transfer of land from public to private ownership. Pressures on the forest resource reached a peak as railways were constructed and unforested areas such as the Great Plains placed demands on the nation's forests for building materials. The Timber Culture Act of 1873 did, however, require settlers in unforested areas to plant trees. By 1878 policy makers realised that agriculture was not the only economic activity that could be supported by the land and turned their attentions to the potential for timber production, but exploitation continued on a large scale. The Timber Trespass Law of 1831, however, halted much of this misuse of the public domain by lumber companies and individuals. Fines and imprisonment were imposed on those who cut timber without authorisation. This marked a change in attitude from one of exploitation to one of conservation. By 1867 legislation was being enacted in numerous states to preserve forest lands and the scientific management of resources began.
1891–1911	Predictions of a timber famine were being made, based on the rapid forest demise of the earlier century. In 1891 the General Revision Act was passed, which under Section 24 (the Forest Reserve Act), gave the President the authority to set aside forest reserves from the public domain. This was the basis on which the US system of national forests was established, beginning with the designation of Yellowstone Park in 1891. By 1907 the number of reserves had increased to 159.
1911–1952	Conflicts arose over the demands of the military during both World Wars. Nevertheless, conservation interests were well served with the establishment of a public domain and the establishment of a Civilian Conservation Corps in 1933. During the Great Depression, the corps deflected labour into forest management.
1952–1980	Numerous important policies were developed during this period relating to environmental protection in general but which led to improved forest management. They include policies on water quality, endangered species, pesticide control and pollution. Although timber production is still the main goal of US forestry, multiple-use policies (including recreation and water management) have led to balanced resource management strategies.
1980–1990	Conflicts of interest arose between groups involved in multiple use, and the Forestry Service received much criticism for its organisation of logging. After identifying stands ready for harvesting, they were leased to private companies to undertake the logging. Much of the wood was exported, often to Japan, at a profit to the company but at a loss to the Forest Service and hence the nation's taxpayers.
1990s	Promotion to the 'New Forestry' by the Society of American Foresters. This follows criticism of the Forest Service for allowing aggressive logging in remaining old-growth forest, e.g. in Oregon and Washington. The 'New Forestry' involves the promotion and protection of forest diversity, harvesting less often and ensuring the presence of wide buffer zones, and encouraging such practices in the private sector as well as the public sector.

Source: based on Bonnicksen (1982) with additions from Nebel and Wright (1996)

and regeneration; the ideal is to maintain equal numbers of trees in each age class, but this is rarely achieved *in toto* (Smith, 1986), especially in old-growth or second-growth forests, where age-class distribution is naturally disposed to older trees. Redressing the balance through thinning can thus take several decades without significant financial return. As a result, most management techniques in

Table 9.3 Silvicultural techniques

1. Intermediate techniques between establishment of seedlings and the harvest

A. The regulation of species composition and growth rates
(a) Release cuttings: the removal of larger trees of competing, undesirable species in stands of saplings or seedlings; this may involve the use of herbicides.
(b) Improvement cuts: the removal of diseased or poor-performing species at the post-sapling stage to improve quality.
(c) Thinning: the density of a stand may be reduced in order to promote growth of the remaining trees; this can be achieved by low/high or mechanical thinning, respectively through cutting lower/higher canopy species or the clearance of strips; thinning may be undertaken periodically, e.g. every 10 years.
(d) Fertilising: this is important to promote growth productivity where some nutrients are deficient, e.g. nitrogen.

2. Prediction of forest growth

Prediction of growth helps determine the volume of the annual cut. Forest-growth models are widely used and are based on observable relationships between growth, age, site, quality and competition.

3. Regeneration

A. Natural regeneration
Natural regeneration occurs in a variety of ways, including advance regeneration. This includes seedlings and saplings that were established before harvesting, sprouting from stumps and regeneration from the seed bank. How effective this is, and how rapidly it occurs, depends on a variety of factors including the quantity and regularity of seed production, germination success, local microclimate, competition and predation by insects and mammals.

B. Artificial regeneration
Artificial regeneration is more manageable and predictable than reliance on the natural process; it also ensures the regeneration of desired species in the desired mix.

(a) Direct seeding by hand or mechanically: this is especially appropriate where it is necessary to seed large areas rapidly and is most successful where logging debris is thin or absent
(b) Planting: this is more effective than direct seeding but is more expensive because of the extra costs involved in the nursery production of seedlings.

4. Harvesting methods depend on the nature of the stands

A. Even-aged stands
(a) Clear-cutting: all trees are removed from a given area in a relatively short time but it is no longer a favoured technique because of adverse aesthetic and environmental impacts such as the acceleration of soil erosion.
(b) Seed-tree methods: these involve the maintenance of scattered mature trees to act as a seed source.
(c) Shelterwood: trees are maintained to provide seed and shelter for new seedlings; once the seedlings are established, the older trees can be removed.
(d) Coppicing: this involves the removal of the upper portions of mature trees so that subsequent sprouts from the old bole can be harvested; it is not widely used.

Using these methods in even-aged stands it is only possible to approach sustained yield by cutting a small proportion of a forest annually; over a period of 50–100 years the entire forest is cut

B. Uneven-aged stands
Uneven-aged stands are managed by selective cutting that involves the harvesting of scattered trees or groups of trees. This approach ensures the stands are maintained and disturbance to the ecosystem is minimal.

Source: based on Lorimer (1982) and Smith (1986)

North American forests approach, rather than achieve, a situation of sustained yield. In general, silvicultural techniques include the regulation of regeneration, species composition and growth; they may be designed to enhance aspects of forestry unrelated to timber production, e.g. improvement of habitats for wildlife. Effective management requires accurate timber inventories, data on productivity and predictions of future growth; it is crucial to determine the age of a stand at harvest, known as

the rotation age. This figure that depends on the species and the time required for optimum growth.

Table 9.3 gives the silvicultural techniques that can improve productivity; all of them influence the structure of forests and are thus agents of environmental change. There are also indirect impacts of silviculture which relate not only to forest management but to the construction of roads and the effect of forest removal on watershed management. The removal of trees constitutes the removal of biomass and nutrients, so it is important for sustainable yield that logging rotations are sufficiently long to allow replenishment. Table 9.3 also shows that herbicides and fertilisers are frequently used in silvicultural practices; these chemicals can lead to a decline in water quality and to declines in aquatic fauna and flora. Road construction can also lead to soil erosion, sometimes exacerbated by the dragging or skidding of logs from the place of cutting to areas where they are loaded onto vehicles for transport. With careful management, however, these problems can be kept to a minimum.

All these practices contribute to environmental change at local and regional scales and have implications for biodiversity. This is illustrated by two studies on the insect faunas of Canadian old-growth (i.e. unlogged/virgin) forest and unlogged spruce forests in northern Sweden. The Canadian study by Spence *et al.* (1996) has shown that nine carabid beetle species are common in pristine forests of montane lodgepole pine, but they are rare or absent from nearby sites undergoing regeneration after harvesting some 27 years ago. This highlights the need for the preservation of uncut stands from which specialist carabids of old-growth forests can eventually expand, as exemplified by the presence of some old-growth carabids in aspen stands that have developed after firing of lodgepole pine. Similar results come from Petterson's (1996) study of spider biodiversity in natural spruce and selectively logged spruce in northern Sweden. The increased species richness of undisturbed spruce stands may reflect a higher diversity of habitats when compared with managed stands, the main difference being lichen encrustation and larger branch size in the undisturbed stands. Selective logging also has an impact on the range of species in the ground flora; increased light availability tends to give a temporary increase in the range of species until succession proceeds. Holien (1996) has also demonstrated that management practices influence the distribution of calcicole crustose lichens in spruce forest in central Norway. Holien's analysis of 100 forest patches in Sor-Trondelag in relation to a range of site and stand characteristics shows that species diversity is higher in old forests and high-altitude forests when compared with those of a young age and at low altitudes. In particular, the presence of old trees and stumps encourages high species diversity.

Afforestation (including reforestation) also brings about environmental change, as illustrated by many site-specific studies undertaken in Europe. Afforestation is widespread; in the Netherlands forest and woodland has increased in extent by nearly 16 per cent since 1981–83, nearly 13 per cent in Switzerland and 12 per cent in the United Kingdom (World Resources Institute, 1996). Afforestation has also occurred on a significant scale over the past decade in Australia, the United States and Canada. There are many reasons for this trend. Firstly, there is a ready market for wood and wood products; secondly, there is a growing recognition that forests and woodlands protect against soil erosion and provide environmental services as well as offering wildlife habitats and recreation facilities; and thirdly, forests provide a sink for carbon dioxide generated from fossil-fuel combustion as well as providing an economic return. The role of afforestation and forest management in carbon sequestration has been discussed by Adger and Brown (1994) and in Apps and Price (1996). In this context afforestation contributes to global environmental regulation through its role as a negative feedback control on atmospheric composition. Mather (1993a) reports that afforestation has caused forest cover in much of western Europe to increase by 2–4 per cent, with an average of 2 per cent for Europe overall. In the United Kingdom and Ireland afforestation respectively amounts to 40 000 and 9000 hectares annually, though in both countries there are concerns about its adverse environmental impact.

A historical perspective is required to give a full understanding of forestry's role as an agent of environmental change in Britain; this is given in Table 9.4, which highlights the role of both the public and private forestry sectors. This forest policy has brought considerable changes in the character of British upland landscapes and their wildlife. For example, the majority of trees planted by the Forestry Commission (FC) in Scotland and Wales are non-native conifers, notably Sitka spruce (*Picea sitchensis*), Norway spruce (*P. abies*) and lodgepole pine (*Pinus contorta*). These species were originally chosen for afforestation programmes because they grow relatively quickly and can tolerate the poor soils that are characteristic of British uplands. Apart from aesthetic changes, involving the blanketing of large areas with uniform stands planted in regular rows, the character of these forests is quite different from ancient woodland. Poor light penetration means that understorey species diversity is much reduced,

Table 9.4 Major developments and policies in UK forestry

Before 1914	Tree planting on private estates; small-scale commercial growing of European larch (*Larix decidua*) and Norway spruce (*Picea abies*); no overall institutional control but the establishment of colonial forest organisation (notably Dehra Dun in India) led to the introduction of forestry courses at Oxford University. The early twentieth-century recommendations, relating to plantation forestry, by these new foresters were largely ignored by the government.
1919–1939	The First World War highlighted the dearth of UK forest resources as the need for timber escalated. Government intervention of felling practices on private estates ensued and a policy of afforestation with the establishment of the Forestry Commission (FC) in 1919. Financial constraints limited the FC to the purchase of marginal land, mainly in the uplands where there was opposition from hill farmers. Grants were also made available for planting by individuals, covering 25 per cent of costs, though such concessions were not taken advantage of on any large scale. In 1992 the Geddes Committee on National Expenditure reported that the FC's policies were uneconomic and it narrowly escaped abolition.
1939–1957	At the start of the Second World War, the UK's forest resources were still in a parlous state and once again necessitated government intervention in the running of private estates. Felling licences were introduced to restrict harvesting for non-strategic purposes. New policies were instituted in 1943 and adopted by the postwar government. The area to be afforested was increased to 1.2 million hectares and a further 0.8 million hectares were to be created by restocking existing woodlands but little consideration was given to environmental factors or to criteria for the selection of existing woods for 'improvement'. The private sector also became increasingly involved in afforestation as high income tax rates in postwar Britain made the already available tax incentives more attractive. Moreover, in the 1940s legislation relating to conservation (Section 4.3.3.) excluded both forestry and agriculture from rigid controls. The 1947 Agriculture Act, which guaranteed prices via deficiency payments, also ensured that afforestation was relegated to the poorest areas, inevitably in the uplands.
1957–1979	Afforestation for strategic purposes thus became incidental to a forestry industry, the case for which was rationalised by the Zuckerman Committee of 1957. This emphasised the role of forestry in maintaining the rural economy through employment opportunities in depressed areas, though integration between forestry and agriculture remained elusive due to separate ownership. Various administrative reorganisations in the FC occurred during the 1960s but it continued to operate on revenue principles rather than on profit maximisation and benefited from government funding. Despite the fact that forestry, as a nationalised industry, did not meet government criteria in terms of rates of financial return it was considered to provide a better social benefit than hill farming, despite fluctuations in labour demands relating to planting–cutting rotation cycle. The 1960s also witnessed increasing opposition to forestry schemes by conservationists which led, for the first time, to official recognition that landscaping advice was needed, albeit subsidiary to timber production. This was promoted by increasing concern about the effect of afforestation on water quality. Although a government cost-benefit study in 1972 highlighted poor social and economic returns, forestry policies remained intact. The replacement of estate duty with capital transfer tax provided even more attractive fiscal conditions for the transfer of wealth from one generation to the next. Declining rates of private afforestation in the mid-1970s were at least partially due to more rigorous, though still legally unenforceable, consultation procedures required by the FC. This decline was halted in 1977 as grants were raised and foresty land was accorded the same privileges as farmland in relation to capital transfer tax. By this time the 1943 target of ca. 2 million hectares had been achieved and justification for a further 1–1.8 million hectares couched in terms of import saving by a Royal Commission report in 1977
1980–1987	Forestry policies continued intact; rates of afforestation were to be maintained on the basis of import savings but were to be encouraged by the private rather than the nationalised sector. FC approval was no longer necessary and some FC forests were sold to private enterprises. As a result, planting rates in the private sector doubled. Of the 25 000 hectares afforested in 1983–84, two-thirds were due to private investors, who by this time had become heavily involved in the afforestation of bare ground rather than the restocking of existing woodlands. High-rate taxpayers thus continued to benefit from various mid-1980s reports that the exchequer was losing vast sums as a result and that returns for the sale of FC forests was not as high as might be expected. No policy changes followed. The Wildlife and Countryside Act of 1981 did little to improve the relationship between forestry and conservation: compensation payments were made available by the NCC to prevent the afforestation of SSSIs and to promote the designation of new SSSIs. By this time attention was being drawn to the fact that UK forestry policies had already resulted in a considerable loss of upland habitats. The forestry versus conservation debate became just as vehement as the agriculture versus conservation debate. In 1985 the FC modified its policies as a result, too late of course to retrieve lost sites of ancient woodland. These new moves involved higher grants for the planting regeneration of broadleaved species and the removal of grants for the replacement of broadleaved woods by conifers. A 1985 amendment to the Wildlife and Countryside Act also charges the FC with conservation duties; its effects are not yet to be seen. Moreover, the role of the private sector in forestry, as discussed in the Commons Public Accounts Committee in 1927, is no longer considered as socially or economically beneficial to the nation in general.
After 1987	In the wake of CAP reorganisation (Section 6.2.3) as a result of agricultural surpluses, it could be argued that forestry has gained a new impetus. The extensification of agriculture may create increased afforestation of land that is marginal for agriculture. On the other hand, fiscal changes in the 1988 budget mean that forestry is no longer as financially advantageous to individual high-rate tax payers. Environmental concerns continued to be voiced in the acidification of lakes and streams.

Source: based on Stewart (1987)

especially under dense stands; the ground flora species tend to be acidophilous and the bird species vary as the forest ages. Species characteristic of open country are initially replaced by those which favour the limited cover provided by young trees; these in turn are replaced by forest dwellers as the forest matures (Bibby, 1987). Lavers and Haines-Young (1993) have analysed the impact of afforestation on upland birds in Britain; they believe that general conclusions are difficult to draw because of the individualistic response of species to afforestation and the characteristics of the afforestation, i.e. extent, character and management. This is also highlighted by the results of Newton *et al.* (1996), whose work on the red kite populations in Wales has shown that afforestation has not had a major impact, though only 16 per cent of the kites' territory has been afforested. However, further afforestation may indeed have a detrimental affect especially if it were to occur at the expense of oak and other hardwoods favoured for nesting, a prime factor in bird distribution. Insect biodiversity may decline after afforestation, as Butterfield *et al.* (1995) have shown for afforested grassland and moorland habitats in northern England. These have lower biodiversity than unafforested areas, mainly because of loss of wetland and open-habitat species. Afforestation may also have an effect on the feeding habits of lagomorph species. Using radio-tracking, Hulbert *et al.* (1996) have shown that female mountain hares prefer pasture and tree plantations more than moorlands. For both hares and rabbits the grazing of pine plantations declined as the plantations aged.

Apart from the effects of afforestation on flora and fauna, there are other environmental implications. For example, the establishment of plantations in British moorlands has affected soils and soil processes as well as the quality and quantity of water yield. The impact of afforestation on soils has varied in relation to soil and substrate types, species composition of the plantation and management, including land preparation techniques. Ploughing, commonly used to prepare upland sites for afforestation, dries out the soil and causes more nutrients to be lost compared with undisturbed habitats. Such losses decline as the plantation matures but they rarely return to predisturbance levels because subsoil is exposed and weathering is promoted. Moreover, the construction of drains tends to reduce the residence time of water in the substrate and aeration is promoted by land preparation techniques; aeration also increases the rate of decomposition of organic matter and the release of nutrients. Where upland blanket peats or raised bogs are afforested, the increased aeration promotes the production of carbon dioxide from stored organic matter; this enters the atmosphere and therefore detracts from the significance of afforestation as a sink for atmospheric carbon dioxide. Few carbon budgets have been determined to establish the dynamics of the situation, though overall it is likely that afforestation results in a carbon gain in the biosphere, at least over decades. What is more at issue, notably in the United Kingdom, is the loss of peatland habitats, especially raised bogs. Forestry practices and timber transport require an expenditure of fossil fuel. In terms of global warming potential (Section 6.5), Karjalainen and Asikainen (1996) have calculated a carbon balance for forestry operations in Finland. On a 20 year timescale the warming effects amount to 1.31×10^6 t (carbon dioxide equivalent). This figure is small when compared with the 30.3×10^6 t in harvested timber.

It has been established that in some regions afforestation can cause acidification of soils and water. Indeed it was long considered that the acidification of many lochs in southwest Scotland was a response to afforestation, though subsequent research on fossil diatoms has shown that sulphurous and nitrous emissions from fossil-fuel burning are the major culprits (Section 6.6). Nevertheless, Rees and Ribbens (1995) have shown that problems of acidification in the Galloway Hills in southwest Scotland have been exacerbated by afforestation. For example, in catchments that have been entirely afforested, the pH of drainage water was 0.7 unit lower than for drainage water from unafforested areas. Rees and Ribbens have also shown that drainage from afforested catchments is associated with low concentrations of calcium, and increased concentrations of aluminium and total organic carbon. The adverse impact of high aluminium concentrations on fish populations has already been referred to in Section 6.6; this problem is also reflected in the following study which showed that salmonid fish populations were reduced in waters of high acidity. The impact on land preparation techniques on drainage quality has been demonstrated by the work of J.D. Miller *et al.* (1996) at an experimental site in Rumster, Caithness, Scotland. Caithness is home to the UK's Flow Country, an extensive area of peatland designated as tundra and rich in wildlife; decisions to afforest substantial areas in the past decade have met widespread criticism. Miller *at al.* have shown that ploughing, to facilitate tree planting on deep peats, and draining, have caused the loss of nitrogenous and sulphurous compounds out of the peatland ecosystem into drainage waters. Within one year this was accompanied by loss of phosphates and potassium salts, originally applied as fertilisers to stimulate tree growth. Increased inputs of marine-derived aerosols occurred through their capture by the tree canopy. Afforestation can also influence water yield, as demonstrated by Sahia and Hall (1996) on the basis of data from the Netherlands. For example, a reduction of 10 per cent cover for coniferous, deciduous and eucalyptus species increased water yields to 20–25 mm, 17–19 mm and 6 mm respectively.

9.3 Forestry and afforestation in the developing world

Today most of the world's forests occur in the boreal or tropical zone (Fig. 6.4); the tropical zone accounts for ca. 50 per cent of the world's forest, including open woodlands as well as evergreen and deciduous moist forests. Almost all of these forests occur in the developing world. Vast areas remain relatively unexplored and, like all other forests, provide ecosystem services that involve biogeochemical cycles and the regulation of atmospheric composition. However, as discussed in Section 8.2 in relation to agriculture, one of the most important environmental issues is the loss of biodiversity, especially in the developing world. The pressures for forest degradation and clearance are considerable. They include the need for fuelwood and logging as well as agriculture. Logging may be undertaken on a sustainable or unsustainable basis and only relatively recently have unsuitable practices become more widespread. Exploitation of tropical forest is also caused, directly and indirectly, by factors emanating from the developed world, notably the high demand for wood and wood products, especially from nations like Japan that have limited forest resources, and because of the need to generate foreign currency to service debts and development projects. Moreover, afforestation programmes are relatively recent innovations in the developing world but they are increasing for a variety of reasons. These include concerns about diminishing forest resources, the mitigation of problems such as soil erosion and desertification (Sections 7.3 and 7.4) and the sequestration of carbon to help counteract industrial development. Both forestry and afforestation bring about environmental change in the developed world for all the same reasons they cause environmental change in the developing world.

Forestry practices are diverse and complex, and even where there are national forest management policies they are often difficult to enforce. Moreover, forest management is not always ecologically sound and is often focused on short-term gains rather than long-term sustainable yield. This is probably comparable with pre-nineteenth century exploitation of forests in the temperate zone. There are many factors involved in the exploitation of forests in the developing world (Section 8.2). Multinational logging companies must take their share of the blame for poor forest management strategies in many developing countries, where little heed is paid to conserving the remaining forest or to the effect of logging and heavy machinery on the soils. Some of these problems arise because there is inadequate information about terrain conditions, species composition, growth rates and tree life cycles, much of which stems from poor planning.

Many of these problems are exemplified by the state of forestry in Nigeria, the subject of a review by Adegbehin (1988). Forest reserves occupy some 9.7 per cent of Nigeria's land area and are most abundant in the lowland rain forest belt in the southern part of the country. These reserves produce most of Nigeria's industrial roundwood, the raw material for various wood-based industries, notably plywood, particle board and match production as well as pulp and paper. This industrial wood accounts for 20 per cent of Nigeria's wood consumption; the other 80 per cent is consumed as fuelwood mainly derived from the more northerly savanna zones. There are now widespread shortages of wood for both purposes, not least because many rainforest areas outside the reserves have been exhausted, areas which supplied about 50 per cent of the country's timber output up until 1960. Deforestation and agricultural encroachment are among the reasons for this, but forest management policies have also been inadequate. As Adegbehin points out, assessment of the resource base itself has been poor and existing inventory data are both incomplete and out of date. In addition, the volume of timber removed in logging operations is either not recorded or inaccurate. Thus there is no closely managed relationship between potential supply and its exploitation. Indeed the available data suggest that the annual rate of roundwood removal, currently ca. $6.75 \times 10^6 \, m^3$, will result in depletion within 15 years despite the initiation of plantation forestry during the 1960s.

Much of this overexploitation is due to poor institutional control. For example, the government, via State Forestry Services and the Federal Department of Forestry, controls and implements regeneration and afforestation schemes but logging is transacted via the private sector with few enforceable controls. This lack of control is illustrated by the fact that until recently operating licences were issued more or less indiscriminately to any company that wanted to set up a sawmill. Such an ad hoc approach is not conducive to efficient use of present resources nor is it appropriate for long-term planning. Already there are sawmills that cannot operate at full capacity because of wood shortages. Improvements in the overall forestry infrastructure, from inventory acquisition through to logging, would at least go some way towards improving resource use. Other mitigating measures could involve the encouragement of private concerns to develop their own plantations; this already occurs on a small scale not only in rainforest regions but also in savanna regions, where *Eucalyptus* is being successfully cultivated to provide fuelwood and poles. The fuelwood shortage in these areas could also be alleviated by the introduction of more efficient wood-burning stoves, and industrial use of wood could be improved by minimising waste. However, eucalyptus may not prove so beneficial in the long term; although its productivity is high, it may reduce the growth of other species in its vicinity, as has been found in many Andean countries of South America.

Poor infrastructure in Nigerian forestry is not the only factor that has transformed the forest landscape. Silvicultural management has been practised since the 1920s, beginning with regulation of the felling cycle then the stimulation of natural regeneration by cutting climbers, removing uneconomic species and canopy thinning in the 1930s, as described by Kio and Ekwebelam (1987). None of these techniques have been particularly successful and in 1944 the tropical shelterwood system (TSS) was introduced, based on management techniques that had already proven successful in parts of the tropical forests of Malaysia. The TSS involved similar techniques to those described above but they were more aggressively applied with specific time periods between each operation. Nevertheless, there was insufficient regeneration of desired species, largely because of the different life cycles, fruiting cycles and density of desired Nigerian species in comparison with the Malaysian forest species on which the methodology was based. (This is another example of the inappropriateness of transposing management and land-use practices from one discrete ecological system to another.) Other problems were also experienced, including the damage of saplings during the logging phase of TSS, which opened the habitat to colonisation by secondary species, and the availability of light as the canopy was opened up encouraged the growth of unwanted climbers and weeds. Thus the lack of fundamental ecological knowledge has played a significant role in the shaping of these forests, and lack of foresight has led to the demise, by deliberate poisoning and girdling, of many species which are now considered commercially viable. TSS was eventually abandoned and subsequent schemes to encourage restocking by enrichment planting of desired species has also been unsuccessful, not only in Nigeria but also in Ghana (Asabere, 1987) and west Africa (Nwoboshi, 1987). In all of these cases the authors recommend that the best way of ensuring future timber resources is to organise forestry on a plantation basis; this will necessitate massive afforestation schemes to combat the losses of natural forest and appropriate infrastructure to ensure survival. As shown in Figure 8.1, current rates of deforestation indicate that Nigeria is likely to lose most of its rainforest by the year 2000, and with a rapidly growing population it needs an immediate solution.

Forest industries also provide a major source of foreign income in Malaysia. According to Tang (1987), the forest sector contributed 14.1 per cent of the total export earnings in 1983, equivalent to 6.8 per cent of the gross national product (GNP). Extensive logging, together with the expansion of agricultural land, will deplete Malaysia's forest resources by the turn of the century unless effective management programmes are established. To this end, Tang reports that some 13.8×10^6 ha of Malaysia's forests have been designated as permanent forest estate (PFE), of which 9.1×10^6 ha are productive forests intended to be managed on the basis of sustained yield. Whether this can be achieved is debatable because the management schemes implemented to date have not been uniformly successful due to the tremendous variation in ecology and terrain. Unless effective schemes can be designed, Malaysia's timber industry may depend on the establishment of plantations, a possibility that has been suggested as the saviour of African forestry (see above). Although the efficient exploitation of Malaysia's forests is beset by problems similar to those of Nigeria, there have been attempts to implement governmental control. This includes governmental allocation of lands for agricultural development as well as controls on wood-harvest volumes. However, these controls are not always strictly implemented and logging operations have sometimes been sanctioned in direct conflict with the needs of indigenous groups. Other studies from Malaysia illustrate the more general effects of logging practices in tropical regions, showing how the destructive effect is far greater than just the removal of trees. Johns (1988) has examined the impact of logging in a part of west Malaysia where the harvest consists of 3.3 per cent of the trees per hectare but which, by damaging saplings and soils, causes an overall loss of nearly 51 per cent of trees. Other indirect effects include the way in which hitherto inaccessible forests are opened up to cultivators following the creation of forest roads; this results in even greater forest destruction, often indiscriminately (Section 8.2). Bussman (1996) has also highlighted the alteration that is taking place in Mount Kenya's forests where there is large-scale selective logging of east African camphor (*Ocotea usambarensis*) for its termite-resistant hardwood. This species is widely dispersed, rarely occurring in pure stands, so logging is difficult. It also means the impact of logging is extensive due to the provision of many roads and tracks. In addition, the cultivation of marihuana is widespread on the southeastern slopes of Mount Kenya, where forest must first be destroyed by slash and burn techniques, and in the cedar forests much illegal logging occurs. All these activities, coupled with poor or non-existent management and lack of implementation of official forest policies, are creating substantial changes in forest and leading to degradation.

More covert reasons may also underlie the inability (or reticence) of government forest agencies in developing countries to enforce regulations. These include the need for foreign currency and the

attraction of foreign companies and their interests in developing economies. Such pressures may be offset by the newly established International Tropical Timber Organisation (ITTO), whose remit is to administer the International Tropical Timber Agreement. As Mather (1990) reports, there are 36 countries producing tropical timber and 33 market countries which are party to this agreement, an agreement which not only aims to regulate the tropical timber trade but to promote sustainable use and conservation of tropical forests. How conservation will be achieved remains to be seen, but it has been noted that a surcharge should be levied on tropical timber, to be paid by the market, which could be used to undertake research into ways of achieving and ensuring tropical forest survival. The ITTO and similar organisations may well become agents of environmental change in the not too distant future.

There is also the corollary that the economic significance of tropical timber will ensure its survival. The economic significance of British woodlands throughout history (Sections 4.6, 5.2 and 5.3) did not, however, ensure their survival to any great extent, nor is there any guarantee that tropical forest plantations will succeed, though as Mather (1990) points out, their prospects are promising. Other factors may also contribute to the survival or demise of tropical forests. For example, will developing nations be able to reduce their reliance on forest resources as other aspects of their industrial economy begin to achieve importance? Mather draws some interesting parallels between the development of agricultural systems and forestry. He believes that the direct exploitation of natural forest, as in many tropical regions, is equivalent to 'hunting–gathering' whereas the managed natural forests of North America, for example, have reached the 'farming' stage. He states that 'if a similar stage can be quickly reached in tropical LDCs (Least Developed Countries), and if plantation forests can be established without serious problems of ecological sustainability or pest outbreaks, then prospects for future timber availability are brighter than the currently rapid rates of deforestation in any tropical countries would at first suggest'. Is there also the possibility that the environmental movement will reduce the demand for such forest products in the developed nations by educating the public on the undesirable environmental consequences of over-exploitation (e.g. the enhanced greenhouse effect and the loss of genetic resources)? Overall the outlook for the sustainable use of tropical forests is not good, despite increasing conservation concerns and widespread government controls. Wood (1990) points out that today only 14 nations house 80 per cent of the world's remaining tropical forests; all of them have high population growth, increasing pressure on existing agricultural systems and substantial foreign debts. Although there is an increasing trend towards plantation forestry, Wood indicates that less than 0.1 per cent of tropical forest is managed for sustainable development.

Although afforestation does not compensate for the loss of natural forest, because of loss of biodiversity and heterogeneity and possible adverse impacts on soils and drainage systems (Section 9.2), the establishment of plantation forestry in the developing world has many advantages. Firstly, it provides wood and wood products for indigenous people as well as providing goods for export; secondly, it provides more effective environmental services than grasslands or scrub, notably soil protection, water regulation and biogeochemical cycling, including carbon storage. Moreover agroforestry, which involves the retention or introduction of trees or shrubs in crop and/or animal production systems, contributes to multicropping, varied products and environmental protection. Agroforestry systems have been described in detail by Nair (1993) and have been examined in Sections 8.3 and 8.4 in relation to the protection of land from soil erosion and desertification. Consequently, they will not be considered here. Changes in the extent of forest plantations in developing countries have been discussed by the World Resources Institute (1996) which states 'FAO estimates that during the 1980s total plantation cover almost doubled in developing countries. The total area planted, however, was only one fifth of the total area of natural forest converted to other uses'. Thus, despite an increase in plantation forestry there remains a considerable overall loss of forest cover in developing countries. The data quoted by the World Resources Institute also indicate that plantation cover in developing countries increased by 88 per cent between 1980 and 1990, a rate much higher than in developed countries (Section 9.2). Table 9.5 gives data for changes in the extent of plantation forestry in the developing world between 1980 and 1990. It shows that the most significant afforestation has occurred in China where 31.83×10^9 ha were afforested, representing an increase of 5.58 per cent per year.

The World Resources Institute (1996) points out that when deforestation is taken into account, this massive afforestation has given rise to a net gain of forest in the Asia–Pacific region. India is another country in which extensive plantation programmes are being implemented.

According to Stewart (1993), the main reason for afforestation in Algeria was the establishment in the mid-1970s of a scheme described as 'a green dam to hold back the Sahara'. This proved to be overambitious and eventually failed. Moreover, Stewart believes that the whole idea was poorly founded, mainly because it deflected effort from the conservation of existing natural forests and because the afforestation, on the southern slopes of the Saharan Atlas, Hodna and Aurés mountains, was too far from the main centres of rural population to provide substantial benefits for rural development. He also states that the plantations are still too young to produce economic quantities of wood. At least he should take heart that they are sequestering carbon! Instead of a 'green dam' the Chinese are engaged in erecting a 'green wall' (Section 8.4). Smil (1993b) has outlined the recent developments in China's forest policy; a great deal of effort has been deployed since 1950 to afforest large areas of a country where there has been a long history of deforestation. Indeed Smil explains how 'official Chinese afforestation claims leave no doubt that since 1950 the country has been engaged in the biggest revegetation effort in human history'. These claims amount to 104×10^6 ha of new plantings between 1949 and 1979, i.e. nearly 3.5×10^6 ha per year.

However, such claims have since been rescinded and the accepted figures are now ca. 30×10^6 ha. Most of this has involved plantation forests, i.e. 66 per cent, with a further 13 per cent in bamboo groves, 6 per cent in shelterbelts, 2 per cent in fuel lots and 13 per cent in timber stands for mixed use. These data represent successful, established afforestation. However, since 1979 China has been engaged in the planting of a Green Great Wall, the objective being to halt desertification in the northern provinces (Section 8.4). Some 8.5×10^6 ha had been planted by 1995. This plus two other large afforestation projects, a shelterbelt 10 km wide in Shandong and slope protection forests along the banks of the Yangtze River, should provide China with an additional 25×10^6 ha by the year 2000. However, planting success rates remain low, though they have risen to 40 per cent from 30 per cent in the days of Mao. According to Jiankun et al. (1996), China's total

Table 9.5 Changes in the extent of plantation forest

	1990 extent (10³ ha)	Percent annual change 1980–1990	Percent annual change in total extent of forest 1980–1990
Algeria	485	6.08	0.88
Senegal	112	118.00	0.52
Mali	14	143.85	0.79
Ethiopia	189	17.49	0.18
Sudan	203	7.58	0.99
Nigeria	151	3.23	0.68
Malawi	126	12.42	1.12
Tanzania	154	12.52	1.13
Madagascar	217	1.65	0.76
Africa[a]	**4 416**	**5.55**	**0.24**
China	31 831	5.58	−0.59
Syria	127	36.01	NA
Bangladesh	235	10.89	NA
India	13 230	32.13	NA
Vietnam	1 470	5.00	NA
Thailand	529	12.55	NA
Indonesia	6 125	11.82	NA
Philippines	203	0.00	NA
Asia[a]	**NA**	**NA**	**NA**
Costa Rica	28	132.86	NA
Honduras	3	101.11	NA
Mexico	109	9.29	NA
Central America and Mexico[a]	**192**	**16.13**	**0.50**
Brazil	4 900	6.63	NA
Paraguay	9	35.00	NA
Peru	184	9.21	NA
Venezuela	253	19.16	NA
Tropical South America[a]	**5 545**	**7.24**	**0.47**

Source: based on World Resources Institute (1996)
[a] The figure for the continent is greater than the column total because the table does not include every country
NA = not available

30×10^6 ha of planted forest, the area planted by 1990 when the last survey was undertaken, will increase to 40×10^6 ha by the year 2000, providing China with 15 per cent forest cover. They also advocate improved management, greater provision of fuelwood plots and conservation of natural and seminatural forest in order to enhance the ability of China's forests to absorb carbon. China's carbon dioxide emissions currently amount to 617.2×10^6 t, some 11.2 per cent of the global total. Indeed Xu (1995) has suggested that if afforestation were to be undertaken on all land available and managed on a permanent rotation basis, the forests would provide a sink for 9.7×10^9 t of carbon; this is equivalent to 16.3 times the total industrial carbon release in 1988, or approximately 15.7 times China's present industrial carbon production.

The issue of global warming has prompted much interest in ways of reducing atmospheric carbon dioxide concentrations, including the role of forest management and afforestation. For example, Brown (1996) has analysed the potential for carbon sequestration in Asia's forests which are currently considered net producers of carbon because of deforestation and degradation. She suggests there is a potential for the establishment of 133×10^6 ha of plantations, including agroforestry, which would provide a sink for 58 per cent of the carbon produced industrially in the region. The major contributor to this would be China as indicated by Xu (1995). Further substantial contributions would include those of India (Ravindranath and Somashekhar, 1995) and Thailand (Wang-wacharakul and Bowonwiwat, 1995). As Table 9.5

illustrates, both India and Thailand have already been increasing their extents of plantation forest. Proposals and prospects for southeast Asia have been discussed by Moura-Costa (1996), who highlights the need for protection of existing forests, including the improvement of logging strategies and afforestation. Examples of afforestation include degraded pasture and the rehabilitation of degraded forest. Appropriate management is required to enhance carbon assimilation; this applies to existing natural forest as well as plantation forest and may include various silvicultural techniques (Table 9.3), e.g. thinning programmes, weeding and nutrient addition (Hoen and Solberg, 1994). Enrichment planting may be practised in degraded forest, i.e. seedlings of required tree species can be planted in logged or disturbed forest then nurtured to ensure their survival to maturity.

Forward planning in relation to forestry and afforestation must also take into account the future needs of the people living in or around the forests and in urban areas, including the demand for wood, wood products and paper. Consumption of these commodities is increasing rapidly in the developing world, and as Dabas and Bhatia (1996) point out, it is becoming increasingly important for developing countries to establish their own production industries instead of relying on imports. Given the favourable growth conditions in tropical regions, tropical plantations are more efficient than temperate plantations at sequestering carbon; as well as providing an economic asset, they have considerable potential to mitigate the effects of global warming. This reflects an approach advocated by Kanowski et al. (1992), who point out that forestry and afforestation must be compatible with the needs of the people (see the earlier comments on Algerian reafforestation); this may involve agroforestry in order to generate food products as well as timber products and fuelwood. They believe that an essential aspect in the development of successful plantation forestry is 'the integration of agroforestry and plantation forestry methodologies and technologies, which requires both social and biological research'. Afforestation thus becomes an important vehicle for rural development; traditional monoculture, e.g. tropical eucalyptus plantations, is inappropriate. Eucalyptus plantations generate wood products but do not provide for other needs. Although there is a general consensus that afforestation in developing countries brings benefits on all scales, local, national and global, Houghton (1996b) warns that rapid global warming may mean that forest mortality and carbon release from soils, peatlands, etc., will negate the effects of forest management and/or make afforestation impossible. Conversely, the additional carbon dioxide generated by industrialisation may promote increased productivity (Section 6.5; Norby et al., 1996).

9.4   The impact of recreational activities

As the standard of living has improved in the developed world it has brought with it a higher disposable income and increased leisure time. The average working week in the United Kingdom is now about 37 hours as compared with 60 hours in 1900. In addition, paid holiday entitlements are generous, having increased from 2 weeks in 1963 to 4 weeks or longer since 1985. Car ownership is also widespread, giving increased access to the countryside, and greater affluence means that finance is available for leisure purposes. These trends are typical of westernised society. In these nations the countryside is increasingly being exploited as a source of recreation and as a venue for sporting events. Where the countryside or environmental features are the foci of interest, environmental change can occur through large influxes of people and the attendant services they require. The very landscape that provides the opportunities for leisure and sporting activities may itself be put at risk in the absence of careful management. The examples which follow illustrate a range of landscape changes created by leisure and sporting activities. Recreation in its broadest sense involves any pleasurable occupation of leisure time. Thus, many recreational activities do not impinge on the physical environment or landscape.

Trampling of vegetation is one of the most widespread environmentally degrading repercussions of recreation and it can lead to excessive soil erosion. Numerous studies have been undertaken in US national parks which illustrate the ecological impact of visitors on camp and picnic sites, nature trails and footpaths. In the Grand Canyon National Park, Arizona, Cole (1986) has investigated 12 high-use and 12 low-use campsites located away from the main tourist access routes in three desert vegetation types consisting of desert scrub, catclaw (*Acacia greggi*) and piñon–juniper (*Pinus edulie–Juniperus osteosperma*) communities. Soil compaction, changes in infiltration rates, soil organic matter, litter and soil moisture content as well as vegetation cover were recorded at all 24 sites. The data revealed that human impact was greatest with initial use; further increase in use levels exhibited a non-linear relationship, illustrating that once a campsite had been established, higher-use levels could be sustained without incurring disproportionate damage. The major discovery concerning vegetation related to campsite core size, which tended to be particularly significant in piñon–juniper areas, where the plant cover is open and the terrain less rugged. Recommended measures to combat this expansion include the planting of cacti in the core periphery and the on-site education of the public via publicity material emphasising the need for core confinement.

This and other studies (e.g. Stohlgren and Parsons, 1986) have led to the conclusion that the most appropriate management strategy for such wilderness areas is to restrict camping to a few designated sites, where activity can be sustained without excessive damage. Cole and Marion (1988) have also examined visitor impact in national parks on the comparatively moist east coast. Looking at three areas, they compared their results with the arid west coast to see whether the same conclusions would apply. All three sites, in the Delaware River catchment, are accessible only by boat; camping takes place in a variety of campsites including camp-grounds, designated primitive sites and undeveloped user-selected sites. In each area Cole and Marion recorded a range of soil and vegetation characteristics from both recreational and undisturbed control sites. Although variations were apparent from site to site in relation to pre-existing environmental characteristics and intensity of use, the study highlighted some general trends with implications for management strategies. Most significant is the fact that the soil and vegetation characteristics on low-use campsites were more similar to those of high-use sites than to control sites, with significant tree damage, reduced tree regeneration, changes in and disappearance of ground flora and compacted soils. User-selected sites were less heavily damaged than designated sites, which were almost completely devegetated, but discarded rubbish was more abundant; this suggests that the provision of disposal facilities at designated sites has proved worthwhile. The two studies, east coast and west coast, led to similar conclusions and

Cole and Marion recommend the provision of designated sites as the most appropriate strategy for minimising the effects of recreation in these wilderness areas.

More recent work by Cole (1995a) has focused on experimental treatments to determine the extent of vegetation disturbance at four sites in four different vegetation types. The results reflect those of the earlier studies insofar as just one night of camping was sufficient to cause obvious damage to the vegetation whatever the type. Moreover, the vegetation at sites where camping was undertaken for four nights was more damaged than where camping occurred for only one night, though the impact was less than twice as intense. The results again indicate that the most appropriate way to minimise damage is to designate specific campsite locations. Cole (1995b) has also shown that, in the weeks immediately after trampling, the response of the vegetation varies between types and with intensity of trampling. Marion and Cole (1996) have examined the response to impact of a range of campsites in the Delaware Water Gap National Recreation Area over a five-year period. In common with other studies, soil and vegetation communities registered the greatest change on initial impact; increased damage occurred with further use but stabilised with continuous use. The results of this study and another study by Cole (1995c) also showed that open grassland vegetation was not as easily damaged as forest vegetation with a herb-rich ground flora. Anderson (1995) has undertaken a similar trampling study to Cole (1995b, c) but has focused on five vegetation communities in a Danish coastal area. Of the species recorded, Anderson found that geophytes (plants which survive the winter because they have subterranean buds) were resistant to trampling by humans, whereas therophytes (annual plants) and hemicryptophytes (plants which die back at the end of the growing season and whose buds are protected by the litter and soil) are susceptible to damage. Of the habitats investigated, the dunes were most vulnerable to damage, followed by coastal grasslands, with salt marsh being most resilient. These findings can be used to formulate management plans in relation to campsite location, period of use and footpath location.

Trampling can also be caused by horses' hooves, as Whinam and Comfort (1996) have discussed in relation to commercial horseriding in the Tasmanian subalpine zone, close to the Tasmanian Wilderness World Heritage Area. Intensive horseriding pressures have here caused soil loss and track construction in several vegetation zones. For example, Whinam and Comfort report that 397 cm^2 of soil per metre width of track has been lost over two years in eucalyptus forest with soil loss of 236 cm^2 per metre width of track occurring in *Gleichenia* moorland. Soil losses were much lower in rainforest, at 35 cm^2 per metre width of track, because of soil redistribution. Excessive use of tracks caused deepening and water accumulation after rainfall; this encouraged the creation of new tracks. Recreational activity can also be the cause of inadvertent but long-lasting impacts related to accidental fires. Heathland, moorlands and open forests are particularly susceptible to fire after sustained dry periods. Especially if high temperatures are maintained, fire will ash the above ground biomass and litter, and it may impair the reproductive potential of plants by reducing the seed bank in the soil and even damage below-ground roots and tubers. Such fires may be caused by discarded cigarettes, broken glass on which the Sun's rays focus or camp fires that run out of control. Littering of the countryside also causes ecological damage, especially to wildlife.

Sporting activities which utilise the landscape are many and varied and may be elements of both recreation and tourism. In the broadest sense, sports are active physical pursuits such as skiing, climbing, sailing and diving. Some of these sports require motor-powered vehicles which can be the major agents of environmental change; sometimes the environmental damage can be substantial, e.g. trail-bike racing. Large-scale sporting events, such as the Winter Olympics, can also bring about environmental change beyond the event itself; the effects extend to the infrastructure needed to support a large number of competitors and spectators.

Edington and Edington (1986) have described the environmental impact of the 'Hare and Hounds' trail-bike race across the Californian Desert between Barstow and Las Vegas. This is an arid, fragile environment that is easily damaged by the mechanical impact of motor-cycle tyres. The race has caused the destruction of some 140 000 creosote bushes (*Larrea divaricata*), 64 000 burro-bushes (*Franseria dumosa*) and 15 000 Mojave yuccas (*Yucca schidigera*) due to mechanical impact and the exposure of plant roots by subsequent soil erosion. Other desert flora and fauna have also been adversely affected by this type of activity as well as the use of dune buggies and jeeps. Soil compaction

by vehicle tyres in recreation areas of the Californian Desert has caused a reduction in the population of annual ephemerals by reducing water penetration that is essential to remove germination inhibitors from the soil seed bank. Changes in plant cover can subsequently influence faunal populations by altering food supplies and habitats.

Sailing, boating and water-skiing can also have adverse effects on aquatic ecosystems both directly and indirectly. In the Norfolk Broads of the United Kingdom, aquatic plants have been damaged by the mechanical action of propellers and boat hulls, which have also accelerated bank erosion. The discharge of sewage from boats has also contributed to eutrophication problems in the region (mainly due to fertiliser run-off from adjacent arable land). The popularity of the area, which attracts some 2000 hired boats and 3000 private craft in the summer months (Crawford, 1985), is gradually causing the impoverishment of what is generally considered one of Britain's most attractive wetland areas. Moreover, numerous reports indicate that sailing and other water-based recreational sports can have adverse affects on waterfowl that abound on many of Britain's artificially created reservoirs. This is a particular problem when sailing is extended into the winter period and can reduce the numbers of overwintering species.

Skiing can also have a significant environmental impact, since provisions for it extend over relatively large areas. This has been discussed by Mosimann (1985) in relation to ski piste construction in the Swiss Alps. His study of more than 200 sites in the cantons of Grison and Valais involved the collection of data on a range of variables relating to relief, the degree of site modification in piste construction such as the amount of levelling, soil characteristics, vegetation cover and its disturbance as well as impact assessment, notably sheet, rill and gully erosion, sediment deposition and landslipping. The relationships between site factors and soil erosion which emerged from this study are given in Table 9.6. This shows that slope form and soil moisture status are particularly influential in determining erosion rates. These factors are also important, along with altitude, in the establishment of new vegetation communities once piste construction has been completed. Revegetation is particularly limited where intensive erosion has occurred at altitudes above 2200 m; here revegetation is also restricted by harsh climatic factors. Between 2000 m and 1600 m vegetation recovery is more rapid but only where erosion is at a

Table 9.6 Relationships between site factors and soil erosion on ski pistes in the cantons of Grison and Valais, Switzerland

A. Most significant factors

- Slope form
- soil moisture status
- Run-off frequency
- Size of catchment

Greatest erosion occurs on long, concave linear hollows with no moisture deficit
Least erosion occurs on ridges

B. Intermediate factors

Shallow soils with low water-retention capacity
Piste length in the range 150–180 m means that erosion rates rise steadily; above 180 m they increase more rapidly
Stone content reduces the amount of compaction during piste construction

C. Least significant factors

Angle of slope has its most marked effect during periods of heavy snowmelt or heavy rainfall

D. Variable factors

Rock type has a limited influence on overall rates of erosion
Gneiss is the most readily eroded rock type, as evidenced by rill and gully development
Schists may be particularly susceptible to landslides
Marls are also subject to relatively high rates of erosion
Other rocks show little variation

Source: based on Mosimann (1985)

minimum and even then it may take between 5 and 8 years after seeding. Below 1600 m revegetation is much more rapid except on poorly drained slopes. Mosimann has also noted that on some revegetated pistes there are sharp boundaries separating well-vegetated and poorly vegetated areas. This occurs because of local variations in soil properties; on the upper part of the slope the soil may be truncated but on the lower slopes the soils are mixed with layers of various depths. Such variations affect nutrient and water transfers within the soil profile, which in turn influence the vegetation communities and the degree of success achieved by reseeding. Mosimann suggests that some of these problems could be overcome if particularly sensitive areas are avoided, especially those above 2000 m, and if the topsoil is restored after levelling, etc. Similar problems have occurred in the Black Forest, as Ries (1996) has reported. At Schauinsland, for example, both soil erosion and solifluction have been intensified by the construction of ski runs and lifts.

Passive sports, such as angling, can also have an environmental impact. Discarded lead shot, lead weights and fishing lines can adversely affect wildlife. Fishing lines can entrap adult birds and if the line is introduced into nests it can cause chick mortality; fishing lines may imperil seals and sea lions by causing lesions which become infected. Discarded lead fishing weights and lead shotgun pellets are a significant cause of game-bird mortality in both North America and the United Kingdom. Significant declines in UK swan populations have been attributed to the ingestion of lead shot and weights, from sediments of lake beds, stream beds and banks which the birds consume as grit. Marschall and Crowder (1996) have examined population changes in brook trout (*Salvelinus fontinalis*) in response to a variety of pressures, including an increase in recreational fishing in southern Appalachian mountain streams. This was found to be minor in relation to other pressures, such as the introduction of an exotic species (e.g. rainbow trout) and acidification, but it still contributed to population decline. Recreational fishing can also deplete stocks of sea fish, as Guastella (1994) has discussed in relation to Durban harbour, South Africa. Here, fish are caught from the shore or from light-tackle boats. Records of catches for the period 1976–1991 show that a gradual decline has occurred, probably through loss of habitat, which could affect the success of juvenile fish, and increasing disturbances by harbour traffic as well as intensive fishing.

Recreational activities in aquatic environments may have indirect effects. For example, Fox *et al.* (1994) have monitored numbers of overwintering pochards (*Aythya ferina*) in the Cotswold Water Park, UK, which comprises over 100 flooded gravel pits. They found that pits with high levels of recreational use had lower bird populations than those with low levels of use. Ecological change can also be brought about inadvertently, as illustrated by the work of Ludwig and Leitch (1996) on the transfer of fish between areas of origin. Their survey of anglers in North Dakota and Minnesota reflects the potential of anglers to disperse fish used for bait from the Mississippi River basin into the Hudson Bay basin via their bait buckets. Algae and macrophytes could also be transported this way. In addition fish populations on which recreational fishing depends may be manipulated by habitat management. Caffrey (1993) has discussed the management of aquatic vegetation in Irish canals used for recreational fishing and has shown that fish biomass is higher when the vegetation cover is between 20 and 70 per cent. The data also show that the architecture of plant species influences fish biomass; higher values occur in species with broad or complex leaves than in species with floating or strap leaves.

There are many instances where wildlife populations have been adversely affected by recreational hunting as well as the obnoxious practice of overhunting particular species to provide wildlife souvenirs. Two examples are the overhunting of African elephants as a source of ivory for ornaments and jewellery and the hunting of black rhinoceroses to obtain their horns, believed to have aphrodisiac properties. However, in relation to recreational hunting, a classic example of overexploitation is the Arabian oryx (*Oryx leucoryx*) in the Middle East. Before the Second World War this species was hunted by Bedouin tribes as part of a traditional subsistence economy. With the development of the oil industry and the introduction of modern vehicles and weapons, hunting parties organised on a large scale rapidly reduced oryx populations to such an extent they were virtually eliminated. The only herds that now inhabit the area are derived from zoo-bred animals which have been reintroduced.

9.5 The impact of tourism

Tourism is a distinctive type of recreation that involves the temporary movement of people to destinations beyond their usual residence along with the activities they pursue. One of the world's fastest growing industries, tourism burgeoned since the Second World War and has grown even more rapidly since the 1960s. Figure 9.2 gives data on increases in international tourist numbers, origins and destinations and reflects the fact that affluent people are becoming increasingly mobile. The factors underpinning these trends have been discussed by Cooper *et al.* (1993) and include improvements in transport, increases in annual leave entitlements and increases in disposable income. Although tourism data show that the major generating and receiving areas are developed nations, especially those of Europe and North America, there has been a marked increase in tourism to developing countries; in many cases this

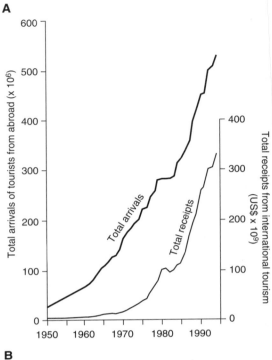

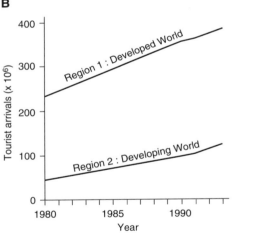

Fig. 9.2 (A) Changes in tourist arrivals from abroad and receipts from tourism since 1950. (B) Distribution of tourist arrivals in relation to developed and developing worlds (based on data in World Tourism Organisation, 1995).

is actively encouraged by governments in order to generate foreign currency. In some developing countries tourism has become the major source of export earnings; tourist income for Nepal and

Kenya accounts for 35.2 per cent and 42.9 per cent respectively (Cater, 1995). Developing nations are also highly susceptible to economic problems that ensue if tourism is depressed through internal strife; Peru and Egypt have respectively suffered from the activities of Shining Path guerillas and Islamic fundamentalists. Wherever tourism is important it is a source of employment as well as wealth generation. It inevitably promotes social as well as environmental change and is a major factor in globalisation.

Although much international tourism is undertaken for the enjoyment of cultural attractions, a high proportion of it is prompted by scenic, wildlife and marine features. Thus tourism and conservation are closely linked; this includes urban as well as ecological conservation. There is also a case for cultural conservation (e.g. King and Stewart, 1996) in the sense that people indigenous to areas attractive to tourism will adapt their traditional lifestyles, or at worst lose them altogether, as they absorb outside influences (see Hunter and Green, 1995). Worst-case examples include the cultural influence of so-called sex tourism in Thailand and the Philippines and the exploitation of child labour in tourist-based industries. As Brohman (1996) points out, developing countries are particularly susceptible to these impacts; their mitigation or avoidance can best be addressed through appropriate planning involving governments and local communities. In this context there are many parallels with forestry and forestry policy in developing countries (Section 9.3). Inadequate planning, poor enforcement of laws and regulations, and lack of involvement of local people tend, in both cases, to generate short-term gains but long-term losses and unsustainable activities. These issues have been discussed by Pagdin (1995) in relation to Nepal, where the development of tourism has occurred rapidly and has not been well planned; it contrasts markedly with recent developments in the Falkland Islands (Riley, 1995) and Bhutan, where numbers of tourists are tightly controlled. Consequently, the concept of sustainable development applies as much to the tourist industry as to agriculture and forestry. This is the underpinning rationale of ecotourism in its true sense. According to Cater (personal communication), this is 'tourism to ecologically and culturally sensitive areas; it reflects the integrity of national and socio-cultural environments and should contribute to environmental conservation, improve local standards of living, provide enlightening and meaningful

experiences for tourist guests and bring long-term benefits to the tourism industry and the local economy'; see also Goodwin (1996). Unfortunately, not all ecotourism has such a sound basis. This is because the prefix 'eco' is being applied to tourism that focuses on landscapes and/or wildlife but which is not necessarily sustainable because of its adverse impact on nature and culture.

Heavy visitor pressures to a particular landscape feature may cause degradation to such an extent that it becomes unattractive; tourism thus becomes self-defeating. Conversely, the closure of a feature to tourism for conservation purposes is undesirable because it defeats the purpose of conservation which is preservation of the feature for the enjoyment (and possibly education) of future generations. Some degree of accommodation between conservation and tourism is essential, involving the management of both people and place.

Successful management policies are employed in Australia's so-called Red Centre, where the major attractions are Ayres Rock and the Olgas. Until the late 1970s there was little control on campsites and motels were constructed at the base of Ayres Rock, giving free access to climbers and walkers. This began to promote excessive erosion and aroused Aboriginal concerns relating to infringements of their sacred sites, which abound in the region. The outcome was the construction of Yulara, a new resort situated 19 km from the rock itself and 37 km from the Olgas (Fig. 9.3). Yulara has a large range of facilities, including two luxury hotels, shops, restaurants, camp-grounds, an information centre and an airstrip. Visitors to the area are encouraged to confine their living requirements to the resort and are bussed out to the various landmarks, where rangers of the Uluru National Park provide tours and prevent entry to certain Aboriginal sites. In this way, footpaths can be opened or closed according to erosional pressures and the Aboriginals' wishes are respected. This is one example where conservation and tourism enjoy a sustainable symbiotic relationship.

In addition, the Australian and Queensland governments have attempted to manage sustainable tourism and recreational activities in the Great Barrier Reef Marine Park. Here the tourist industry is worth at least $300 million annually; $100 million comes from recreational fishing and the rest from resort accommodation, etc., and boat hire (the data are for 1987–88 and are derived from A.R.M. Young, 1996). The park occupies $345 \times 10^3 \, km^2$ and is divided into four sections; each has a zoning plan

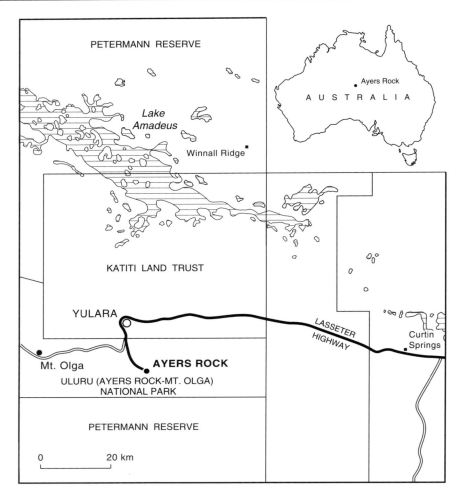

Fig. 9.3 Organisation of tourism facilities in Australia's Uluru National Park.

which indicates the choice of activities and where they can be performed. Activities include sea sports and various types of fishing; the zones include areas for general use, habitat protection, scientific research and preservation only. No activities are allowed in the preservation zone and traditional fishing is the only activity allowed in the research zone; even then it requires a permit. Nevertheless, the construction of marinas, breakwaters, etc., plus waste discharges and reef damage caused by boats, etc., take their toll. The issue of waste discharge is especially significant because sewage effluent causes cultural eutrophication (Section 7.5) to which coral reefs are particularly susceptible along with increased turbidity caused by sediment discharge

from building sites in resort areas. The zoning system is designed to confine damage as well as to minimise it. In fact, Marion and Rogers (1994) propose that containment of recreational activities in coral reef areas is a vital management strategy, just as it is in the National Parks of the United States (Section 9.4). Their study, based on the Virgin Islands National Park, showed that containment, the restriction of high-impact uses, the provision of educational programmes and information, and an enforcement system for park rules are the most appropriate management options.

Another example of ecotourism is practised in the Galápagos Islands, where management strategies have been designed to minimise tourist impact. The

Galápagos Islands combine the attractions of natural beauty with a unique and fearless wildlife. The island archipelago is situated in the Pacific Ocean ca. 1000 km west of Ecuador and it is famous as the source of inspiration for Charles Darwin's theory of evolution. The islands, comprising volcanic cones and rocks on the Galápagos Platform, itself comprising coalesced volcanics high above the surrounding sea floor, began to form ca. 10 K years ago due to the extrusion of lava, etc., from the Earth's crust. Consequently, the islands have never been joined to the mainland and their isolation has given rise to a large number of endemic organisms, including 229 plants and an unknown but considerable number of animals, e.g. the Galápagos penguin (*Spheniscus mendiculus*), the famous giant tortoise (*Geochelone elephantopus*) and several species of the lava lizard (*Tropidurus* spp.). Details of the geology, fauna and flora are given in Jackson (1993). The wildlife is the prime tourist attraction, especially the fauna, whose fearlessness often makes it highly visible. Indeed in view of the turbulent history of the Galápagos Islands, it is surprising that so much wildlife survives and, most surprisingly, that it retains its extrovert nature. Factors such as disputed ownership, exploitation of the giant tortoise by four centuries of buccaneers, the introduction of donkeys, goats and rats, and the introduction of agriculture, a twentieth-century decision of the Ecuadorian government, have all militated against the survival of the unique Galápogean environment. The introduction of agriculture especially seems to be an uneconomic and palpably unwise development (Adsersen, 1989). There are currently, I write as a visitor to the Galápagos in 1996, plans under review by Ecuador's president for increased development in the islands, because they generate so much wealth for the country. If these plans are implemented they will undoubtedly cause changes to such a fragile environment.

Tourism is currently regulated and visitors are only allowed on to the islands with a licensed, fully trained guide. Most visitors are not accommodated on the islands but on boats that range in size from cruise liners with ca. 90 passengers to small fishing vessels that carry ca. 6. Passengers from boats that cannot reach the shore, the larger vessels, are ferried ashore by panga. Once on shore there are strict Galápagos Park rules (Table 9.7) and tourists are confined to well-marked trails. Littering and leaving food are strictly forbidden, the aim being to prevent

Table 9.7 Rules for preservation in the Galápagos National Park, Ecuador

1. No plant, animal or remains of them (shells, bones pieces of wood) or other natural objects should be removed or disturbed.
2. Be careful not to transport any live material to the islands, or from island to island.
3. Do not take any food to the uninhabited islands.
4. Animals may not be touched or handled.
5. Animals may not be fed.
6. Do not startle or chase any animal from its resting or nesting spot.
7. Do not leave the areas which are designated as visiting sites.
8. Litter of all types must be kept off the islands.
9. Do not buy souvenirs or objects made of plants or animals of the islands.
10. Camping anywhere in Galápagos without a permit is against the law.
11. When camping, do not build campfires.
12. Do not paint names or graffiti on rocks.
13. All groups which visit the National Park must be accompanied by a qualified guide approved by the National Park.
14. The National Park is divided into different zones to facilitate its management and administration.
15. Do not hesitate to show your conservationist attitude and remind others of their responsibilities.

Source: abridged from a document of park rules provided by a shipping company

alien plants and animals colonising the islands. Similarly, the throwing of materials overboard is forbidden. Despite these efforts, and despite the high cost of Galápagos tours which limits numbers, there are still those who ignore the rules and the guides. The containment of tourists in boats reduces the tourist impact considerably, especially in relation to resort development and sewage, etc. However, there are hotels on the larger islands and the towns have expanded considerably since the 1970s. According to Carrasco, quoted in Jackson (1993), there were 6000 international visitors in 1972; by 1990 this had risen to 25 000, generating $100 million per year. There are now ca. 10 000 Ecuadorians visiting the islands annually. The massive increase in tourists has attracted much migration from the mainland. This is uncontrolled and the population has risen from ca. 5000 in the late 1970s to 10 000 today; these people need homes, food and water. The role of tourism as an agent of environmental change may thus extend beyond the impact of the tourists themselves. The impact of both residents and tourists will undoubtedly increase if development plans proceed despite the obvious fragility of the islands with their unique wildlife.

Tourism has produced significant changes in a wide range of contexts and in a wide range of environments. For example, Lukashina *et al.* (1996) have reported on pollution in the Sochi region of Russia's Black Sea coast. This is a popular resort area for Russians but air quality and beach quality have declined because of poor emission controls on motor vehicles and the rapid increase in tourist numbers giving rise to poorly planned resort expansion. In addition the Black Sea itself is becoming increasingly polluted, notably through eutrophication (Section 7.5) from sewage, urban run-off and agriculture. This study focuses on the deterioration of environmental quality whereas Haralambopoulos and Pizam (1996) focus on the way tourism has changed social conditions on the Greek island of Samos. Their survey of residents in the resort town of Pythagorion pinpointed several negative impacts of tourism despite its overall economic benefit. These included vandalism, sexual harassment, increased incidences of drug dependency, affrays, and high shop prices. The survey also showed that those directly involved with tourism welcomed its expansion whereas those with little direct economic interest were less enthusiastic. This study probably reflects the attitudes of resort residents elsewhere in the Mediterranean, where tourism has expanded substantially since the 1960s.

Another social problem that may be exacerbated by tourism in some regions is rural depopulation. Where tourism is concentrated in coastal resorts, as in many Mediterranean countries, there is likely to be an influx of people from rural areas, especially the young, leaving behind an ageing population. In many Mediterranean countries this has caused the abandonment of traditional lifestyles based on agropastoral systems. However, in recent years some of these countries have begun to develop alternative types of tourism in rural areas. This is exemplified by developments in Portugal, where there has been a trend away from the mass tourism characteristic of the 1960s and 1970s and the development of specialised tourism. According to Cavaco (1995), this comprises the revival of spa resorts and new, activity-based tourism involving hunting holidays. This diversification of tourism is partly a response to problems of rural development. Moreover, in other parts of Europe on-farm tourism is becoming increasingly commonplace and is usually associated with walking or sporting activities. This is a response to changes in the Common Agricultural Policy (CAP), changes that involve set-aside to reduce certain types of agricultural productivity and the resulting requirement of farmers to diversify their sources of income (Benjamin, 1994). This will give rise to a new range of impacts on Europe's rural areas.

The impact of tourism on Nepal is in direct contrast to the examples cited above. Here tourist attractions include eight of the world's highest mountain peaks, notably Everest and Annapurna. Since the late 1950s especially, when Nepal opened its borders to outsiders, the numbers of trekkers and mountain-climbing groups have increased markedly and, as is the case in many developing countries, this has been encouraged as a source of foreign income. The endemic fuelwood crisis is now being aggravated by the fuelwood requirements of tourists (Gurung and De Coursey, 1994). Inefficient stoves are kept burning day and night to provide heat and food for the tourist lodges in Sagarmatha, the Mount Everest National Park. This situation may well be aggravated by the construction of an airstrip above the Sherpa village of Namche Bazar and the renovation of a Japanese-run hotel. This village, situated along the trail from Lukla to Everest Base Camp, has undergone tremendous changes as the number of tourists to Nepal has risen from less than 10 000 in 1960 to ca. 270 000 in 1991 and there are plans to increase that number to ca. 1 million by the year 2000. Approximately 25 per cent of these tourists will visit Sagarmatha and, although this is welcomed as a source of employment, there are drawbacks.

Apart from the increased pressures on fuelwood, rubbish disposal is disfiguring the landscape, especially along the trail from Lukla, through Namche Bazar, to Everest Base Camp. Climbing expeditions into the peaks themselves have also left behind piles of rubbish. This problem is compounded by the fact that in such a high Alpine environment decomposition rates are very slow so that discarded materials may persist for several years if not decades. Moreover, expeditions tend to congregate in a few favoured areas that provide rapid access to peaks as well as adequate water supplies and shelter from natural hazards like avalanches. Add to this the recent rapid growth in the numbers of expeditions, climbers and trekkers, and it is not surprising that the problem has reached massive proportions, despite the availability of new access sites from Tibet, whose borders were opened to climbers in 1980. The harsh terrain and the inhospitable weather conditions mean that climbers subordinate rubbish disposal to their mountaineering goals and their own survival; the major problem areas are the base camps. Additional studies on tourism and its impact on Nepal include those of Wells (1993) and Stevens (1993). Wells has highlighted the problem of Nepal's parks and reserves, which are major tourist attractions. Although they generate $27 million annually in tourist receipts, they receive only $1 million in direct income, only 20 per cent of their running costs. Thus they are not well funded. This low investment means that low-level management is inadequate and degradation has occurred in some areas. Thus long-term sustainability is being sacrificed for short-term gains. On the positive side, Stevens (1993) considers that standards of living have risen as a result of tourism, especially in the Mount Everest region, but they have not produced significant cultural change.

Tourism has also had a significant impact on many national parks in Africa; visitors now come on safari holidays, the modern descendants of the big-game hunts of the late nineteenth century. Hunting is no longer allowed and tourists visit these national parks to view and photograph the wildlife in their natural habitats. Most of these parks are located in savanna regions, and in many of them tourists regularly drive off the existing game-viewing tracks into the dry terrain; this has generated concern that increased tourist traffic will detrimentally alter the environment. Onyeanusi (1986) has looked at the Masai Mara National Reserve in Kenya, where more than 40 000 visitors each year generate considerable off-road traffic in order to obtain close-up views of large carnivores like the cheetah. (*Acinonyx jubatus*), leopard (*Panthera pardus*) and lion (*P. leo*).

Such activity is necessitated by the provision of only 420 km of game-viewing in a reserve of 1673 km^2 and the fact there are relatively low populations of the carnivores, and individual carnivores do not always locate themselves in convenient spots! Onyeanusi's study involved an assessment of the impact of vehicle tyres on off-road tracks in a specific area of what is predominantly grassland savanna. As would be expected, the amount of damage increased reduction in vegetation cover was caused by initial vehicle runs. The study also shows that grasslands with a high above-ground biomass lose their standing crop more rapidly than those with a lower biomass. Extrapolating his results to the entire nature reserve, Onyeanusi estimated that the actual loss of biomass

was low. Although the visual impact of the extra trackways implied that significant damage was taking place, the actual ecological impact was relatively small. Nevertheless, the impact is detrimental and could be minimised by the provision of additional primary designated trackways. Increased tourist traffic could exacerbate this problem; further data has been collected by the Economic Intelligence Unit (1991).

9.6   The environmental implications of biotechnology

9.6.1   Agriculture

Biotechnology is a general term that embraces aspects of biochemistry, microbiology and chemical engineering, which in combination have harnessed the capacities of living organisms, especially micro-organisms and tissue cultures, to undertake specific processes with agricultural, medical, industrial and environmental applications. Indeed agriculture itself is a form of biotechnology since it involves the manipulation of whole organisms through selection and breeding programmes. Modern biotechnology is based more on manipulation at the cellular level than at the organism level, and the relatively recent innovation of genetic engineering has added another dimension at the genetic level. Since its inception in the early 1970s, genetic engineering has represented a major scientific advance because of the potential it provides to manipulate organisms ranging from the most simple life forms, such as bacteria, to the most complex, including *Homo sapiens sapiens*. The applications in agriculture are legion.

Nevertheless, misgivings have been expressed in relation to two aspects of the science. Firstly, much research focuses on human embryos and on animal embryos; this raises ethical questions hitherto unprecedented and which are contentious. Although the research is still in its early stages, the ability to manipulate life in this way requires clearly defined, enforceable legislation. Secondly, there are controversial issues relating to the control of genetically modified organisms (GMOs) and their release into the environment. It will require a cautious approach if humanity is to heed the lessons of earlier experiments when organisms have been introduced into new environments, either deliberately or inadvertently, often with detrimental consequences. Examples from Australia and New Zealand are given in Section 5.9.

Such a precautionary approach to the potential of genetic engineering is to be applauded in the wake of the dilemmas that have been created by the harnessing of nuclear power, a science developed in the 1940s to meet the weapons requirements of the Second World War. Promoted thereafter as an energy source both cheap and clean, neither claim has actually come true, and 50 years later there is considerable unease over its potential to annihilate the world and the long-lasting toxicity of its waste (Section 5.4.3). Although nuclear power and genetic engineering may appear unconnected, they do have much in common; the term 'nuclear' relates to a nucleus, a characteristic of atoms and cells which are respective building blocks of matter and living organisms. Nuclear power can be harnessed by splitting the nuclei of atoms that comprise the basis of all matter; genetic engineering attempts to change the nature of DNA a component of the nuclei of all living cells which is able to transfer characteristics from one generation to the next. Whatever the drawbacks of genetically engineered organisms ultimately prove to be, and as yet there is little evidence available, there are a number of possibilities that could prove to be environmentally beneficial. Some of these potentialities will be discussed below in relation to agriculture and in Section 9.6.2 in relation to other aspects of biotechnology such as environmental remediation.

The applications of biotechnology in agriculture include crop improvement, the control of pests and diseases, and the enhancement of nitrogen fixation. The basic aim is to improve the channelling of energy and nutrients into the crop. In the past this has been achieved by manipulating the environment but modern biotechnology allows it to be achieved by modifying the crop. Plant breeding programmes now use several techniques for crop improvement (Chrispeels and Sadava, 1994), producing stocks of plants with advantageous characteristics such as disease resistance; these are shown in Figure 9.4. For example, it is possible to exploit the capacity of many plants for sexual or vegetative reproduction, in order to produce stocks from cuttings. Stocks of potatoes, apples and strawberries are produced in this way as well as an array of garden plants. Plants can also be produced using tissue culture techniques, as discussed by Lindsey and Jones (1989), wherein complete plants are produced from tissue cells, even single cells, grown

in vitro. The applications of tissue culture range from the relatively simple production of large numbers of plants by cloning to the sophisticated production of new species and useful chemicals, as well as facilitating genetic engineering (see reviews in Brown and Thorpe, 1995; Pauls, 1995).

The ability of plants to regenerate from a few cells is related to the characteristic known as totipotency. This means that each individual cell has the ability to develop into any kind of cell the organism will require for survival. It is characteristic of both plant and animal cells; totipotency in animal cells is lost at an early stage in embryonic development but in plants all cells maintain this ability throughout the life span of the individual. In theory it is possible to produce new plants from any group of cells in an appropriate growth medium. In practice most plant production using tissue culture utilises cells that are known as callus, a mass of undifferentiated cells which are produced to repair damage, or cells from leaves, roots or stems, although it is not yet possible to regenerate all crops plants in this way.

Tissue culture has facilitated the production of several disease-resistant crops by sidestepping the tendency of plants to pass infections between generations via seeds, tillers or tubers. For example, it is now possible to produce cassava seedlings, in which a mosaic virus usually affects the parental line, by culturing unaffected cells that can be propagated in sterile conditions. This would be impossible using simple cuttings and such disease-resistant strains would be an obvious advantage to many farming systems in Africa, where cassava is a major crop, especially if disease resistance can be combined with high productivity. The cloning of plants using tissue culture is also being used to produce improved crop strains and it has considerable potential for the improvement of oil palms, coconuts and tea, all of which are important economic crops in many developing countries. Examples of crops which have been improved through tissue culture include cassava (*Manihot esculenta* Crantz) now available in a virus-free type (Hershey, 1993), millet (*Setaria italica*) in salt-tolerant form (Jing-Fen *et al.*, 1993) and many others (Chrispeels and Sadava, 1994; Lindsey and Jones, 1989). The technique is particularly important for producing seedlings of trees that are commercially valuable because it facilitates the production of clones of elite trees, hence, bypassing the time required for trees to mature. Oil palms (*Elaeis guineensis*) in particular are a commercial

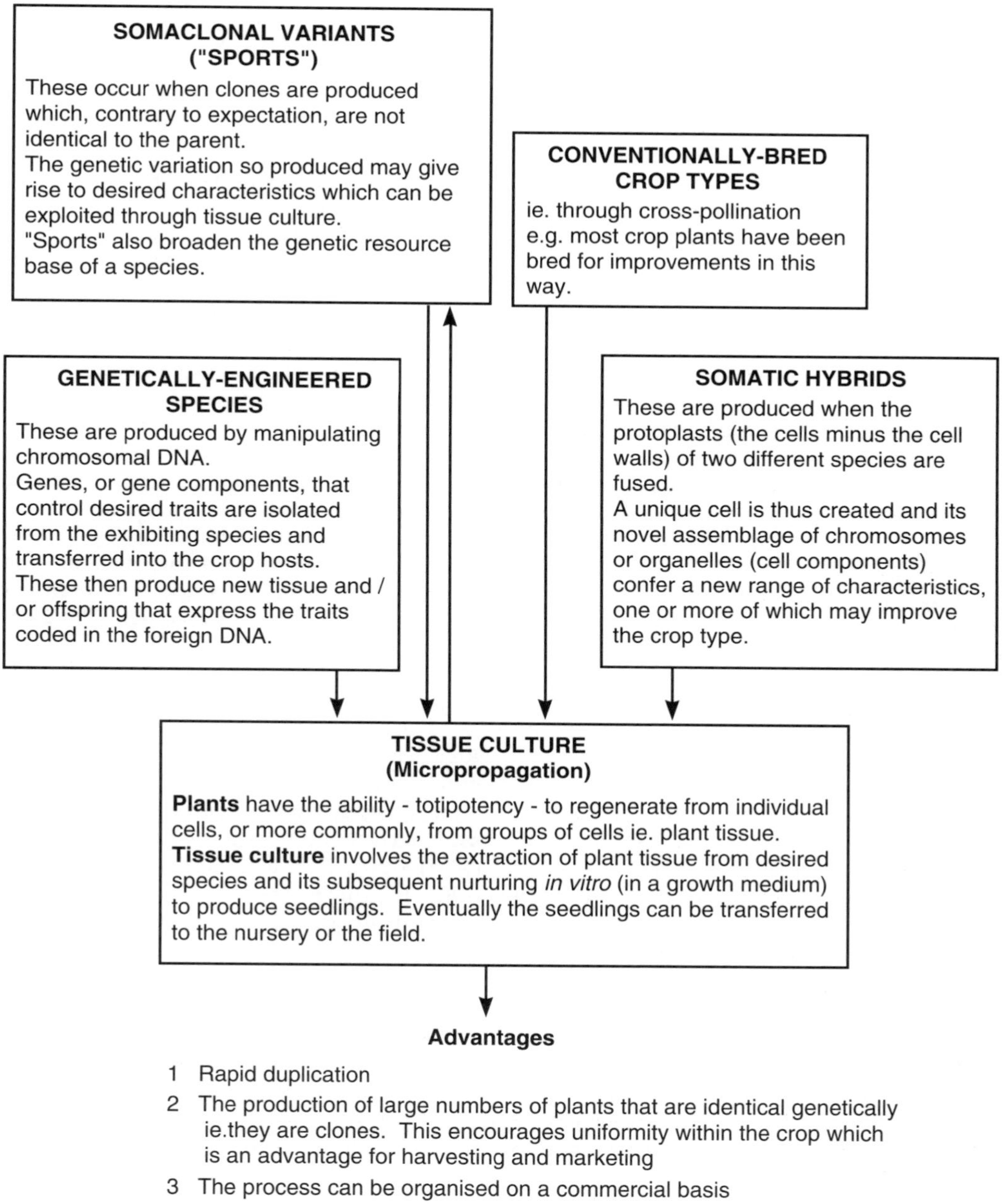

```
┌─────────────────────────────────┐
│   SOMACLONAL VARIANTS           │
│      ("SPORTS")                 │
│ These occur when clones are     │
│ produced which, contrary to     │
│ expectation, are not            │        ┌──────────────────────────┐
│ identical to the parent.        │        │  CONVENTIONALLY-BRED      │
│ The genetic variation so        │        │    CROP TYPES            │
│ produced may give rise to       │        │ ie. through cross-       │
│ desired characteristics which   │        │ pollination              │
│ can be exploited through        │        │ e.g. most crop plants    │
│ tissue culture.                 │        │ have been bred for        │
│ "Sports" also broaden the       │        │ improvements in this      │
│ genetic resource base of a      │        │ way.                     │
│ species.                        │        └──────────────────────────┘
└─────────────────────────────────┘
```

SOMACLONAL VARIANTS ("SPORTS")

These occur when clones are produced which, contrary to expectation, are not identical to the parent.
The genetic variation so produced may give rise to desired characteristics which can be exploited through tissue culture.
"Sports" also broaden the genetic resource base of a species.

CONVENTIONALLY-BRED CROP TYPES

ie. through cross-pollination
e.g. most crop plants have been bred for improvements in this way.

GENETICALLY-ENGINEERED SPECIES

These are produced by manipulating chromosomal DNA.
Genes, or gene components, that control desired traits are isolated from the exhibiting species and transferred into the crop hosts.
These then produce new tissue and / or offspring that express the traits coded in the foreign DNA.

SOMATIC HYBRIDS

These are produced when the protoplasts (the cells minus the cell walls) of two different species are fused.
A unique cell is thus created and its novel assemblage of chromosomes or organelles (cell components) confer a new range of characteristics, one or more of which may improve the crop type.

TISSUE CULTURE (Micropropagation)

Plants have the ability - totipotency - to regenerate from individual cells, or more commonly, from groups of cells ie. plant tissue.
Tissue culture involves the extraction of plant tissue from desired species and its subsequent nurturing *in vitro* (in a growth medium) to produce seedlings. Eventually the seedlings can be transferred to the nursery or the field.

Advantages

1 Rapid duplication
2 The production of large numbers of plants that are identical genetically ie.they are clones. This encourages uniformity within the crop which is an advantage for harvesting and marketing
3 The process can be organised on a commercial basis

Fig. 9.4 Ways in which crop plants can be modified or duplicated (from Mannion, 1995e).

crop produced by tissue culture for many years. Many vegetable and fruit crops, including potatoes and strawberries, are similarly replicated to produce large numbers of virus-free seedlings for horticultural enterprises. According to Brown and Thorpe (1995), tissue culture can now be used to replicate most crops, though the procedures for cereals and many woody species are more complex than for other dicotyledenous crops.

Somaclonal variation (Fig. 9.4) is another

attribute of plants that can be exploited to produce new crop plants. This is a characteristic that may occur in cloned species; clones or 'sports' derived from the parent plant are not always genetically identical and those that differ provide additions to the gene pool of the species. The sports may be disappointing in their performance when compared with the parent or they may have new advantages which can be exploited through tissue culture. Somaclonal variation has been recognised in many plant species and has been exploited to produce improved strains. Examples include potato (*Solanum tuberosum* L.) bred with resistance to the late-blight fungus (*Phytophthora infestans*), which caused the calamitous potato famine in Ireland during the 1840s (Douches and Jastrzebski, 1993). Sugar-cane has also been modified in this way (Malhotra, 1995). In addition somatic hybridisation, described in Figure 9.4, is used for crop improvement; tomato (*Lycopersicum esculentum*), tobacco (*Nicotiania tabacum*) and potato have all been improved this way (Kalloo, 1993; Pelletier, 1993). Nowadays, genetic engineering is increasingly providing opportunities for crop improvement; this is considered below, following additional applications of conventional biotechnology, pest control, etc., because these attributes may be incorporated into crop plants, either potentially or in reality.

In relation to pest and disease control, agricultural systems may incorporate a form of 'biological warfare' where one organism, a biopesticide, is introduced into the system to reduce the population of other organisms. The most important commercial biopesticide currently comprises preparations of a bacterium called *Bacillus thuringiensis* Bt. There are several varieties of this organism which act as insect pathogens because they produce insecticidal chemicals (reviewed by Marrone, 1994). According to Wood Mackenzie (1995), the biopesticide market was worth $130 million in 1994, of which Bt bioinsecticides accounted for 92 per cent. Bt pesticides thus represent 'big business' and a business that is growing rapidly. Bts have the advantage of being environmentally benign and they are relatively easy to produce and apply (Rappaport, 1992). Cotton is one of the most important crops to which Bt (Bt var. *kurstaki*) is applied; its use to combat the cotton bollworm has prompted genetic engineering of the crop with genes of the bacterium (see below). However, there is concern that resistance is beginning to develop in some lepidoptera (the insect

group to which the bollworm belongs). Apart from highlighting the capacity of insects to develop pesticide resistance to pesticides, and the ever present battle between insect and farmer, this suggests the commercial life of crops with engineered Bt genes may be relatively short-lived.

Other biopesticides include those based on fungi (mycopesticides) and viruses. According to Wainwright (1992), there are more than 400 fungi that attack insects and mites, which implies there is considerable potential for the production of mycopesticides (see also Leathers *et al.*, 1993). Several commercial products are already available, along with mycoherbicides, mycofungicides and myconematicides (Chrispeels and Sadava, 1994; Mannion, 1995a). Examples of mycopesticides are *Beauveria bassiana*, to combat the Colorado potato beetle, and *Metarhizium anisopliae*, produced in Brazil to kill froghoppers in sugar-cane. Three commercial mycoherbicides are Collego®, Devine®, and Casst®. Collego is a pathogen of the northern joint velch (*Vicia* sp.), a pest of soya bean and rice crops in the southern states of the United States; Devine kills milkweed (*Asclepias* sp.), a pest of Floridan citrus groves; and Casst targets sickle pod (*Arabis canadensis*) and coffee senna (*Cassia occidentalis*), weeds of soya bean and peanut crops. Research is also under way on the use of baculoviruses as biopesticides (Shuler, 1995). Bts, mycopesticides and baculovirus products can be used as components of integrated pest management (IPM) strategies, which also involve chemical and cultural approaches to pest control. IPMs are designed to curtail the establishment of insect resistance to chemical and biological pesticides and to minimise environmental impact (Leslie and Cuperus, 1993).

Crop improvement and biopesticides represent two aspects of the application of biotechnology to agriculture. Attention has also focused on the enhancement of soil nitrogen (as nitrate) through the use of nitrogen-fixing bacteria. The absence of sufficient available nitrogen in soils is one of the most significant limiting factors on crop productivity and this is why artificial nitrate fertilisers are so widely applied. However, as discussed in Sections 7.5 and 8.5, the use of artificial fertilisers can contribute to the cultural eutrophication of aquatic environments. As an alternative, cultures of the bacterium *Rhizobium* spp. can be applied to enhance nitrogen availability. This bacterium is naturally occurring in a symbiotic relationship with leguminous plant species, including several crop types, e.g. alfalfa, peas and soya bean. Moreover, in some parts of the world, increased productivity has been achieved by the inoculation of soils and/or seeds with *Rhizobia*. There are many species of *Rhizobia*; individually host specific, they fix nitrogen from the atmosphere only when present in root nodules and they show considerable variation in their nitrogen-fixing ability. These characteristics mean that a range of preconditions must be met before the addition of *Rhizobium* to the agricultural system is effective in increasing productivity (Catroux and Amarger, 1992).

The efficiency of nitrogen fixation by *Rhizobium* spp. is also determined to a large extent by the amount of nitrogen salts in the soil, so the application of nitrate fertiliser will actually inhibit bacterial nitrogen fixation. The options thus include the development of *Rhizobium* spp. that are active even when nitrates are abundant, or to reduce fertiliser application. From an environmental viewpoint it is preferable to reduce the amount of fertiliser in order to minimise the potential effects of eutrophication. *Rhizobium* seeding, either by inoculation into soil or legume seeds, has been successful in North America, where soya beans are widely grown in soils without any naturally occurring rhizobial bacteria. This is chiefly because the soya bean and its naturally occurring symbiotic bacteria are natives of the Far East. This implies that to obtain increased productivity from non-indigenous legumes it may well be advantageous to accompany their introduction with some rhizobial strains. The hypothesis is borne out by Primrose's (1991) observations on the use of *Rhizobia* inocula in Australia and New Zealand, especially successful at increasing the productivity of alfalfa and clover. These crops were introduced from Europe and are grown on soils with no indigenous *Rhizobia* and into which cultures are injected annually as the bacteria do not survive all year round. The addition of other bacteria and fungi may also prove beneficial for nitrogen enhancement (Mannion, 1995a).

All these aspects of biotechnology can be altered by genetic engineering (Fig. 9.4). This is also known as recombinant DNA technology, gene cloning and *in vitro* genetic manipulation. It involves the manipulation of deoxyribonucleic acid (DNA), the basic chromosomal unit that exists in all cells and which contains genetic information that is passed on to future generations. Plant and animal breeding

programmes have utilised a crude form of genetic engineering by crossing or interbreeding related species in the anticipation the offspring will contain preferred characteristics. In the case of plants the end product may be a species that is relatively drought resistant, a high carbohydrate yielder or easy to harvest. In the case of animals the objective may be high meat, wool or milk production. Virtually all the plants and animals that comprise the Earth's agricultural systems have been manipulated in this way. Such breeding programmes, in conjunction with the use of fertilisers and crop-protection chemicals, have given rise to what is commonly known as the Green Revolution of the twentieth century. Now, however, it is possible to identify specific gene components that contain the relevant hereditary information. Moreover, the technology is available to transpose that part of the chromosome into a related species which will then produce offspring bearing the particular characteristic. As Marx (1989) has discussed, the science of genetic engineering has been developed since the 1970s, but its origins really began in the mid-nineteenth century, when Charles Darwin and Gregor Mendel established the principles of heredity. Subsequent developments, including the isolation of cell components, the recognition of DNA in the cell nucleus as the purveyor of hereditary information and the determination of its structure in the 1950s by Watson and Crick, have all paved the way for the genetic engineering of the 1990s and beyond.

Recombinant DNA is produced by combining DNAs from different sources, discussed in detail by Chrispeels and Sadava (1994) and Christou (1995). The foreign DNA is inserted into a vector; the foreign DNA contains the genetic information necessary to confer the target characteristic, and the vector must be recognisable to the host in order for replication to occur. The vector may be a plasmid (extrachromosomal molecules of DNA that exist in many bacteria) or a phage (normally a bacterial virus). Another option is the direct transfer of DNA, which may be accomplished in various ways. This was first undertaken by Paul Berg and his research group at Stanford University in California, who joined genes from the bacterium *Escherichia coli* with DNA from simian virus 40 (SV40). Using similar procedures it is now possible to engineer organisms that can produce relatively large quantities of useful substances. Some of the resulting medical applications include the production of interferon, a potential antiviral and anticancer agent that until recently was only available in very small quantities by extracting it directly from human cells. It is now also possible to manufacture human insulin, a substance essential for diabetics who have become immune to animal-derived insulin, and growth hormones to treat children with growth deficiencies. Moreover, gene therapy is becoming available to combat a number of hereditary diseases (Pappas, 1996).

Genetic engineering has already been used in agriculture to produce many commercial crops with specific advantages. These include cotton with engineered resistance to the bollworm, soya bean with herbicide resistance and the Flavr Savr® tomato with delayed ripening to reduce wastage during transport. The objective of genetic engineering in crop plants is to improve productivity. This may take many forms. Productivity may be increased by enhancing the ability of crop plants to withstand pests, diseases and competitors i.e. the biological factors that impair crop growth. Alternatively, productivity may be increased by altering the capacity of crop plants to tolerate environmental conditions that are not conducive to optimum crop growth, e.g. high salinity, drought and length of growing season. These are the main environmental constraints on crop growth. Genetic engineering offers opportunities to modify crops in hitherto unlikely, if not impossible, ways in order to produce so-called designer crops. Agriculture has traditionally focused more on environmental manipulation than on crop modification. However, in relation to environmental change, genetic engineering may prove a double-edged sword. This is considered below in the context of the advantages and disadvantages of biotechnology and genetic engineering (Fig. 9.5).

The agrochemical company Monsanto marketed its transgenic cotton for the first time in 1995–96. The cotton was engineered to contain a gene from Bt (see above) to promote integral pest control. Considerable success was achieved in field trials; in its first season the high success rate has not been entirely replicated but the 'new' cotton has nevertheless raised productivity. However, the onset of resistance to the Bt insecticide in some insects, may mean that the usefulness of such engineered crops is short-lived. Even so, the engineering of pest resistance in crops is a major objective of the seed companies and the agrochemical industry. Although the transfer of Bt genes to crop plants is a prime

Table 9.8 Advantages of genetically engineered insect resistance in crop plants

- Continuous protection is provided irrespective of insect life cycles, season or weather conditions; this a major advantage over chemical pesticides.
- Provided transgenic seeds are marketed at a reasonable price, the costs are lower than if chemical pesticides are used; there is no need for repeated applications, which require labour.
- The costs of bringing a transgenic crop to the marketplace are much lower than those for developing a new chemical pesticide.
- The entire crop plant is protected, including underground parts.
- The protection is provided *in situ* and there is no possibility of contamination of the wider environment.
- Protection is target-specific, so beneficial insects remain unaffected; consequently, there is less disruption of food chains and food webs than occurs with conventional pesticides. However there may be other ecological disadvantages such as transmission of the specific gene to wild relatives (see text).
- The active factor is biodegradable and there is virtually no possibility of it becoming concentrated in the environment; but the resulting gene products could be toxic to animals or humans.
- Pesticide residues are absent, so transgenic crops may be preferred by consumers; it should be a legal requirement for consumers to be informed about gene products through appropriate labelling.
- Engineered pesticide resistance overcomes the increasing resistance of insect pests to conventional pesticides; the advantage may be short-lived.
- The reduction in the use of chemical pesticides will decrease the amount of fossil-fuel energy intput to high-technology agricultural systems.
- Transgenic crops may become important for integrated pest management strategies.

Source: based on Gatehouse *et al.* (1992)

target (Wearing and Hokkanen, 1994), it is possible that genes from viruses which attack insects may also be manipulated (Leishy and Van Beek, 1992; Thacker, 1993/94). The insecticidal potency of Bts and viruses, or their genes, could also be enhanced through genetic engineering. Rice is the main source of food for ca. 33 per cent of the world's population, and improved insect resistance is vital (Zapataarias *et al.*, 1995); rice transformed with Bt genes is already undergoing field trials at the International Rice Research Institute in the Philippines (Bennett, 1995). There are many advantages of engineered resistance to insects in crop plants, as detailed in Table 9.8. Indeed there appear to be few disadvantages, though it is possible that insect resistance, and other traits mentioned below, may eventually become established in the wild relatives of the engineered crops, causing problems for beneficial insect populations. Moreover, the insect toxin produced must be non-toxic or rapidly degradable if the crop is to be consumed by animals or humans. It is therefore necessary to assess these and other risks before marketing (e.g. Purrington and Bergelson, 1995; Sumida, 1996; Kasanmoentalib, 1996).

Diseases are also responsible for considerable crop losses and genetic engineering is being used to develop resistant cultivars. The chief causes of crop diseases are bacteria, viruses and fungi. Viruses are especially difficult to treat in the field, so inbred resistance is particularly advantageous. Work is in hand on potato, tobacco, tomato, alfalfa and melon (reviewed in Mannion 1995a). Resistance to potato viruses X and Y, and potato leaf-roll virus can be engineered in the potato as well as resistance to the late-blight fungus (Moffat, 1992). Much effort is being expended on the development of virus resistance in rice (Bennett, 1995); examples of viral diseases include tungro and ragged stunt. Panopoulos *et al.* (1996) have reviewed progress in the engineering of crop resistance to bacteria. This may be achieved in several ways, including the introduction of bacterial genes into crop plants or the introduction of genes from other plants or insects.

However, the greatest current investment in transgenic crop research is focused on the promotion of herbicide resistance, especially in the major cereals. The main advantage of herbicide resistance is that broad-spectrum chemical herbicides can still be used. Thus the crop remains unimpaired and its competitors (for light and nutrients) can be eliminated. Moreover, many agrochemical companies are broadening their product ranges by acquiring seed companies so they can market both transgenic seeds and the herbicides. Examples of herbicides to which resistance is being engineered are glufosinate, glyphosate, biomoxynil and imazaquin, and targeted crops include rape, soya bean, maize, sugar beet, wheat and rice (Wood Mackenzie, 1994; Dunwell, 1995). However, the monopolisation of specific markets involving both seed and matched herbicide marketing is not necessarily advantageous, especially in relation to price control and the purchasing ability of farmers in the developing world.

This issue is part of the wider consideration of the power relations of biotechnology, the moral concerns surrounding the patenting of genes, which severely limits their use by anyone besides the patent holders, and the question of technology transfer between the developed world and the developing world. The developing world is particularly important as the search for new genetic resources intensifies and is concentrated there. Who should benefit, the gene hosts or the technologists that can turn the genes into resources? How should the benefits be apportioned? Such considerations are also related to the use, abuse and conservation of biodiversity (Sections 7.2 and 8.2). The debate has been reviewed in Shiva (1993), Lacy (1995), Crouch (1995), Sasson (1996) and Reid (1996). It was a major focus of the Earth Summit held in Rio de Janeiro in 1992 and is addressed in the United Nations Convention on Biological Diversity (Lacy, 1995; Section 10.2.4). Moreover, this debate relates to biotechnology as a whole, not just agricultural biotechnology.

Other aspects of agriculture to which genetic engineering can be applied include the manipulation of chemicals produced within crop plants. Such chemicals often have considerable value, as in the case of speciality oils. Genetic engineering can capitalise on the ability of specific plants to produce these substances by improving their productivity and quality, or it can alter plants to produce alien substances, so they behave as bioreactors (Goddijn and Pen, 1995). Rape is one of the plants being genetically modified to alter the oil it produces. There is considerable potential for the production of biomass fuels; rapeseed oil has already been used to power buses in the United Kingdom and Zeneca have engineered it to produce Biopol, a biodegradable plastic. In addition Spelman (1994) has reviewed the possibilities for the engineered production of therapeutic substances for use in human health care. There is also considerable potential for engineering tolerance to various types of environmental stress, such as drought and salinity (reviewed in Mannion, 1995a). However, the development of such attributes may encourage the spread of agriculture into areas now considered marginal and which constitute the world's remaining natural ecosystems (see below and Fig. 9.5). The possibilities of genetically engineering the bacteria *Rhizobia* to enhance their ability to fix nitrogen have been examined by Mytton and Skøt (1993). There is potential to engineer the bacteria themselves to improve performance and eventually the main crop plants may be engineered to fix their own nitrogen directly. This would reduce or eliminate the need for artificial fertilisers; the advantages would be similar to those given in Table 9.8 for engineered insect resistance, and in relation to the wider environment the risk of cultural eutrophication would be reduced (Sections 7.5 and 8.5).

The potential of genetic engineering for animal-based agriculture is also considerable. Firstly, feed crops can be engineered to contain increased minerals which lead to an increase in secondary (animal) productivity. In this way, wood and meat production can be improved. Milk production can be improved through the use of genetically engineered growth hormones, the somatotropins, and animal health can benefit from engineered viruses. Many of the developments in these fields have been reviewed by Robinson and McEvoy (1993). In addition, animal reproduction can be manipulated to increase productivity, notably through the production of several embryos, i.e. clones, from one fertilised egg, as Deboer et al. (1995) have discussed in relation to dairy cattle. It is also now possible to produce transgenic species of livestock though this raises many ethical issues (Mepham, 1993; Thompson, 1993). Just as contentious are the actual and potential uses of animals as bioreactors, i.e. genetically engineering animals to produce substances for use in human health care, and as sources of organs for transplanting into humans. For example, human haemoglobin can be produced by transgenic pigs (A. Sharma et al., 1994). The production of genetically engineered fish appears to raise fewer public concerns, though as Pullin (1996) points out, there is the possibility of gene exchange and competition with related wild species and competition in a similar way to transgenic plants and their wild relatives (see above).

As the foregoing discussion implies, there are considerable advantages and disadvantages of modern biotechnology, especially genetic engineering in agriculture. Those that relate to the environment are shown in Figure 9.5. On the positive side, increased productivity from existing agricultural systems should contribute to conservation by diminishing the need for additional land to be cultivated as the world's population continues to grow. Improved energy transfer via disease-resistant and pest-resistant crops and livestock, as well as enhanced nitrogen availability, could in principle reduce the inputs of fossil fuels to the world's agricultural systems and thus curtail outputs of heat-trapping and acid-forming gases which cause pollution (Sections 6.5 and 6.6). Overall the sustainability of agricultural systems could be improved substantially and nutritional levels of much of the world's population improved. However, it is equally possible that few of the advantages could become reality. For example, the production of drought-resistant and pest-resistant species could lead to even greater expansion of agricultural systems, further reducing the remaining natural ecosystems with concomitant loss of biodiversity. Escaped transgenic crops could compete with wild species and engineered genes may escape to create 'superweeds', causing extinctions and reducing the gene pool. These issues have been discussed by Schmitt and Linder (1994), Raybould and Gray (1994) and many others; work to date indicates that trials of each transgenic species should be carried out to ascertain the risks (see discussion in Kling, 1996).

Many of the real and potential cultural advantages and disadvantages of biotechnology have been mentioned above. Disadvantages include monopolies over seeds and chemicals by transnational companies whose business is to make profits, and gene appropriation from developing countries with little or no recompense. This amounts to 'genetic imperialism' and comprises the appropriation of genes from the wild, including genes from species not yet established as crops or products, i.e. pharmaceuticals, and from plant and germplasm collections. Where there is co-operation between developing countries and agribusiness in the developed world, this has been advantageous, as has been demonstrated by Peel (1996) in relation to the Saskawa–Global 2000 project, which has led to crop yields being improved by 100 per cent in several countries in west Africa. Moreover, it has been argued that biotechnology is likely to provide the best, and possibly the only, way forward for agriculture in the developing world (e.g. Nene, 1996), including the least-developed countries, which are heavily dependent on agriculture, as is the case in sub-Saharan Africa (Okafor, 1994).

9.6.2 Other applications

There are many applications of biotechnology besides agriculture, and they too have implications for environmental quality. Biotechnology is widely applied in the treatments of industrial effluents and sewage, and Section 7.5 has already considered the possibilities of denitrification of water intended for domestic consumption. Any such processes that can improve environmental quality are to be welcomed,

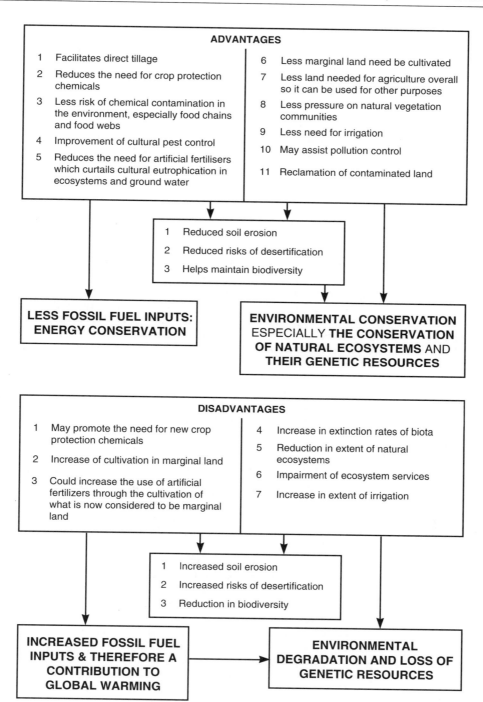

Fig. 9.5 Biotechnology and the environment: advantages and disadvantages (from Mannion, 1997a).

provided they themselves do not lead to inadvertent environmental deterioration. Biotechnology for environmental management can be classified into four categories: waste and wastewater treatment (including sewage), resource recovery and recycling, pollution monitoring and abatement, and renewable energy resources.

The treatment of industrial, agricultural and domestic wastes that are reduced to sludge may be undertaken using various digestion processes. The processes involved are complex and have been discussed by Nebel and Wright (1996). The aerobic treatment of sewage and waste is well established in several industries, e.g. food, chemical and pulp and paper industries; it involves the microbial decomposition of dissolved organic compounds. The chief aims are to reduce the solid volume of waste, to minimise offensive odours and to remove any pathogens. Like anaerobic digestion (see below), the aerobic processing of waste material produces methane, which can be used as a biogas fuel. The processes and technologies in aerobic waste treatment have been detailed by Nebel and Wright (1996) and Jackson and Jackson (1995) who point out that the treatment plants must be designed according to the waste in question. Sewage treatment is divided into three stages. Preliminary treatment removes debris and grit; the debris is collected against a screen then raked away; the grit is allowed to settle out. Primary treatment, the next stage, removes particles of organic material; the sewage is passed very slowly through large tanks and the organic particles collect at the bottom. This accounts for up to 50 per cent of the total organic matter. Oils and fats can also be skimmed off the surface. The organic sediment and the oily skimmings comprise raw sludge, which can be treated anaerobically (see below). The remaining effluent then undergoes secondary treatment. Secondary treatment is also known as biological treatment because micro-organisms are encouraged to feed on the organic matter by providing them with oxygen. If the reactions and energy transfers are allowed to proceed uninterrupted, the end products are carbon dioxide and water enriched with minerals. However, in order to avoid cultural eutrophication through the release of this mineral-rich effluent into drainage systems, a further step in wastewater treatment is usually undertaken. Known as biological nutrient removal, it also involves organisms, this time to reduce the concentrations of nitrogen and phosphate. Moreover, sewage effluent can be used to generate biomass, from secondary treatment, and this can be used as animal or human food (see below).

The sludge produced from primary treatment can be digested anaerobically. This involves bacteria that can break down organic molecules such as proteins, lipids and polysaccharides (such as cellulose and starch) in the absence of oxygen. Anaerobic reactions take place in a variety of digesters designed to deal with specific types of waste products. Municipal sewage works have used digesters to treat sewage since the early 1900s; the biogas they produce is on average 70 per cent methane and 30 per cent carbon dioxide, and it can be used to power the digestion plant itself. The digested sludge can be used as fertiliser. Biogas produced in this way is not widely used as an alternative energy source in the developed world, but is now widely used in China, where several million small-scale digesters have been constructed to produce fuel for cooking and lighting at a local level. There are also analogies between this type of energy production and the decomposition of organic rubbish in landfill sites, where anaerobic conditions promote the often hazardous production of methane (Section 6.8). The process of digestion is completed in about five weeks, and since any pathogens have been removed, the treated sludge can be used as liquid fertiliser or converted into sludge cake by removing the water. This can also be used as fertiliser.

The biotechnology involved in resource recovery and recycling involves the harnessing of organisms to concentrate useful substances, usually metals (Jordan et al., 1996). Mineral extraction using organisms is sometimes known as biomining and its use is increasing as a means of extracting minerals from ores or from tailings. It harnesses the natural ability of organisms, especially bacteria, to break down mineral rock. In particular, bacterial leaching is used to extract metals from sulphide ores. It involves the oxidation of less soluble sulphides to more soluble sulphates, thereby generating solutions enriched in the required metals, solutions from which they can be easily extracted. The bacteria involved in this process are known as chemolithotrophs because they have the ability to produce energy for growth from oxidation of inorganic sulphurous compounds. According to Woods and Rawlings (1989), the most important organisms are *Thiobacillus ferrooxidans*, *T. thiooxidans*, *Leptospirillum ferrooxidans* and several *Sulfolobus* spp. which not only utilise inorganic substances as their energy source but also survive in highly acidic conditions and work at high temperatures. The chief metals that are being recovered using biomining are copper and uranium, though it is possible to obtain cobalt, nickel, zinc, lead and gold. Details of copper, uranium and gold recovery are given in Rawlings and Silver (1995).

Biomining has many advantages over the conventional recovery of minerals. Firstly, where near-surface ores have been depleted, deep-seated ores may be extracted by *in situ* leaching, obviating the need to undertake extensive and environmentally damaging excavation (Section 6.2). Secondly, low-grade sources of minerals and waste dumps can be efficiently exploited. Leachate must, however, be carefully controlled to avoid contamination of surface and below-ground drainage, though further economies can be made by recycling the leaching solution. If this were to be applied on a large scale, the widespread contamination of land and water by natural mine-leaching processes could be considerably reduced as could the input of fossil-fuel energy. One example from the United States is the Kennecott Chino Mine in New Mexico, which has a daily yield of cement copper of 45–50 t (Woods and Rawlings, 1989). The cement copper, which is produced from the leachate by the addition of scrap iron, can then be refined by smelting.

These processes may be used to remove impurities such as arsenides which detract from the value of mineral concentrates. Biomining may also be useful in the desulphurisation of coal, helping to combat the acidification of ecosystems (Section 6.6) due to fossil-fuel combustion. Moreover, Goldstein *et al.* (1993) have demonstrated that some species of bacteria, *Pseudomonas cepacia* E37 and *Erwinia herbicola*, can be harnessed to extract phosphates from rock phosphate ore. Eventually it may become possible to use bacteria to concentrate metals that are dispersed in the environment. One such possibility is the use of the bacterium *Pedomicrobium*, involved in the deposition of placer gold (Mann, 1992; Southam and Beveridge, 1994). Tebo (1995) has reviewed the potential of marine bacteria for metal precipitation. Rawlings and Silver (1995) also point out that genetic engineering has considerable application in this field by developing improved strains of the relevant bacteria with enhanced efficiency, which can operate on an increased range of substrates and under a variety of environmental conditions.

The propensity of certain types of bacteria to scavenge metals could also be employed to recover metals from wastewater and sewage. Other micro-organisms may also be used in this way, e.g. fungi and algae. Incorporating them into treatment plants that already rely on biological processing (see above) may be cost-effective and environmentally beneficial if sufficient volumes of useful substances can be obtained. The possibilities have been discussed by Ross (1988), who refers to the ability of the fungi *Rhizopus arrhizus* and *Penicillium chrysogenum* to concentrate uranium, and the algae *Chlorella regulis* and *C. vulgaris* to concentrate uranium and gold respectively. Moreover, Dwivedy and Mathur (1995) report that bacteria have been used to remove radium and manganese from effluent produced by uranium processing. As well as recycling valuable substances, reducing the pressure on primary resources, these organisms may prove effective in mitigating pollution due to high concentrations of heavy metals.

Pollution of the environment can take many forms and vary in intensity. Biotechnology can contribute to both pollution detection and monitoring as well as its mitigation. Indeed the use of bacteria in sewage treatment (see above) could be considered as a form of pollution mitigation. Biosensors can be used to detect and monitor pollution. This involves the use of an organism, or its incorporation into a device, to provide a measure of pollution levels. Biosensors are many and varied; their applications go beyond environmental problems and into medical diagnostics (Hall, 1990). A wide range of environmental biosensors can be used in many different contexts. For example, multicellular organisms such as trout can be used to indicate water quality. However, recent developments in biosensor production involve single-celled organisms or single cells derived from multicellular organisms. According to Bains (1993), most of this new generation of biosensors employ bacteria that behave as sensors through the inhibition of their physiological processes. This affects carbon dioxide production from respiration as well as changes in pH, cell function, etc. An example is the measurement of biological oxygen demand (BOD). This is important because BOD is a measure of the pollution potential of organic-rich effluents, e.g. where the activity of micro-organisms is high the oxygen demand is high and can therefore be depleted rapidly; Sections 7.5 and 8.5 examine at the cultural eutrophication of aquatic ecosystems, and Jackson and Jackson (1995) explain how BOD is determined. Many biosensors have been developed to detect pesticides and their residues. For example, Rainina *et al.* (1996) have developed a transgenic strain of the bacterium *Escherichia coli* to detect organophosphorus neurotoxins and related chemical warfare agents, and Bains (1992) reports that various types of algae are being used to detect herbicides that work by inhibiting photosynthesis.

Pollution mitigation strategies are increasingly incorporating the services of organisms, especially micro-organisms. This is sometimes known as bioremediation and is a rapidly developing field of investigation and application. In some respects the use of buffer zones, comprising riparian vegetation communities, to protect lakes and rivers from excess nitrate and phosphate derived from arable land is a form of bioremediation, preventative rather than reactive. The riparian vegetation acts as a sink for these nutrients, which are therefore prevented from causing cultural eutrophication (Section 7.5). The use of higher plants to effect remediation is also known as phytoremediation and is a form of ecological engineering (Ernst, 1996). Cunningham and Lee (1995) state that 'phytoremediation is defined as the use of green plants to remove, contain, or render harmless environmental contaminants'. Phytoremediation may sometimes be a component of land reclamation strategies (Section 6.4). In the case of soil remediation, plants can be used to improve soils contaminated with metals and organic substances. The remediation of metal-contaminated soils exploits the ability of specific plants to absorb metals in concentrations that would prove toxic to other plants. Phytoremediative plants have developed the ability to absorb, translocate and store metals so they do not inhibit metabolic processes; they are sometimes known as 'hyperaccumulator' plants (Cunningham and Lee, 1995) and can accumulate up to 3 per cent dry weight of heavy metals and as much as 25 per cent dry weight in sap.

Metals to which such tolerances have been recorded include zinc, nickel, cobalt and substrates and soils, including mine tailings, with high accumulations of these metals can be stabilised using a covering of the relevant hyperaccumulators. It may even become possible to extract the metals commercially from the plant tissue. Moreover, genetic engineering could be employed to improve the efficiency of hyperaccumulators and to develop new strains. An example of this has been reported by Rugh *et al.* (1996), who have transferred a gene for mercury accumulation from a bacterium into

Arabidopsis, a species of cress. Under laboratory conditions this proved effective for mercuric chloride uptake with subsequent conversion to metallic mercury. Metallic mercury is less hazardous than the mercuric chloride and it evaporates from the plant's surfaces. Rugh *et al.* conclude that if plants with mercury-accumulating genes were planted along the edges of Florida's wetlands, there would be much less ecological damage caused by drainage contaminated with mercury compounds from fungicides and bactericides formerly used in Florida's citrus groves. They also state that silver, gold, copper and cadmium could be removed using a similar strategy. These metals are not volatile, unlike mercury, so there is potential for harvesting. Moreover, Pollard and Baker (1996) have reported on the capacity of *Thlaspi caerulescens* (a brassica) to accumulate zinc and the potential that it offers for bioremediation. Brown *et al.* (1994) have reported that the species can also concentrate cadmium, and Banuelos (1996) has examined the potential of several plants (including several species of saltbush, rape and tall fescue) to concentrate selenium and boron. It is suggested that a vegetation community which includes many of these species could be a useful management tool for agricultural soils in parts of the western United States, where selenium and boron concentrations are naturally but dangerously high. In a similar way, many plants are tolerant of organic compounds. Section 9.6.1 has already examined the development of crops with engineered resistance to herbicides. This relies on the fact that such plants have the physiological ability to break down herbicides into non-toxic components. As Cunningham and Lee (1995) point out, this reflects crop selectivity and it is this characteristic which can be exploited to develop bioremedial strategies for soils, etc., contaminated with organic compounds. Genetic engineering also provides the opportunity to enhance the ability of specific plants to break down organic chemicals; this could be achieved by incorporating appropriate genes from micro-organisms.

Micro-organisms are also providing opportunities for bioremediation and have been used in many environments to alleviate a wide range of pollution problems. Certain micro-organisms, usually bacteria, have the ability to degrade hazardous substances. Polychlorinated biphenyls (PCBs) are degraded by several strains of *Rhizobia* (Damaj and Ahmad, 1996), *Sphingomonas paucimobilis* SS86 decomposes γ-HCH (Senoo *et al.*, 1996), *Acinetobacter calcoaceticus* breaks down diesel and heating oil (Marin *et al.*, 1995) and *Pseudomonas* spp. (*P. alcaligenes*, *P. mendocina*, *P. putida* biovar. B. and *P. stutzeri*) can decompose methyl violet and its precursors, dimethylaniline and phenol (Sarnaik and Kanekar, 1995). Bacteria can also be used to decontaminate groundwater and there are even species which will decompose the explosive, trinitrotoluene (TNT), and mustard gas. Many bacteria also concentrate metals. For example, Leblanc *et al.* (1996) have shown that *Thiobacillus*-type bacteria concentrate arsenic. In a stream draining an acid lead–zinc mine at Gard, France, the bacteria cause ferric arsenate and arsenic sulphate to accumulate close to the mine, limiting the pollution downstream. In addition, the ability of the bacterium *Desulfovibrio desulfuricans* M6 to degrade sulphur compounds could be used to desulphurise fossil fuels (Kim *et al.*, 1995). Fungi as well as bacteria can be employed in bioremediation procedures. For example, Paszczynski and Crawford (1995) have reviewed information on the white rot fungus (*Phanerochaete chrysosporium*) which has a unique enzyme system that is capable of degrading the complex polymer lignin; white rot is also known to degrade several xenobiotic chemicals and is thus a potential bioremediation agent. *P. sordida* can perform similar tasks (Glasser and Lamar, 1995). Both species have been used to degrade pentachlorophenol and creosote in soils.

Perhaps the most well-known aspect of bioremediation concerns oil-spill mitigation. Such measures were used as part of the clean-up operation after the massive oil spill from the Exxon Valdez off Prince William Sound, Alaska, in 1989. This type of bioremediation in the marine environment has been reviewed by Swannell *et al.* (1996), who point out that the biodegradation process can be accelerated by the addition of nutrients, e.g. mineral fertilisers, which encourage naturally occurring oil-decomposing bacteria to multiply. Concerns have, however, been voiced about the addition of non-indigenous microflora and their efficacy. These issues have been addressed on an experimental basis by Venosa *et al.* (1996). Their monitoring of plots treated with nutrients, plots treated with a microbial inoculum from the area and plots with no additions following a deliberate oil-spill along Delaware Bay, in the United States, showed that plots with no additions recovered more slowly than the other two plots. However, there was little difference between plots treated with microbial inocula and those treated

with nutrients, and Venosa *et al.* observe that the comparatively rapid amelioration of untreated plots was probably due to abundant nitrate. On the basis of a literature review, Atlas (1995) also concludes that seeding with micro-organisms is not especially effective, but in the case of fertiliser application, as in the treatment of the Exxon Valdez spill, rates of biodegradation increased by at least a factor of 3. Dolt *et al.* (1995) have shown that the rate of recovery of soils polluted with hydrocarbons does not increase if they are seeded with bacteria. It is likely that genetic engineering could be used to improve the ability of bacteria and other organisms to carry out remediation. Although this might be advantageous for pollution mitigation, the release of such transgenic organisms may cause other, possibly worse, environmental problems by altering biogeochemical cycles. The release of transgenic organisms must be tightly controlled (Stephenson and Warnes, 1996).

Modern biotechnology has also been applied to the development of new food and fuel sources. In relation to food energy, Litchfield (1989) has described examples of how aerobic sewage treatment can be manipulated to produce single-cell proteins (SCPs). These are the agglomerated, and subsequently dried, cells of single-celled micro-organisms like algae, yeasts, fungi and certain bacteria, which can be cultivated on an industrial scale for the production of protein that can be used for either human or animal consumption. SCPs can be produced in two ways: by harnessing organisms that photosynthesise and by using biotechnological methods that involve the provision of non-photosynthetic organisms with oxygen, nutrients such as carbon, nitrogen and phosphorus, and energy. Bacteria and algae are the most commonly used photosynthesising organisms in SCP production and are best cultivated in shallow ponds where temperatures are high and an adequate nutrient supply is present. For example, the blue-green alga *Spirulina maxima* has been produced on a commercial basis from Lake Texcoco in Mexico, and in common with the products of similar projects in Hawaii, Taiwan, Israel and Thailand, the dried algae are sold as health foods. Israel has successfully combined sewage treatment with SCP production; the sewage provides the nutrients and the year-round high light intensities promote rapid algal growth. The resulting biomass is thermally dried to remove pathogens and the final product is used as an animal feed.

SCP production from non-photosynthetic organisms has also reached the stage of commercial availability, mainly as animal feed. Pruteen® is one such product, developed by ICI in the 1960s and early 1970s, mainly used as an alternative to soyabean meal. The process involves the bacterium *Methylophilus methylotrophus*, which takes methanol as its energy source, and the provision of nutrients such as ammonia (to provide nitrogen), phosphorus, calcium and potassium. According to Litchfield (1989), the dried product consists of 72 per cent protein, but although it is an acceptable supplement to animal feed, the worldwide drop in the price of soya bean meal put an end to its commercial production. However, technology such as this is used to produce a wide variety of other food products for both animal and human consumption, such as yeasts, flavouring, gums and vitamins (Crueger and Crueger, 1990). Of the products currently available for human consumption, Quorn® is perhaps the best-known SCP. It consists of fungal mycelia from the fungus *Fusarium graminearum*, produced by fermentation using glucose as the energy source and ammonia as the nitrogen source. The end product resembles meat insofar as it is fibrous and can be flavoured and/or coloured accordingly. Compared with lean raw beefsteak, Quorn has a protein content of ca. 47 per cent, nearly 20 per cent less than the beefsteak, but it contains half the fat and is rich in fibre (Angold *et al.*, 1989). Indeed Primrose (1991) has suggested that SCPs could be produced to supplement poor diets in developing countries, where protein consumption is low and agricultural productivity is low.

Several other SCPs are under development. Feed is required for the increase in aquaculture, e.g. fish and shrimp farming, including feed derived from microalgae. The production of microalgal biomass has been discussed by Gladue and Maxey (1994), who examine various production systems, including the use of outdoor ponds (mentioned earlier in relation to *Spirulina maxima*) and fermentation systems which use sugars as an energy source. One such example is the marine alga *Brachiomonas submarine* var. *pulsifera* (Tsavalos and Day, 1994), and the commercial applications of microalgae in general have been discussed by Radmer and Parker (1994). Some types of organisms can also be harnessed to act as bioreactors in order to produce useful products, e.g. speciality chemicals.

Concerns about pollution from fossil-fuel use,

especially acidification (Section 6.6), and the importance of establishing renewable energy sources have prompted the development of biomass fuels. Ethanol is the best-known biomass fuel; it was developed in Brazil because of the abundance of sugar-cane, not to satisfy environmental considerations. Today ethanol provides some 30 per cent of Brazil's fuel requirements and there are plans to increase its use for the generation of domestic electricity. Ethanol can also be produced from a variety of other source materials, including wood pulp; Sweden makes ethanol from wood pulp, a by-product of its forestry industry. In the United States ethanol-based fuel is produced from sweet sorghum. According to Henk and Linden (1994), sorghum is particularly suitable for ethanol production because it produces high yields of fermentable sugars and cellulose. In Europe the production of biomass energy is increasing, partly because of set-aside and the use of resulting land for biomass-oil crops and partly because of environmental concerns. The possibilities and prospects have been examined by Easterly and Burnham (1996) and Hall and House (1995). One such example is rapeseed oil, which is currently being evaluated in several pilot schemes in the United Kingdom. Poola *et al.* (1994) have also drawn attention to the potential of orange and eucalyptus oil, which may be suitable for spark-ignition engines. Moreover, Braun (1996) reports on the production of biodiesel from algal biomass in the Netherlands.

All of the applications of biotechnology discussed in this section could potentially be improved by genetic engineering.

9.7 Conclusion

Although the topics discussed in this chapter are diverse, none can be lightly dismissed as a significant agent of environmental change. In a temporal context, the state of modern forestry represents the culmination of forest and woodland resource use that probably began with the earliest hominids. Recreation, tourism and sport are products of the industrial era, and especially the result of increasing leisure time since the Second World War. Biotechnology, in the form of plant and animal breeding, also has a long history, beginning with the first agriculturists ca. 10 K years BP. However, recent developments in genetic engineering are opening up possibilities that have no historical parallels, so there is no baseline data against which to predict any future environmental changes they may cause. For all of these agents of environmental change, economic and political considerations are primary factors in their operation and future development. Unfortunately, environmental considerations often play a subservient role to economic and political expediency and the concept of long-term sustainability is rarely accorded more than cosmetic attention.

Several dichotomies arise out of forestry in the developed world. In North America foresight and management have promoted a vigorous sustainable wood industry, despite the early pillages of European settlers in the colonial era. In contrast woodland exploitation in Britain occurred through history and prehistory, leaving behind a landscape largely bereft of natural woodland and where a mantle of alien conifers dominates many upland areas. There is increasing demand in the developed world for wood and wood products, placing more pressures on indigenous forests. Land and fuelwood are scarce in many developing countries, which also need to earn foreign currency; this combination of circumstances has prompted wood harvesting at a hitherto unprecedented rate and often under inadequate management. Forest survival is thus threatened on a large scale. The irony of this is reflected in forest destruction on the one hand and the apparent success of agroforestry (social forestry) as a means of sustainable agricultural development on the other hand. Dilemmas such as these have no obvious, readily implemented solutions, nor are they conducive to gene pool conservation or the amelioration of the enhanced greenhouse effect.

The environmental impacts of recreation, tourism and sport are insignificant in relation to those posed by forest demise. Nevertheless, all three activities can adversely affect the environment and if uncontrolled they can result in the destruction of landforms, fauna and flora that first motivated the activity. Planning and management, if the political and economic will is there to implement them, can effect sustainable recreation and tourism. This is essential if such industries are to continue as a source of income in developed and developing countries. Furthermore, all three activities could and should work to the advantage of environmental conservation by encouraging people into new landscapes and thus heightening their environmental awareness, so they themselves become part of the environmental protection movement. Education is probably the best

form of conservation because public opinion influences politicians, as exemplified by the environmental debates of the 1980s and 1990s that figured so prominently in the media.

Finally, biotechnology has played a significant role in improving productivity via traditional plant and animal breeding programmes and in controlling environmental quality via wastewater treatment, and it is now beginning to feature in mineral extraction. Biotechnology has, however, brought civilisation to the verge of another environmental experiment in the context of the release of genetically engineered organisms. The questions that must be asked relate to the promises it presents, especially in the context of world food supplies and the alleviation of environmental pollution, and the caution it demands in relation to the potential ecological havoc it may wreak. Both proponents and protesters are right to present their views and it is incumbent on politicians to ensure that adequate regulatory controls provide as large a degree of environmental protection as is possible when dealing with the relatively unknown.

Further reading

Apps, M.J. and Price, D.J. (eds.) (1996) *Forest Ecosystems, Forest Management and the Global Carbon Cycle.* Springer-Verlag, Berlin.

Burns, P. and Holden, A. (1995) *Tourism: a New Perspective.* Prentice Hall, London.

di Castri, F. and Younès, T. (eds.) (1996) *Biodiversity, Science and Development: Towards a New Partnership.* CAB International, Wallingford, and the International Union of Biological Sciences, Paris.

Glick, B.R. and Pasternak, J.J. (1994) *Molecular Biotechnology: Principles and Applications of Recombinant DNA.* ASM Press, Washington DC.

Hunter, C. and Green, H. (1995) *Tourism and the Environment: A Sustainable Relationship.* Routledge, London.

Persley, G.J. (ed.) (1996) *Biotechnology and Integrated Pest Management.* CAB International, Wallingford.

Reid, W.V., Laird, S., Meyer, C.A., Gámez, R., Sittenfield, A., Janzen, D.H., Gollin, M.A. and Juma, C. (1993) *Biodiversity Prospecting: Using Genetic Resources for Sustainable Development.* World Resources Institute, USA, Instituto Nacional de Biodiversidad, Costa Rica, Rainforest Alliance, USA, and African Centre for Technology Studies, Kenya.

Sharma, N.P. (ed.) (1992) *Managing the World's Forests: Looking for Balance Between Conservation and Development.* Kendall/Hunt Publishers, Dubuque, IA.

CHAPTER 10

Conclusion and prospect

10.1 Introduction

During the past $2-3 \times 10^6$ years the Earth has experienced large-scale environmental changes which relate to the alternation of cold and warm stages, or glacial and interglacial periods. The evidence from terrestrial deposits, ocean cores and ice cores indicates there have been at least 17 such cycles, each lasting approximately 120 K years, with the warm part of the cycle occupying between 10 and 20 K years. Thus, climatic change has been highly significant in promoting environmental change by influencing energy flows and biogeochemical cycling, both of which affect the biota and are affected by the biota. Against this continuum of environmental change, human evolution and dispersal have occurred, culminating in the evolution of *Homo sapiens sapiens* and the subsequent spread of the species to all but the most inhospitable parts of the globe. As the current interglacial became established, early Neolithic civilisations began to develop technology that subsequently changed the face of the Earth. The domestication of plants and animals and the initiation of permanent agriculture provided the basis for organised, planned food production, supplementing and often replacing the uncertainty of hunting and gathering food-procurement strategies and providing the opportunity for permanent settlement. From these beginnings modern agricultural systems have developed, leading to the modification and replacement of interglacial ecosystems and the alteration of Earth-surface processes. Small-scale industrial activities, often associated with agriculture and metal production, also developed as an increasing range of natural resources were exploited, culminating in the Industrial Revolution of the eighteenth century. Primarily based in Europe, industrialisation rapidly spread to what are now the developed nations, and constituted another potent agent of environmental change. Industrialisation is currently spreading to developing nations, where it is considered a prerequisite to achieving the relatively high standards of living that characterise developed countries.

There are two major agents of environmental change in the current interglacial. The first is climate, and in view of palaeoenvironmental evidence (Chapter 2), there is every reason to believe that another ice age or cold period will ensue in the next 5–10 millennia. The second agent of change is *H. sapiens sapiens*, a relatively recent development in terms of the evolution of the Earth's biota but one that has already altered the Earth's surface to such an extent that in many areas it is no longer possible to identify the natural components of the interglacial environment. The tools of *H. sapiens sapiens* are chiefly agriculture and industry, both of which are primarily influenced by science and technology, and controlled by energy inputs along with economic and political expediency.

Although it is difficult to present a summary of all the topics discussed in this book, other than to state the obvious in relation to the dynamism of planet Earth, it is even more difficult to provide a prognosis as to the fate of the environment and the fate of the human race. Although humans are often perceived as the ultimate controllers of their environment, it is clear that knowledge has not yet progressed to the state where effective control is possible. This includes scientific and technological expertise as well as its translation into equitable local, national and international policies through political agencies. For example, the world's energy problems, which include the provision of an adequate supply of food energy as well as fuel energy have not yet been solved. There have also been many intentional modifications of Earth-surface processes, often environmentally detrimental; they have occurred as a result of past and present resource use and it seems that society is either unable or unwilling to combat them. Technological developments are usually Janus-faced; they cause unforeseen detrimental environmental repercussions. This is due to a variety of factors, including the power relations of technology and its wealth-generating capacity which tend to encourage short-term goals or gains and eschew long-term sustainability. There is also a failure to learn from past mistakes – history does indeed repeat itself – and thus to implement the precautionary principle, as well as inadequate recognition of the economics–ecology paradigm. Moreover, the global family is increasing rapidly and any population increases mean additional pressure on a diminishing resource base.

What prognoses can be offered in the light of these facts? Inevitably, predictions can only be based on past experience coupled with realistic projections relating to technological potential. Even then it is wise to be somewhat hesitant! There is the pessimistic stance which espouses the attitude that people are self-destructive, and by modifying or destroying the life-support systems on which they rely, they are condemning *H. sapiens sapiens* to extinction. Conversely, there is the optimistic view, whose protagonists believe that scientific and technological development will open up new opportunities to meet the increasing pressures of a growing global population. Difficulties in interpreting the fossil record mean it is virtually impossible to be certain that there is a case for pessimism. The fossil record is unequivocal in documenting that extinctions have happened in the past but it is rarely possible to determine precisely why they have occurred. Nevertheless, the fossils do reveal something significant: life on Earth has evolved in tandem with changes in the composition of the atmosphere, each influencing the other in a complex interrelationship. This is the crux of the Gaia hypothesis (Section 1.3).

The optimistic view is much easier to justify insofar as there are numerous precedents throughout history which testify to the efficiency of *H. sapiens sapiens* as an inventor of artefacts that can expand the resource base. Increased food production during the past 10 K years is a case in

point. The production of food surpluses, especially in what are now the developed nations, has led to a situation wherein the vast majority of the population are no longer engaged in food production but nevertheless enjoy a high degree of food security. Nineteenth- and twentieth-century developments in the discovery and exploitation of new fossil-fuel resources and their application to food production have also facilitated the support of much larger populations than those of the preceding historic period. The recent developments in biotechnology illustrate that innovation is continuing. The real questions are whether or not such innovations can proceed at the same pace as the changes in the environment, so they can counteract potential threats to survival, and whether or not these innovations themselves and thus generate positive feedback.

Photosynthesis and climate dynamics, the two basic mechanisms that sustain life on earth are still not fully understood. Through their energy transfers they control the underpinning functions of all Earth-surface and atmospheric processes, as well as human activities. Despite the sophistication of civilisation, its food energy requirements must still be met almost totally from photosynthesising organisms, mainly green plants; and their energy production is still largely determined by the environment. Understanding the complex process of photosynthesis, possibly improving it and even finding a way to mimic it may be one key to ensuring the future of society. The subject of climate dynamics is another expedient matter if even the most conservative predictions about the impact of the enhanced greenhouse effect are to be heeded. The world is already experiencing global warming (Fig. 6.3) but the relationship between atmospheric carbon dioxide concentrations, aerosols and climatic change remains unclear. Palaeoenvironmental investigations have much to offer in the search for a solution. This shows how necessary it is to turn to the past, as well as understanding current atmospheric processes, in order to provide answers for the future.

There is little doubt that technological innovations generate positive feedback in Earth–atmosphere interrelationships and thus cause environmental change. Fossil-fuel burning associated with industrialisation, and biomass burning associated with forest clearance for agriculture are the main causes of the enhanced greenhouse effect; acidification is a consequence of fossil-fuel burning, and cultural eutrophication, salinisation and excessive soil erosion are all consequences of agricultural innovations. Present and future innovations must be treated with caution. What impact genetically engineered organisms will have on the environment and on natural flora and fauna remains to be seen. In many respects this is a bigger unknown factor than the enhanced greenhouse effect since there are, locked in the geological record, some precedents for the impact of changing carbon dioxide concentrations that provide a limited basis for prediction. There are no such precedents for the impact of genetically engineered organisms. Genetic manipulation holds much promise for the future, but it is unlikely it will be free of hazards, and the ability to control evolution in this way is indeed a powerful tool that must be treated with the respect it deserves, including a soundly based international legislative infrastructure. Precaution should be the watchword.

Environmental history records that civilisation has already created a wide range of potent instruments of environmental change. Their effects will have to be confronted in the ensuing decades. This is a formidable task requiring a wide range of both social and technological strategies. It is not enough to identify local, national and global impacts or the agents and rates of change. This is just the first stage in understanding the problems; the next stage is to take concerted action. Shades of despair and glimmers of hope accompany the approach of the twenty-first century. Political instability due to loss of food security, as the enhanced greenhouse effect exacts its toll on food production systems, may undermine all that is good about planet Earth by initiating even more wars than there are today; nuclear war could eliminate civilisation completely. Conversely, since the early 1980s there has been increasingly widespread concern over environmental issues. This has pervaded the entire spectrum of political activity. Will this lead to unprecedented international co-operation? Will environmental issues steal the role of the unidentified extraterrestrials, up to now the only common threat to humanity, and probably imagined at that? If so, then perhaps the environmental changes that have so far occurred to the detriment of the Earth's life-support systems have not occurred in vain, provided the positive feedback has not passed the point of no return, promoting changes in the Earth's atmosphere that are inevitable and unsuitable for the continuance of the human race.

10.2   Environmental factors: a series of perspectives

The data hereto presented reflect the dynamic state of planet Earth and the all-pervading stamp of humanity. The past 2 to 3×10^6 years have been characterised by the natural oscillations of glacial and interglacial periods. Indeed the interglacial periods are short-lived, so it could be concluded that the equilibrium state of the Earth is colder than average temperatures today, by ca. 8–$10\,°C$. Moreover, there is the question as to when the next ice age will occur since some $10\,K$ years of the present interglacial have already elapsed. It is possible that anthropogenically induced warming may postpone or even eliminate further ice ages.

In tandem with the other widespread culturally induced changes that affect soils, the biota and water, this poses a vital question: Is the Earth at an ecological threshold? Conversely, is Gaia (Section 1.3) sufficiently robust to adjust to such major perturbations in energy flows and biogeochemical cycles so that negative feedback predominates and the status quo is maintained? Much of what has been reported in this text reflects the tenor of published literature, which dwells on adverse human impacts. This may give a false impression of the state of planet Earth but there is a consensus that global warming is now occurring (Section 6.5) and abundant evidence for environmental degradation. Such developments are a consequence of humanity's past and present use of resources and they contribute to society's potential for survival.

To separate people from environment is impossible; all resources derive from the environment (both biotic and abiotic components) and humanity persists as a result of resource manipulation. This has varied temporally and spatially within the constraints of the physical environment. Agriculture, for example, is limited by climatic and soil characteristics. Without sustained food production, it would be impossible to have mineral exploitation and industrialisation. However, where it exists, and especially in developed countries, food security has been achieved at considerable environmental cost, and in the developing world there is often the double problem of too little food and environmental degradation. However, these situations have arisen as a result of cultural rather than environmental factors. Economy and ecology have become estranged, even antagonistic. Industrialisation has been a major factor in wealth generation but there has been a price to pay; it has exacted an ecological toll. The use of fossil fuels is a key component of the people/environment relationship and links people with Gaia because of the impact of fossil-fuel use on the atmosphere.

10.2.1 The Quaternary period

It is well established that the Quaternary is characterised by glacial and interglacial periods. This oscillation of warm and cold periods caused the growth and decay of polar ice caps, ups and downs of sea level and major changes in the configuration of the Earth's biomes. These glacial and interglacial stages were not internally uniform; the glacial stages often comprised two or more stadials, between which interstadials developed, and during the last interglacial there is evidence from the GRIP ice core, as yet uncorroborated, for an unstable climate and for a 'Little Ice Age' during the present interglacial (Chapter 2). The palaeoenvironmental record also indicates that the switch from an ice age to an interglacial is relatively rapid, suggesting the climatic system can alter rapidly. Moreover, faunal assemblages changed remarkably from one interglacial to another, especially in the temperate zone, and a major extinction event occurred as the last ice age ended. Whether this was naturally or culturally induced is a matter for speculation.

The wealth of palaeoenvironmental evidence from an array of sources has contributed to solving the controversial question as to why climate changes. The cyclical nature of the stratigraphic record coupled with the high degree of correlation between marine, terrestrial and ice cores has led to the reinstatement of the astronomical theory, refined in the 1920s by Milutin Milankovitch (Section 2.6). This theory involves the control of climate by periodic changes in the Earth's orbit around the Sun and changes in various characteristics of the movement of the Earth's axis (Fig. 2.6). Other factors are also likely to be important, e.g. volcanic eruptions, sunspot cycles, orogenic uplift. Moreover, the composition of the atmosphere is now understood to amplify the effects of Milankovitch cycles. For example, the atmosphere of the last ice age had 25 per cent less carbon dioxide and 50 per cent less methane than the pre-industrial Holocene. This testifies to the significance of the greenhouse effect and the importance of the global biogeochemical cycle of carbon in climate regulation. How Milankovitch cycles and the carbon biogeochemical cycle interact is, however, a matter for conjecture. The involvement of the global carbon biogeochemical cycle, as well as reinforcing the Gaia hypothesis (Section 1.3) and its emphasis on the reciprocation between atmospheric composition and life, has opened a debate on the relative importance of the pools and the flux rates. During glacial periods did the carbon from the atmospheric pool become incarcerated in the terrestrial pool, i.e. the biota and soils, or in the oceans? Consequently, do either or both of these pools act as a buffer through changes in primary productivity? Answers to such questions are vital in order to comprehend how these components of the carbon cycle will alter as global fossil-fuel consumption continues to increase, emitting previously sequestered carbon to the atmosphere. In other words, it is essential to identify negative and positive feedback and thresholds in the carbon cycle.

Although the Quaternary period has been characterised by considerable natural environmental change, it has also witnessed the emergence of another powerful agent, the hominids, and eventually *Homo sapiens sapiens* (Fig. 4.2). It is interesting to speculate that the environmental changes caused by tectonic uplift ca. 5×10^6 years ago or more, i.e. the uplift of the Himalayas, contributed to human evolution in some way. Whether it did or not remains speculative, as does the detail concerning early relationships between hominids and their environment. Nor is it clear why numerous hominid species became extinct, despite the ability of some to migrate into Europe and Asia, well beyond their centre of origin in Africa. The controversies surrounding the origin of modern humans and their spread have been discussed in Sections 4.2 and 4.3. Of particular importance is the record of hominid and human abilities to manipulate the environment and resources through ingenuity and technology. This led to what has to be described as a major turning-point in natural and cultural environmental history, the domestication of plants and animals and the inception of agricultural systems. This was a momentous development; various natural systems became control systems as humans learnt to manipulate energy flows and biogeochemical cycles. Thereafter, and on the basis of the advantages generated by agriculture, additional technologies developed. These facilitated further human advancement.

10.2.2 The impact of industrialisation

The emergence of fuel-powered urban-industrial systems, a process which began with the Industrial Revolution, generated new agents of environmental change, e.g. large-scale mineral extraction and fossil-fuel consumption. Moreover, there was a shift in the distribution of population from rural areas to urban centres. Together with industrialisation, this produced increasing volumes of waste products.

Mineral extraction created environmental change by disfiguring landscapes and polluting drainage networks. In many cases the science and technology to combat such problems is already available and major advances are being made in bioremediation (Section 9.6.2). What is often lacking is the legal infrastructure to ensure that exploiting companies undertake reclamation. However, such impacts tend to be local in contrast to the global impact of fossil-fuel use. Its most significant repercussions are acidification, a regional problem in the northern hemisphere, and the enhanced greenhouse effect which is global. The release of carbon from the world's forests as deforestation occurs is also contributing to the enhanced greenhouse effect, as are CFCs, heat-trapping gases that destroy the Earth's shield of stratospheric ozone.

The enhanced greenhouse effect has been discussed in detail in Section 6.5 and since it has only just been established that global warming is occurring, further discussion will be reserved until Section 10.3.1, which is concerned with the climatic future. Although the impact of global warming is not yet widely apparent, there is abundant evidence for the acidification of ecosystems. The nitrous and sulphurous emissions produced by fossil burning combine with water in the atmosphere to produce acids, which have detrimental effects on aquatic and terrestrial plant and animal communities. In some areas acidification began some 200 years ago but it was not recognised until the 1960s, when it was also established that the acidic pollution could be transported considerable distances from its point of origin. Once the links had been established, politicians eventually responded by establishing initiatives to develop mitigating policies (Section 6.6). In Europe and North America such policies have reduced acidic emissions and there is evidence for ecosystem recovery. Further advances in reducing acid emissions could be made, though it is ironic that recent research has shown how a reduction in acid pollution in the northern hemisphere could amplify global warming (Section

6.5). The aerosols produced as acid rain is generated act as cloud condensation nuclei and therefore increase cloudiness. In turn this radiates heat back into space. In addition the newly industrialising countries, notably in Asia, are increasingly generating acid rain as well as increasing their contribution to the pool of carbon dioxide in the atmosphere. This will enhance global warming and alter the pattern of acid rain deposition. Similar problems relate to CFCs. CFCs destroy stratospheric ozone (Section 6.7) and although there are international policies to reduce their consumption, the ozone shield will take a long time to recover. In the meantime an illicit trade in CFCs has begun to flourish and the demands from developing countries is increasing.

Fuel-powered urban-industrial systems generate not only pollution but vast amounts of varied waste products. The disposal and treatment of such wastes promotes environmental change (Section 6.8). For example, sewage promotes cultural eutrophication in aquatic ecosystems whereas domestic waste disposal uses precious land, is aesthetically unattractive and possibly dangerous if methane is produced. Some types of waste are intrinsically hazardous, e.g. dioxins, nuclear waste and PCBs, and are expensive as well as difficult to destroy. Today biotechnology is increasingly being used to mitigate pollution problems and to degrade hazardous waste (Section 9.6.2). Moreover, recycling programmes are becoming increasingly widespread in order to conserve both energy and materials.

The development and maintenance of fuel-powered urban-industrial systems is thus exacting a considerable toll on the environment. The most significant impact is due to fossil-fuel consumption and the capacity this confers on human communities to support continued population increases and industrial development. The legacy of fossil-fuel consumption has yet to be fully realised.

10.2.3 The impact of agriculture

Agriculture and human populations have enjoyed a synergistic relationship since the first domestication of plants and animals 10 K years ago, and possibly even earlier. The range of domesticated biota has increased through time and human communities have developed many ways of manipulating energy flows and biogeochemical cycles. The extent of human ingenuity employed in the manipulation of ecosystems and the resulting range of agroecosystems are quite remarkable. Some agricultural systems may rely almost entirely on solar energy whereas others are characterised by a massive fossil-fuel energy subsidy. Whatever form it takes, agriculture is a major agent of environmental change, possibly the most significant agent.

The nature and organisation of agricultural systems are responses to cultural stimuli which operate within the constraints of the physical environment. The stimuli may include population growth rates, availability of markets, the need to generate foreign currency and the desire for food security, which relates to political superiority. The responses may include increased energy inputs through scientific and technological developments, e.g. pesticides and fertilisers, an increase in land clearance and a decrease in the length of fallow periods. Whatever the characteristics of agricultural systems, they replace natural ecosystems and almost always cause a reduction in biodiversity. The reduction in the extent of natural ecosystems and their fragmentation are causing a high rate of extinction that is unparalleled in the geological record. This means there is a diminution of gene pools just at the time when science is learning how to manipulate the basis of life. This represents loss of opportunities for the future, including the loss of gene products that could potentially be pharmaceuticals or agrochemicals. Moreover, the demise of natural ecosystems may well be impairing the capacity of the biosphere to regulate biogeochemical cycles, including the carbon cycle, with all the significance that holds for the regulation of global climate (Sections 7.2 and 8.2).

In the developed world, especially in Europe, agriculture has caused the removal of a significant proportion of the natural vegetation (see data in Section 1.5). Soil erosion and desertification have ensued in some regions. However, the 'industrialisation' of agriculture, involving large inputs of fossil fuels through mechanisation, fertiliser and pesticide use, has generated further environmental change, as well as contributing to the enhanced greenhouse effect (Section 10.2.2). Water quality and aquatic ecosystems have been adversely affected by agricultural run-off rich in nitrates and phosphates. This nutrient-rich run-off damages aquatic food chains and food webs, and it impairs water quality through cultural eutrophication, with implications for human health (Section 7.5). In addition, the use of pesticides and herbicides has caused floral and fauna changes, bringing some species close to extinction. The prodigious capacity of insects to develop resistance to chemical pesticides has meant there is a continual need for new pesticides. On a positive note, and to counteract the fact that DDT is given an undeserved share of publicity, lessons have been learnt from past mistakes and today's pesticide industry is highly regulated and environmentally aware. Biotechnological developments may reduce the need for chemical pesticides in the future (Section 9.6.1), though such developments may themselves cause environmental change.

The most significant impacts of agriculture in the developing world are deforestation, desertification, salinisation and soil erosion. These are reactions which become manifest in the physical environment as cultural factors initiate change. Environmental degradation has ensued in many arid and semiarid regions as traditional shifting cultivation and nomadic herding have been replaced by permanent cultivation and settled pastoral agriculture. Moreover, there are many examples of the adverse impact of innovations designed to improve productivity, e.g. irrigation and its role in soil salinisation. Although remedial measures to combat some of these problems are available, many developing nations do not have the technical or financial ability to institute them, and even where foreign aid is available reclamation schemes have not always been successful. Nor do mitigating schemes contribute to the solution of underlying problems such as high rates of population growth and inequitable land ownership.

The impact of agriculture on the Earth's surface has undoubtedly been immense. Agricultural systems in Europe, North America, etc., and their high rates of productivity have in the past provided a firm basis on which industrialisation has proceeded and improved standards of living have been achieved. Many developing nations are proceeding along a parallel path, but for others their high population growth rates and extensive land degradation are slowing the development process.

10.2.4 The impact of social and political factors

People bring about environmental change, therefore the process intensifies or accelerates as populations increase. Indeed population change is often invoked as a reason for environmental and cultural change in prehistory. Rarely can such conjecture be supported because of inadequate knowledge about population numbers. However, in a modern context it is established that environmental change is related to population growth but the relationship is by no means straightforward because some people use more resources than others. Figure 10.1 shows the changes in population by continent and for selected countries. It reflects the rapidly increasing populations of developing countries and the recent much lower, sometimes negative, rates of change in North America and Europe. Population growth of such magnitude as occurred since 1950 has undoubtedly produced environmental change. However, resource use in the United Kingdom is an order of magnitude higher than in India, and two orders of magnitude higher than in Burkina Faso. There is no single index than can be used to reflect absolute resource use per capita but data on energy consumption represent a crude measure of reliance on resources. Data for selected countries and regions are given in Table 10.1. Thus, in relation to carbon dioxide output to the atmosphere, the contribution per capita from developing countries is very low when compared with developed countries. Even in China, where development is occurring rapidly, per capita fossil-fuel consumption is still considerably below the levels for the United States or the United Kingdom. However, the data given in Table 10.1 reflect the fact that fossil-fuel consumption is increasing rapidly in many developing countries, e.g. China and India, which have more than doubled their consumption in the past 20 years. This will be referred to again in Section 10.3.4.

Environmental change has also been initiated through unjust environments. Before the Middle Ages it is difficult to determine with certainty whether systems of land tenure of land appropriation caused people either to migrate to new areas, where they cleared land for agriculture, or to employ unsustainable practices such as overgrazing. For the period since 1500 there are many examples of the creation of unjust environments which have prompted environmental degradation. The expansion of Europe between the sixteenth and nineteenth centuries is a case in point. European expansion into the Americas, Africa and Australasia brought profound changes not only to the environments of these continents but also within indigenous cultures (Sections 5.7 to 5.9). Moreover,

the legacy of this colonialism is still apparent today in many countries. In much of Latin America, for example, land ownership remains mostly in the hands of the elite 10 per cent of the population; this has created a duel society comprising the rich and the poor. The poor are often forced through poverty and necessity to exploit natural resources unsustainably, e.g. the 'shifted' cultivators of Latin America's rainforests (Section 8.2).

Imperialism is a means of exerting power which can be manifest in many ways. Colonialism represents power over people and land (the resource base); its principal characteristic is the appropriation of resources by the powerful from the powerless. Today imperialism is less blatant but no less unjust. Ecological imperialism could today be interpreted as the usurpation of biomass and genes from the South (developing nations) by the North (developed nations). Agricultural produce, forest products, etc., find ready markets in the North but the nations of the South cannot always feed themselves. The North therefore appropriates the sunshine, soil and water of the South, and in the search for 'useful' genes to fuel the biotechnology industry (Section 9.6), genetic imperialism is becoming a possibility (see below in relation to the United Nations Environment and Development Summit in Rio de Janeiro, 1992). Technological imperialism is yet another form of powermongering: the North is technologically advanced whereas the South aspires to technological advancement. Assistance and technology transfer are not readily dispensed from the North to the South, even where technologies to mitigate environmental degradation are concerned. Such developments are not necessarily conscious policies on the part of companies or governments in the developed world. The underpinning rationale is wealth generation for shareholders and citizens who are not necessarily aware of unjust environments or the North's involvement in their creation. Moreover, those who are aware often plead their powerlessness in the face of such a massive issue; they often respond the same way to the domestic agenda. Environmental issues do not inspire personal responsibility.

On a positive note, environmental awareness has increased substantially during the past three decades and many activist groups have been formed, e.g. Greenpeace and Friends of the Earth. These developments have contributed to the emergence of environmental issues on the agenda of local, national and international politics. Reference was made to the Montreal Protocol to curb the use of

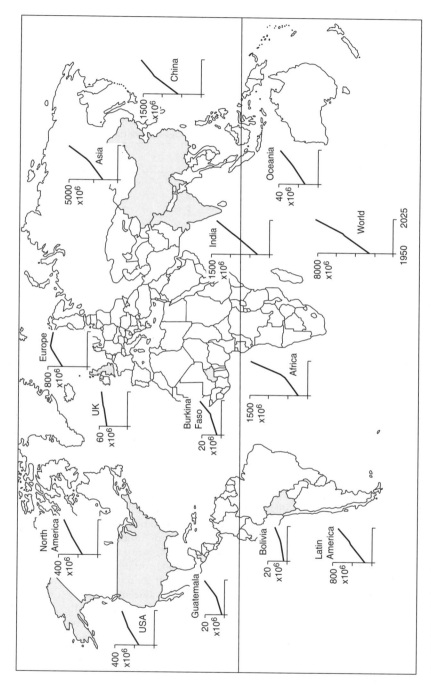

Fig. 10.1 Population change by continent and for selected countries since 1950 along with projections for 2025 (based on World Resources Institute, 1996).

Table 10.1 Data on energy consumption, 1973–1993

A. Energy consumption per continent and per capita				
	Total consumption in 1993 (PJ)	Percent change since 1973	Per capita consumption in 1973 (GJ)	Percentage change since 1973
Africa	8 805	144	13	41
Asia[a]	95 679	185	28	92
Latin America	16 300	~90	35	~30
North America	90 947	~30	318	~ −3
Europe	108 523	90	148	73
Oceania	4 595	93	166	44

B. Energy consumption per capita for selected countries				
	Total consumption in 1993 (GJ)	Percent change since 1973		
USA	317	−7		
UK	164	7		
China	25	110		
India	10	128		
Bolivia	12	56		
Burkina Faso	1	150		
Guatemala	7	4		

Source: based on World Resources Institute (1996)
[a] Some 31% derives solely from China

CFCs in Section 6.7 and international measures to reduce acidic emissions were examined in Section 6.6. Table 10.2 gives a summary of the most notable developments. Especially influential were the United Nations Conferences of 1972 (the Stockholm Conference) and 1992 (often known as the Earth Summit) along with the 1984 report of the World Commission on Environment and Development (WCED); often known as the Brundtland Report, it was published in 1987. The 1972 conference raised public awareness and was the first international debate on environmental issues. WCED crystallised and publicised the basic human rights that each individual should be guaranteed, and highlighted the principles of sustainable development and intergenerational security. The Earth Summit was attended by more than 100 world leaders, who discussed environmental problems of the present and future through the foci of conventions on biodiversity, climatic change, forest principles and Agenda 21 (Table 10.3). These events have increased general awareness and given rise to government agencies with environmental remits. The policies of the Earth Summit were intended to help the environment and the disadvantaged, but it is debatable how much tangible improvement has occurred or will occur in the next few years, as they are implemented.

10.3 Environmental factors: their future impacts

In the past, social and economic factors have brought about environmental change by interacting with the biosphere and its physical, chemical and biological factors; this will continue into the future. However, the complexity of environmental change means it is difficult to make predictions about future trends that are reliable. Earth-surface processes may react in unexpected ways to anthropogenic stimuli, so even the best models may be found lacking. Nevertheless, any consideration of future environmental change must involve climate, industry and agriculture.

Prediction of future climatic change is imprecise but the Earth is now experiencing a warming trend caused by anthropogenic emissions of heat-trapping gases. The trend will continue as emissions increase, though international agreements now exist to curb such emissions and this will affect the rate at which global warming ensues. A related statistic is the pace of industrialisation in developing nations and how much reliance continues to be placed on fossil fuels. Nuclear power as an alternative energy source may be environmentally benign, insofar as it does not produce heat-trapping gases or acidic emissions, but there are other hazards which can generate catastrophic

Table 10.2 Major international agreements and conferences since 1970

1971	Convention on Wetlands of International Importance Especially as Waterfowl Habitat (Ramsar Convention)
1972	Convention concerning the protection of the World Cultural and Natural Heritage (World Heritage Convention) United Nations Conference on the Human Environment, Stockholm Greenpeace founded First European Ecology Party/Green Party founded in Switzerland Values Party (campaigning on environmental issues) founded in New Zealand and was the first such party to enter candidates for national elections
1973	Convention on International Trade in Endangered Species (CITES) Green Party founded in the United Kingdom
1974	World Food Conference World Population Plan of Action
1975	International Environmental Education Programme
1977	United Nations Conference on Desertification, Nairobi
1979	Convention on Long-Range Transboundary Air Pollution (LRTAP) Launch of World Climate Impact Programme
1980	German Green Party founded Launch of World Conservation Strategy
1982	United Nations Convention on the Law of Sea (LOS)
1984	Founding of '30 per cent Club' to curb acidic emissions – Helsinki protocol signed in 1985 World Commission on Environment and Development – so-called Brundtland Report published in 1987
1985	Tropical Forestry Action Plan (TFAP) Convention on the Protection of the Ozone Layer signed in Vienna
1987	Protocol on Substances that Deplete the Ozone Layer signed in Montreal (Montreal Protocol)
1989	Convention on the Control of Transboundary Movements of Hazardous Wastes (Basel Convention)
1991	World Conservation Strategy for the 1990s
1992	United Nations Conference on Environment and Development, Rio de Janeiro (Earth Summit) United Nations Commission on Sustainable Development established

Source: based on Tolba *et al.* (1992) and McCormick (1995)

environmental change. In addition agriculture may become increasingly 'industrialised' in developing nations as they strive to generate sufficient food for internal consumption and export commodities. The use of biotechnology, as in the case of so many earlier technologies, may bring unexpected environmental change.

All of these developments will be influenced by population growth, the balance of political power on a global basis, local and national systems of land tenure, the presence of poverty or plenty and a host of factors related to social organisation and power relations. Moreover, international agreements such as those initiated at the Earth Summit of 1992 (Table 10.2) will alter the course and intensity of environmental change. In the next few decades technological advances will be important and it is likely there will be a reduction in the number of people directly involved with the land as increasing numbers turn to industry for their

livelihoods. These people will be no less dependent on the environment for resources, and although much can be written about the uncertainties surrounding environmental change, it can be said with absolute conviction that the state of the environment will remain of paramount importance.

10.3.1 The future of the climate

Palaeoenvironmental evidence (Chapters 2 and 3) indicates that the Quaternary period has experienced comparatively rapid environmental change, when measured on the geological timescale. Interglacial periods last ca. 15 to 20 K years, and since the last ice age ended ca. 12 K years ago, it is not unreasonable to suppose the Earth will eventually cool. This is the long-term perspective. However, current records indicate the occurrence of anthropo-

Table 10.3 Major agreements of the 1992 Earth Summit (United Nations Conference on Environment)

Framework Convention on Climate Change
This was signed by more than 150 nations. Its aim is to provide an international framework within which future actions can be taken. The main climate problem was recognised as global warming and discussions ensued concerning reductions in carbon dioxide emissions, i.e. to reduce to 1990 levels by 2000. The EU declined to ratify the treaty; individual countries set their own targets and lengthened the time available to reach them. Developing countries expected developed countries to set the pace and to fund the cuts for both groups of countries. This treaty, although being diluted by ambiguous composition, has at least established the value of the precautionary principle since, in 1992, the evidence for anthropogenic global warming was equivocal.

The Convention of Biological Diversity
The focus of this treaty, which came into force in 1994, is the conservation of the Earth's biota through the protection of species and habitats or ecosystems. The debate polarised developing and developed countries insofar as the developing countries house more of the world's biodiversity (where it is experiencing many threats; see Section 8.2) than the developed countries, which are skilled in biotechnology (Section 9.6). The convention was signed by 153 countries, a major exception being the United States because of fears that US biotechnology could be threatened. The United States has since capitulated.

Statement on Forest Principles
This represents a considerably diluted (and legally unenforcible) consensus on forest management which resulted from earlier (1990–91) attempts to formulate a global forest convention. The principles assert the sovereign right of individual countries to profit from their forest resources but advocate this should take place within a framework of forest protection, conservation and management.

The Rio Declaration on Environment and Development
This has its origins in the UN conference on the Human Environment in Stockholm, 1972 (Table 10.2). It established basic guidelines for the attitude of individuals and nations to the environment and development.

Agenda 21
This is an action plan for sustainable development. It established tasks, targets, costs, etc., for actions that combine economic development and environmental protection. In many nations, e.g. the United Kingdom, the suggestions of Agenda 21 have been incorporated into local and national development strategies. It led to the establishment of the UN Commission on Sustainable Development.

genic warming (Section 6.5). It is not without the bounds of possibility that this warming is delaying or counteracting the onset of the next ice age. For example, the 'Little Ice Age' which began in the fourteenth century may have been halted by rising concentrations of carbon dioxide in the atmosphere, caused by widespread deforestation in Europe. There is little evidence for periods similar to the 'Little Ice Age' in pre-Holocene interglacial deposits, though it remains to be seen whether the climatic swings in the Eemian section of the Greenland GRIP core (Section 2.4) are confirmed as real, or as artefacts of ice deformation or sampling. These possibilities highlight the need to elucidate the environmental changes that occurred in pre-Holocene interglacial periods, in order to determine the nature and rates of ecosystem response to natural agents of environmental change. This will provide baseline data on natural systems, their response rates and stimuli. Moreover, palaeoenvironmental investigations are essential if the operation of the global carbon cycle is to be understood, especially in the context of climate regulation and the role of the terrestrial and marine biota.

Palaeoenvironmental studies contribute to the prediction of climate changes in two other ways. Firstly, reconstructions of climate and environments from palaeoenvironmetal data can be used to test mathematical models, thereby helping to validate them. Any discrepancies can be investigated, allowing refinements to be made to both the mathematical models and the palaeoenvironmental methods of reconstruction. Secondly, palaeoenvironmental studies indicate that the change from one stable climatic regime to another may occur rapidly, even over decades and certainly over centuries. Perturbations to the present atmosphere through the addition of greenhouse gases may thus induce rapid rather than gradual change. GCMs vary considerably in their predictions of global warming. This highlights the complexity of the climate system and the inadequacy of current scientific understanding. However, there is general agreement that high latitudes will experience greater warming than low latitudes (Section 6.5). Consequently, the release of vast stores of carbon from the tundra peatlands and boreal zone will reinforce global warming. Moreover, if there are further reductions in acidic emissions, which at current levels cause a counteracting cooling effect, global warming will be accelerated. The forecasting of the magnitude of reinforcing factors is almost impossible for a variety of reasons, including legislation to curb acidic emissions, accelerating industrialisation in developing countries which will produce increased acidic emissions, and the reaction of the feedback mechanisms within the carbon, sulphur and nitrogen cycles.

It is difficult to predict the extent, intensity and spatial variation of global warming, but it is even more difficult to predict its impact on Earth-surface processes, ecosystem distribution and characteristics, agricultural productivity and population distribution. For some nations the advantages of an extended growing season will be considerable, for others the loss of agricultural and silvicultural interests will be economic disadvantages. Moreover, many island nations are concerned that rising sea levels will at best impair their tourist industries and at worst engulf their atolls, etc., creating environmental refugees.

The only certainty is that the future climate is uncertain. This also means that many other aspects of the future cultural and environmental attributes of planet Earth are uncertain. Environmental determinism is indeed alive and well! Changes in the resource base may give rise to changes in the distribution of political power, providing opportunities for some and reducing the possibilities for others.

10.3.2 The future effects of industrialisation

Industrialisation will continue to have a major influence on the environment, especially through the continued use of fossil fuels. Even if renewable and environmentally benign fuels are developed in the long term, fossil fuels will remain the major sources of energy for the next three or four decades at least. Moreover, the decline in fossil-fuel consumption that characterises the developed world will be outweighed by rapidly increasing fossil-fuel consumption in the newly industrialising countries of the developing world. The early stages of industrialisation are proving energy intensive, as occurred in Europe in the eighteenth and nineteenth centuries, and there are few controls of gaseous or particulate emissions. Consequently, carbon dioxide emissions will continue to increase, exacerbating the

enhanced greenhouse effect. Moreover, acid deposition will increase in regions hitherto unaffected. The aspirations of developing countries to raise standards of living and reduce foreign debts, etc., through export-generated income are likely to take priority over international agreements to limit heat-trapping gases and acidic emissions. Changes in energy policy and energy efficiency are necessary to reduce or avoid these problems as well as the economic problems associated with reliance on imported fossil fuels, notably oil from politically unstable regions such as the Middle East. However, there is much debate as to which alternative energy sources warrant intensive development. Hydroelectric power, nuclear power, wind power, wave power and biomass fuels all have disadvantages, and solar power is difficult to harness for large-scale consumption.

Other aspects of environmental change caused by industrialisation include numerous forms of pollution generated by the deliberate and inadvertent release of waste products. Apart from causing environmental degradation (and sometimes posing health risks), this constitutes a waste of resources. Many waste products and effluents could be treated to minimise their impact on drainage systems, etc., and to effect the recycling of useful substances. For example, recent developments in biotechnology (Section 9.6.2) reflect the capacity of many living organisms to undertake such tasks. It is thus likely that in the next few decades many kinds of organisms will be harnessed to recycle heavy metals, decompose hazardous wastes and xenobiotic chemicals, prevent the cultural eutrophication of sensitive wetlands and drainage systems and possibly even to recycle and/or concentrate radioactive substances such as uranium. Environmental bioremediation is also likely to become big business in the wake of environmental legislation and environmental taxes (Section 10.3.4), and will generate significant amounts of wealth.

10.3.3 The future effects of agriculture

Speculation as to how agriculture will affect the global environment in the next few decades is made particularly difficult because of problems associated with predicting the impact of global warming on agricultural systems (Section 6.5.2). Whatever doubts there may be about the impact of global warming, the fact remains there will be continued global population growth until the middle of the twenty-first century; world population is likely to grow from its current 5.7×10^9 to 10×10^9 before stabilising. (Fig. 10.1 gives projections until 2025.) Consequently, there will be continuous pressure to increase the productivity of existing agricultural systems and increasing pressure to transform the world's remaining natural ecosystems into agricultural systems. However, the situation is complex because of the spatial variations in the distribution of agricultural productivity, and hence food security, and rates of population growth. The highest degree of food security occurs mainly in the developed world, as shown in Figure 10.2, which gives details of food trade. Rates of population growth in the developed world are low or even negative, and there is a food surplus of as much as 10–50 per cent. The converse applies in much of the developing world (Fig. 10.3), where population growth rates are between 2 and 3 per cent and will continue to be high for some time (Fig. 10.4). In these countries, populations are expected to double in ca. 25 years, not only increasing pressure on agricultural systems but on resources in general. Moreover, the need to generate foreign income to repay debts and to finance the development process are likely to increase demands on primary industries, notably agriculture and forestry, to produce export goods.

These population patterns have considerable implications for environmental change in general, and for agriculture in particular there are two more or less opposite global trends. Firstly, in the developed world there is a trend towards extensification due to the production of surplus foodstuffs. This is manifest as set-aside policies and diversification of farm activities into leisure, tourism and forestry. Conversely in the developing world there is a trend towards agricultural intensification as the demands for food and export goods rise. Increasing agricultural productivity in the developing world can be achieved either by bringing new lands into cultivation, at the expense of natural ecosystems, and/or by increasing the productivity of existing agricultural systems. The first option would not be in the interests of long-term sustainability because of the ensuing loss of biodiversity (a loss of genetic resources) and impairment of biogeochemical exchanges (Sections 7.2 and 8.2) with implications for the regulation of global climate. It would also increase soil erosion and desertification. Moreover, real gains in productivity could be low insofar as

there is a consensus that the best land is already being cultivated. Thus few gains are likely to be made from farming marginal land.

Increasing the productivity of existing agricultural land is the preferred option. Indeed it has been argued that high-technology agriculture (high-yield farming) has not only fed the world's population, and fuelled its past growth, but also has a role in the conservation of natural habitats. One such argument is presented in Table 10.4. Many of these issues also relate to the likely future role of biotechnology in agriculture, as shown in Figure 9.5. Inevitably there are advantages and disadvantages of high-technology agriculture, some of which apply to the additional inputs it involves, e.g. pesticides and fertilisers (Chapters 7 and 8), and some of which apply to its power relations. Power relations involve the control of technology and globally there is a polarisation between the developing and developed worlds. The expertise tends to be concentrated in the developed world but it is most needed in the developing world. Although there are many disadvantages of high-technology agriculture, there is much that could be improved to minimise the adverse impacts. This requires legislation, policing, education and technology transfer. It also requires implementation of the precautionary principle. In relation to biotechnology there are many potential disadvantages (Fig. 9.5) but with adequate controls many of them could be avoided. Possibly one of the biggest challenges facing agriculture is to reduce fossil-fuel inputs, which not only contribute to global warming but also cause the impairment of water resources when used to produce nitrate fertilisers.

Although Table 10.4 may represent an extreme view of the role of high-technology agriculture in promoting cultural and environmental well-being, it is clearly here to stay. Perhaps the more important issue is the control of the technology and its implementation, rather than the technology itself.

10.3.4 The future effects of social and political factors

Rates of population growth and their spatial variations (Figs 10.3 and 10.4) are a vital ingredient in determining the pattern of future global environmental change. However, as stated in Section 10.2.4, it is not sufficient to consider absolute numbers of people alone, because the environmental

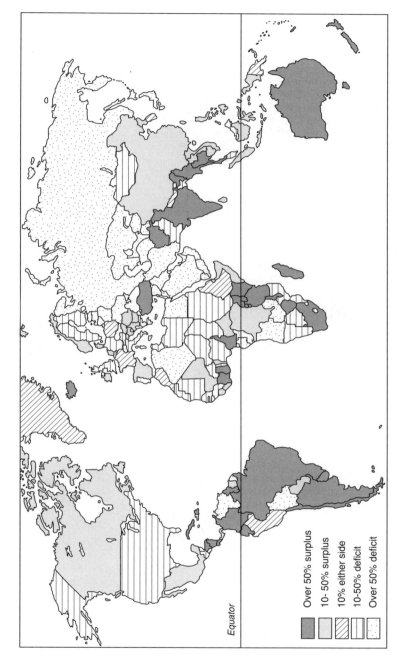

Fig. 10.2 Balance of trade in food products as a percentage of total trade in food products (based on Philips Atlas of the World, 1992).

Equator

Over 50% surplus
10- 50% surplus
10% either side
10-50% deficit
Over 50% deficit

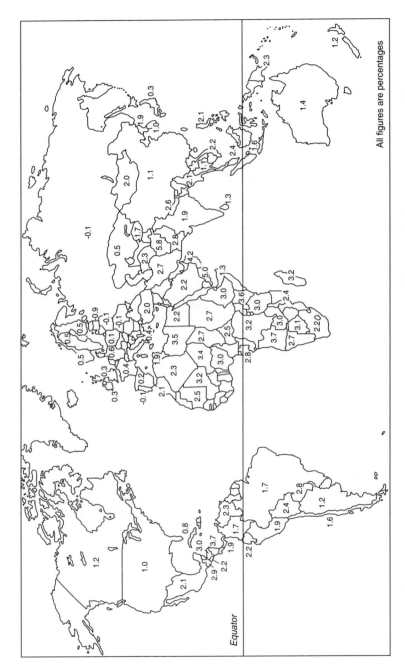

Fig. 10.3 Average annual rates of population change for 1990–1995 (based on World Resources Institute, 1996).

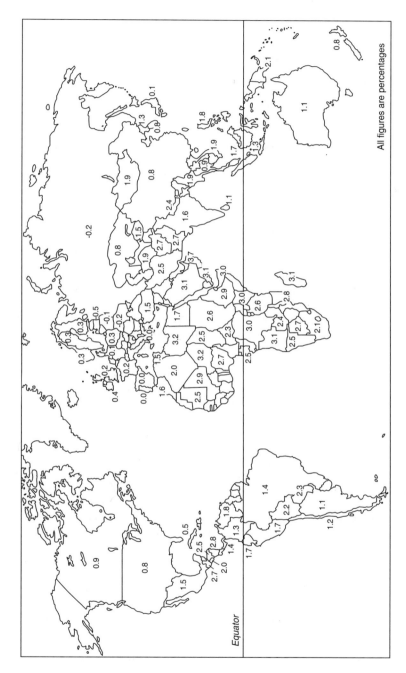

Fig. 10.4 Average annual rates of population change for 2000–2025 (based on World Resources Institute, 1996).

Table 10.4 The case for high-yield (industrialised) agriculture

Reasons why high-yield (industrialised) agriculture is environmentally and culturally beneficial
- High-yield agriculture is essential for the conservation of remaining natural ecosystems and wildlife.
- High-yield plantation forestry is essential for the preservation of existing natural forests.
- Increased crop yields, hence enhanced food security, does not encourage increased population growth but appears to cause a significant decline in population growth rates.
- The methods employed to produce high yields are safe for human consumption and for the environment because of stringent registration procedures, e.g. for pesticides.
- The food produced by high-yield agriculture is cheaper than organically produced food.
- Organic farming is culturally and environmentally a high-risk venture because of unforeseen natural toxins and the requirement for additional hectarage to compensate for reduced productivity per unit area. Moreover, poor farmers have large families which ultimately require more land and cause deforestation.

Strategies to feed increasing world population based on high-yield agriculture
- Improvements are required in crop and tree yields, especially for agriculture in the developing world. The application of many aspects of biotechnology is particularly important to improve the productivity of existing crops and to develop new crops and new strains of existing crops by exploiting gene pools.
- Use of only the best agricultural land is essential to ensure high productivity and facilitate the abandonment of marginal land to wildlife and/or recreation. The use of pesticides, etc., will thus be limited spatially.

Source: Avery (1995)

impact of any given individual relates to their level of resource use. As developing countries improve their standards of living, the environmental cost of individuals will intensify; the rapidly growing populations of these nations will thus exert a greater environmental impact than they do now. Increased demands will both arise from and lead to agricultural change and industrial development. Although the aspirations of developing countries, in terms of living standards can be estimated on the basis of those currently enjoyed by developed nations, it is not so easy to gauge the aspirations of the developed nations themselves because they depend on new, even undiscovered, resources. Insofar as the developed world has experienced a demographic transition and population numbers have stabilised, along with an energy transition whereby energy consumption per capita is stabilising or even decreasing, it is interesting to speculate about a resource-use transition. Will this also reach a plateau? Conversely, the developing world may experience increased pressure on its resources as its populations crave exotic products, woods, plants, etc., and as tourism increases, promoting globalisation. Will there be continued appropriation of its soils, minerals, biomass, etc.? Or will yet another transition be reached as its populations reserve increasing volumes of their own resources for internal consumption?

In view of the recognition that local environmental change eventually adds up to the transformation of the global environment, there is the question of responsibility for promoting change.

One reason why environmental change occurs is because individuals live in environments that are unjust and which force and/or limit the range of options. Systems of land tenure, leasing and product marketing are often unjust. In that they favour wealth generation for a few, such systems are unlikely to disappear. It is, however, heartening that education, media interest, tourism and information technology all promote environmental awareness and the exchange of expertise. All are likely to remain important in the future. The deepening involvement of politics with environmental issues at all levels from the local to the global is encouraging. The Earth Summit of 1992, the UNEP conference in Rio de Janeiro, was a landmark in environmental politics. Its outcome has been somewhat disappointing in terms of practical achievements but it represents the acknowledgement by political parties of all persuasions that the environment is all-important. Rio is unlikely to be the only Earth summit. Its repercussions will be felt well into the twenty-first century and in the next few years it is likely there will be a strengthening of Rio's Climate Treaty (Table 10.3) or possibly a new treaty; either option will involve wide participation to limit emissions of heat-trapping gases and the introduction of specific targets. Precursors to this involve national policies, many of which are already in place to control emissions in various ways.

Political and institutional reactions to environmental change are likely to include a wide range of environmental taxes, increasing the penalties for breaches of environmental legislation and improved enforcement. In the 25 countries of the Organisation for Economic Co-operation and Development (mainly European countries plus North America, Japan, Mexico, Australia and New Zealand) there are already hundreds of environmental taxes which involve many aspects of the environment ranging from fuel consumption to waste disposal. Some nations, including Denmark, Finland, the Netherlands, Norway and Sweden, have already introduced a carbon tax and impose some of the highest rates of value-added tax on domestic motor fuels. The objective is to reduce fossil-fuel consumption. Similar taxes will be imposed elsewhere, as will emission controls on vehicles and industry to improve air quality. Legislation to control factors such as land use, deforestation and waste disposal will increase in the developing world and enforcement agencies will become more effective.

It is, however, important that at least some elements of the precautionary principle should become associated with environmental planning and legislation. This requires acceptance of the fact that current assessment of environmental issues is imperfect, that prediction is usually flawed and that Gaia works in mysterious ways. Such admissions themselves point to a need for precaution, as do the recorded repercussions of past and current technologies (including agriculture). Precaution requires experimentation, etc., including risk assessment, on the part of developers to ascertain what environmental impacts will be caused by the proposed innovations. An example of the current application of the precautionary principle is given in Figure 7.10, which details the tests required for the registration of new pesticides. Employment of the precautionary principle is essential to establish whether or not new technologies will cause irreversible environmental damage. It could be argued that sustainable development is impossible without the widespread exercise of this principle.

10.4 Envoi

What is certain is that environmental change will continue to occur but what is uncertain is the direction and rate of change at all scales from the local to the global. In the short term it seems likely that global warming will ensue and the resulting regimes of temperature and precipitation will alter the patterns of the world's biomes and agricultural systems. The warming trend will not be globally uniform; high latitudes will warm more than low latitudes but beyond that the predictions vary. Although rising sea levels due to thermal expansion of the oceans and melting of polar ice caps will have adverse impacts on many low-lying nations, changes in the climatic constraints on agriculture may provide opportunities and advantages for others. Much hinges on the rapidity with which climatic change occurs and the capacity for adjustment that exists within environmental and cultural systems. This highlights the reciprocity between environmental systems and their political, social and economic factors, especially manifest through food security. If the environmental systems change then feedback mechanisms inevitably produce adjustments in the way people live and interact; there are significant implications for power relations that control standards of living and wealth distribution.

Global warming is not the only agent of future environmental change. Rapidly increasing population numbers in many developing countries will exacerbate existing fuelwood and food shortages, and it is unlikely that rates of deforestation will diminish in the next decade as increased agricultural land is created. Consequently, soil erosion and desertification are likely to increase. These are grave environmental problems that represent significant loss of vital resources. They are not conducive to sustainable development. Of particular significance is the irreversible loss of biodiversity. Moreover, these problems are often exacerbated through external pressures created by wealthy nations, e.g. ready markets for biomass products. The agents of environmental change thus become global but the effects begin locally then escalate into global issues, e.g. deforestation and its impact on the carbon biogeochemical cycle.

On an optimistic note, there are some encouraging signs which imply that civilisation is at least beginning to recognise the necessity of achieving a symbiotic and sustainable relationship with the environment. This is manifest in the technology available to reverse some of the undesirable effects of human activities. Examples include the reclamation of derelict land, pollution abatement and measures to combat acidification, salinisation, desertification and soil erosion. The late twentieth century has witnessed important advances in environmental improvement and conservation but the challenge remains monumental. Moreover, the challenges involve each and every person, not just environmental scientists. Reducing population growth rates must be a major target; this will require changing attitudes towards the family unit and towards the status and education of women. Even today almost all methods of fertility control are directed only at women. Although the relationship between population numbers and environmental impact is not straightforward, because of different levels of resource use associated with varied standards of living, population control will reduce environmental pressures and increase the possibility of sustainable development. For people in the developed world the target should involve a reduction in resource use, perhaps coupled with increased efficiency of resource use. This again emphasises the responsibility of the individual in the achievement of global sustainability. Moreover, the mutual dependence of the developing and developed worlds is highlighted by the synergy between biotechnology and biodiversity; the technology is a facility of developed nations but it relies to a large extent on the genetic resources in the biodiversity of the developing nations. The developed nations will therefore contribute to the conservation of biodiversity, provided there is a just procedure for allocating benefits to the developing countries. In turn, the preservation of biodiversity will maintain effective biogeochemical cycles, and thus contribute to the regulation of global climate.

Another cause for optimism is the increasing significance of the environment as a political issue at all levels and for all shades of political activity. Some of this has been prompted by the increased media coverage of environmental problems and the ease with which information is now transmitted around the globe. The establishment of international agreements to curb CFC and acidic emissions represents formal recognition of the associations between people and the environment, the internationality of environmental issues and the need for collective mitigation measures. The road to and from Rio has been rocky but the resulting impetus must not be lost; complacency must not be allowed to return. The lessons from the past also highlight the need for precautions when developing new technologies. This fact is well illustrated by the current debates over biotechnology.

However, taking precautions does not mean the abandonment of all research into environmental change; in fact, the need for improved understanding suggests that research should intensify. It is important for palaeoenvironmental research to continue to focus on past climatic changes: how they are caused, their rates and their impacts. Without such baseline data any simulation will remain speculative and subject to wide margins of error. The accuracy of predictive models also needs to be improved; this will require a improved appreciation of how the basic life-support systems of planet Earth operate and how they interrelate with the biota. Regulation of these systems could become feasible through land-use policies. Lessons from the past must be heeded; this is part of a precautionary approach insofar as past technologies have caused deliberate and inadvertent environmental change, often through a lack of forethought and the stimulus of short-term gains. Avoidance should become the environmental watchword and the challenge of the new millennium in order to promote cultural and environmental well-being. It would be appropriate to adopt as a motto what Pearl S. Buck said in her novel *The Good Earth*, published in 1931: 'It was true that all their lives depended upon the Earth'. Moreover, for posterity there are some apposite lines from a First World War memorial, 'For your tomorrow, we gave our today.' It must not become 'For our today we took your tomorrow.'

Further reading

Auty, R.M. (1995) *Patterns of Development: Resources, Policy and Economic Growth*, Edward Arnold, London.

Brown, L.R. *et al.* (1996) *State of the World 1996.* Earthscan, London.

Chatterjee, P. and Finger, M. (1994) *The Earth Brokers.* Routledge, London.

Cooper, D.E. and Palmer, J.A. (eds.) (1995) *Just Environments.* Routledge, London.

Johnston, R.J., Taylor, P.J. and Watts, M.J. (eds.) (1995) *Geographies of Global Change: Remapping the World in the Late Twentieth Century,* Blackwell, Oxford.

McCormick, J. (1995) *The Global Environmental Movement.* J. Wiley, Chichester.

Organisation for Economic Co-operation and Development (1995) *Environmental Taxes in OECD Countries.* OECD, Paris.

Watson, R.T., Zinyowera, M.C. and Moss, R.H. (eds.) (1996) *Climate Change 1995: Impacts, Adaptations and Mitigation of Climate Change.* Cambridge University Press, Cambridge.

References

Abbatt, J.P.D. and Molina, M.J. (1993) Status of stratospheric ozone depletion. *Annual Review of Energy and the Environment* **18**, 1–29.

Abramovitz, J.N. (1996) Sustaining freshwater ecosystems. In L.R. Brown (ed.) *State of the World 1996*. Earthscan, London, pp. 60–77.

Abrol, I.P. and Sehgal, J.L. (1994) Degraded lands and their rehabilitation in India. In D.J. Greenland and I. Szabolcs (eds) *Soil Resilience and Sustainable Land Use*. CAB International, Wallingford, pp. 129–44.

Adamson, D.A. and Fox, M.D. (1982) Change in Australian vegetation since European settlement. In J.M.B. Smith (ed.) *A History of Australasian Vegetation*. McGraw-Hill, Boston, pp. 109–46.

Adegbehin, J.O. (1988) Meeting the increasing wood demand from the Nigerian forests. *Journal of World Forest Resource Management* 3, 31–46.

Adger, W.N. and Brown, K. (1994) *Land Use and the Causes of Global Warming*. J. Wiley, Chichester.

Adsersen, H. (1989) The rare plants of the Galápagos islands and their conservation. *Biological Conservation* **47**, 49–77.

Agassi, M. (ed.) (1996) *Soil Erosion, Conservation and Rehabilitation*. Marcel Dekker, New York.

Ahmad, S., Aslam, M. and Shafiq, M. (1996) Reducing water seepage from earthen ponds. *Agricultural Water Management* 30, 69–76.

Aiello, L.C. (1993) The fossil evidence for modern human origins in Africa: a revised view. *American Anthropologist* **95**, 73–96.

Aiello, L.C. (1994) Thumbs up for our early ancestors. *Science* 265, 1540–41.

Aitchison, E. (1996) Methane generation from UK landfall sites and its use as an energy source. *Energy Conversion and Management* 37, 1111–16

Akagi, H., Malm, O., Kinjo, Y., Harada, M., Branches, F.J.P. Pfeiffer, W.C. and Kato, H. (1995) Methylmercury pollution in the Amazon, Brazil. *Science of the Total Environment* **175**, 85–95.

Aksu, A.E., Yasar, D., Mudine, P.J. and Gillespie, H. (1995) Late glacial–Holocene paleoclimatic and paleoceanographic evolution of the Aegean Sea– micropaleonological and stable isotope evidence. *Marine Micropaleontology* **25**, 1–28.

Aladin, N.V., Plotnikov, I.S. and Potts, W.T.W. (1995) The Aral Sea desiccation and possible ways of rehabilitating and conserving its northern part. *Environmetrics* **6**, 17–29.

Alaouimhamdi, M. and Aleya, L. (1995) Assessment of the eutrophication of Al-Massira Reservoir (Morocco) by means of a survey of the biogeochemical balance of phosphate. *Hydrobiologia* **297**, 75–82.

Alayev, E.B., Badenkov, Y.P. and Karavaeva, N.A. (1990) The Russian Plain. In B.L. Turner II, W.C. Clark, R.W. Kates, J.F. Richards, J.T. Mathewes and W.B. Meyer (eds) *The Earth Transformed by Human Action*. Cambridge University Press, Cambridge, pp. 543–60.

Alcock, S., Cherry, J.F. and Davis, J.L. (1994) Intensive survey, agriculture practice and the classical landscape of Greece. In I. Morris (ed.) *Classical Greece: Ancient Histories and Modern Archaeologies*. Cambridge University Press, Cambridge, pp. 137–70.

Alcock, S.E. (1993) *Graecia Capta*. Cambridge University Press, Cambridge.

Alexandratos, N. (ed.) (1995) *World Agriculture: towards 2010: an FAO study*. J. Wiley, Chichester.

Ali, J. and Mirza, M.I. (1993) Space technology in the mapping of desertification in Pakistan. *Advances in Space Research*. **13**, 103–6.

Allaglo, L.K., Areba, A., D'Almeida, N.C., Gu-Konu, E.Y., Kounetsron, K. and Seddoh, K.F. (1987) Togo, its geopotential and attempts for land-use planning – a case study. In P. Arndt and G.W. Lüttig (eds) *Mineral Resources Extraction, Environmental Protection and Land-use Planning in the Industrial and Developing Countries*. E. Schweizerbart'sche Verlagsbuchhandlung, Stuttgart, pp. 243–70.

Allegre, J.C. and Cassel, D.K. (1996) Dynamics of soil physical properties under alternative systems to slash and burn. *Agriculture Ecosystems and Environment* **58**, 39–48.

Allegre, J.C. and Rao, M.R. (1996) Soil and water conservation by contour hedging in the humid tropics of Peru. *Agriculture Ecosystems and Environment* **57**, 17–25.

Allen, R.C. (1991) The two English agricultural revolutions, 1450–1850. In B.M.S Campbell and M. Overton (eds) *Land, Labour and Livestock: Historical Studies in European Agricultural Productivity.* Manchester University Press, Manchester, pp. 237–54.

Allison, T.D., Moeller, R.E. and Davis, M.B. (1986) Pollen in laminated sediments provides evidence for a mid-Holocene forest pathogen outbreak. *Ecology* 67, 1101–05.

Alsaleh, I.A. and Taylor, A. (1994) Lead concentration in the atmosphere and soil of Riyadh, Saudi Arabia. *Science of the Total Environment* 141, 261–67.

Alsulaimi, J., Viswanathan, M.N., Naji, M. and Sumait, A. (1996) Impact of irrigation on brackish groundwater lenses in Northern Kuwait. *Agricultural Water Management* 31, 75–90.

Amman, B., Lotter, A.F., Eicher, U., Gaillard, M.J., Wohlfarth, B., Haeberli, W., Lister, G., Maisch, M., Niessen, F. and Schlüchter, C. (1994) The Würmian late-glacial in lowland Switzerland. *Journal of Quaternary Science* 9, 119–25.

Amonooneizer, E.H., Nyamah, D. and Bakiamoh, S.B. (1996) Mercury and arsenic pollution in soil and biological samples around the mining town of Obuasi, Ghana. *Water, Air and Pollution* 91, 363-373.

An, Z. (1994) China during the Neolithic. In S.J. De Laet, A.H. Dani, J.L. Lorenzo and R.B. Nunoo (eds) *History of Humanity, Volume I: Prehistory and the Beginnings of Civilization.* Routledge, London, and UNESCO, Paris, pp. 482–89.

Andersen, J. (1993) Beetle remains as indicators of the climate in the Quaternary. *Journal of Biogeography* 20, 557–62.

Andersen, S. T. (1966) Interglacial vegetation succession and lake development in Denmark. *Palaeobotanist* 15, 117–27.

Anderson, D. (1987) Managing the forest: the conservation of Lembus Forest, Kenya, 1904–1963. In D. Anderson and R. Grove (eds) *Conservation in Africa: People, Politics and Practice.* Cambridge University Press, Cambridge, pp. 249–68.

Anderson, P.M. and Brubaker, L.B. (1994) Vegetation history of northcentral Alaska: a mapped summary of Late-Quaternary pollen data. *Quaternary Science Reviews* 13, 71–92.

Anderson, R.S. (1993) A 35,000 year vegetation and climate history from Potato Lake, Mogollon Rim, Arizona. *Quaternary Research* 40, 351–59.

Anderson, T.W. and MacPherson, J.B. (1994) Wisconsinan late-glacial environmental change in Newfoundland: a regional review. *Journal of Quaternary Science* 9, 171–78.

Anderson, U.V. (1995) Resistance of Danish coastal vegetation types to human trampling. *Biological Conservation* 71, 223–30.

Andrews, J.T. (1994) Wisconsinan late-glacial environmental change on the southeast Baffin Shelf, southeast Baffin Island and northern Labrador. *Journal of Quaternary Science* 9, 179–83.

Andrews, P. (1982) Hominoid evolution. *Nature* 295, 185–86.

Andriessen, P.A.M., Helmens, K.F., Hooghiemstra, H., Riezebos, P.A. and van der Hammen, T. (1994) Absolute chronology of the Pliocene–Quaternary sediment sequence of the Bogota area, Colombia. *Quaternary Science Reviews* 12, 483–501.

Andrieu, V., Huang, C.C., O'Connell, M. and Paus, A. (1993) Late-glacial vegetation and environment in Ireland: first results from four western sites. *Quaternary Science Reviews* 12, 681–705.

Anglinbrown, B., Armourbrown, A. and Lalor, G.C. (1995) Heavy metal pollution in Jamaica I: survey of cadmium, lead and zinc concentrations in the Kintyre and Hope Flat district. *Environmental Geochemistry and Health* 17, 51–56.

Angold, R., Beech, G. and Taggart, J. (1989) *Food Biotechnology.* Cambridge University Press, Cambridge.

Anthony, D., Telegin, D.Y. and Brown, D. (1991) The origin of horseback riding. *Scientific American* 265, 44–48.

Apps, M.J. and Price, D.T. (eds) (1996) *Forest Ecosystems, Forest Management and the Global Carbon Cycle.* Springer-Verlag, Berlin.

Archer, J. (1994) Policies to reduce nitrogen loss to water from agriculture in the United Kingdom. *Marine Pollution Bulletin* 29, 444-449.

Archibold, O.W. (1995) *Ecology of World Vegetation.* Chapman and Hall, London.

Arkhipov, S.A., Bespaly, V. G., Faustova, M.A., Glushkova, O.Y., Isayeva, L.L. and Velichko, A.A. (1986a) Ice-sheet reconstructions. *Quaternary Science Reviews* 5, 475–88.

Arkhipov, S.A., Isayeva, L.L., Bespaly, V.G. and Glushkova, O.Y. (1986b) Glaciation of Siberia and North-east USSR. *Quaternary Science Reviews* 5, 463–74.

Asabere, P.K. (1987) Attempts at sustained yield management in the tropical high forests of Ghana. In F. Mergen and J.R. Vincent (eds) *Natural Management of Tropical Moist Forests.* Yale University School of Forestry and Environmental Studies, New Haven, CT, pp. 47–70.

Ashworth, A.C., Markgraf, V. and Villagran, C. (1991) Late Quaternary climatic history of the Chilean channels based on fossil pollen and beetle analyses, with an analysis of the modern vegetation and pollen rain. *Journal of Quaternary Science* 6, 279–91.

Assadourian, C.S. (1992) The colonial economy: the transfer of the European system of production to New Spain and Peru. *Journal of Latin American Studies* 24(suppl.), 55–68.

Atkinson, I.A.E. and Cameron, E.K. (1993) Human influence on the terrestrial biota and biotic communities of New Zealand. *Trends in Ecology and Evolution* 8, 447–51.

Atkinson, T.C., Briffa, K.R. and Coope, G.R. (1987) Seasonal temperatures in Britain during the past 22,000 years, reconstructed using beetle remains. *Nature* **325**, 587–92.

Atlas, R.M. (1995) Petroleum biodegradation and oil spill bioremediation. *Marine Pollution Bulletin* **31**, 178–82.

Attfield, R. (1994) *Environmental Philosophy: Principles and Prospects*. Avebury, Aldershot.

Aucott, M. and Winka, M. (1996) Findings and recommendations of the New Jersey mercury emissions standard setting task force. *Journal of Hazardous Materials* **47**, 103–17.

Aumonier, S. (1996) The greenhouse gas consequences of waste management – identifying preferred options. *Energy Conversion and Management* **37**, 1117–22.

Avery, D.T. (1995) *Saving the Planet with Pesticides and Plastic*. Hudson Institute, Indianapolis IN.

Avnir, D., Ludlow, D.K. and Steinberger, E.H. (1995) On replenishing the depleted ozone by its photogeneration in the stratosphere. *Energy Sources* **17**, 495–501.

Awad, M.A., Nada, A.A., Hamza, M.S. and Froehlich, K. (1995) Chemical and isotopic investigation of grounwater in Tahta region, Sohag – Egypt. *Environmental Geochemistry and Health* **17**, 147–153.

Awadallah, R.M., Soltan, M.E. and Rashed, M.N. (1996) Relationship between heavy metals in mud sediments and beach soil of the River Nile. *Environment International* **22**, 253–58.

Axtell, J. (1992) *Beyond 1492: Encounters in Colonial North America*. Oxford University Press, Oxford.

Badve, R.M., Kumaran, K.P.N. and Rajhekhar, C. (1993) Eutrophication of Lonar Lake, Maharashtra. *Current Science* **65**, 347–51.

Bahn, P.G. (1994) Time for a change. *Nature* **367**, 511–12.

Bains, W. (1992) Sensors for a clean environment. *Bio/Technology* **10**, 515–18.

Bains, W. (1993) *Biotechnology from A to Z*. Oxford University Press, Oxford.

Baker, R.G. (1984) Holocene vegetational history of the western United States. In H.E. Wright Jr (ed.) *Late-Quaternary Environments of the United States* (2 vols). University of Minnesota Press, Minneapolis MN, and Longman, London, Vol. 2. *The Holocene*, pp. 109–27.

Baker, R.J., Vandenbussche, R.A., Wright, A.J., Wiggins, L.E., Hamilton, M.J., Reat, E.P., Smith, M.H., Lomakin, M.D. and Chesser, R.K. (1996) High levels of genetic change in rodents of Chernobyl. *Nature* **380**, 707–8.

Bakewell, P. (1987) Mining. In L. Bethell (ed.) *Colonial Spanish America*. Cambridge University Press, Cambridge, pp. 203–49.

Balopoulos, E.T. and Friligos, N.C. (1994) Water circulation and eutrophication in the northwestern Aegean Sea – Thermaikos Gulf. *Toxicological and Environmental Chemistry* **41**, 155–67.

Banda, A.Z., Maghembe, J.A., Ngugi, D.N. and Chome, V.A. (1994) Effect of intercropping maize and closely spaced *Leucaena* hedgerows on soil conservation and maize yield on a steep slope at Ntcheu, Malawi. *Agroforestry Systems* **27**, 17–22.

Bandi, H.-G. (1994) Palaeolithic and Mesolithic art in Europe. In S.J. De Laet, A.H. Dani, J.L. Lorenzo and R.B. Nunoo (eds) *History of Humanity, Volume I: Prehistory and the Beginnings of Civilization*. Routledge, London, and UNESCO, Paris, pp. 234–41.

Banuelos, G.S. (1996) Managing high levels of boron and selenium with trace element accumulator crops. *Journal of Environmental Science and Health A* **31**, 1179–96.

Barber, K.E., Chambers, F.M., Dumayne, L., Haslam, C.J., Maddy, D. and Stoneman, R.E. (1994) Climatic change and human impact in North Cumbria: peat stratigraphic and pollen evidence from Bolton Fell Moss and Walton Moss. In J. Boardman and J. Walden (eds) *The Quaternary of Cumbria: Field Guide*. Quaternary Research Association, Oxford, pp. 20–54.

Barber, R.G. (1995) Soil degradation in the tropical lowlands of Santa Cruz, Eastern Bolivia. *Land Degradation and Rehabilitation* **6**, 95–107.

Bard, E., Hamelin, B. and Fairbanks, R.G. (1990) U-Th ages obtained by mass spectroscopy in corals from Barbados: sea level during the past 130,000 years. *Nature* **346**, 456–58.

Barker, G. (1985) *Prehistoric Farming in Europe*. Cambridge University Press, Cambridge.

Barlow, B.A. (1994) Phytogeography of the Australian region. In R.H. Groves (ed.) *Australian Vegetation*, 2nd edn. Cambridge University Press, Cambridge, pp. 3–35.

Barmawidjaja, D.M., Vanderzwaan, G.J., Jorissen, F.J. and Puskaric, S. (1995) 150 years of eutrophication in the northern Adriatic sea. *Marine Geology* **122**, 367–84.

Barnola, J.M., Raynaud, D., Korotkevich, Y.S. and Lorius, C. (1987) Vostok ice core provides 160,000-year record of atmospheric CO_2. *Nature* **329**, 408–14.

Barnosky, A.D. (1986) 'Big game' extinction caused by Late Pleistocene climatic change: Irish elk (*Megaloceras giganteus*) in Ireland. *Quaternary Research* **25**, 128–35.

Barnosky, A.D. (1994) Defining climate's role in ecosystem evolution: clues from Late Quaternary mammals. *Historical Biology* **8**, 173–90.

Bartley, D.D, and Chambers, C. (1992) A pollen diagram, radiocarbon ages and evidence of agriculture on Extwistle Moor, Lancashire. *New Phytologist* **121**, 311–20.

Bar-Yosef, O. (1994) The contributions of Southwest Asia to the study of the origin of modern humans. In M.H. Nitecki and D.V. Nitecki (eds) *Origins of Anatomically Modern Humans*. Plenum, New York, pp. 23–66.

Bassett, M.G. (1985) Towards a 'common language' in stratigraphy. *Episodes* **8**, 87–92.

Battarbee, R.W. (1994) Diatoms, lake acidification and the Surface Water Acidification Program (SWAP) – a review. *Hydrobiologia* **274**, 1–7.

Battarbee, R.W., Allott, T.E.H., Juggins, S., Kreiser, A.M., Curtis, C. and Harriman, R. (1996) Critical loads of acidity to surface waters: an empirical diatom-based palaeolimnological model. *Ambio* **25**, 366–69.

Battle, M., Bender, M., Sowers, T., Tans, P.P., Butler, J.H., Elkins, J.W., Ellis, J.T., Conway, T., Zhang, N., Lang, P. and Clarke, A.D. (1996) Atmospheric gas concentrations over the past century measured in air from firn at the South Pole. *Nature* **383**, 231–35.

Bazzaz, F.A. and Fajer, E.D. (1992) Plant life in a CO_2-rich world. *Scientific American* **266**, 68–74.

Beaumont, P. (1989) *Drylands: Environmental Management and Development.* Routledge, London.

de Beaulieu, J.-L., Andrieu, V., Ponel, P., Reille, M and Lowe, J.J. (1994) The Weichselian Late-glacial in southwestern Europe (Iberian Peninsula, Pyrenees, Massif Central, northern Italy). *Journal of Quaternary Science* **9**, 101–107

Bechmann, R. (1990) *Trees and Man: the forest in the Middle Ages* (translated by Katharyn Dunham). Paragon House, New York.

Becker, B. (1993) An 11,000 year German oak and pine dendrochronology for radiocarbon calibration. *Radiocarbon* **35**, 201–13.

Beckerman, W. (1995) *Small is Stupid: Blowing the Whistle on the Greens.* Duckworth, London.

Beer, J., Blinov, A., Bonani, G., Finkel, R.C., Hofmann, H.J., Lechmann, B., Oeschger, H., Sigg, A., Schwander, J., Staffelbach, T.J. Stauffer, B., Suter, M. and Wölfi, W. (1990) Use of [10]Be in polar ice to trace the 11-year cycle of solar activity. *Nature* **347**, 164–66.

Beerling, D.J., Birks, H.H. and Woodward, F.I. (1995) Rapid late-glacial atmospheric CO_2 changes reconstructed from the stomatal density record of fossil leaves. *Journal of Quaternary Science* **10**, 379–84.

Behrendt, H. and Boekhold, A. (1993) Phosphorus saturation in soils and groundwaters. *Land Degradation and Rehabilitation* **4**, 233–43.

Bell, M. (1989) Environmental archaeology as an index of continuity and change in the medieval landscape. In M. Aston, D. Austin and C. Dyer (eds) *The Rural Settlements of Medieval England.* Blackwell, Oxford, pp. 269–86.

Bell, P.R.F. and Elmetri, I. (1995) Ecological indicators of large-scale eutrophication in the Great Barrier Reef lagoon. *Ambio* **24**, 208–15.

Bender, M., Sowers, T., Dickson, M.-L., Orchardo, J., Grootes, P., Mayewski, P.A. and Meese, D.A. (1994) Climate correlations between Greenland and Antarctica during the past 100,000 years. *Nature* **372**, 663–66.

Bender, M., Ellis, T., Tans, P., Francey, R. and Lowe, D. (1996) Variability in the O_2/N_2 ratio of Southern Hemisphere air, 1991–1994. Implications for the carbon cycle. *Global Biogeochemical Cycles* **10**, 9–21.

Benjamin, C. (1994) The growing importance of diversification activities for farm households. *Journal of Rural Studies* **10**, 331–42.

Bennett, J. (1995) Biotechnology and the future of rice production. *GeoJournal* **35**, 333–35.

Bennett, K.D. (1988) Holocene pollen stratigraphy of central East Anglia, England, and comparison of pollen zones across the British Isles. *New Phytologist* **109**, 237–53.

Bennett, K.D. and Humphry, R.W. (1995) Analysis of late-glacial and Holocene rates of vegetational change at two sites in the British Isles. *Review of Palaeobotany and Palynology* **85**, 263–87.

Bennett, K.D., Fossit, J.A., Sharp, M.J. and Switsur, V.R. (1990a) Holocene vegetational and environmental history at Loch Lang; South Vist, Western Isles, Scotland. *New Phytologist* **114**, 281–98.

Bennett, K.D., Simonson, W.D. and Peglar, S.M. (1990b) Fire and man in post-glacial woodlands of eastern England. *Journal of Archaeological Science* **17**, 635–42.

Benton, L.J. and Mannion, A.M. (1995) *The post-glacial vegetation history of the Lashford Lane fen catchment, Oxfordshire, UK.* Department of Geography, University of Reading Geographical Papers B 39, 43 pp.

Benz, B.F. and Iltis, H. (1990) Studies in archaeological maize. I: the 'wild' maize from San Marcos cave re-examined. *American Antiquity* **55**, 500–11.

Berdnikov, K. (1995) Draft of the Povolje's Wood of the 17th century as a source of historical monitoring of the environment. In I.G. Simmons and A.M. Mannion (eds) *The Changing Nature of the People–Environment Relationship: Evidence from a Variety of Archives,* Proceedings of the International Geographical Union Commission on Historical Monitoring of Environmental Changes Meeting, Přihrazy, Czech Republic. Department of Social Geography and Regional Development, Charles University, Prague, pp. 27–31.

Berger, W.H., Killingley, J.S. and Vincent, E. (1987) Time scale of the Wisconsin/Holocene transition: oxygen isotope record in the western equatorial Pacific. *Quaternary Research* **28**, 295–06.

Berglund, B.E. (ed.) (1986) *Handbook of Holocene Palaeohydrology and Palaeoecology.* J. Wiley, Chichester.

Berglund, B.E., Björck, S., Lemdahl, G., Bergsten, H., Nordberg, K. and Kolstrup, E. (1994) Late Weichselian environmental change in southern Sweden and Denmark. *Journal of Quaternary Science* **9**, 127–32.

Bergman, C.A. (1993) The development of the bow in Western Europe: a technological and functional perspective. In G.L. Peterkin, H.M. Bricker and P.M. Mellars (eds) *Hunting and Animal Exploitation in the Later Palaeolithic and Mesolithic of Eurasia.* American Anthropological Association, pp. 95–105.

Berry, L., Lewis, L.A. and Williams, C. (1990) East African Highlands. In B.L. Turner II, W.C. Clark, R.W. Kates, J.F. Richards, J.T. Matthews and W.B. Meyer (eds) *The Earth as Transformed by Human Action.* Cambridge University Press, Cambridge, pp. 533–42.

Beuneche, D. (1996) Backfitting French nuclear power plants – integration by lots. *ATW International Zeitschrift Fur Kernenergie* **41**, 33–34.

Bewick, D. (1994) The mobility of pesticides in soil – studies to prevent groundwater contamination. In H. Börner (ed.) *Pesticides in Ground and Surface Water.* Springer-Verlag, Berlin. pp. 57–86.

Bibby, C.J. (1987) Effects of management of commercial conifer plantations on birds. In J.E.G. Good (ed.) *Environmental Aspects of Plantation Forestry in Wales.* Institute of Terrestrial Ecology, Monkswood, pp. 70–75.

Biot, Y. and Lu, X.X. (1995) Loss of yield caused by soil erosion on sandy soils in the UK. *Soil Use and Management* **11**, 157–62.

Birks, H.H., Paus, A., Svendsen, J.I., Torbjørn, A., Mangerud, J. and Landvik, J.Y. (1994) Late Weichselian environmental change in Norway, including Svalbard. *Journal of Quaternary Science* **9**, 133–45.

Birks, H.J.B. (1986) Late-Quaternary biotic changes in terrestrial and lacustrine environments, with particular reference to north-west Europe. In Berglund B.E. (ed.) *Handbook of Holocene Palaeoecology and Palaeohydrology.* J. Wiley, Chichester, pp. 3–65.

Birks, H.J.B. and Peglar, S.M. (1979) Interglacial pollen spectra from Sel Ayre, Shetland. *New Phytologist* **83**, 559–75.

Biswas, A.K. (1993) Land resources for sustainable agricultural development in Egypt. *Ambio* **22**, 556–60.

Blackstock, T.H., Stevens, J.P., Howe, E.A. and Stevens, D.P. (1995) Changes in the extent and fragmentation of heathland and seminatural habitats between 1920–22 and 1987–88 in the Llŷn Peninsula, Wales, UK. *Biological Conservation* **72**, 33–44.

Blaikie, P. (1986) Natural resource use in developing countries. In R.J. Johnston and P.J. Taylor (eds) *A World in Crisis?* Blackwell, Oxford, pp. 107–26.

Blaikie, P. and Brookfield, H. (1987) *Land Degradation and Society.* Methuen, London.

Blair, J. (1994) *Anglo-Saxon Oxfordshire.* Oxfordshire Books, Oxford, and Alan Sutton, Stroud.

Blaschke, P.M., Trustrum, N.A. and DeRose, R.C. (1992) Ecosystem processes and sustainable land use in New Zealand steeplands. *Agriculture, Ecosystems and Environment* **41**, 153–78.

Blette, V.L. and Newton, R.M. (1996) Application of the integrated lake watershed acidification study model to watershed liming in Woods Lake, New York. *Biogeochemistry* **32**, 363–83.

Blumler, M.A. and Byrne, R. (1991) The ecological genetics of domestication and the origins of agriculture. *Current Anthropology* **32**, 23–54.

Blunden, J. (1985) *Mineral Resources and Their Management.* Longman, London.

Blunden, J. and Curry, N. (1996) Analysing amenity and scientific problems: the Broadlands, England. In P. Sloep and A. Blowers (eds) *Environmental Problems as Conflicts of Interest.* Arnold, London, pp. 37–61.

Blunden, J. and Turner, G. (1985) *Critical Countryside.* BBC, London.

Blunier, T., Chappellaz, J., Schwander, J., Stauffer, B. and Raynaud, D. (1995) Variations in atmospheric methane concentration during the Holocene epoch. *Nature* **374**, 46–49.

Blytt, A. 1876. *Essays on the Immigration of the Norwegian Flora During Alternating Rainy and Dry Periods.* Cammermeye, Christiana (now Oslo).

Boardman, J. and Evans, R. (1994) Soil erosion in Britain: a review. In R.J. Rickson (ed.) *Conserving Soil Resources: European Perspectives.* CAB, Wallingford, pp. 3–12.

Bobovnikova, T., Dibtseva, A., Mitroshkov, A. and Pleskachevskaya, G. (1993) Ecological assessment of a region with PCB emissions using samples of soil, vegetation and brest milk -a case study. *Science of the Total Environment* **140**, 357–64.

Bodhankar, N. and Chatterjee, B. (1994) Pollution of limestone aquifer due to urban waste disposal around Raipur, Madhya Pradesh, India. *Environmental Geology* **23**, 209–13.

Body, R. (1987) *Red or Green For Farmers (and The Rest of Us)?* Broad Leys Publishing, Saffron Walden, UK.

Boggess, C.F., Flaig, E.G. and Fluck, R.C. (1995) Phosphorus budget-basin relationships for Lake Okeechobee tributary basins. *Ecological Engineering* **5**, 143–62.

Bohncke, S.J.P. (1993) Late-glacial environmental changes in the Netherlands: spatial and temporal patterns. *Quaternary Science Reviews* **12**, 707–17.

Boix, C., Calvo, A., Imeson, A.C., Schoorl, J.M., Soto, S., Tiemessen, I. (1995) Properties and erosional response of soils in a degraded ecosystem in Crete (Greece). *Environmental Monitoring and Assessment* **37**, 79–92.

Bonavia, D. and Grobman, A. (1989) Andean maize: its origins and domestication. In D.R. Harris and G.C. Hillman (eds) *Foraging and Farming: the Evolution of Plant Exploitation.* Unwin Hyman, London, pp. 456–70.

Bond, G., Broecker, W., Johnsen, S., McManus, J., Labeyrie, L., Jouzel, J. and Bonani, G. (1993) Correlations between climate records from North Atlantic sediments and Greenland ice. *Nature* **365**, 143–47.

Bond, J. (1994) Forest, chases, warrens and parks in Medieval Wessex. In M. Aston and C. Lewis (eds) *The Medieval Landscape of Wessex.* Oxbow Monograph 46, pp. 115–58.

Bond, W.J., Stock, W.D. and Hoffman, M.T. (1994) Has the Karoo spread? A test for desertification using carbon isotopes from soils. *South African Journal of Science* **90**, 391–97.

Bonnefille, R. and Mohammed, U. (1994) Pollen-inferred climatic fluctutations in Ethiopia during the last 3000 years. *Palaeogeography, Palaeoclimatology, Palaeoecology* **109**, 331–43.

Bonnichsen, R. and Schneider, A.L. (1995) Roots. *The Sciences* **35**, 26–31.

Bonnicksen, T.M. (1982) The development of forest policy in the United States. In R.A. Young (ed.) *Introduction to Forest Science*. J. Wiley, New York, pp. 7–36.

Bonzongo, J.C.J., Heim, K.J., Chen, Y.A., Lyons, W.B., Warwick, J.J., Miller, J.R. and Lechler, P.J. (1996) Mercury pathways in the Carson river – Lahontan Reservoir, Nevada, USA. *Environmental Toxicology and Chemistry* **15**, 677–83.

Boserup, E. (1965) *The Conditions of Agricultural Growth: the Economics of Agrarian Change Under Population Pressure*. Aldine, Chicago.

Bosman, A.F., Hooghiemstra, H. and Cleef, A.M. (1994) Holocene mire development and climatic change from a high Andean *Plantago rigida* cushion mire. *The Holocene* **4**, 233–43.

Boutron, C.F., Candelone, J.P. and Hong, S.M. (1995) Greenland snow and ice cores – unique archives of large-scale pollution of the troposphere of the northern hemisphere by lead and other heavy metals. *Science of the Total Environment* **161**, 233–41.

Bowcock, A.M., Ruiz-Linares, A., Tomfohrde, J., Minch, E., Kidd, J.R. and Cavalli-Sforza, L.L. (1994) High resolution of human evolutionary trees with polymorphic microsatellites. *Nature* **368**, 455–57.

Bowdler, S. (1993) Sunda and Sahul: a 30 K yr culture area. In M.A. Smith, M. Spriggs and B. Frankhauser (eds) *Sahul in Review: Pleistocene Archaeology in Australia, New Guinea and Island Melanesia*. Department of Prehistory, Research School of Pacific Studies, Australian National University, Occasional Papers in Prehistory 24, pp. 60–61.

Bowen, D.Q. (1991) Time and space in the glacial sediment systems of the British Isles. In J. Ehlers, P.L. Gibbard and J. Rose (eds) *Glacial Deposits in Great Britain and Ireland*. A.A. Balkema, Rotterdam, pp. 3–11.

Bowen, D.Q. (1994) The Pleistocene of North West Europe. *Science Progress Oxford* **76**, 209–23.

Bowen, D.Q., Richmond, G.M., Fullerton, D.S., Šibrava, V., Fulton, R.J. and Velichko, A.A. (1986a) Correlation of Quaternary glaciations in the northern hemisphere. *Quaternary Science Reviews* **5**, 509–10.

Bowen, D.Q., Rose, J., McCabe, A.M. and Sutherland, D.G. (1986b) Correlation of Quaternary glaciations in England, Ireland, Scotland and Wales. *Quaternary Science Reviews* **5**, 299–340

Bowen, D.Q., Hughes, S., Sykes, G.A. and Miller, G.H. (1989) Land-sea correlations in the Pleistocene based on isoleucine epimerisation in non-marine molluscs. *Nature* **340**, 49–51.

Boyden, S. (1987) *Western Civilisation in Biological Perspective*. Clarendon Press, Oxford.

Boyle, A.E. (1990) Quaternary deepwater paleoceanography. *Science* **249**, 863–70.

Boynton, W.R., Garber, J.H., Summers, R. and Kemp, W.M. (1995) Inputs, transformations, and transport of nitrogen and phosphorus in Chesapeake Bay and selected tributaines. *Estuaries* **18**, 285–314.

Bradbury, J.P. (1975) Diatom stratigraphy and human settlement in Minnesota, *Geological Society of America Special Paper* No. 171.

Bradley, D.G., Machugh, D.E., Cunningham, P. and Loftus, R.T. (1996) Mitochondrial diversity and the origins of African and European cattle. *Proceedings of the National Academy of Sciences of the United States of America* **93**, 5131–35.

Bradley, R.S. and Jones, P.D. (1993) 'Little ice age' summer temperature variations: their nature and relevance to recent global warming trends. *The Holocene* **3**, 367–76.

Bradshaw, A.D. and Chadwick, M.J. (1980) *The Restoration of Land*. Blackwell, Oxford.

Brain, C.K. and Sillen, A. (1988) Evidence from Swartkrans cave for the earliest use of fire. *Nature* **336**, 464–66.

Brand-Miller, J.C. and Colagiuri, S. (1994) The carnivore connection: dietary carbohydrate in the evolution of NIDDM. *Diabetologia* **37**, 1280–86.

Brassell, S.C., Eglinton, G., Marlowe, I.T., Pflaumann, U. and Sarnthein, M. (1986) Molecular stratigraphy: a new tool for climate assessment. *Nature* **320**, 129–33.

Braun, A.R. (1996) The reuse and fixation of CO_2 in chemistry, algal biomass and fuel substitutions in the traffic sector. *Energy Conservation and Management* **37**, 1229–34.

Brereton, R., Bennett, S. and Mansergh, I. (1995) Enhanced greenhouse climate change and its potential effect on selected fauna of southeastern Australia – a trend analysis. *Biological Conservation* **72**, 339–54.

Bridgland, D. (1994) *The Quaternary History of the Thames Valley*. Chapman and Hall, London.

Briffa, K.R., Jones, P.D., Schweingruber, F.H., Shiyatov, S.G. and Cook, E.R. (1995) Unusual 20th-century summer warmth in a 1,000 year temperature record from Siberia. *Nature* **376**, 156–59.

Broadus, J.M. (1993) Possible impacts of, and adjustments to, sea-level rise: the case of Bangladesh and Egypt. In R.A.Warrick, E.M. Barrow and T.M.L. Wigley (eds) *Climate and Sea-Level Change: Observations Projections and Implications*. Cambridge University Press, Cambridge, pp. 263–75.

Broberg, O. (1988) Delayed nutrient responses to the liming of Lake Gårdsjön, Sweden. *Ambio* **17**, 22–27.

Brody, D.J., Pirkle, J.L., Kramer, R.A., Flegal, K.M., Matté, T.D., Gunter, K.W. and Paschal, D.C. (1994) Blood lead levels in the US population: phase 1 of the 3rd national health and nutrition examination survey (NHANES III, 1988 to 1991). *Journal of the American Medical Association* **272**, 277–83.

Broecker, W.S. (1994) Massive iceberg discharges as triggers for global climate change. *Nature* **372**, 421–24.

Broecker, W.S. and van Donk, J. (1970) Insolation changes, ice volumes and the ^{18}O record in deep-sea cores. *Reviews of Geophysics and Space Physics* **8**, 169–97.

Brohman, J. (1996) New directions in tourism for third-world development. *Annals of Tourism Research* **23**, 48–70.

Bromage, T.G. and Schrenk, F. (1995) Biogeographic and climatic basis for a narrative of early hominid evolution. *Journal of Human Evolution* **28**, 109–14.

Bromage, T.G., Schrenk, F. and Zonneveld, F.W. (1995) Palaeoanthropology of the Maláwi Rift: an early hominid mandible from the Chiwondo Beds, northern Malawi. *Journal of Human Evolution* **28**, 71–108.

Brook, E.J., Sowers, T. and Orchardo, J. (1996) Rapid variations in atmospheric methane concentration during the past 110,000 years. *Science* **273**, 1087–91.

Brown, D.C.W. and Thorpe, T.A. (1995) Crop improvement through tissue culture. *World Journal of Microbiology and Biotechnology* **11**, 409–15.

Brown, L.R. (1994) Fertilizer use keeps dropping. In L.R. Brown, H. Kane and D.M. Roodman (eds) *Vital Signs 1994*. Norton, New York, pp. 42–43.

Brown, S. (1993) Tropical forests and the global carbon cycle – the need for sustainable land-use patterns. *Agriculture Ecosystems and Environment* **46**, 31–44.

Brown, S. (1996) Mitigation potential of carbon dioxide emissions by management of forests in Asia. *Ambio* **25**, 273–78.

Brown, S.L., Chaney, R.L., Angle, J.S. and Baker, A.J.M. (1994) Phytoremediation potential of *Thlaspi caerulescens* and bladder campion for zinc-contaminated and cadmium-contaminated soils. *Journal of Environmental Quality* **23**, 1151–57.

Brubaker, C.L. and Wendel, J.F. (1994) Reevaluating the origin of domesticated cotton (*Gossypium hirsutum* Malvaceae) using nuclear restriction fragment length polymorphisms (RFLPs). *American Journal of Botany* **81**, 1309–26.

Brugam, R.B. (1978) Pollen indicators of land-use change in southern Connecticut. *Quaternary Research* **9**, 349–62.

Brunet, M., Beauvilain, A., Coppens, Y., Heintz, E., Moutaye, A.H.E. and Pilbeam, D. (1995) The first australopithecine 2,500 kilometres west of the Rift Valley (Chad). *Nature* **378**, 273–75.

Brynildsen, L.I., Selnaes, T.D., Strand, P. and Hove, K. (1996) Countermeasures for radiocesium in animal products in Norway after the Chernobyl accident – techniques, effectiveness and cost. *Health Physics* **70**, 665–72.

Budd, P., Gale, D., Ixer, R.A.F. and Thomas, R.G. (1994) Tin sources for prehistoric bronze production in Ireland. *Antiquity*, **68**, 518–24.

Budd, P.M., Pollard, A.M., Thomas, R.G. and Williams, P.A. (1992) The early development of metallurgy in the British Isles. *Antiquity* **66**, 677–86.

Bunting, M.J. (1994) Vegetation history of Orkney, Scotland. Pollen records from small basins in west Mainland. *New Phytologist* **128**, 771–92.

Burkle, L.H. (1993) Late Quaternary interglacial stages warmer than present. *Quaternary Science Reviews* **12**, 825–31.

Burney, D.A., Brook, G.A. and Cowart, J.B. (1994) A Holocene pollen record for the Kalahari Desert of Botswana from a U-series dated speleothem. *The Holocene* **4**, 225–32.

Bush, M.B. (1993) An 11400 year palaeoecological history of a British Chalk grassland. *Journal of Vegetation Science* **4**, 47–66.

Bush, M.B., and Hall, A.R. (1987) Flandrian *Alnus*: expansion or immigration? *Journal of Biogeography* **14**, 479–81.

Bush, M.B., Piperno, D.R., Colinvaux, P.A., De Oliveira, P.E., Krissek, L., Miller, M.C. and Rowe, W.E. (1992) A 14,300-yr palaeoecological profile of a lowland tropical lake in Panama. *Ecological Monographs* **62**, 251–75.

Bussman, R.W. (1996) Destruction and management of Mount Kenya's forests. *Ambio* **25**, 314–17.

Butlin, N.G. (1993) *Economics of the Dreamtime. A Hypothetical History*. Cambridge University Press, Cambridge.

Butterfield, J., Luff, M.L., Baines, M. and Eyre, M.D. (1995) Carabid beetle communities as indicators of conservation potential in upland forests. *Forest Ecology and Management* **79**, 63–77.

Byrd, B.F. (1994) From early humans to farmers and herders – recent progress on key transitions in southwest Asia. *Journal of Archaeological Research* **2**, 221–53.

Caffrey, J.M. (1993) Aquatic plant management in relation to Irish recreational fisheries development. *Journal of Aquatic Plant Management* **31**, 162–68.

Cai, S., Yue, L., Shang, Q. and Nordberg, G. (1995) Cadmium exposure among residents in an area contaminated by irrigation water in China. *Bulletin of the World Health Organization* **73**, 359–67.

Campbell, B.M.S. and Overton, M. (1993) A new perspective on medieval and early modern agriculture: six centuries of Norfolk farming c.1250–c.1850. *Past and Present* **141**, 38–105.

Campbell, I.D. and McAndrews, J.H. (1993) Forest disequilibrium caused by rapid Little Ice Age cooling. *Nature* **366**, 336–38.

Cann, R.L. (1993) Human dispersal and divergence. *Trends in Ecology and Evolution* **8**, 27–31.

Cann, R.L., Rickards, O. and Lum, J.K. (1994) Mitochondrial DNA and human evolution: our one lucky mother. In M.H. Nitecki and D.V. Nitecki (eds) *Origins of Anatomically Modern Humans*. Plenum Press, New York, pp.135–48.

Cann, R.L., Stoneking, M. and Wilson, A.C. (1987) Mitochondrial DNA and human evolution. *Nature* **325**, 31–36.

Carbonell, E., Bermudez de Castro, J.M., Arsuaga, J.L., Diez, J.C., Rosas, A., Cuenca-Bescós, G., Sala, R.,

Mosquera, M. and Rodriguez, X.P. (1995) Lower Pleistocene hominids and artifacts from Atapuerca – TD6 (Spain). *Science* **269**, 826–30.

Carney, H.J. (1982) Algal dynamics and trophic interactions in the recent history of Frains lake. *Ecology* **63**, 1814–26.

Carney, H.J., Binford, M.W., Marin, R.R. and Goldman, C.R. (1993) Nitrogen and phosphorus dynamics and retention in ecotones of Lake Titicaca, Bolivia/Peru. *Hydrobiologia* **251**, 39–47.

Carreira, J.A. and Niell, F.X. (1995) Mobilization of nutrients by fire in a semiarid scrubland ecosystem of southern Spain. *Arid Soil Research and Rehabilitation* **9**, 73–89.

Carter, F.W. and Turnock, D. (eds) (1993) *Environmental Problems in Eastern Europe*. Routledge, London.

Carvalho, L. and Moss, B. (1995) The current status of a sample of English sites of special scientific interest subject to eutrophication. *Aquatic Conservation - Marine and Freshwater Ecosystems* **5**, 191–204.

Castaneda, A.R. and Bhuiyan, S.I. (1996) Groundwater contamination by ricefield pesticides and some influencing factors. *Journal of Environmental Science and Health A* **31**, 83–99.

Cater, E. (1995) Environmental contradictions in sustainable tourism. *Geographical Journal* **161**, 21–28.

Catroux, G. and Amarger, N. (1992) *Rhizobia* as soil inoculants in agriculture. In J.C. Fry and M.J. Day (eds) *Release of Genetically Engineered and Other Microorganisms*. Cambridge University Press, Cambridge, pp. 1–13.

Cavaco, C. (1995) Tourism in Portugal – diversity, diffusion, and regional and local development. *Tijdschrift voor Economische en Sociale Geografie.* **86**, 64–71.

Cavalli-Sforza, L.L. and Cavilli-Sforza, F. (1995) *The Great Human Diasporas*. Addison-Wesley, Reading, MA, USA.

Cavalli-Sforza, L.L., Menozzi, P. and Piazza, A. (1993) Demic expansions and evolution. *Science* **259**, 639–46.

Cavalli-Sforza, L.L., Menozzi, P. and Piazza, A. (1994) *The History and Geography of Human Genes*. Princeton University Press, Princeton, NJ.

Chadwick, H. (1994) Envoi: on taking leave of antiquity. In J. Boardman, J. Griffin and O. Murray (eds) *The Oxford History of the Classical World*. Oxford University Press, Oxford, pp. 807–28.

Chang, T.T. (1989) Domestication and spread of the cultivated rices. In D.R. Harris and G.C. Hillman (eds) *Foraging and Farming: the Evolution of Plant Exploitation*. Unwin Hyman, London, pp. 408–17.

Chapman, J. (1989) Demographic trends in neothermal south-east Europe. In C. Bonsall (ed.) *The Mesolithic in Europe*. Edinburgh University Press, Edinburgh, pp. 500–515.

Chapman, J. and Müller, J. (1990) Early farmers in the Mediterranean basin: the Dalmation evidence. *Antiquity* **64**, 127–34.

Chapman, J. and Shiel, R. (1993) Social change and land use in prehistoric Dalmatia. *Proceedings of the Prehistoric Society* **59**, 61–104.

Chapman, R. (1995) Urbanism in Copper and Bronze Age Iberia? *Proceedings of the British Academy* **86**, 29–46.

Chappell, J. and Shackleton, N.J. (1986) Oxygen isotopes and sea level. *Nature* **324**, 137–40.

Chappellaz, J., Barnola, J.M., Raynaud, D., Korotkevich, Y.S. and Lorius, C. (1990) Ice-core record of atmospheric methane over the last 160,000 years. *Nature* **345**, 127–31.

Chappellaz, J., Blunier, T., Raynaud, D., Barnola, J.M., Schwander, J. and Stauffer, B. (1993) Synchronous changes in atmospheric CH_4 and Greenland climate between 40 and 8 k yr BP. *Nature* **366**, 443–45.

Charbonneau, R. and Kondolf, G.M. (1993) Land-use change in California, USA – nonpoint source water quality impacts. *Environmental Management* **17**, 453–60.

Charles, C.D. and Fairbanks, R.G. (1992) Evidence from Southern Ocean sediments for the effect of North Atlantic deep-water flux on climate. *Nature* **355**, 416–19.

Charles, C.D., Lynchstieglitz, J., Ninnemann, U.S. and Fairbanks, R.G. (1996) Climate connections between the hemispheres revealed by deep sea sediment – ice core correlations. *Earth and Planetary Science Letters* **142**, 19–27.

Charles, D.F. (1990) Effects of acid deposition on North American lakes: palaeolimnological evidence from diatoms and chrysophytes. *Philosphical Transactions of the Royal Society of London* **B317**, 403–12.

Charles, D.F. and Smol, J.P. (1994) Long-term chemical changes in lakes – quantitative inferences from biotic remains in the sediment record. *Advances in Chemistry Series* **237**, 3–31.

Charlson, R.J. and Heintzenberg, J. (eds) (1995) *Aerosol Forcing of Climate*. J. Wiley, Chichester.

Charlson, R.J., Lovelock, J.E. Andreae, M.O. and Warren, S.G. (1987) Oceanic phytoplankton, atmospheric sulphur, cloud albedo and climate. *Nature* **326**, 655–61.

Chartres, C. (1987) Australia's land resources at risk. In A. Chisholm and R. Dumsday (eds) *Land Degradation: Problems and Policies*. Cambridge University Press, Cambridge, pp. 7–26.

Chavaillon, J. (1994) Africa during the Lower Palaeolithic and the first settlements. In S.J. De Laet, A.H. Dani, J.L. Lorenzo and R.B. Nunoo (eds) *History of Humanity. Volume I: Prehistory and the Beginnings of Civilization*. Routledge, London and UNESCO, Paris, pp. 35–43.

Chen, F., Wu, R., Pompei, D. and Oldfield, F. (1995) Magnetic property and particle size variations in the Late Pleistocene and Holocene parts of the Dadongling loess section near Xining, China. *Quaternary Proceedings* **4**, 27–40.

Chen, J.H., Curran, H.A., White, B. and Wasserburg, G.J. (1991) Precise chronology of the last interglacial period: $^{234}U-^{230}Th$ data from fossil and coral reefs in the Bahamas. *Geological Society of America Bulletin* **103**, 82–97.

Chiew, F.H.S., Whetton, P.H., McMahon, T.A. and Pittock, A.B. (1995) Simulation of the impacts of climate change on runoff and moisture in Australian catchments. *Journal of Hydrology* **167**, 121–47.

Chiew, F.H.S. and McMahon, T.A. (1993) Detection of trend and change in annual flow of Australian rivers. *International Journal of Climatology* **13**, 643–53.

Childe, V.G. (1936) *Man makes himself*. Watts, London.

Chisci, G. (1990) Soil erosion versus desertification in the semi-arid Mediterranean environment. In J.L. Rubio and R.J. Rickson (eds) *Strategies to Combat Desertification in Mediterranean Europe*. Commission of the European Communites, Luxembourg, pp. 132–47.

Chlachula, J. (1996) Geology and Quaternary environments of the first preglacial Palaeolithic sites found in Alberta, Canada. *Quaternary Science Reviews* **15**, 285–313.

Choi, J.Y. and Koo, J.K. (1993) The monthly impact of nonpoint sources to Hoedong Lake, Korea. *Water Science and Technology* **28**, 137–42.

Chorley, R.J. and Kennedy, B.A. (1971) *Physical Geography: A Systems Approach*. Prentice-Hall, Englewood Cliffs NJ.

Chrispeels, M.J. and Sadava, D.E. (1994) *Plants, Genes, and Agriculture*. Jones & Bartlett, Boston, MA.

Christopherson, R.W. (1994) *Geosystems: An Introduction to Physical Geography*. Macmillan, New York.

Christou, P. (1995) Strategies for variety-independent genetic transformation of the important cereals, legumes and woody species utilizing particle bombardment. *Euphytica* **85**, 13–27.

Chung, Y.S., Kim, T.K. and Kim, K.H. (1996) Temporal variation and cause of acidic precipitation from the monitoring network in Korea. *Atmospheric Environment* **30**, 2429–35.

Ciais, P., Tans, P.P., Troher, M., White, J.W.C. and Francey, R.J. (1995) Large northern hemisphere terrestrial CO_2 sink indicated by the C-13/C-12 ratio of atmospheric CO_2. *Science* **269**, 1098–1102.

Clapp, B.W. (1994) *An Environmental History of Britain*. Longman, London.

Clapperton, C. (1979) Glaciation in Bolivia before 3.27 M yr. *Nature* **277**, 375–77.

Clapperton, C. (1993) *Quaternary Geology and Geomorphology of South America*. Elsevier, Amsterdam.

Clark, G. (1991) Yields per acre in English agriculture, 1250–1860: evidence from labour inputs. *Economic History Review* **44**, 445–60.

Clark, J.G.D. (1954) *Excavations at Star Carr*. Cambridge University Press, Cambridge.

Clark, J.G.D. (1972) *Star Carr: A Case Study in Bioarchaeology*. Addison-Wesley, Reading, MA.

Clark, J.S. and Robinson, J. (1993) Paleoecology of fire. In P.J. Crutzen and J.G. Goldammer (eds) *Fire in the Environment: the Ecological, Atmospheric and Climatic Importance of Vegetation Fires*. J. Wiley, New York, pp. 193–214.

Clark, S. and Edwards, A.J. (1994) Use of artificial reef structures to rehabilitate reef flats degraded by coral mining in the Maldives. *Bulletin of Marine Science* **55**, 724–44.

Clarke, A.G. (1986) The air. In R.E. Hester (ed.) *Understanding Our Environment*. Royal Society of Chemistry, London, pp. 71–118.

Clarke, R.J. and Tobias, P.V. (1995) Sterkfontein member 2 foot bones of the oldest South African hominid. *Science* **269**, 521–24.

Clayton, K. (1995) Predicting sea-level rise and managing the consequences. In T. O'Riordan (ed.) *Environmental Science for Environmental Management*. Longman, Harlow, pp. 165–84.

Cleere, H. (1976) Some operating parameters for Roman ironworks. *Bulletin of the Institute of Archaeology, London*, **13**, 233–46.

Cleveland, C.B. (1996) Mobility assessment of agrichemicals – current laboratory suggestions for future directions. *Weed Technology* **10**, 157–68.

Clutton-Brock, J. (1992) The process of domestication. *Mammal Review* **22**, 79–85.

Clutton-Brock, J. (1995) Origins of the dog: domestication and early history. In J. Serpell (ed.) *The Domestic Dog: Its Evolution, Behaviour and Interactions with People*. Cambridge University Press, Cambridge, pp.7–20.

Clutton-Brock, J. and Noe-Nygaard, N. (1990) New osteological and C-isotope evidence on Mesolithic dogs: companions to hunters and fishers at Star Carr, Seamer Carr and Kongemose. *Journal of Archaeological Science* **17**, 643–53.

Coates, D.R. (1987) Subsurface impacts. In K.J. Gregory and D.E. Walling (eds) *Human Activity and Environmental Processes*. J. Wiley, Chichester, pp. 271–304.

Cocking, E.C. (1989) Plant cell and tissue culture. In J.L. Marx (ed.) *A Revolution in Biotechnology*. Cambridge University Press, Cambridge, pp. 119–29.

Cole, D.N. (1986) Recreational impacts on back country campsites in Grand Canyon National Park, USA. *Journal of Environmental Management* **10**, 651–59.

Cole, D.N. (1995a) Disturbance of natural vegetation by camping – experimental applications of low-level stress. *Environmental Management* **19**, 405–16.

Cole, D.N. (1995b) Experimental trampling of vegetation. 1: Relationship between trampling intensity and vegetation response. *Journal of Applied Ecology* **32**, 203–14.

Cole, D.N. (1995c) Experimental trampling of vegetation. 2. Predictors of resistance and resilience. *Journal of Applied Ecology* **32**, 215–24.

Cole, D.N. and Marion, J.L. (1988) Recreation impacts in some riparian forests of the eastern United States. *Journal of Environmental Management* **12**, 99–107.

Coles, B. and Coles, J. (1986) *Sweet Track to Glastonbury*. Thames and Hudson, London.

Colhoun, E.A. (1996) Application of Iversen's glacial–interglacial cycle to interpretation of the last late glacial and Holocene vegetation history of western Tasmania. *Quaternary Science Reviews* **15**, 557–80.

Colhoun, E.A., Van Degeer, G., Fitzsimons, S.J. and Heusser, L.E. (1994) Terrestrial and marine records of the last glaciation from Western Tasmania–do they agree? *Quaternary Science Reviews* **13**, 293–300.

Colhoun, E.A., van der Geer, G., Fitzsimons, S.J. and Heusser, L.E. (1984) Terrestrial and marine records of the last glaciation from western Tasmania: do they agree? *Quaternary Science Reviews* **13**, 293–300.

Colhoun, E.A., Mabin, M.C.G, Adamson, D.A. and Kirk, R.M. (1992) Antarctic ice volume and contribution to sea-level fall at 20,000 yr BP from raised beaches. *Nature* **358**, 316–19.

Colinvaux, P. (1987) Amazon diversity in the light of the paleoecological record. *Quaternary Science Reviews* **6**, 93–114.

Collis, J. (1984) *The European Iron Age*. Batsford, London.

Coltori, M. *et al.* (1982) Reversed magnetic polarity at an early palaeolithic site in central Italy. *Nature* **300**, 173–76.

Conacher, A. and Conacher, J. (1988) The exploitation of the soils. In R.L. Heathcote (ed.) *The Australian Experience: Essays in Australian Land Settlement and Resource Management*. Longman, Melbourne, pp. 127–38.

Conzen, M.P. (ed) (1990) *The Making of the American Landscape*. Unwin Hyman, Boston.

Coope, G.R. (1977) Fossil coleopteran assemblages as sensitive indicators of climatic changes during the Devensian (last) cold stage. *Philosophical Transactions of the Royal Society of London* **B280**, 313–40.

Coope, G.R. (1986) Coleoptera analysis. In Berglund, B.E. (ed.) *Handbook of Holocene Palaeoecology and Palaeohydrology*. J. Wiley, Chichester, pp. 703–13.

Coope, G.R. and Lister, A.M. (1987) Late-glacial mammoth skeletons from Condover, Shropshire, England. *Nature* **330**, 472–74.

Cooper, C., Fletcher, J., Gilbert, D. and Wanhill, S. (1993) *Tourism: Principles and Practice*. Longman, Harlow.

Cope, C. (1991) Gazelle hunting strategies in the southern Levant. In O. Bar-Yosef and F.R. Valla (eds) *The Natufian Culture in the Levant*. Ann Arbor, MI, pp. 341–58.

Coppens, Y. and Geraads, D. (1994) The period of *Homo habilis* and *Homo erectus*: an overview. In S.J. De Laet, A.H. Dani, J.L. Lorenzo and R.B. Nunoo (eds) *History of Humanity, Volume I: Prehistory and the Beginnings of Civilization*. Routledge, London, and UNESCO, Paris, pp. 27–34.

Covey, C., Sloan, L.C. and Hoffert, M.I. (1996) Paleoclimate data constraints on climate sensitivity – the paleocalibration method. *Climatic Change* **32**, 165–84.

Cowell, C.M. (1995) Presettlement Piedmont forests: patterns of composition and disturbance in Central Georgia. *Annals of the Association of American Geographers* **85**, 65–83.

Cowie, J.W. and Bassett, M.G. (1989) Global Stratigraphic chart with geochronometric and magnetostratigraphic calibration. *Episodes* **12** (2) (suppl.).

Cowley, T.J. and Kim, K.-Y. (1994) Milankovitch forcing of the last interglacial sea level. *Science* **265**, 1566–68.

Cowling, R.M., MacDonald, I.A.W. and Simmons, A.T. (1996) The Cape Peninsula, South Africa – physiographical, biological and historical background to an extraordinary hot-spot of biodiversity. *Biodiversity and Conservation* **5**, 527–50.

Crabtree, P.J. (1994) Animal exploitation in East Anglian villages. In J.Rackham (ed.) *Environment and Economy in Anglo-Saxon England*. Council for British Archaeology Research Report 89, pp. 40–54.

Craig, J.R., Vaughan, D.J. and Skinner, B.J. (1996) *Resources of the Earth. Origin, Use and Environmental Impact*, 2nd edn. Prentice Hall, Upper Saddle River, NJ.

Crawford, P. (1985) *The Living Isles*. BBC Publications, London.

Croll, J. (1864) On the physical cause of the change of climate during geological epochs. *Philosophical Magazine* **28**, 121–37.

Croll, J. (1875) *Climate and Time*. Appleton, New York.

Cronon, W. (1983) *Changes in the Land: Indians, Colonists and the Ecology of New England*. Hill and Wang, New York.

Crosby, A.W. (1986) *Ecological Imperialism: the Biological Expansion of Europe, 900–1900*. Cambridge University Press, Cambridge.

Crouch, M.L. (1995) Biotechnology is not compatible with sustainable agriculture. *Journal of Agricultural and Environmental Ethics* **8**, 98–111.

Crowley, G.M. and Kershaw, A.P. (1994) Late Quaternary environmental change and human impact around Lake Bolac, western Victoria, Australia. *Journal of Quaternary Science* **9**, 367–77.

Crowley, T.J. and Kim, K.-Y. (1994) Milankovitch forcing of the last interglacial sea level. *Science* **265**, 1566–68.

Crueger, W. and Crueger, A. (1990) *Biotechnology: A Textbook of Industrial Microbiology* (English translation). Sinauer Associates, Sunderland, MA.

Crutzen, P.J. (1996) My life with O_3, NO_x, and other YZO_x compounds (Nobel Lecture). *Angewandte Chemie* **35**, 1758–77 (international edition in English).

Cumming, B.F., Davey, K.A., Smol, J.P. and Birks, H.J.B. (1994) When did acid-sensitive Adirondack Lakes (New York, USA) begin to acidify and are they still acidifying? *Canadian Journal of Fisheries and Aquatic Sciences* **51**, 1550–68.

Cunliffe, B. (1994) Iron Age societies in Western Europe and beyond, 800–140 BC. In B. Cunliffe (ed.) *The Oxford Illustrated Prehistory of Europe*. Oxford University Press, Oxford, pp. 336–72.

Cunliffe, B. (1995) *Danebury: An Iron Age Hillfort in Hampshire*. Council for British Archaeology Research Report 102.

Cunningham, S.D. and Lee, C.R. (1995) Phytoremediation: plant-based remediation of contaminated soils and sediments. In H.D. Skipper and R.F. Turco (eds) *Bioremediation. Science and Applications*. Soil Science Society of America, American Society of Agronomy and Crop Science Society of America, Madison, WI, pp. 145–56.

Currant, A.P. (1986) Man and the Quaternary interglacial faunas of Britain. In S.N. Collcut (ed.) *The Palaeolithic of Britain and its Nearest Neighbours*. Department of Archaeology and Prehistory, University of Sheffield, pp. 50–52.

Currie, R.G. (1995) Luni-solar 18.6 and solar cycle 10–11 year signal in Chinese dryness–wetness indices. *International Journal of Climatology* **15**, 497–515.

Currie, R.G., Wyatt, T. and O'Brien, D.P. (1993) Deterministic signals in European fish catches, wine harvests, sea-level, and further experiments. *International Journal of Climatology* **13**, 665–87.

Curtis, J.H. and Hodell, D.A. (1996) Climate variability on the Yucatan Peninsula (Mexico) during the past 3500 years, and implications for Maya cultural evolution. *Quaternary Research* **46**, 37–47.

Cwynar, L.C., Levesque, A.J., Mayle, F.E. and Walker, I. (1994) Wisconsinan late-glacial environmental change in New Brunswick: a regional synthesis. *Journal of Quaternary Science* **9**, 161–64.

Dabas, M. and Bhatia, S. (1996) Carbon sequestration through afforestation: role of tropical industrial plantations. *Ambio* **25**, 327–30.

Dahl, T.E. (1990) *Wetland Losses in the United States 1780s to 1980s*. US Department of the Interior, Fish and Wildlife Service, Government Printing Office, Washington, DC.

Damaj, M. and Ahmad, D. (1996) Biodegradation of polychlorinated biphenyls by *Rhizobia* – novel finding. *Biochemical and Biophysical Research Communications*. **218**, 908–15.

Damm, D., Grandjean, P., Lyngbye, T., Trillingsgaard, A. and Hansen, O.N. (1993) Early lead exposure and neonatal jaundice – relation to neurobehavioral performance at 15 years of age. *Neurotoxicology and Teratology* **15**, 173–81.

Daniels, W.L. and Zipper, C.E. (1995) Improving coal surface mine reclamation in the Central Appalachian Region. In J. Cairns. Jr (ed.) *Rehabilitating Damaged Ecosystems*. 2nd edn. Lewis Publishers, Boca Raton FL, pp. 187–217.

Danilin, M.Y., Sze, N.D., Ko, M.K.W., Rodriguez, J.M. and Prather, M.J. (1996) Bromine–chlorine coupling in the Antarctic ozone hole. *Geophysical Research Letters* **23**, 153–56.

Dansgaard, W. (1987) Ice core evidence of abrupt climatic change. In Berger, W.H. and Labeyrie, L.D. (eds) *Abrupt climatic change*. D. Reidel, Dordrecht, 223–33.

Dansgaard, W., Johnsen, S.J., Clausen, H.B., Dahl-Jensen, D., Gundestrup, N.S., Hammer, C.U., Hvidbjerg, C.S., Steffensen, J.P., Sveinbjørnsdottir, A.E., Jouzel, J. and Bond, G. (1993) Evidence for general instability of past climate from a 250-Kyr ice-core record. *Nature* **364**, 218–20.

Darwin, C. (1859) *On The Origin of Species*. Murray, London.

David, F. (1993) Altitudinal variation in the response of the vegetation to late-glacial climatic events in the northern French alps. *New Phytologist* **125**, 203–20.

Davis, J.H. (1985) The pyrethroids: an historical introduction. In J.P. Leahy (ed.) *The Pyrethroid Insecticides*. Taylor & Francis, London, pp. 1–41.

Davis, G.R. (1990) Energy for planet Earth. *Scientific American* **2631**, 54.

Davis, M.B. (1976) Erosion rates and land-use history in southern Michigan. *Environmental Conservation* **3**, 139–48.

Davis, M.B. (1981) Outbreaks of forest pathogens in forest history. *Proceedings of the IV International Palynological Conference (1976–1977)* **3**, 216–27.

Davis, M.B. (1984) Holocene vegetational history of the eastern United States. In H.E. Wright Jr. (ed.) *Late-Quaternary Environments of the United States* (2 Vols). University of Minnesota Press, Minneapolis, and Longman, London, Vol. 2, *The Holocene*, pp. 166–81.

Davis, O.K. and Sellers, W.D. (1987) Contrasting climatic histories for western North America during the early Holocene. *Current Research in the Pleistocene* **4**, 87–89.

Davis, S.J.M. (1987) *The Archaeology of Animals*. Batsford, London.

Davis, W.M. (1909) The geographical cycle. In D.W. Johnson (ed.), *Geographical Essays by William Morris Davis*. Ginn, Boston, USA, pp. 254–56.

Day, S.P. (1993) Post-glacial vegetational history of the Oxford region. *New Phytologist* **119**, 445–70.

Day, S.P. (1993) Woodland origin and 'ancient woodland indicators': a case-study from Sidlings Copse, Oxfordshire, UK. *The Holocene* **3**, 45–53.

Day, S.P. and Mellars, P.A. (1994) 'Absolute' dating of Mesolithic human activity at Star Carr, Yorkshire: new palaeoecological studies and identification of the 9600 BP radiocarbon 'plateau'. *Proceedings of the Prehistoric Society* **60**, 417–22.

Dayan, T. (1994) Early domesticated dogs of the Near East. *Journal of Archaeological Science* **21**, 633–40.

Dayan, T. and Simberloff, D. (1995) Natufian gazelles: proto-domestication reconsidered. *Journal of Archaeological Science*. **22**, 671–75.

Dazhong, W. (1993) Soil erosion and conservation in China. In D. Pimentel (ed.) *World Soil Erosion and Conservation*. Cambridge University Press, Cambridge, pp. 63–85.

D'Costa, D.M. and Kershaw, A.P. (1995) A Late

Pleistocene and Holocene pollen record from Lake Terang, Western Plains of Victoria, Australia. *Palaeogeography, Palaeoclimatology, Palaeoecology* **113**, 57–67.

Dean, W.R.J., Hoffman, M.T., Meadows, M.E. and Milton, S.J. (1995) Desertification in the semiarid Karoo, South Africa – review and reassessment. *Journal of Arid Environments* **30**, 247–64.

DeAngelis, M., Barker, N.I. and Petrov, V.N. (1987) Aerosol concentrations over the last climatic cycle (160 K yr) from an Antarctic ice core. *Nature* **325**, 318–21.

Dearing, J. (1994) Reconstructing the history of soil erosion. In N. Roberts (ed.) *The Changing Global Environment*. Blackwell, Oxford, pp. 242–61.

de Beaulieu, J.-L., Andrieu, V., Ponel, P., Reille, M. and Lowe, J.J. (1994) The Weichselian late-glacial in southwestern Europe (Iberian Peninsula, Pyrenees, Massif Central, northern Apennines). *Journal of Quaternary Science* **9**, 101–7.

Deboer, I.J.M., Brom, F.W.A. and Vorstenbosch, J.M.G. (1995) An ethical evaluation of animal biotechnology – the clones in dairy-cattle breeding. *Animal Science* **61**, 453–63.

Decker-Walters, D., Walters, T., Cowan, W. and Smith, B.D. (1993) Isozymic characterisation of wild populations of *Cucurbita pepo*. *Journal of Ethnobiology* **13**, 55–72.

Dedatta, S.K. (1995) Nitrogen transformations in wetland rice ecosystems. *Fertilizer Research* **42**, 193–203.

Dehn, M. (1995) An evaluation of soil conservation techniques in the Ecuadorian Andes. *Mountain Research and Development* **15**, 175–82.

Dejong, F. and Kowalchuk, T.E. (1995) The effect of shelterbelts on erosion and soil properties. *Soil Science* **159**, 337–45.

De Laet, S.J. (1994a) From the beginnings of food production to the first states. In S.J. De Laet, A.H. Dani, J.L. Lorenzo and R.B. Nunoo (eds) *History of Humanity, Volume I: Prehistory and the Beginnings of Civilization*. Routledge, London, and UNESCO, Paris, pp. 366–76.

De Laet, S.J. (1994b) Europe during the Neolithic. In S.J. De Laet, A.H. Dani, J.L. Lorenzo and R.B. Nunoo (eds) *History of Humanity, Volume I: Prehistory and the Beginnings of Civilization*. Routledge, London, and UNESCO, Paris, pp. 490–500.

Delascuevas, C. and Pueyo, J. (1995) The influence of mineralogy and texture in the water content of rock-salt formations – its implication in radioactive waste disposal. *Applied Geochemistry* **10**, 317–27.

Delcourt, H.R. and Delcourt, P.A. (1991) *Quaternary Ecology. A Paleoecological Perspective*. Chapman and Hall, London.

Delson, E. (1989) Oldest Eurasian stone tools. *Nature* **340**, 96.

Dempster, J.P. (1987) Effects of pesticides on wildlife and priorities for the future. In K.J. Brent and R.K. Atkins

(eds) *Rational Pesticide Use*. Cambridge University Press, Cambridge, pp. 17–25.

Denton, G.H. and Hendy, C.H. (1994) Younger Dryas age advance of Franz-Josef glacier in the Southern Alps of New Zealand. *Science* **264**, 1434–37.

de Ronde, J.G. (1993) What will happen to the Netherlands if sea-level rise accelerates? In R.A. Warrick, E.M. Barrow and T.M.L. Wigley (eds) *Climate and Sea-Level Change: Observations, Projections and Implications*. Cambridge University Press, Cambridge, pp. 322–35.

DeRouffignac, C., Bowen, D.Q., Coope, G.R., Keen, D.H., Lister, A.M., Maddy, D., Robinson, J.E., Sykes, G.A. and Walker, M.J.C. (1995) Late Middle Pleistocene interglacial deposits at Upper Strensham, Worcestershire, England. *Journal of Quaternary Science* **10**, 15–31.

Derwent, R.G. (1996) The influence of human activities on the distribution of hydroxyl radicals in the troposphere. *Philosophical Transactions of the Royal Society of London* **A354**, 501–31.

Desmet, J., Hofman, G., Vanderdeelen, J., Vanmeirvenne, M. and Baert, L. (1996) Phosphate enrichment in the sandy loam soils of West Flanders, Belgium. *Fertilizer Research* **43**, 209–15.

Devernal, A., Hillairemarcel, C. and Bilodeau, G. (1996) Reduced meltwater outflow from the Laurentide ice margin during the Younger Dryas. *Nature* **381**, 774–77.

Dhir, R.P. (1993) Problem of desertification in arid zone of Rajasthan – a view. *Annals of Arid Zone* **32**, 79–88.

Diaz, H.F. and Anderson, C.A. (1995) Precipitation trends and water consumption related to population in the southwestern United States – a reassessment. *Water Resources Research* **31**, 713–20.

Diederichsen, A. and Hammer, K. (1995) Variation of cultivated flax (*Linum usitatissimum* L. subsp. *usitatis-simum*) and its progenitor pale flax (subsp. *angustifolium* (Huds) Thell). *Genetic Resources and Crop Evolution* **42**, 263–72.

Dillehay, T.D. and Collins, M.B. (1988) Early cultural evidence from Monte Verde in Chile. *Nature* **332**, 150–52.

Dilsaver, L.M. and Colten, C.E. (1992) *The American Environment: Interpretations of Past Geographies*. Rowman and Littlefield, Lenham, MD.

Ding, Z.,Yu, Z., Rutter, N.W. and Liv, T. (1994) Towards an orbital time scale for Chinese loess deposits. *Quaternary Science Reviews* **13**, 39–70.

Ding, Z.L., Rutter, N., Han, J.T. and Liv, T.S. (1992) A coupled environmental system formed at about 2.5 Ma over eastern Asia. *Palaeogeography, Palaeoclimatology and Palaeoecology* **94**, 223–42.

Dinham, B. (1993) *The Pesticide Hazard*. Zed Books, London.

Dinham, B. (ed.) (1995) *The Pesticide Trail: the Impact of Trade Controls on Reducing Pesticide Hazards in Developing Countries*. Pesticides Trust, London.

Dixit, S.S. and Smol, J.P. (1994) Diatoms as indicators in

the Environmental Monitoring and Assessment Program, Surface Waters (EMAP–SW). *Environmental Monitoring and Assessment* **31**, 275–306.

Dobson, A.P. (1996) *Conservation and Biodiversity.* W.H. Freeman, New York.

Dodson, J.R. De Salis, T., Myers, C.A. and Sharp, A.J. (1994) A thousand years of environmental change and human impact in the alpine zone of Mt. Kosciusko, New South Wales. *Australian Geographer* **25**, 77–87.

Doebley, J. (1990) Molecular evidence and the evolution of maize. *Economic Botany.* **44**, supplement 3, 6–27.

Dogheim, S.M., Mohamed, E.Z., Alla, S.A.G., Elsaied, S., Emel, S.Y., Mohamed, E.Z. and Fahmy, S.M. (1996) Monitoring of pesticide residues in human milk, soil, water and food samples collected from Kafr El Zayat Governorate. *Journal of AOAC International* **79**, 111–16.

Dokken, T.M. and Hald, M. (1996) Rapid climatic shifts during isotope stages 2–4 in the polar North Atlantic. *Geology* **24**, 599–602.

Dolt, W., Feidieker, D., Steiof, M., Becker, P.M. and Kampfer, P. (1995) Comparison of *ex situ* and *in situ* techniques for bioremediation of hydrocarbon polluted soils. *International Biodeterioration and Biodegradation* **35**, 301–16.

Donner, J. (1995) *The Quaternary History of Scandinavia.* Cambridge University Press, Cambridge.

Doolittle, W.E. (1992) Agriculture in North America on the eve of contact: a reassessment. *Annals of the Association of American Geographers* **82**, 386–401.

Döös, B.R. (1994) Environmental degradation, global food production and risk for large-scale migrations. *Ambio* **23**, 124–30.

Dorale, J.A., Gonzalez, L.A., Reagan, M.K., Pickett, D.A., Murrell, M.T. and Baker, R.G. (1992) A high-resolution record of Holocene climate change in speleothem calcite from Cold Water Cave, northeast Iowa. *Science* **258**, 1626–30.

Douches, D.S. and Jastrzebski, K. (1993) Potato. In G. Kalloo and B.O. Bergh (eds) *Genetic Improvement of Vegetable Crops.* Pergamon, Oxford, pp. 605–44.

Downey, A., Atkinson, R., Fraser, P. and Shanklin, J.D. (1996) The Antarctic ozone hole (1994) *Australian Meteorological Magazine* **45**, 123–29.

Downton, M.W. (1995) Measuring tropical deforestation – development of the methods. *Environmental Conservation* **22**, 229–34.

Dragovich, D. and Patterson, J. (1995) Condition of rehabilitated coal mines in Hunter Valley, Australia. *Land Degradation and Rehabilitation* **6**, 29–39.

Dregne, H., Kassas, M. and Rozanov, B. (1991) A new assessment of the world status of desertification. *Desertification Control Bulletin* **20**, 6–18.

Dregne, H.E. (1995) Erosion and soil productivity in Australia and New Zealand. *Land Degradation and Rehabilitation* 6, 71-78.

Dubrova, Y.E., Nesterov, V.N., Krouchinsky, N.G., Ostapenko, V.A., Neumann, R., Neil, D.L. and

Jeffreys, A.J. (1996) Human minisatellite mutation rate after the Chernobyl accident. *Nature* **380**, 683–86.

Dudgeon, D. (1995) River regulation in southern China – ecological implications for conservation and environmental management. *Regulated Rivers Research and Management* **11**, 35–54.

Dudley, N. (1990) *Nitrates: the Threat to Food and Water.* Green Print, London.

Dugan, P. (ed.) (1993) *Wetlands in Danger.* Mitchell Beazley, London.

Dulvy, N.K., Stanwell-Smith, D., Darwall, W.R.T. and Horrill, C.J. (1995) Coral Mining at Mafia Island, Tanzania – a management dilemma. *Ambio* **24**, 358–65.

Dumayne, L. (1993) Invader or native? Vegetation clearance in northern Britain during Romano-British time. *Vegetation History and Archaeobotany* **2**, 29–36.

Dumayne, L. and Barber, K.E. (1994) The impact of the Romans on the environment of northern England: pollen data from three sites close to Hadrian's Wall. *The Holocene* **4**, 165–73.

Dunwell, J. (1995) Transgenic cereal crops. *Chemistry and Industry* **18**, 730–33.

Duplessy, J.C., Delibrias, G., Turon, J.L., Pujol, C. and Duprat, J. (1981) Deglacial warming of the northeastern Atlantic ocean: correlation with the palaeoclimatic evolution of the European continent. *Palaeogeography, Palaeoclimatology, Palaeoecology* **35**, 121–44.

Duplessy, J.C., Labeyrie, L., Arnold, M., Paterne, M., Duprat, J. and van Weering, T.C.E. (1992) Changes in surface salinity of the North Atlantic Ocean during the last deglaciation. *Nature* **358**, 485–88.

Dupont, L.M. (1993) Vegetation zones in NW Africa during the Brunhes chron reconstructed from marine palynological data. *Quaternary Science Reviews* **12**, 189–202.

Dutton, A., Fasham, P.J., Jenkins, D.A., Caseldine, A.E. and Hamilton-Dyer, S. (1994) Prehistoric copper mining on the Great Orme, Llandudno, Gwynedd. *Proceedings of the Prehistoric Society.* **60**, 245–86.

Dwivedy, K.K. and Mathur, A.K. (1995) Bioleaching – our experience. *Hydrometallurgy* **38**, 99–109.

Dyer, J.M. (1995) Assessment of climatic warming using a model of forest species migration. *Ecological Modelling* **79**, 199–219.

Eagles, B. (1994) The archaeological evidence for settlement in the fifth to the seventh centuries AD. In M. Aston and C. Lewis (eds) *The Medieval Landscape of Wessex.* Oxbow Monograph 46, pp. 13–32.

Earth Report (1988) *The Earth Report: Monitoring The Battle for Our Environment.* Mitchell Beazley, London.

Easterly, J.L. and Burnham, M. (1996) Overview of biomass and waste fuel resources for power production. *Biomass and Bioenergy* **10**, 79–92.

Ebohon, O.J. (1996) The scope and limits of sustainable development in Africa built environment sector. *International Journal of Sustainable Development and World Ecology* **3**, 1–12.

Economic Intelligence Unit (1991) Managing tourism and the environment: a Kenyan case study. *Travel and Tourism Analyst* **2**, 78–87.

Eden, S.E. (1994) Using sustainable development: the business case. *Global Environmental Change* **4**, 160–67.

Edington, J.M. and Edington, M.A. (1986) *Ecology, Recreation and Tourism.* Cambridge University Press, Cambridge.

Edwards, K. (1993) Soil erosion and conservation in Australia. In D. Pimentel (ed.) *World Soil Erosion and Conservation.* Cambridge University Press, Cambridge, pp. 147–69.

Edwards, K.J. and Berridge, J.M.A. (1994) The late Quaternary history of Loch a' Bhogaidh, Rinns of Islay SSSI, Scotland. *New Phytologist* **128**, 749–70.

Edwards, K.J. and Hirons, K.R. (1984) Cereal pollen grains in pre-elm decline deposits: implications for the earliest agriculture in Britain and Ireland. *Journal of Archaeological Science* **11**, 71–80.

Edwards, L., Burney, J. and Dehaan, R. (1995) Researching the effects of mulching on cool period soil erosion control in Prince Edward Island, Canada. *Journal of Soil and Water Conservation* **50**, 184–87.

Edwards, R.L., Beck, J.W., Burr, G.S., Donahue, D.J., Chappell, J.M.A., Bloom, A.L., Druffel, E.R.M. and Taylor, F.W. (1993) A large drop in atmospheric ^{14}C/^{12}C and reduced melting in the Younger Dryas, documented with ^{230}Th ages of corals. *Science* **260**, 962–68.

Eglinton,G., Bradshaw, S.A., Rosell, A., Sarnthein, M., Pflaumann, V. and Tiedemann, R. (1992) Molecular record of secular sea surface temperature changes on 100-year timescales for glacial terminations I, II and IV. *Nature* **356**, 423–26.

Ehlers, J., Gibbard, P.L. and Rose, J. (1991) Glacial deposits of Britain and Europe: General Overview. In: J. Ehlers, P.J. Gibbard and J. Rose (eds) *Glacial Deposits in Great Britain and Ireland.* A.A. Balkema, Rotterdam, pp. 493–501.

Ehrlich, P.R. and Wilson, E.O. (1991) Biodiversity studies: science and policy. *Science* **253**, 758–62.

Eide, L.K., Beyer, I.K. and Jansen, E. (1996) Comparison of Quaternary interglacial periods in the Iceland Sea. *Journal of Quaternary Science* **11**, 115–24.

Ekholm, P. (1994) Bioavalability of phosphorus in agriculturally loaded rivers in southern Finland. *Hydrobiologia* **287**, 179–94.

Ekman, S.R. and Scourse, J.D. (1993) Early and Middle Pleistocene pollen stratigraphy from British Geological Survey borehole 81/26, Fladen Ground, central North Sea. *Review of Palaeobotany and Palynology* **79**, 285–95.

Elassward, R.M. (1995) Agricultural prospects and water resources in Libya. *Ambio* **24**, 324–27.

Elenga, H., Schwartz, D. and Vincens, A. (1994) Pollen evidence of late Quaternary vegetation and inferred climate changes in Congo. *Palaeogeography, Palaeoclimatology, Palaeoecology* **109**, 345–56.

Ellefson, P.V. and Ek, A.R. (1996) Privately initiated forestry and forest products research and development – current status and future challenges. *Forest Products Journal* **46**, 37–43.

Elliott, P., Shaddick, G., Kleinschmidt, I., Jolley, D., Walls, P., Beresford, J. and Grundy, C. (1996) Cancer incidence near municipal solid waste incinerators in Great Britain. *British Journal of Cancer* **73**, 702–10.

Ellis, S., Taylor, D. and Masood, K.R. (1993) Land degradation in northern Pakistan. *Geography* **78**, 84–87.

Eltahir, E.A.B. (1996) Role of vegetation in sustaining large-scale atmospheric circulations in the tropics. *Journal of Geophysical Research – Atmospheres* **101**, 4255–68.

Elvingson, P. (1995) More trees than ever damaged. *Acid News* **5**, 6.

Ely, L.L., Enzel, Y., Baker, V.R. and Cayan, D.E. (1993) A 5000-year record of extreme floods and climate change in the southwestern United States. *Science* **262**, 410–12.

Emeis, K.-C., Anderson, D.M., Doose, H., Kroon, D. and Schulz-Bull, D. (1995) Sea-surface temperatures and the history of monsoon upwelling in the northwest Arabian sea during the last 500,000 years. *Quaternary Research* **43**, 355–61.

Emontspohl, A.-F. (1995) The northwest European vegetation at the beginning of the Weichselian glacial (Brørup and Odderade interstadials) – new data for northern France. *Review of Palaeobotany and Palynology* **85**, 231–42.

Enell, M. and Fejes, J. (1995) The nitrogen load to the Baltic Sea – present situation, acceptable future load and suggested source reduction. *Water Air and Soil Pollution* **85**, 877–82.

Engel, T. and Frey, W. (1996) Fuel resources for copper smelting in antiquity in selected woodlands in the Edom highlands to the Wadi-Arabah, Jordan. *Flora* **191**, 29–39.

Engstrom, D.R. and Swain, E.B. (1986) The chemistry of lake sediments in time and space. *Hydrobiologia* **143**, 37–44.

Engstrom, D.R., Swain, F.B. and Kingston, J.C. (1985) A palaeoliminological record of human disturbance from Harvey's Lake, Vermont: geochemistry, pigments and diatoms. *Freshwater Biology* **15**, 261–88.

Ernst, W.H.O. (1996) Bioavailability of heavy metals and decontamination of soil by plants. *Applied Geochemistry* **11**, 163–67.

Esser, G. (1995) Contribution of monsoon Asia to the carbon budget of the biosphere, past and future. *Vegetatio* **121**, 175–88.

Evans, J.G. (1993) The influence of human communities on the English chalklands from the Mesolithic to the Iron Age: the molluscan evidence. In F.M. Chambers (ed.) *Climate Change and Human Impact on the Landscape: Studies in Palaeoecology and Environmental Archaeology.* Chapman and Hall, London, pp. 147–56.

Evans, J.G., Limbrey, S.J., Máté, I. and Mount, R. (1993) An environmental history of the Upper Kennet Valley, Wiltshire, for the last 10,000 years. *Proceedings of the Prehistoric Society* **59**, 139–95.

Evans, L.T. (1993) *Crop Evolution, Adaptation and Yield.* Cambridge University Press, Cambridge.

Evans, R. (1990) Soils at risk of accelerated erosion in England and Wales. *Soil Use and Management* **6**, 125–31.

Evans, R. (1993) Extent, frequency and rates of rilling of arable land in localities in England and Wales. In S. Wicherek (ed.) *Farm Land Erosion in Temperate Plains Environment and Hills.* Elsevier, Amsterdam, pp. 177–90.

Evans, R. (1995) Some methods of directly assessing water erosion of cultivated land – a comparison of measurements made on plots and in fields. *Progress in Physical Geography* **19**, 115–29.

Evans, R.G, Girgin, B.N., Chenoweth, J.F. and Kroger, M.W. (1995) Surge irrigation with residues to reduce soil erosion. *Agricultural Water Management* **27**, 283–97.

Evelyn, J. (1664) *Silva: or a discourse of forest-trees*, 5th edn. J. Walthoe *et al.,* London.

Fang, J.-Q. (1991) Lake evolution during the past 30,000 years in China, and its implications for environmental change. *Quaternary Research* **36**, 37–60.

Farid, M.S.M., Atta, S., Rashid, M., Munnink, J.O. and Platenburg, R. (1993) Impact of the reuse of domestic waste water for irrigation on groundwater quality. *Water Science and Technology* **27**, 147–57.

Farman, J.L., Gardiner, B.G. and Shanklin, J.D. (1985) Large losses of total ozone reveal seasonal ClO_x/NO_x interaction. *Nature* **315**, 207–10.

Fearnside, P.M. (1993) Deforestation in Brazilian Amazon – the effect of population and land tenure. *Ambio* **22**, 537–45.

February, E.C. (1994) Rainfall reconstruction using wood charcoal from 2 archaeological sites in South Africa. *Quaternary Research* **42**, 100–107.

Fernandes, H.M., Franklin, M.R., Veiga, L.H.S., Freitas, P. and Gomiero, L. (1996) Management of uranium mill tailings – geochemical processes and radiological risk assessment. *Journal of Environmental Radioactivity* **30**, 69–95.

Fiedel, S.J. (1992) *Prehistory of the Americas*, 2nd edn. Cambridge University Press, Cambridge.

Field, M.H., Huntley, B. and Müller, H. (1994) Eemian climate fluctuations observed in a European pollen record. *Nature* **371**, 779–83.

Field, M.M. (1993) Plant macrofossils from the Lower Channel sediments at Marsworth, Buckinghamshire, UK. *New Phytologist* **123**, 195–201.

Finlayson, C.M. (1991) Australia and Oceania. In I.M. Finlayson and M. Moser (eds) *Wetlands.* Facts on File, Oxford, pp. 179–208.

Finnecy, E.E. and Pearce, K.W. (1986) Land contamination and reclamation In R.E. Hester (ed.) *Understanding Our Environment.* Royal Society of Chemistry, London, pp. 172–225.

Flannery, K.V. (1986) *Guilá Naquitz.* Academic Press, New York.

Flavin, C. (1996) Facing up to the risks of climate change. In L.R. Brown (ed.) *State of the World 1996.* Earthscan, London, pp. 21–39.

Fleischer, S., Anderson, G., Brodin, Y., Dickson, W., Herrmann, J. and Muniz, I. (1993) Acid water research in Sweden – knowledge for tomorrow. *Ambio* **22**, 258–63.

Flint, E.P. (1994) Changes in land use in south and Southeast Asia from 1880 to 1980 – a data base prepared as part of a coordinated research program on carbon fluxes in the tropics. *Chemosphere* **29**, 1015–62.

Flood, J. (1995) *Archaeology of the Dreamtime*, 3rd edn. Angus and Robertson, Sydney.

Flower, R.J., Pippey, B., Rose, N.L., Appleby, P.G. and Batterbee, R.W. (1994) Palaeolimnological evidence for the acidification and contamination of lakes by atmospheric pollution in western Ireland. *Journal of Ecology* **82**, 581–96.

Flowers, T.J. and Yeo, A.R. (1995) Breeding for salinity resistance in crop plants – where next? *Australian Journal of Plant Physiology.* **22**, 875–84.

Foell, W., Green, C., Amann, M., Bhaltacharya, S., Carmichael, G., Chadwick, M., Cinderby, S., Hangland, T., Hettelingh, J.P., Hordijk L., Kuylenstierna, J., Shah, J., Shrestha, R., Streets, D. and Zhao, D. (1995) Energy use, emissions, and air-pollution reduction stratrgies. *Water Air and Soil Pollution* **85**, 2277–82.

Fojt, W. and Harding, M. (1995) 30 years of change in the vegetation communities of 3 valley mires in Suffolk, England. *Journal of Applied Ecology* **32**, 561–77.

Foley, R. (1987) *Another Unique Species: Patterns in Human Evolutionary Ecology.* Longman, London.

Foner, H.A. (1993) Lead pollution in Israel. *Water Science and Technology* **27**, 253–62.

Fontes, J.C., Mélières, F., Gibert, E., Qing, L. and Gasse, F. (1993) Stable isotope and radiocarbon balances of Kio Tibetan Lakes (Sumxi Co, Longmu Co) from 13,000 BP. *Quaternary Science Reviews* **12**, 875–87.

Food and Agiculture Organisation (1993) *Marine Fisheries and the Law of the Sea: A Decade of Change.* Fisheries Circular 853, FAO, Rome, pp. 34–35.

Foote, A.L., Pandey, S. and Krogman, N.T. (1996) Processes of wetland loss in India. *Environmental Conservation* **23**, 45–54.

Forman, S.L., Oblesby, R., Markgraf, V. and Stafford, T. (1995) Palaeoclimatic significance of late Quaternary eolian deposition on the Piedment and High Plains, Central United States. *Global and Planetary Change* **11**, 35–55.

Fossitt, J.A. (1994) Late-glacial and Holocene vegetation history of Western Donegal, Ireland. *Biology and*

Environment: Proceedings of the Royal Irish Academy **94B**, 1–31.

Fox, A.D., Jones, T.A., Singleton, R. and Agnew, A.D.Q. (1994) Food supply and the effects of recreational disturbance on the abundance and distribution of wintering pochard on a gravel pit complex in Southern Britain. *Hydrobiologia* **280**, 253–61.

Francaviglia, R.V. (1992) Mining and landscape transformation. In L.M. Dilsaver and C.E. Colten (eds) *The American Environment: Interpretations of Past Geographies*. Rowman and Littlefield, Lanham, MD, pp. 89–114.

Francou, B. Mourguiart, P and Fournier, M. (1995) Glacier advance during the Younger Dryas in the Andes of Peru. *Comptes Rendus de L'Academic des Sciences, Serie II*, **320**, 593–99.

Franzen, L.G. (1994) Are wetlands the key to ice-age cycle enigma? *Ambio* **23**, 300–308.

Frayer, D.W., Wolpoff, M.H., Thorne, A.G., Smith, F.H. and Pope, G.G. (1993) Theories of modern human origins: the paleontological test. *American Anthropologist* **95**, 14–50.

Fridriksson, S. (1987) Plant colonisation of a volcanic island, Surtsey, Iceland. *Arctic and Alpine Research* **19**, 425–31.

Friends of the Earth (1995) *Environment Watch: Western Europe*. Friends of the Earth, London.

Fritts, H.C. (1976) *Tree Rings and Climate*. Academic Press, New York.

Fritz, G.J. (1995) New dates and data on early agriculture: the legacy of complex hunter-gatherers. *Annals of the Missouri Botanical Garden* **82**, 3–15.

Fronval, T., Jansen, E., Bloemendal, J. and Johnsen, S. (1995) Oceanic evidence for coherent fluctuations in Fennoscandian and Laurentide ice sheets on millennium timescales. *Nature.* **374**, 443–46.

Fryrear, D.W. (1995) Soil losses by wind erosion. *Soil Science Society of America Journal* **59**, 668–72.

Fuchs, A. and Leuenberger, M.C. (1996) Delta ^{18}O of atmospheric oxygen measured on the GRIP ice core document stratigraphic disturbances in the lowest 10 per cent of the core. *Geophysical Research Letters* **23**, 1049–52.

Fuji, N. (1986) Global correlation on the paleoclimatic changes between Lake Biwa sedimentary evidence and other marine and terrestrial records. *Proceedings of the Japanese Academy* **62**(B,1), 1–4.

Fullen, M.A. and Mitchell, D.J. (1994) Desertification and reclamation in north-central China. *Ambio* **23**, 131–35.

Fumihito, A., Miyake, T., Sumi, S.I., Takada, M., Ohno, S. and Kondo, N. (1994) One subspecies and the red junglefowl (*Gallus gallus gallus*) suffices as the matriarchic ancestor of all domestic breeds. *Proceedings of the National Academy of Sciences of the United States of America* **91**, 12505–9.

Funder, S., Hjort, C. and Landrik, J.Y. (1994) The last glacial cycles in East Greenland, an overview. *Boreas* **23**, 283–93.

Gabunia, L. and Vekua, A. (1995) A Plio Pleistocene hominid from Dmanisi, East Georgia, Caucasia. *Nature* **373**, 509–12.

Gade, D.W. (1996) Deforestation and its effects in highland Madagascar. *Mountain Research and Development* **16**, 101–16.

Gajewski, K., Winkler, M.G. and Swain, A.M. (1985) Vegetation and fire history from three lakes with varved sediments in northwestern Wisconsin (USA). *Review of Palaeobotany and Palynology* **44**, 277–92.

Galinat, W.C. (1992) Evolution of corn. *Advances in Agronomy* **47**, 203–31.

Galvin, J., Stephenson, J., Wlodarczyk, J., Loughran, R. and Waller, G. (1993) Living near a lead smelter – an environmental health risk assessment in Boolaroo and Argenton, New South Wales. *Australian Journal of Public Health* **17**, 373–78.

Gamble, C. (1986) *The Palaeolithic Settlement of Europe*. Cambridge University Press, Cambridge.

Gao, Y. and Bradshaw, A.D. (1995) The containment of toxic wastes 2: metal movement in leachate and drainage at Parc lead – zinc mine, North Wales. *Environmental Pollution* **90**, 379–82.

Garašanin, M. (1994) The Balkan Peninsula and South-East Europe. In S.J. De Laet, A.M. Dani, J.L. Lorenzo and R.B. Nunoo (eds) *History of Humanity, Volume I: Prehistory and the Beginnings of Civilization*. Routledge, London, and UNESCO, Paris, pp. 527–39.

Garciaoliva, F., Lugo, R.M. and Maass, J.M. (1995) Long-term net soil erosion as determined by Cs-137 redistribution in an undisturbed and perturbed tropical deciduous forest ecosystem. *Geoderma* **68**, 135–47.

Gascoyne, M. (1996) The geochemical environment of nuclear fuel waste disposal. *Canadian Journal of Microbiology* **42**, 401–9.

Gatehouse, A.M.R., Boulter, D. and Hilder, V.A. (1992) Potential of plant-derived genes in the genetic manipulation of crops for insect resistance. In A.M.R. Gatehouse, V.A. Hilder and D. Boulter (eds) *Plant Genetic Manipulation for Crop Protection*. CAB International, Wallingfod, pp. 155–181.

Gates, W.L., Henderson-Sellers, A., Boer, G.J., Folland, C.K., Kitoh, A., McAvaney, B.J., Semazzi, F., Smith, N., Weaver, A.J. and Zeng, Q.-C. (1996) Climate models – evaluation. In J.T. Houghton, L.G. Meira Filho, B.A. Callander, N. Harris, A. Kattenberg and K. Maskell (eds) *Climate Change 1995: The Science of Climate Change*. Cambridge University Press, Cambridge, pp. 228–84.

Gaudzinski, S. (1995) Wallertheim revisited: a re-analysis of the fauna from the Middle Palaeolithic site of Wallertheim (Rheinhessen/Germany). *Journal of Archaeological Science* **22**, 51–66.

Gaynor, J.D. and Findlay, W.I. (1995) Soil and phosphorus loss from conservation and conventional tillage in corn production. *Journal of Environmental Quality* **24**, 734–41.

Gehrke, C., Johanson, U., Callaghan, T.V., Chadwick, D. and Robinson, C.H. (1995) The impact of enhanced ultraviolet-B radiation on litter quality and decomposition processes in *Vaccinium* leaves from the sub-Arctic. *Oikos* 72, 213-222.

Geikie, A. 1863. On the phenomena of the glacial drift of Scotland. *Transactions of the Geological Society of Glasgow* **1**, 1–190.

Geikie, J. 1874. *The Great Ice Age*. W. Ibister, London.

Gepts, P. (1990) Biochemical evidence bearing on the domestication of *Phaseolus* (Fabaceae) beans. In P. Bretting (ed.) New Perspectives on the origin and evolution of New World domesticated plants. *Economic Botany* **44**, (suppl.), 28–38.

Gerasimidis, A. and Athanasiadis, N. (1995) Woodland history of northern Greece from the mid Holocene to recent time based on evidence from peat pollen profiles. *Vegetation History and Archaeobotany* **4**, 109–16.

Ghosh, T.K. (1994) Environmental impact analysis of desertification through remote sensing and land-based information system. *Journal of Arid Environments.* **25**, 141–50.

Gibbons, A. (1993) Geneticists trace the DNA trail of the first Americans. *Science* **259**, 312–13.

Gicheru, P.T. (1994) Effects of residue mulch and tillage on soil moisture conservation. *Soil Technology.* **7**, 209–20.

Gilbert, G.K. (1890) Lake Bonneville. *US Geological Society Monograph* **1**, 1–438.

Gilland, B. (1993) Cereals, nitrogen and population: an assessment of global trends. *Endeavour, New Series* **17**, 84–88.

Girling, M.A. and Greig, J. (1985) A first fossil record for *Scolytus scolytus* (F.)(elm bark beetle): its occurrence in elm decline deposits from London and the implications for neolithic elm disease. *Journal of Archaeological Science* **12**, 347–51.

Girling, R. and Jackman, B. (1989) Nothing to do with you? *Sunday Times Colour Supplement.* 26 Feb, 1989.

Gitay, H., Wilson, J.B. and Lee, W.G. (1996) Species redundancy: a redundant concept? *Journal of Ecology* **84**, 121–24.

Giusberti, G. and Peretto, C. (1991) Evidence of intentional animal marrow-bone breakage in the Palaeolithic site of La Pineta di Isernia, Molise, Italy. *Anthropologie* **95**, 765–78.

Gladue, R.M. and Maxey, J.E. (1994) Microalgal feeds for aquaculture. *Journal of Applied Phycology* **6**, 131–41.

Glantz, M.H., Rubinstein, A.Z. and Zonn, I. (1993) Tragedy in the Aral Sea basin. *Global Environmental Change* **3**, 174–98.

Glasser, J.A. and Lamar, R.T. (1995) Lignin-degrading fungi as degraders of pentachlorophenol and creosote in soil.

In H.D. Skipper and R.F. Turco (eds) *Bioremediation: Science and Applications*. Soil Science Society of America, American Society of Agronomy and Crop Science Society of America, Madison WI, pp. 117–33.

Gleick, J. (1987) *Chaos: The Making of a New Science*. Viking, New York.

Glennie, P.D. (1990) Industry and towns 1500–1730. In R.A. Dodgshon and R.A. Butlin (eds) *An Historical Geography of England and Wales*. Academic Press, London, 2nd edn., pp. 199–222.

Glennie, P.D. (1991) Measuring crop yields in early modern England. In B.M.S. Campbell and M. Overton (eds) *Land and Livestock: Historical Studies in European Agricultural Productivity*. Manchester Universty Press, Manchester, pp. 255–85.

Gnamus, A., Horvat, M. and Stegnar, P. (1995) The mercury content among deer and browsed foliage as a means of ascertaining environmental pollution of the mining regions of Idrija – a case study from Slowenia. *Zeitschrift fur Jagdwissenschaft* **41**, 198–208.

Goddijn, D.J.M. and Pen, J. (1995) Plants as bioreactors. *Trends in Biotechnology* **13**, 379–87.

Godwin, H. (1975) *History of the British flora*, 2nd edn. Cambridge University Press, Cambridge.

Goebal, T. and Arsenov, M. (1995) Accelerator radiocarbon dating of the initial Upper Palaeolithic in southeast Siberia. *Antiquity* **69**, 349–57.

Goede, A. (1994) Continuous early last glacial palaeoenvironmental record from a Tasmanian speleothem based on stable isotope and minor element variations. *Quaternary Science Reviews* **13**, 283–91.

Goede, A., McDermott, F., Hawkesworth, C., Webb, J. and Finlayson, B. (1996) Evidence of Younger Dryas and neoglacial cooling in a Late Quaternary paleotemperature record from a speleothem in Victoria, Australia. *Journal of Quaternary Science* **11**, 1–7.

Goldstein, A.H., Rogers, R.D. and Mead, G. (1993) Mining by microbe. *Bio/Technology* **11**, 1250–54.

Goodwin, H. (1996) In pursuit of ecotourism. *Biodiversity and Conservation* **5**, 277–91.

Gophen, M., Ochumba, P.B.O. and Kaufman, L.S. (1995) Some aspects of the perturbation in the structure and biodiversity of the ecosystem of Lake Victoria (East Africa). *Aquatic Living Resources* **8**, 27–41.

Goslar, T., Kuc, T., Ralska-Jasiewiczowa, M., Rózánski, K., Arnold, M., Bard, E., van Geel, B., Pazdur, M.F., Szeroczyska, K., Wicik, B., Wieckowski, K. and Walanus, A. (1993) High-resolution lacustrine record of the last glacial/Holocene transition in central Europe. *Quaternary Science Reviews* **12**, 287–94.

Gott, J.R. III (1993) Implications of the Copernican principle for our future prospects. *Nature* **363**, 315–19.

Gould, A.B., Hendrix, J.W. and Ferriss, R.S. (1996) Relationship of mycorrhizal activity to time following reclamation of surface mine land in western Kentucky 1: propagule and spore population densities. *Canadian Journal of Botany* **74**, 247–61.

Gowlett, J.A.J. (1988) A case of developed oldowan in the acheulian? *World Archaeology* **20**, 13–26.

Graedel, T.E. and Crutzen, P.J. (1993) *Atmospheric Change: An Earth System Perspective*. W.H. Freeman, New York.

Grainger, A. (1990) *The Threatening Desert: Controlling Desertification*. Earthscan, London.

Granéli, E. and Haraldson, C. (1993) Can increased leaching of trace metals from acidified areas influence phytoplankton growth in coastal waters? *Ambio* **22**, 308–11.

Granli, T. and Bockman, O.C. (1995) Nitrous oxide (N_2O) emissions from soils in warm climates. *Fertilizer Research* **42**, 159–63.

Graveland, J. and Vanderwal, R. (1996) Decline in snail abundance due to soil acidification causes eggshell defects in forest passerines. *Oecologia* **105**, 351–60.

Graves, J. and Reavey, D. (1996) *Global Environmental Change: Plants, Animals and Communities*. Longman, Harlow.

Green, C.P. plus 8 others (1984) Evidence of two temperate episodes in Late Pleistocene deposits at Marsworth, UK. *Nature* 309, 778–81.

Green, F.J. (1994) Cereals and plant foods: a re-assessment of the Saxon economic evidence from Wessex. In J. Rackham (ed.) *Environment and Economy in Anglo-Saxon England*. Council for British Archaeology Research Report 89, pp. 83–88.

Green, M.B., Hartley, G.S. and West, T.F. (1987) *Chemicals for Crop Improvement and Pest Management*. 3rd edn. Pergamon, Oxford.

Greig, J. (1992) The deforestation of London. *Review of Palaeobotany and Palynology* **73**, 71–86.

Grewal, S.S., Juneja, M.L., Singh, K. and Singh, S. (1994) A comparison of 2 agroforestry systems for soil, water and nutrient conservation on degraded land. *Soil Technology* **7**, 145–53.

Griffiths, I.LL. (1994) *The Atlas of African Affairs*. 2nd edn. Routledge, London.

Grigg, D. (1987) The Industrial Revolution and land transformation. In M. Wolman and F.G.A. Fournier (eds) *Land Transformation in Agriculture*. J., Chichester, pp. 79–109.

Grigg, D. (1989) *English Agriculture: An Historical Perspective*. Blackwell, Oxford.

Grigg, D. (1992) *The Transformation of Agriculture in the West*. Blackwell, Oxford.

Grigg, D. (1995) The nutritional transition in Western Europe. *Journal of Historical Geography* **21**, 247–61.

Grimm, E.C, Jacobson, G.L., Watts, W.A., Hansen, B.C.S. and Maasch, K.A. (1993) A 50,000- year record of climate oscillations from Florida and its temporal correlation with the Heinrich events. *Science* **261**, 198–200.

Grootes, P., Stuiver, M., White, J.W.C., Johnsen, S. and Jouzel, J. (1993) Comparisons of oxygen isotope records from the GISP2 and GRIP Greenland ice cores. *Nature* **366**, 552–54.

Grove, A.T. (1993) Africa's climate in the Holocene. In T. Shaw, P. Sinclair, B. Andah and A. Okpoko (eds) *The Archaeology of Africa*. Routledge, London, pp. 32–42.

Grove, J.M. (1988) *The Little Ice Age*. Methuen, London.

Grove, R. (1987) Early themes in African conservation: the Cape in the nineteenth century. In D. Anderson and R. Grove (eds) *Conservation in Africa*. Cambridge University Press, Cambridge, pp. 21–39.

Groves, C.P. (1994) The origin of modern humans. *Interdisciplinary Science Reviews* **19**, 23–34.

Gruhn R. and Bryan, A.L. (1984) The record of Pleistocene megafaunal extinctions at Taima-taima, northern Venezuela. In P.S. Martin and R.G. Klein (eds) *Quaternary Extinctions: A Prehistoric Revolution*. University of Arizona Press, Tucson AZ, pp.128–37.

Guastella, L.A.M. (1994) A quantitative assessment of recreational angling in Durban Harbour, South Africa. *South African Journal of Marine Science* **14**, 187–203.

Guiot, J., Harrison, S.P. and Prentice, I.C. (1993) Reconstruction of Holocene precipitation patterns in Europe using pollen and lake-level data. *Quaternary Research* **40**, 139–49.

Gumbricht, T. (1993) Nutrient removal processes in freshwater submersed macrophyte systems. *Ecological Engineering* **2**, 1–30.

Gunn, J.M. and Keller, W. (1990) Biological recovery of an acid lake after reductions in industrial emissions of sulphur. *Nature* **345**, 431–33.

Guo, D.-S. (1995) Hongshan and related cultures. In S.M. Nelson (ed.) *The Archaeology of Northeast China*. Routledge, London, pp. 21–64.

Gurung, C.P. and DeCoursey, M. (1994) The Annapurna Conservation Area Project: a pioneering example of sustainable tourism. In E. Cater and G. Lowman (eds) *Ecotourism: a sustainable option?* J. Wiley, Chichester, and the Royal Geographical Society, London, pp. 177–94.

Gutierrez, J.R. (1993) The effect of water, nitrogen, and human-induced desertification on the structure of ephemeral plant communities in the Chilean coastal desert. *Revista Chilena de Historia Natural* **66**, 337–44.

Hader, D.P., Worrest, R.C., Kumar, H.D. and Smith, R.C. (1995) Effects of increased solar ultraviolet radiation on aquatic ecosystems. *Ambio* **24**, 174–80.

Hagmann, J. (1996) Mechanical soil conservation with contour ridges: cure for, or cause of, rill erosion. *Land Degradation and Development* **7**, 145–60.

Hall, D.O. and House, J.I. (1995) Biomass – a modern and environmentally acceptable fuel. *Solar Energy Materials and Solar Cells*. **38**, 521–42.

Hall, E.A.H. (1990) *Biosensors*. Open University Press, Milton Keynes.

Hall, R. (1994) *English Heritage Book of Viking Age York*. Batsford, London, and English Heritage, London.

Hall, R.I. and Smol, J.P. (1996) Paleolimnological assessment of long-term water quality changes in south central Ontario lakes affected by cottage development and acidification. *Canadian Journal of Fisheries and Aquatic Sciences* **53**, 1–17.

Hall, V.A. (1990) Recent lanscape history from a Co. Down lake deposit. *New Phytologist* **115**, 377–83.

Hall, V.A., Pilcher, J.R. and McCormac, F.G. (1993) Tephra-dated lowland landscape history of the north of Ireland, AD 750–1150. *New Phytologist* **125**, 193–202.

Hallam, J.S., Edwards, B.J.N., Barnes, B. and Stuart, A.J. (1973) The remains of a Late Glacial elk with associated barbed points from High Furlong, near Blackpool, Lancashire. *Proceedings of the Prehistoric Society* **39**, 100–128.

Hammarlund, D. and Buchardt, B. (1996) Composite stable isotope records from a Late Weichselian lacustrine sequence at Graenge, Lolland, Denmark – evidence for Allerød and Younger Dryas environments. *Boreas* **25**, 8–22.

Hammitt, J.K., Jain, A.K., Adams, J.L. and Wuebbles, D.J. (1996) A welfare-based index for assessing environmental effects of greenhouse-gas emissions. *Nature* **381**, 301–3.

Hannah, L., Lohse, D., Hutchinson, C., Carr, L.L. and Lankerani, A. (1994) A preliminary inventory of human disturbance of world ecosystems. *Ambio* **23**, 246–50.

Hansen, B.C.S. (1995) A review of the late glacial pollen records from Ecuador and Peru with reference to the Younger Dryas event. *Quaternary Science Reviews* **14**, 853–65.

Hansen, B.C.S., Seltzer, G.O. and Wright, H.E. Jr (1994) Late Quaternary vegetational change in the central Peruvian Andes. *Palaeogeography, Palaeoclimatology, Palaeoecology* **109**, 263–85.

Hao, W.M. and Ward, D.E. (1993) Methane production from global biomass burning. *Journal of Geophysical Research – Atmospheres* **98**, 20657–61.

Haralambopoulos, N. and Pizam, A. (1996) Perceived impacts of tourism – the case of Samos. *Annals of Tourism Research* **23**, 503–26.

Hare, R. (1954) *Pomp and Pestilence: Infectious Disease, its Origins and Conquest.* Victor Gollancz, London.

Harlan, J.R. (1992) *Crops and Man.* American Society of Agronomy and Crop Science Society of America, Madison WI.

Harlan, J.R. (1994) Plant domestication: an overview. In S.J. De Laet, A.H. Dani, J.L. Lorenzo and R.B. Nunoo (eds) *History of Humanity, Volume I: Prehistory and the Beginnings of Civilization.* Routledge, London, and UNESCO, Paris, pp. 377–88.

Harris, C. (1987) France in North America. In R.D. Mitchell and P.A. Groves (eds) *North America: the Historical Geography of a Changing Continent.* Hutchinson, London, pp. 65–92.

Harris, P.A. and Zuberer, D.A. (1993) Subterranean clover enhances production of coastal bermudagrass in the revegetation of lignite mine spoil. *Agronomy Journal* **85**, 236–41.

Harrison, R.M. (1993) A perspective on lead pollution and health 1977–1992. *Journal of the Royal Society of Health* **113**, 142–48.

Harrison, R.P. (1992) *Forests: The Shadow of Civilization.* University of Chicago Press, Chicago.

Harrison, S.P. and Digerfeldt, G. (1993) European lakes as palaeohydrological and palaeoclimatic indicators. *Quaternary Science Reviews* **12**, 233–48.

Hart, B.T. and Lake, P.S. (1987) Studies of heavy metal in pollution in Australia with particular emphasis on aquatic systems. In T.C. Hutchinson and K.M. Meena (eds) *Lead, Mercury, Cadmium and Arsenic in the Environment.* J. Wiley, Chichester, pp. 187–216.

Hartz, N. and Milthers, V. (1901) Det senglaciale ler i Alleørd Teglvaerksgrav. *Meddelelser Dansk Geologisk Forening* **8**, 31–59.

Harvey, D. (1993) The nature of environment: dialectics of social and environmental change. In R. Milliband and L. Panitch (eds) *Socialist Register*, pp. 1–51.

Hassan, F.A. (1993) Town and village in ancient Egypt: ecology, society and urbanization. In T. Shaw, P. Sinclair, B. Andah and A. Okpoko (eds) *The Archaeology of Africa: Food, Metals and Towns.* Routledge, London, pp. 551–69.

Hathaway, J.S. (1993) Alar: the EPA's mismanagement of an agricultural chemical. In D. Pimentel and H. Lehmann (eds) *The Pesticide Question: Environment, Economics and Ethics.* Chapman and Hall, New York, pp. 337–43.

Havens, K., Aumen, N.G., James, R.T. and Smith, V. (1996) Rapid ecological changes in a large subtropical lake undergoing cultural eutrophication. *Ambio* **25**, 150–55.

Havens, K.E., Flaig, E.G., James, R.T., Lostal, S. and Muszick, D. (1996) Results of a program to control phosphorus discharges from dairy operations in south-central Florida, USA. *Environmental Management* **20**, 585–93.

Hawkes, J.G. (1990) *The Potato: Evolution, Biodiversity and Genetic Resources.* Belhaven Press, London.

Hawkes, J.G. (1991) The evolution of tropical American root and tuber crops with special reference to potatoes. In J.G. Hawkes, R.N. Lester, M. Nee and N. Estrada (eds) *Solonaceae III: Taxonomy, Chemistry, Evolution.* Royal Botanic Gardens, Kew, for the Linnaean Society of London, pp. 347–56.

Hawkins, J.P. and Roberts, C.M. (1993) Effects of recreational scuba diving on coral reefs – trampling on reef-flat communities. *Journal of Applied Ecology* **30**, 25–30.

Haycock, N.E., Pinay, G. and Walker, C. (1993) Nitrogen retention in river corridors: European perspective. *Ambio* **22**, 340–46.

Hays, J.D., Imbrie, J. and Shackleton, N.J. (1976) Variations in the Earth's orbit: pacemaker of the ice ages. *Science* **194**, 1121–32.

Hazeldine, R.S. and McKeown, C. (1995) A model approach to radioactive waste disposal at Sellafield. *Terra Nova* **7**, 87–95.

Heathcote, R.L. (1983) *The Arid Lands: Their Use and Abuse*. Longman, London.

Heathcote, R.L. (1987) Land. In G. Davison, J.K. McCarty and A. McLeary (eds) *Australians in 1888*. Fairfax, Syme and Weldon, Sydney, pp. 49–67.

Heathcote, R.L. (1994) *Australia*, 2nd edn. Longman, Harlow.

Hebda, R.J. (1995) British Columbia vegetation and climatic history with focus on 6 ka BP. *Geographie Physique et Quaternaire* **49**, 55–79.

Hecky, R.E., Bugenyi, F.W.B., Ochumba, P., Talling, J.F., Mugidde, R., Gophen, M. and Kaufman, L. (1994) Deoxygenation of the deep water of Lake Victoria, East Africa. *Limnology and Oceanography* **39**, 1476–81.

Hedeager, L. (1992) *Iron Age Societies: From Tribe to State in Northern Europe, 500 BC to AD 700*. Blackwell, Oxford.

Hegerl, G., van Storch, H., Hasselman, K., Santer, B.D., Cubasch, U. and Jones, P.D. (1996) Detecting greenhouse-gas-induced climate change with an optimal fingerprint method. *Journal of Climate* **9**, 2281–2306.

Hegerl, G.C. and Cubasch, U. (1996) Greenhouse gas induced climate change. *Environmental Science and Pollution Research* **3**, 99–102.

Heine, J.T. (1993) A re-evaluation of the evidence for a Younger Dryas climatic reversal in the Tropical Andes. *Quaternary Science Reviews* **12**, 769–79.

Heinrich, H. (1988) Origin and consequences of cyclic ice rafting in the northeast Atlantic Ocean during the past 130,000 years. *Quaternary Research* **29**, 142–52.

Henk, L.L. and Linden, J.C. (1994) Silage processing of forage biomass to alcohol fuel. *ACS Symposium Series* **566**, 391–410.

Henle, K., Poschlod, P., Margules, C. and Settele, J. (1996) Species survival in relation to habitat quality, size, and isolation: summary conclusions and future directions. In J. Settele, C. Margules, P. Poschlod and K. Henle (eds) *Species Survival in Fragmented Landscapes*. Kluwer Academic, Dordrecht, pp. 373–81.

Henriksen, A., Lien, L., Tracaen, T.S., Sevaldrud, I.S. and Brakke, D.F. (1988) Lake acidification in Norway – present and predicted chemical status. *Ambio* **17**, 259–66.

Henry, D.O. (1989) *From Foraging to Agriculture: The Levant at the End of the Ice Ages*. University of Pennsylvania Press, Philadelphia PA.

Herman, J.R., Newman, P.A., and Larko, D. (1995) Meteor – 3/TOMS observations of the 1994 ozone hole. *Geophysical Research Letters* **22**, 3227-29.

Hershey, C.H. (1993) Cassava *Manihot esculenta* Crantz. In G. Kalloo and B.O. Bergh (eds) *Genetic Improvement of Vegetable Crops*. Pergamon, Oxford, pp. 669–91.

Heusser, C.J. (1993) Late-glacial of southern South America. *Quaternary Science Reviews* **12**, 345–50.

Heusser, C.J. (1994) Paleoindians and fire during the late Quaternary in southern South America. *Revista Chilena de Historia Natural* **67**, 435–43.

Heusser, L.E. and van de Geer, G. (1994) Direct correlation of terrestrial and marine paleoclimatic records from four glacial–interglacial cycles – DSDP site 594 southwest Pacific. *Quaternary Science Reviews* **13**, 273–82.

Higgitt, D.L. and Rowan, J.S. (1996) Erosion assessment and administration in subtropical China. A case study from Fujian Province. *Land Degradation and Development* **7**, 1–10.

Higham, N. (1992) *Rome, Britain and the Anglo-Saxons*. Seaby, London.

Hill, I.R. (1985) Effects on non-target organisms in terrestrial and aquatic organisms. In J.P. Leahy (ed.) *The Pyrethroid Insecticides*. Taylor & Francis, London, pp. 151–262.

Hill, J.D. (1995) The Pre-Roman Iron Age in Britain and Ireland (ca. 800 BC to AD 100): an overview. *Journal of World Prehistory* **9**, 47–98.

Hillman, G.C. and Davies, M.S. (1990) Measured domestication rates in wild wheats and barley under primitive cultivation, and their archaeological implications. *Journal of World Prehistory*. **4**, 157–222.

Hillman, G.C., Colledge, S.M. and Harris, D.R. (1989) Plant-food economy during the Epipalaeolithic period at Tell Abu Hureyra, Syria: dietary diversity, seasonality, and models of exploitation. In D.R. Harris and G.C. Hillman (eds) *Foraging and Farming: the Evolution of Plant Domestication*. Unwin Hyman, London, pp. 240–68.

Hingley, R. (1993) Society in Scotland from 700 BC to AD 200. *Proceedings of the Society of Antiquaries for Scotland* **122**, 7–53.

Hinton, D.A. (1994) The archaeology of eighth- to eleventh century Wessex. In M. Aston and C. Lewis (eds) *The Medieval Landscape of Wessex*. Oxbow Monograph 46, pp. 33–46.

Hjelmar, O. (1996) Disposal stratigies for municipal solid waste incineration residues. *Journal of Hazardous Materials* **47**, 345–68.

Ho, C.-K. and Li, Z.-W. (1987) Paleolithic subsistence strategies in North China. *Current Research in the Pleistocene* **4**, 7–9.

Hobbs, R.J. and Hopkins, A.J.M. (1990) From frontier to fragments: European impact on Australia's vegetation. *Proceedings of the Ecological Society of Australia* **16**, 93–114.

Hoen, F.H. and Solberg, B. (1994) Potential and economic efficiency of carbon sequestration in forest biomass through silvicultural management. *Forest Science* **40**, 429–51.

Hoffecker, J.F., Powers, W.R. and Goebel, T. (1993) The colonization of Beringia and the peopling of the New World. *Science* **259**, 46–53.

Hoffman, M.T., Bond, W.J. and Stock, W.D. (1995) Desertification of the eastern Karoo, South Africa –

conflicting paleoecological, historical, and soil isotopic evidence. *Environmental Monitoring and Assessment* **37**, 159–77.

Holden, T.G., Hather, J.G. and Watson, J.P.N. (1995) Mesolithic plant exploitation at the Roc del Migdia. *Journal of Archaeological Science* **22**, 769–78.

Holdridge, L.R. (1967) *Life Zone Ecology*. Tropical Science Center, San José.

Holien, H. (1996) Influence of site and stand factors on the distribution of crustose lichens of the calcicales in a suboceanic spruce forest area in Central Norway. *Lichenologist* **28**, 315–30.

Holmes, K.J. and Ellis, J.H. (1996) Potential environmental impacts of future halocarbon emissions. *Environmental Science and Technology* **30**, A348–55.

Hong, H., Xu, L., Zhang, L., Chen, J.C., Wong, Y.S. and Wan, T.S.M. (1995) Environmental fate and chemistry of organic pollutants in the sediment of Xiamen Harbour and Victoria Harbour. *Marine Pollution Bulletin* **31**, 229–36.

Hong, S., Candelone, J.P. and Boutron, C.F. (1994) Greenland ice history of the pollution of the atmosphere of the northern hemisphere for lead during the last 3 millennia. *Analusis* **22**, M38–40.

Hong, S., Candelone, J.P., Patterson, C.C. and Boutron, C.F. (1996) History of ancient copper smelting pollution during Roman and Medieval times recorded in Greenland ice. *Science* **272**, 246–49.

Hong, S.M., Candelone, J.P., Patterson, C.C. and Boutron, C.F. (1994) Greenland ice evidence of hemispheric lead pollution 2–3 millennia ago by Greek and Roman civilizations. *Science* **265**, 1841–43.

Hooghiemstra, H. (1984) *Vegetation and Climatic History of the High Plain of Bogotá, Colombia: A continuous Record of the last 3.5 million years*. Cramer, Vaduz.

Hooghiemstra, H. (1988) The orbital-tuned marine oxygen isotope record applied to the Middle and Late Pleistocene pollen record of Funza (Columbian Andes). *Palaeogeography, Palaeoclimatology,Palaeoecology* **66**, 9–17.

Hooghiemstra, H. (1989) Quaternary and upper-Pliocene glaciations and forest development in the tropical Andes: evidence from a long high-resolution pollen record from the sedimentary basin of Bogotá, Colombia. *Palaeogeography, Palaeoclimatology, Palaeoecology* **72**, 11–26.

Hooghiemstra, H. and Ran, E.T.H. (1994) Late and middle Pleistocene climatic change and forest development in Colombia: pollen record Funza II (2–158 m core interval). *Palaeogeography, Palaeoclimatology, Palaeoecology* **109**, 211–46.

Hooghiemstra, H., Melici, J.L., Berger, A. and Shackleton, N.J. (1993) Frequency spectra and paleoclimatic variability of the high resolution 30–1450 Ka Funza I pollen record (Eastern Cordillera, Colombia). *Quaternary Science Reviews* **12**, 141–56.

Hope, G. and Tulip, J. (1994) A long vegetation history from lowland Jaya, Indonesia. *Palaeogeography, Palaeoclimatology, Palaeoecology* **109**, 385–98.

Horai, S., Kondo, R., Nakagawahattori, Y., Hayashi, S., Sonoda, S. and Tajima, K. (1993) Peopling of the Americas, founded by 4 major lineages of mitochondial-DNA. *Molecular Biology and Evolution* **10**, 23–47.

Horn, D.R. (1988) *Ecological Approach to Pest Management*. Elsevier, London.

Hornblower, S. (1994) Greece: the history of the classical period. In J. Boardman, J. Griffin and O. Murray (eds) *The Oxford History of the Classical World*. Oxford University Press, Oxford, pp. 124–55.

Horowitz, A. (1989) Continuous pollen diagrams for the last 3.5 M.Y. from Israel: vegetation, climate and correlation with the oxygen isotope record. *Palaeogeography, Palaeoclimatology, Palaeoecology* **72**, 63–98.

Horseman, S.T. and McEwen, T.J. (1996) Thermal constraints on disposal of heat-emitting waste in argillaceous rocks. *Engineering Geology* **41**, 5–16.

Horton, B. (1995) Geographical distribution of changes in maximum and minimum temperature. *Atmospheric Research* **37**, 101–18.

Horton, D.R. (1984) Red Kangaroos: last of the Australian megafauna. In P.S. Martin and R.G. Klein (eds) *Quaternary Extinctions: A Prehistoric Revolution*. University of Arizona Press, Tucson AZ, pp. 639–80.

Hosaka, K. (1995) Successive domestication and evolution of the Andean potato as revealed by chloroplast DNA restriction endonuclease analysis. *Theoretical and Applied Genetics* **90**, 356–63.

Houghton, J. (1994) *Global Warming: The Complete Briefing*. Lion Publishing, Oxford.

Houghton, J.T., Jenkins, G.J. and Ephraums, J.J. (eds) (1990) *Climate Change: the IPCC scientific assessment*. Cambridge University Press, Cambridge.

Houghton, J.T., Callander, B.A. and Varney, S.K. (eds) (1992) *Climate Change 1992: The Supplementary Report to the IPCC Scientific Assessment*. Cambridge University Press, Cambridge.

Houghton, J.T., Meira Filho, L.G., Bruce, J., Hoesung Lee, Callender, B.A., Haites, E.N. and Maskell, K. (eds) (1995) *Climate Change 1994: Radiative Forcing of Climate Change and an Evaluation of the IPCC 1592 Emission Scenarios*. Cambridge University Press, Cambridge.

Houghton, J.T., Meiro Filho, I.G., Callendar, B. A., Harris, N., Kettenburg, A. and Maskell, K. (eds) (1996) *Climate Change 1995: The Science of Climate Change*. Cambridge University Press, Cambridge.

Houghton, R.A. (1995) Land-use change and the carbon cycle. *Global Change Biology* **1**, 275–87.

Houghton, R.A. (1996a) Land-use change and terrestrial carbon: the temporal record. In M.J. Apps and D.T. Price (eds) *Forest Ecosystems, Forest Management and the Global Carbon Cycle*. Springer-Verlag, Berlin, pp. 117–34.

Houghton, R.A. (1996b) Converting terrestrial ecosystems from sources to sinks of carbon. *Ambio* **25**, 267–72.

Hours, F. (1994) Western Asia in the period of *Homo habilis* and *Homo erectus*. In S.J. De Laet, A.J. Dani, J.L. Lorenzo and R.B. Nunoo (eds) *History of Humanity, Volume I: Prehistory and the Beginnings of Civilization*. Routledge, London, and UNESCO, Paris, pp. 62–77.

House, M.R. (1995) Orbital forcing timescales: an introduction. In M.R. House and A.S. Gale (eds) *Orbital Forcing Timescales and Cyclostratigraphy*. Geological Society, London, pp. 1–18.

Housley, R.A. (1988) The environmental context of Glastonbury Lake Village. *Somerset Levels Papers* **14**, 63–82.

Howells, G. (1990) *Acid Rain and Acid Waters*. Ellis Horwood, London.

Hu, F.S., Brubaker, L.B. and Anderson, P.M. (1995) Postglacial vegetation and climate change in the northern Bristol Bay region, southwest Alaska. *Quaternary Research* **43**, 382–92.

Hubbard, K.G. and Floresmendoza, F.J. (1995) Relating United States crop land use to natural resources and climate change. *Journal of Climate* **8**, 329–35.

Huber, M.E. (1994) An assessment of the status of the coral reefs of Papua New Guinea. *Marine Pollution Bulletin* **29**, 69–73.

Hublin, J.J., Spoor, F., Braun, M., Zonneveld, F. and Condemi, S. (1996) A late Neanderthal associated with Upper Palaeolithic artifacts. *Nature* **381**, 224–26.

Hughes, J.D. (1982) Deforestation, erosion and forest management in ancient Greece and Rome. *Journal of Forest History* **26**, 60–75.

Hughes, J.D. (1992) Sustainable agriculture in ancient Egypt. *Agricultural History* **66**, 12–22.

Hughes, M.K., Xiangding, W., Xuemei, S. and Garfin, G.M. (1994) A preliminary reconstruction of rainfall in North-central China since AD 1600 from tree-ring density and width. *Quaternary Research* **42**, 88–99.

Hulbert, I.A.R., Iason, G.R. and Racey, P.A. (1996) Habitat utilization in a stratified upland landscape by 2 lagomorphs with different feeding strategies. *Journal of Applied Ecology* **33**, 315–24.

Hulme, M. (1996) Recent climatic change in the world's drylands. *Geophysical Letters* **23**, 61–64.

Humphries, C.J. and Fisher, C.T. (1994) The loss of Bank's Legacy. *Philosophical Transactions of the Royal Society of London* **B344**, 3–9.

Hunter, C. and Green, H. (1995) *Tourism and the Environment*. Routledge, London.

Huntley, B. (1990) European post-glacial forests: compositional changes in response to climatic change. *Journal of Vegetation Science* **1**, 507–18.

Hurni, H. (1993) Land degradation, famine, and land resource scenarios in Ethiopia. In D. Pimentel (ed.) *World Soil Erosion and Conservation*. Cambridge University Press, Cambridge, pp. 27–61.

Hurt, P. (1994) *A Conspiracy of Optimism*. University of Nebraska Press, NE.

Hutton, J. (1795) *Theory of the Earth*. William Creech, Edinburgh.

Igarashi, Y. (1994) Quaternary forest and climate history of Hokkaido, Japan. *Quaternary Science Reviews* **13**, 335–44.

Imasu, R., Suga, A. and Matsuno, T. (1995) Radiative effects and halocarbon global warming potential of replacement compounds for chlorofluorocarbons. *Journal of the Meteorological Society of Japan* **73**, 1123–36.

Ingólfsson, O. and Norddahl, H. (1994) A review of the environmental history of Iceland, 13,000-9000 yr BP. *Journal of Quaternary Science* **9**, 147–50.

Ingram, R.G., Wang, J., Lin, C., Legendre, L. and Fortier, L. (1996) Impact of freshwater on a Sub-Arctic coastal ecosystem with seasonal sea ice (southeastern Hudson Bay, Canada) I: interannual variability and predicted global warming on river plume dynamics and sea ice. *Journal of Marine Systems* **7**, 221–31.

Innes, J.L. (1992) Forest decline. *Progress in Physical Geography* **16**, 1–64.

Isager, S. and Skydsgaard, J.E. (1992) *Ancient Greek Agriculture*. Routledge, London.

Islebe, G.A., Hooghiemstra, H., Brenner, M., Curtis, J.H. and Hodell, D.A. (1996) A Holocene vegetation history from lowland Guatemala. *The Holocene* **6**, 265–71.

Islebe, G.A., Hooghiemstra, H. and van der Borg, K. (1995) A cooling event during the Younger Dryas chron in Costa Rica. *Palaeogeography, Palaeoclimatology, Palaeoecology* **117**, 73–80.

Ivens, G.W. (1993) *The UK Pesticide Guide*, 6th edn. CAB International, Wallingford, and British Crop Protection Council, Farnham.

Iversen, J. (1941) Landnam i Danmarks stenalder. *Danmarks Geologiske Undersøgelse Series IV*, **66**, 20–68.

Iversen, J. (1944) *Viscum, Hedera* and *Ilex* as climate indicators. *Geologiske Förenigens Stockholm Förhandlingar* **66**, 463–83.

Iversen, J. (1956) Forest clearance in the stone age. *Scientific American* **194**, 36–41.

Iversen, J. (1958) The bearing of glacial and interglacial epochs on the formation and extinction of plant taxa. *Uppsala Universiteit Årsk* **6**, 210–15.

Iwata, H., Tanabe, S., Sakai, N., Nishimura, A. and Tatsukawa, R. (1994) Geographical distribution of persistent organochlorines in water and sediments from Asia and Oceania, and their implications for global redistribution from lower latitudes. *Environmental Pollution* **85**, 15–23.

Jackson, A.R.W. and Jackson, J.M. (1995) *Environmental Science*. Longman, Harlow.

Jackson, M.H. (1993) *Galápagos: A Natural History*, 2nd edn. University of Calgary Press, Calgary.

Jaglan, M.S. and Qureshi, M.H. (1996) Irrigation development and its environmental consequences. *Environmental Management* **20**, 323–36.

Jagtap, T., Chavan, V.S. and Untawale, A.G. (1993) Mangrove ecosystems of India: a need for protection. *Ambio* **22**, 252–54.

James, I.N. and James, P.M. (1989) Ultra-low-frequency variability in a simple atmospheric circulation model. *Nature* 342, 53-55.

Jameson, M.H., Runnels, C.N. and van Andel, T.H. (1994) *A Greek Countryside: the Southern Argolid from Prehistory to the Present Day.* Stanford University Press, Stanford, CA.

Jamieson, D. (1994) Global environmental justice. In R. Attfield and A. Belsey (eds) *Philosophy and the Natural Environment.* Royal Institute of Philosophy Supplement 36. Cambridge University Press, Cambridge, pp. 199–210.

Jamieson, T.F. (1865) On the history of the last geological changes in Scotland. *Quarterly Journal of the Geological Society of London* 21, 161–95.

Janiot, L.T., Sericano, J.L. and Roses, O.E. (1994) Chlorinated pesticide occurrence in the Uruguay River (Argentina–Uruguay). *Water Air and Soil Pollution* **76**, 323–31.

Jansson, M., Anderson, R., Berggren, H. and Leonardson, L. (1994) Wetlands and lakes as nitrogen traps. *Ambio* **23**, 320–25.

Jarvis, P.J. (1993) Environmental changes. In R.W. Furness and J.J.D. Greenwood (eds) *Birds as Monitors of Environmental Change.* Chapman and Hall, London, pp. 42–85.

Jelinowska, A., Tucholka, P., Gasse, F. and Fontes, J.C. (1995) Mineral magnetic record of environment in late Pleistocene and Holocene sediments, Lake Manas, Xinjiang, China. *Geophysical Research Letters* **22**, 953–56.

Jenkins, D.G. (1987) Was the Pliocene-Pleistocene boundary placed at the wrong stratigraphic level? *Quaternary Science Reviews* **6**, 41–42.

Jessen, K. (1949) Studies in Late Quaternary deposits and flora-history of Ireland. *Proceedings of the Royal Irish Academy* **52B**, 85–290.

Jessen, K. and Milthers, V. (1928) Stratigraphical and palaeontological studies of inter-glacial fresh water deposits in Jutland and northwest Germany. *Danmarks Geologiske Undersøgelse, Series II,* **48**, 1–379.

Jiang, Y.B., Yung, J.L. and Zurek, R.W. (1996) Decadal evolution of the Antarctic ozone hole. *Journal of Geophysical Research – Atmospheres* **101**, 8985–99.

Jiankun, H., Aling, Z. and Yong, Y. (1996) Technology options for CO_2 mitigation in China. *Ambio* **25**, 249–53.

Jing-Fen, J., Wei, L. and Xiang-Shan, Z. (1993) Selection of NaCl-tolerant variant of *Sectaria italica* via tissue culture technique and its physiological and biochemical characteristics. In C.B. You, Z.L. Chen and Y. Ding (eds) *Biotechnology in Agriculture.* Kluwer Academic, Dordrecht, pp. 309–12.

Johanson, D.C. and White, T.D. (1979) A systematic assessment of early African hominids. *Science* **203**, 321–30.

Johns, A.D. (1988) Economic development and wildlife conservation in Brazilian Amazonia. *Ambio* **17**, 302–6.

Johnsen, S.J., Clausen, H.B., Dansgaard, W., Gundestrup, N.S., Hammer, C.V. and Tauber, H. (1995) The Eem stable isotope record along the GRIP ice core and its interpretation. *Quaternary Research* **32**, 117–24.

Johnsen, S.J., Clausen, H.B., Dansgaard, W., Fuhrer, K., Gundestrup, N., Hammer, C.U., Iversen, B., Jouzel, J., Stauffer, B. and Steffensen, J.P. (1992) Irregular glacial interstadials recorded in a new Greenland ice core. *Nature* **359**, 311–13.

Johnsen, S.J., Clausen, H.B., Dansgaard, W., Gundestrup, N.S., Hammer, C.V. and Tauber, H. (1995) The Eem stable isotope record along the GRIP ice core and its interpretation. *Quaternary Research* **43**, 117–24.

Johnson, D. and Lewis, L.A. (1995) *Land Degradation: Creation and Destruction.* Blackwell, Oxford.

Johnson, H.B. (1987) Portuguese settlement. In L. Bethell (ed.) *Colonial Brazil.* Cambridge University Press, Cambridge, pp. 1–38.

Johnson, J.B., Montgomery, M., Thompson, G.E., Wood, G.C., Sage, P.W. and Cooke, M.J. (1996) The influence of combustion-derived pollutants on limestone deterioration 2: the wet deposition of pollutant species. *Corrosion Science* **38**, 267–78.

Jolly, D., Bonnefille, R. and Roux, M. (1994) Numerical interpretation of a high resolution Holocene pollen record from Burundi. *Palaeogeography, Palaeoclimatology, Palaeoecology* **109**, 357–70.

Jones, A.E. and Shanklin, J.D. (1995) Continued decline of total ozone over Halley, Antarctica. *Nature* **376**, 409–11.

Jones, G., Straker, V. and Davis, A. (1991) Early medieval plant use and ecology. In A. Vince (ed.) *Aspects of Saxon-Norman London, 2: Finds and Environmental Evidence.* London and Middlesex Archaeological Society Special Paper 12, 347–85.

Jones, J.S. (1991) Farming is in the blood. *Nature* **351**, 97–98.

Jones, M. (1996) Plant exploitation. In T.C. Champion and J.R. Collis (eds) *The Iron Age in Britain and Ireland: Recent Trends.* J.R. Collis, Sheffield, pp. 29–40.

Jones, P.D. (1995) Recent variations in mean temperature and the diurnal temperature range in the Antarctic. *Geophysical Research Letters* **22**, 1345–48.

Jones, P.D., Marsh, R., Wigley, T.M.L. and Peel, D.A. (1993) Decadal timescale links between Antarctic Peninsula ice-core oxygen-18, deuterium and temperature. *The Holocene* **3**, 14–26.

Jones, P.D., Briffa, K.R. and Schweingruber, F. H. (1995) Tree-ring evidence of the widespread effects of explosive volcanic eruptions. *Geophysical Research Letters* **22**, 1333–36.

Jones, R.L. and Keen, D.H. (1993) *Pleistocene Environments in the British Isles*. Chapman and Hall, London.

Jordan, M.A., McGuiness, S. and Phillips, C.V. (1996) Acidophilic bacteria – their potential mining and environmental applications. *Minerals Engineering* **9**, 169–81.

Joseph, L., Moritz, C. and Hugall, A. (1995) Molecular support for vicariance as a source of diversity in rainforest. *Proceedings of the Royal Society of London* **B260**, 177–82.

Joshi, D.C. and Dhir, R.P. (1994) Amelioration of soils irrigated with sodic water in the arid region of India. *Soil Use and Management* **10**, 30–34.

Joshi, R.V. (1994) Southern Asia in the period of *Homo habilis* and *Homo erectus*. In S.J. De Laet, A.H. Dani, J.L. Lorenzo and R.B. Nunoo (eds) *History of Humanity, Volume I: Prehistory and the Beginnings of Civilization*. Routledge, London, and UNESCO, Paris, pp. 78–85.

Jouzel, J. Lorius, C., Petit, J.-R., Genthon, C. Barkhov, N.I., Kotlyakov, V.M. and Petrov, V.M. (1987) Vostok ice core: a continuous isotope temperature record over the last climate cycle (160,000 years). *Nature* **329**, 403–8.

Jouzel, J., Petit, J.R. and Raynaud, D. (1990) Palaeoclimatic information from ice cores: the Vostok records. *Transactions of the Royal Society of Edinburgh: Earth Sciences* **81**, 349–55.

Jung, M.C. and Thornton, I. (1996) Heavy metal contamination of soils and plants in the vicinity of a lead–zinc mine, Korea. *Applied Geochemistry* **11**, 53–59.

Justic, D., Rabalais, N.N. and Turner, R.E. (1995) Stoichiometric nutrient balance and origin of coastal eutrophication. *Marine Pollution Bulletin* **30**, 41–46.

Kahlown, M.A. and Hamilton, J.R. (1994) Status and prospects of karez irrigation. *Water Resources Bulletin* **30**, 125–48.

Kahlown, M.A. and Hamilton, J.R. (1996) Sailaba irrigation practices and prospects. *Arid Soil Research and Rehabilitation* **10**, 179–91.

Kalb, J.E. (1995) Fossil elephantoids, Awash paleolake basins, and the Afar triple junction, Ethiopia. *Palaeogeography, Palaeoclimatology, Palaeoecology* **114**, 357–68.

Kalloo, G. (1993) Tomato. In G. Kalloo and B.O. Bergh (eds) *Genetic Improvement of Vegetable Crops*. Pergamon, Oxford, pp. 587–604.

Kanowski, P.J., Savill, P.S., Adland, P.G., Burley, J., Evans, J., Palmer, J.R. and Woods, P.G. (1992) Plantation forestry. In N.P. Sharman (ed.) *Managing the World's Forests: Looking for Balance between Conservation and Development*. Kendall/Hunt, Dubuque, IA, pp. 375–401.

Kappelle, M. and Juarez, M.F. (1995) Agroecological zonation along an altitudinal gradient in the montane belt of the Los Santos Forest Reserve in Costa Rica. *Mountain Research and Development* **15**, 19–37.

Karjalainen, T. and Asikainen, A. (1996) Greenhouse gas emissions from the use of primary energy in forest operations and long-distance transportation of timber in Finland. *Forestry* **69**, 215–28.

Kasanmoentalib, S. (1996) Science and values in risk assessment – the case of deliberate release of genetically-engineered organisms. *Journal of Agricultural and Environmental Ethics* **9**, 42–60.

Kassas, M. (1995) Desertification: a general review. *Journal of Arid Environments* **30**, 115–28.

Kauppi, L., Pietilainen, O.P. and Knuuttila, S. (1993) Impacts of agricultural nutrient loading on Finnish watercourses. *Water Science and Technology* **28**, 461–71.

Kaushalya, R. (1992) Monitoring the impact of desertification in western Rajasthan using remote sensing. *Journal of Arid Environments* **22**, 293–304.

Kebin, Z. and Kaiguo, Z. (1989) Afforestation for sand fixation in China. *Journal of Arid Environments* **16**, 3–10.

Keeling, C.D., Whorf, T.P., Wahlen, M. and Vanderplicht, J. (1995) Interannual extremes in the rate of rise of atmospheric carbon dioxide since 1980. *Nature* **375**, 660–70.

Keeling, R.F., Piper, S.C. and Heimann, M. (1996) Global and hemispheric CO_2 sinks deduced from changes in atmospheric O_2 concentration. *Nature* **381**, 218–21.

Keigwin, L.D., Curry, W.B., Lehman, S.J. and Johnsen, S. (1994) The role of the deep ocean in North Atlantic climate change between 70 and 130 K yr ago. *Nature* **371**, 323–26.

Keleher, C.J. and Rahel, F.J. (1996) Thermal limits to salmonid distributions in the Rocky Mountain Region and potential habitat loss due to global warming – a geographic information system (GIS) approach. *Transactions of the American Fisheries Society* **125**, 1–13.

Kellomaki, S., Hanninen, H. and Kolstrom, M. (1995) Computations on frost damage to Scots Pine under climatic warming in boreal conditions. *Ecological Applications* **5**, 42–52.

Kelso, G.K. (1994) Palynology in historical rural-landscape studies: Great Meadows, Pennsylvania. *American Antiquity* **59**, 359–72.

Kerley, G.I.H., Knight, M.H. and Dekock, M. (1995) Desertification of subtropical thicket in the Eastern Cape South Africa – are there alternatives? *Environmental Monitoring and Assessment* **37**, 211–30.

Kershaw, A.P. (1994) Pleistocene vegetation of the humid tropics of northeastern Queensland, Australia. *Palaeogeography, Palaeoclimatology, Palaeoecology* **109**, 399–412.

Kershaw, A.P. and Nanson, G.C. (1993) The last full glacial cycle in the Australian region. *Global and Planetary Change* **7**, 1–9.

Khair, K., Haddad, F. and Fattouh, S. (1992) The effects of overexploitation on coastal aquifers in Lebanon, with special reference to saline intrusion. *Hydrogeology* 3, 349–62.

Khan, A.H. (1994) History of decline and present status of natural tropical forest in Punjab. *Biological Conservation* 67, 205–10.

Khan, F.K. (1991) *A Geography of Pakistan*. Oxford University Press, Oxford.

Khoshoo, T.N. and Tejwani, K.G. (1993) Soil erosion and conservation in India (status and policies) In D. Pimentel (ed.) *World Soil Erosion and Conservation*. Cambridge University Press, Cambridge, pp. 110–45.

Khotinsky, N.A. (1993) Anthropogenic changes in the landscapes of the Russian Plain during the Holocene. *Grana* (suppl. 2), 70–74.

Kim, B.H., Kim, H.Y., Kim, T.S. and Park, D.H. (1995) Selectivity of desulfurization activity of *Desulfovibrio desulfuricans* M6 on different petroleum products. *Fuel Processing Technology* 43, 87–94.

Kimbel, W.H., Johnson, D.C. and Rak, Y. (1994) The first skull and other new discoveries of *Australopithecus afarensis* at Hadar, Ethiopia. *Nature* 368, 449–51.

King, D.A. and Stewart, W.P. (1996) Ecotourism and commodification – protecting people and places. *Biodiversity and Conservation* 5, 293–305.

Kinghorn, A.D. and Seo, E.-K. (1996) Cultivating the pharmacopoeia. *Chemtech*, July 1996, 46–54.

Kio, P.R.O. and Ekwebelam, S.A. (1987) Plantations versus natural forests for meeting Nigeria's wood needs. In F. Mergen and J.R. Vincent (eds) *Natural Management of Tropical Moist Forests*. Yale University School of Forestry and Environmental Studies, New Haven CT, pp. 149–76.

Kiome, R.M. and Stocking, M. (1995) Rationality of farmer perception of soil erosion – the effectiveness of soil conservation in semiarid Kenya. *Global Environmental Change – Human and Policy Dimensions* 5, 281–95.

Kirchhoff, V.W.J.H., Schuch, N.J., Pinheiro, D.K. and Harris, J.M. (1996) Evidence of an ozone hole perturbation at 30 degrees south. *Atmospheric Environment* 30, 1481.

Kittel, T.G.F. *et al.* (1995) The VEMAP integrated database for modeling United States ecosystem/vegetation sensitivity to climate change. *Journal of Biogeography* 22, 857–62.

Klein, R.G. (1992) The archaeology of modern humans. *Evolutionary Anthropology* 1, 5–14.

Klein, R.G. (1994) The problem of modern human origins. In M.H. Nitecki and D.V. Nitecki (eds) *Origins of Anatomically Modern Humans*. Plenum, New York, pp. 3–21.

Klima, B. (1994) The period of *Homo sapiens sapiens* to the beginnings of food production. In S.J. De Laet, A.H. Dani, J.L. Lorenzo and R.B. Nunoo (eds) *History of Humanity, Volume I: Prehistory and the Beginnings of Civilization*. Routledge, London, and UNESCO, Paris, pp. 176–84.

Kling, J. (1996) Could transgenic supercrops one day breed superweeds? *Science* 274, 180–81.

Kliot, N. (1994) *Water Resources and Conflict in the Middle East*. Routledge, London.

Knotkova, D., Boschek, P. and Kreislova, K. (1995) Effect of acidification on atmospheric corrosion of structural metals in Europe. *Water Air and Soil Pollution* 85, 2661–66.

Koch, J.M. and Ward, S.C. (1994) Establishment of understorey vegetation for rehabilitation of bauxite-mined areas in the jarrah forest of Western Australia. *Journal of Environmental Management* 41, 1–15.

Kohlmaier, G.H., Hager, G.H., Nadler, A., Wurth, G. and Lüdeke, M.K.B. (1995) Global carbon dynamics of higher latitude forests during anticipated climate change – ecophysiological versus biomes migration view. *Water Air and Soil Pollution* 82, 455–64.

Koptsik, G. and Mukhina, J. (1995) Effects of acid deposition on acidity and exchangeable cations in podzols of the Kola Peninsula. *Water Air and Soil Pollution* 85, 1209–14.

Kozlov, M.V., Haukioja, E., Bakhtiarov, A.V. and Stroganov, D.M. (1995) Heavy metals in birch leaves around a nickel–copper smelter at Monchegorsk, Northwestern Russia. *Environmental Pollution* 90, 291–99.

Kramer, B. and Becker, B. (1993) German oak and pine ^{14}C calibration, 7200–9439 BC. *Radiocarbon* 35, 125–35.

Kramer, K. (1995) Phenotypic plasticity of the phenology of 7 European tree species in relation to climatic warming. *Plant Cell and Environment* 18, 93–104.

Kremenetski, C.V. (1995) Holocene vegetation and climate history of southwestern Ukraine. *Review of Palaeobotany and Palynology* 85, 289–301.

Krivanek, C.S. (1996) Mercury control technologies for MWCS – the unanswered questions. *Journal of Hazardous Materials* 47, 119–36.

Kubiakmartens, L. (1996) Evidence for possible use of plant foods in Palaeolithic and Mesolithic diet from the site of Calowanie in the central part of the Polish Plain. *Vegetation History and Archaeobotany* 5, 33–38.

Kudrass, H.R., Erienkeuser, H., Vollbrecht, R. and Weiss, W. (1994) Global nature of the Younger Dryas cooling event inferred from oxygen isotope data from Sulu Sea cores. *Nature* 349, 406–9.

Kuhry, P., Hooghiemstra, H., van Geel, B. and van der Hammen, T. (1993) The El Abra stadial in the Eastern Cordillera of Columbia (South America). *Quaternary Science Reviews* 12, 333–43.

Kuiper, H.A. (1996) The role of toxicology in the evaluation of new agrochemicals. *Journal of Environmental Science and Health B* 31, 353–63.

Kumar, M. and Bhandari, M.M. (1993) Human use of the sand dune ecosystem in the semiarid zone of the Rajasthan Desert of India. *Land Degradation and Rehabilitation* 4, 21–36.

Kumar, N., Gwiazda, R., Anderson, R.F. and Froelich,

P.N. (1993) ^{231}Pa/^{230}Th ratios in sediments as a proxy for past changes in Southern Ocean productivity. *Nature* **362**, 45–48.

Kunz, M.L. and Reanier, R.E. (1994) Paleoindians in Beringia: evidence from Arctic Alaska. *Science* **263**, 660–62.

Kurtén, B. and Anderson, E. (1980) *Pleistocene Mammals of North America*. University of Columbia Press, New York.

Kuzmin, Y.V. and Chernuk, A.V. (1995) Human impact on environment in the Neolithic Bronze Age in southern Primorye (far eastern Russia). *The Holocene* **5**, 479–84.

Kvasov, D.D. and Blazhchishin, A.I. (1978) The key to sources of the Pliocene and Pleistocene glaciation is at the bottom of the Barents Sea. *Nature* **273**, 138–40.

Labitzke, K. and van Loon, H. (1988) Associations between the 11 year solar cycle, the QBO and the atmosphere, part I: the troposphere and stratosphere in the northern hemisphere in winter. *Journal of Atmospheric Terrestrial Physics* **50**, 197–206.

Lacy, W.B. (1995) The global plant genetic resources system – a competition-cooperation paradox. *Crop Science* **35**, 335–45.

Ladizinsky, G. (1987) Pulse domestication before cultivation. *Economic Botany* **41**, 60–65.

Lagerås, P. and Sandgren, P. (1994) The use of mineral magnetic analyses in identifying Middle and Late Holocene agriculture – a study of peat profiles in Småland, Southern Sweden. *Journal of Archaeological Science* **21**, 687–97.

Lagudah, E.S. and Appels, R. (1992) Wheat as a model system. In G.P. Chapman (ed.) *Grass Evolution and Domestication*. Cambridge University Press, Cambridge, pp. 225–65.

Lal, R. (1993) Soil erosion and conservation in West Africa. In D. Pimentel (ed.) *World Soil Erosion and Conservation*. Cambridge University Press, Cambridge, pp. 7–25.

Lal, R. (1994) Soil erosion by wind and water: problems and prospects. In R. Lal (ed.) *Soil Erosion Research Methods*. Soil and Water Conservation Society, Ankeny, and St Lucie Press, Delray Beach FL, pp. 1–9.

Lal, R. (1996a) Deforestation and land-use effects on soil degradation and rehabilitation in Western Nigeria I: soil physical and hydrological properties. *Land Degradation and Development* **7**, 19–45.

Lal, R. (1996b) Deforestation and land-use effects on soil degradation and rehabilitation in Western Nigeria II: soil chemical properties. *Land Degradation and Development* **7**, 87–98.

Lal, R. (1996c) Deforestation and land-use effects on soil degradation and rehabilitation in Western Nigeria III: runoff, soil erosion and nutrient loss. *Land Degradation and Development* **7**, 99–119.

LaMarche, V.C., Graybill, D.A., Fritts, H.C. and Rose, M.R. (1984) Increasing atmospheric carbon dioxide: tree ring evidence for growth enhancement in natural vegetation. *Science* **225**, 1019–21.

Lamb, H.H. (1995) *Climate, History and the Modern World.*, 2nd edn. Routledge, London.

Lambeck, K. and Nakada, M. (1992) Constraints on the age and duration of the last interglacial period and on sea-level variations. *Nature* **357**, 125–28.

Lambert, D. (1987) *The Cambridge Guide to Prehistoric Man*. Cambridge University Press, Cambridge.

Lamberti, G.A. and Berg, M.B. (1995) Invertebrates and other benthic features as indicators of environmental change in Juday Creek, Indiana. *Natural Areas Journal* **15**, 249–58.

Lanpo, J. (1985) China's earliest palaeolithic assemblages. In W. Rukang and J.W. Olsen (eds) *Palaeoanthropology and Palaeolithic Archaeology in the People's Republic of China*. Academic Press, London, pp. 135–45.

Lanpo, J. and Weiwen, H. (1985) The palaeolithic of China. In W. Rukang and J.W. Olsen (eds) *Palaeoanthropology and Palaeolithic Archaeology in the People's Republic of China*. Academic Press, London, pp. 211–23.

Lara, A. and Villalba, R. (1993) A 3620-year temperature record from *Fitzroya cupressoides* tree rings in southern South America. *Science* **260**, 1104–6.

Larney, F.J., Izaurralde, R.C., Janzen, H.H., Olson, B.M., Solberg, E.D., Lindwall, C.W. and Nyborg, M. (1995) Soil erosion – crop productivity relationships for 6 Alberta soils. *Journal of Soil and Water Conservation* **50**, 87–91.

Larsen, E., Sejrup, H.P., Johnsen, S.J. and Krudsen, K.L. (1995) Do Greenland ice cores reflect NW European interglacial climate variations? *Quaternary Research* **43**, 125–32.

Lash, S. and Urry, J. (1994) *Economies of Signs and Space*. Sage Publications, London.

Lavers, C. and Haines-Young, R. (1993) The impact of afforestation on upland birds in Britain. In C. Watkins (ed.) *Ecology Effects of Afforestation: Studies in the History and Ecology of Afforestation in Western Europe*. CAB International, Wallingford, pp. 127–52.

Lawrence, G. and Vanclay, F. (1992) Agricultural production and environmental degradation in the Murray–Darling Basin. In G. Lawrence, F. Vanclay and B. Furze (eds) *Agriculture, Environment and Society: Contemporary Issues for Australia*. Macmillan, Melbourne, pp. 33–59.

Lawton, R. (1990) Population and Society 1730–1914. In R.A. Dodgshon and R.A. Butlin (eds) *An Historical Geography of England and Wale.*, 2nd edn. Academic Press, London, pp. 285–321.

Lawton, R. and Pooley, C.G. (1992) *Britain 1740–1950: An Historical Geography*. Edward Arnold, London.

Layton, R., Foley, R. and Williams, E. (1991) The transition between hunting and gathering and the specialised husbandry of resources. *Current Anthropology*. 32, 255-74.

Leahy, J.P. (1985) Metabolism and environmental

degradation. In J.P. Leahy (ed.) *The Pyrethroid Insecticides*. Taylor & Francis, London, pp. 263–342.

Leakey, M. D. (1979) *Olduvai Gorge: My Search for Early Man*. Collins, London.

Leakey, M.G., Feibel, C.S., McDougall, I. and Walker, A.C. (1995) New four-million-year-old hominid species from Kanapoi and Allia Bay, Kenya. *Nature* **376**, 565–71.

Leakey, R. and Lewin, R. (1992) *Origins Reconsidered: In Search of What Makes Us Human*. Little, Brown and Co., London.

Leathers, T.D., Gupta, S.C. and Alexander, N.T. (1993) Mycopesticides – status, challenges and potential. *Journal of Industrial Microbiology* **12**, 69–75.

Leblanc, M., Achard, B., Othman, D.B., Luck, J.M., Bertrandsarfati, J. and Personne, J.C. (1996) Accumulation of arsenic from acid mine waters by ferruginous bacterial accretions (stromatolites). *Applied Geochemistry* **11**, 541–44.

Leemans, R. (1996) Biodiversity and global change. In K.J. Gaston (ed.) *Biodiversity: A Biology of Numbers and Difference*. Blackwell, Oxford, pp. 367–87.

Lees, B.G., Hayne, M. and Price, D. (1993) Marine transgression and dune initiation on Western Cape – Northern Australia. *Marine Geology* **114**, 81–89.

Legge, A.J. and Rowley-Conwy, P.A. (1989) Some preliminary results of a re-examination of the Star Carr fauna. In C. Bonsall (ed.) *The Mesolithic in Europe*. John Donald, Edinburgh, pp. 225–30.

Legrand, M.R., Delmas, R.J. and Charlson, R.J. (1988a) Climate-forcing implications from Vostok ice-core sulphate data. *Nature* **334**, 418–20.

Legrand, M.R., Lorius, C., Barkov, N.I. and Petrov, V.N. (1988b) Vostok (Antarctica) ice core: atmospheric chemistry changes over the last climate cycle (160,000 years). *Atmospheric Environment* **22**, 317–31.

Lehman, H. (1993) Values, ethics and the use of synthetic pesticides in agriculture. In D. Pimentel and H. Lehman (eds) *The Pesticide Question: Environment, Economics and Ethics*. Chapman and Hall, New York, pp. 347–79.

Leigh, D.S. and Feeney, T.P. (1995) Palaeochannels indicating wet climate and lack of response to lower sea level, southeast Georgia. *Geology* **23**, 687–90.

Leishy, D.J. and van Beek, N. (1992) Baculoviruses: possible alternatives to chemical insecticides. *Chemistry and Industry* (6th April), 250–54.

Lemly, A.D., Finger, S.E. and Nelson, M.K. (1993) Sources and impacts of irrigation drainwater – contaminated arid wetlands. *Environmental Toxicology and Chemistry* **12**, 2265–79.

Lenihan, J.M. and Neilson, R.P. (1993) A rule-based formation model for Canada. *Journal of Biogeography* **20**, 615–28.

Lenihan, J.M. and Neilson, R.P. (1995) Canadian vegetation sensitivity to projected climatic change at three organisation levels. *Climatic Change* **30**, 27–56.

Lent, R.M., Herczeg, A.L., Welch, S. and Lyons, W.B.

(1992) The history of metal pollution near a lead smelter in Spencer Gulf, South Australia. *Toxicological and Environmental Chemistry* **36**, 139–53.

Leroy, S. and Dupont, L. (1994) Development of vegetation and continental aridity in northwestern Africa during the Late Pliocene: the pollen record of ODP Site 658. *Palaeogeography, Palaeoclimatology, Palaeoecology* **109**, 295–316.

Leslie, A.R. and Cuperus, G.W. (eds) (1993) *Successful Implementation of Integrated Pest Management for Agricultural Crops*. Lewis, Boca Raton FL.

Leung, G.Y. (1996) Reclamation and sediment control in the middle Yellow River valley. *Water International* **21**, 12–19.

Levesque, A.J., Mayle, F.E., Walker, I.R. and Cwynar, L.C. (1994) The amphi-Atlantic oscillation: a proposed late-glacial climatic event. *Quaternary Science Reviews* **12**, 629–43.

Lewin, R. (1987) Africa: cradle of modern humans. *Science* **237**, 1292–95.

Lewin, R. (1993a) *Human Evolution: An Illustrated Introduction*. Blackwell, Oxford.

Lewin, R. (1993b) *The Origin of Modern Humans*. W.H. Freeman, New York.

Lewis, L.A. and Nyamulinda, V. (1996) The critical role of human activities in land degradation in Rwanda. *Land Degradation and Development* **7**, 47–55.

Lewis, M.E., Roberts, C.A. and Manchester, K. (1995) Comparative study of the prevalence of maxillary sinusitis in Later Medieval urban and rural populations in Northern England. *American Journal of Physical Anthropology* **98**, 497–506.

Leyden, B.W. (1995) Evidence of the Younger Dryas in Central America. *Quaternary Science Reviews* **14**, 833–39.

Leyden, B.W., Brenner, M., Hodell, D.A. and Curtis, J.H. (1994) Orbital and internal forcing of climate on the Yucatan Peninsula for the past ca. 36 Ka. *Palaeogeography, Palaeoclimatology, Palaeoecology* **109**, 193–210.

Libert, B. (1995) *The Environmental Heritage of Soviet Agriculture*. CAB International, Wallingford.

Lijklema, L. (1995) Development and eutrophication - experiences and perspectives. *Water Science and Technology* **31**, 11–15.

Likens, G.E., Driscoll, C.T. and Buso, D.C. (1996) Long-term effects of acid rain: response and recovery of a forest ecosystem. *Science* **272**, 244–46.

Lindquist, O. (1995) Environmental impact of mercury and other heavy metals. *Journal of Power Sources* **57**, 3–7.

Lindquist, S. and Tengberg, A. (1993) New evidence of desertification from case studies in Northern Burkina Faso. *Geografiska Annaler Series A* **75**, 127–35.

Lindsey, K. and Jones, M.G.K. (1989) *Plant Biotechnology in Agriculture*. Open University Press, Milton Keynes.

Lindskog, P. (1995) What did the Cape Verde Islands look like at the time of colonisation around 1460? On the causes of land degradation and desertification of an archipelago. In I.G. Simmons and A.M. Mannion (eds) *The Changing Nature of the People–Environment Relationship: Evidence from a Variety of Archives.* Proceedings of the International Geographical Union Commission on Historical Monitoring of Environmental Changes Meeting, Příhrazy, Czech Republic. Department of Social Geography and Regional Development, Charles University, Prague, pp. 27–31.

Litchfield, J.H. (1989) Single-cell proteins. In J.L. Marx (ed.) *A Revolution in Biotechnology.* Cambridge University Press, Cambridge, pp. 71–81.

Litchfield, M.H. (1985) Toxicity to mammals. In J.P. Leahy (ed.) *The Pyrethroid Insecticides.* Taylor & Francis, London, pp. 99–150.

Livett, E.A., Lee, J.A. and Tallis, J.H. (1979) Lead, zinc and copper analysis of British blanket peats. *Journal of Ecology* **67**, 865–91.

Livett, E.A. (1988) Geochemical monitoring of atmospheric heavy metal pollution: theory and application. *Advances in Ecological Research* **18**, 65–177.

Lober, D.J. (1996) Municipal solid waste policy and public participation in household resource reduction. *Waste Management and Research* **14**, 125–43.

Loftus, R.T., MacHugh, D.E., Ngre, L.O., Balaim, D.S., Badi, A.M. and Bradley, D.G. (1994) Mitochondrial genetic-variation in European, African and Indian cattle populations. *Animal Genetics.* **25**, 265–71.

Lonsdale, M. and Braithwaite, R. (1988) The scrub that conquered the bush. *New Scientist* **120**(1634), 52–55.

Lorenzo, J.L. (1994) The origins of humanity in America. In S.J. De Laet, A.H. Dani, J.J. Lorenzo and R.B. Nunoo (eds) *History of Humanity, Volume I: Prehistory and the Beginnings of Civilization.* Routledge, London, and UNESCO, Paris, pp. 290–96.

Lorimer, C.G. (1982) Silviculture. In R.A. Young (ed.) *Introduction to Forest Science.* J. Wiley, New York, pp. 209–34.

Lorius, C. and Oeschger, H. (1994) Palaeo-perspectives: reducing uncertainties in global change? *Ambio* **23**, 30–36.

Lorius, C., Jouzel, J., Ritz, C., Merlivat, L., Barkov, N.I., Korotkevitch, Y.S. and Kotlyakov, V.M. (1985) A 150,000 year climatic record from Antarctic ice. *Nature* **316**, 591–96.

Lorius, C., Raisbeck, G., Jouzel, J. and Raynaud, D. (1989) Long-term environmental records from Antarctic ice cores. In J. Oeschger and C.C. Langway Jr (eds) *The Environmental Record in Glaciers and Ice Sheets.* J. Wiley, Chichester, pp. 343–61.

Loubere, P. (1988) Gradual Late Pliocene onset of glaciation: a deep-sea record from the northeast Atlantic. *Palaeogeography, Palaeoclimatology, Palaeoecology* **63**, 327–34.

Lovejoy, C.O. (1981) The origin of man. *Science* **211**, 341–50.

Lovelock, J.E. (1972) Gaia as seen through the atmosphere. *Atmospheric Environment* **6**, 579–80.

Lovelock, J.E. (1991) *Gaia: The Practical Science of Planetary Medicine.* Gaia Books Limited, London.

Lowe, J.J., Ammann, B., Birks, H.H., Björck, S., Coope, G.R., Cwynar, L., de Beaulieu, J.-L., Mott, R.J., Peteet, D.M. and Walker, M.J.C. (1994) Climatic changes in areas adjacent to the North Atlantic during the last glacial–interglacial transition (14–9 ka BP): a contribution to IGCP-253. *Journal of Quaternary Science* **9**, 185–98.

Lowe, J.J., Coope, G.R., Sheldrick, C., Harkness, D.D. and Walker, M.J.C. (1995) Direct comparison of UK temperatures and Greenland snow accumulation rates, 15,000-12,000 yr ago. *Journal of Quaternary Science* **10**, 175–80.

Loy, T.H. Spriggs, M. and Wickler, S. (1992) Direct evidence for human use of plants 28,000 years ago: starch residues on stone artifacts from the northern Solomon Islands. *Antiquity* **66**, 898–912.

Lozhkin, A.V. (1993) Geochronology of late Quaternary events in northeastern Russia. *Radiocarbon* **35**, 429–33.

Lozhkin, A.V. and Anderson, P.M. (1995) The last interglaciation in Northeast Siberia. *Quaternary Research* **43**, 147–58.

Ludwig, H.R. and Leitch, J.A. (1996) Interbasin transfer of aquatic biota via anglers' bait buckets. *Fisheries* **21**, 14-18.

Ludwig, J.A. and Tongway, D.J. (1995) Desertification in Australia – an eye to grass roots and landscapes. *Environmental Monitoring and Assessment* **37**, 231–237.

Ludwig, K.R., Szabo, B.J., Moore, J.G. and Simmons, K.R. (1991) Crustal subsidence rate off Hawaii determined from ^{234}U/^{238}U ages of drowned coral reefs. *Geology* **19**, 171–74.

Lukashina, N.S., Amirkhanov, M.M., Anisimov, V.I. and Trunev, A. (1996) Tourism and Environmental degradation in Sochi, Russia. *Annals of Tourism Research* **23**, 654–65.

Lundberg, J. and Ford, D.C. (1994) Late Pleistocene sea level change in the Bahamas from mass spectrometric U-series dating of submerged speleothem. *Quaternary Science Reviews* **13**, 1–14.

Lyell, C. (1833) *Principles of Geology.* John Murray, London.

Lyle, M. (1988) Climatically forced organic carbon burial in equatorial Atlantic and Pacific oceans. *Nature* **335**, 529–32.

Lyons, T.J., Smith, R.C.G. and Xinmei, H. (1996) The impact of clearing for agriculture on the surface energy budget. *International Journal of Climatology* **16**, 551–58.

McCaffrey, S.C. (1993) Water, politics and international law. In P.H. Gleick (ed.) *Water in Crisis: A Guide to the World's Fresh Water Resources.* Oxford University Press, Oxford, pp. 92–104.

McCalpin, J.P. (1992) Glacial geology of the Upper

Wairau Valley, Marlborough, New Zealand. *New Zealand Journal of Geology and Geophysics* **35**, 211–22.

McCarthy, M.R. (1995) Archaeological and environmental evidence for the Roman impact on vegetation near Carlisle, Cumbria. *The Holocene* **5**, 491–95.

McChesney, C.J., Koch, J.M. and Bell, D.T. (1995) Jarrah forest restoration in Western Australia – canopy and topographic effects. *Restoration Ecology* **3**, 105–10.

McCluskey, M. (1993) Note on the fragmentation of primary rainforest. *Ambio* **22**, 250–51.

McComb, A.J. and Davis, J.A. (1993) Eutrophic waters of southwestern Australia. *Fertilizer Research* **36**, 105–14.

McCormick, J. (1995) *The Global Environmental Movement*, 2nd edn. J. Wiley, Chichester.

McCorriston, J. and Hole, F.A. (1991) The ecology of seasonal stress and the origins of agriculture in the Near East. *American Anthropologist* **93**, 46–69.

MacDonald, G.M. and McLeod, T.K. (1996) The Holocene closing of the ice-free corridor – a biogeographical perspective. *Quaternary International* **32**, 87–95.

McGarry, D. (1993) Degradation of soil structure. In G.H. McTainsh and W.C. Boughton (eds) *Land Degradation Processes in Australia*. Longman, Melbourne, pp. 271–305.

McGlone, M.S. and Basher, L.R. (1995) The deforestation of the upper Awatere catchment, Inland Kaikoura Range, Marlborough, South Island, New Zealand. *New Zealand Journal of Ecology* **19**, 53–66.

McGlone, M.S. and Neall, V.E. (1994) The late Pleistocene and Holocene vegetation history of Jaranaki, North Island, New Zealand. *New Zealand Journal of Botany* **32**, 251–69.

McGuffie, K., Henderson-Sellers, A., Zhang, H., Durbridge, T.B. and Pitman, A.J. (1995) Global climate sensitivity to tropical deforestation. *Global and Planetary Change* **10**, 97–128.

McLaughlin, A. and Mineau, P. (1995) The impact of agricultural practices on biodiversity. *Agriculture Ecosystems and Environment* **55**, 201–12.

McManus, J.F., Bond, G.C., Broecker, W.S., Johnsen, S., Labeyrie, L. and Higgins, S. (1994) High-resolution climate records from the North Atlantic during the last interglacial. *Nature* **371**, 326–29.

McNeal, B.J., Stanley, C.D., Graham, W.D., Gilreath, P.R., Downey, D. and Creighton, J.F. (1995) Nutrient loss trends for vegetable and citrus fields in west-central Florida. *Journal of Environmental Quality* **24**, 95–100.

McNeill, J.R. (1992) *The Mountains of the Mediterranean World. An Environmental History*. Cambridge University Press, Cambridge.

MacNeish, R.S. (1992) *The Origins of Agriculture and Settled Life*. University of Oklahoma Press, Norman and London.

Maddox, J. (1995) Sustainable development unsustainable. *Nature* **374**, 305.

Magee, J.W., Bowler, J.M., Miller, G.H. and Williams, D.L.G. (1995) Stratigraphy, sedimentology, chronology and palaeohydrology of Quaternary lacustrine deposits at Madigan Gulf, Lake Eyre, South Australia. *Palaeogeography, Palaeoclimatology, Palaeoecology* **113**, 3–42.

Magny, M. and Ruffaldi, P. (1995) Younger Dryas and early Holocene lake-level fluctuations, Jura Mountains, France. *Boreas* **24**, 155–72.

Maher, B.A., Thompson, R. and Zhou, L.P. (1994) Spatial and temporal reconstructions of changes in the Asian palaeomonsoon – a new mineral magnetic approach. *Earth and Planetary Science Letters* **125**, 461–71.

Maisels, C.K. (1990) *The Emergence of Civilization*. Routledge, London.

Malhotra, S.D. (1995) Biotechnology and sugarcane. *International Sugar Journal* **97**, 160–63.

Malm, O., Castro, M.B., Bastos, W.R., Branches, F.J.P., Guimaraes, J., Zuffo, C.E. and Pfeiffer, W.C. (1995) An assessment of Hg pollution in different gold mining areas in Amazon Brazil. *Science of the Total Environment* **175**, 127–40.

Maltby, E. and Immirzi, C.P. (1993) Carbon dynamics in peatlands and other wetland soils: regional and global perspectives. *Chemosphere* **27**, 999–1023.

Mangerud, J., Andersen, S.T., Berglund, B.E. and Donner, J.J. (1974) Quaternary stratigraphy of Norden, a proposal for terminology and classification. *Boreas* **3**, 109–28.

Mann, S. (1992) Bacteria and the Midas touch. *Nature* **357**, 358–59.

Mannion, A.M. (1986) Vegetation succession and climax. In R.D. Thompson, A.M. Mannion, C.W. Mitchell, M. Parry and J.R.G. Townshend *Processes in Physical Geography*. Longman, London, pp. 302–15.

Mannion, A.M. (1989) Palaeoecological evidence for environmental change during the last 200 years. I. Biological data. *Progress in Physical Geography* **13**, 23–46.

Mannion, A.M. (1992a) Environmental change: lessons from the past. In A.M. Mannion and S.R. Bowlby (eds), *Environmental Issues in the 1990s*. J. Wiley, Chichester, pp. 39–59.

Mannion, A.M. (1992b) Acidification and eutrophication. In A.M. Mannion and S.R. Bowlby (ed.) *Environmental Issues in the 1990s*. J. Wiley, Chichester. pp. 177–95.

Mannion, A.M. (1994) The new environmental determinism. *Environmental Conservation* **21**, 7–8.

Mannion, A.M. (1995a) *Agriculture and Environmental Change: Temporal and Spatial Dimensions*. J. Wiley, Chichester.

Mannion, A.M. (1995b) Environmental quality during the last 500 years: palaeoenvironmental evidence. In I.G. Simmons and A.M. Mannion (eds) *The Changing Nature of the People - Environment Relationship: Evidence from a Variety of Archives*. Proceedings of the International Geographical Union Commission on Historical Monitoring of Environmental Changes Meeting, Příhrazy, Czech Republic, published by

Department of Social Geography and Regional Development, Charles University, Prague, Czech Republic, pp. 19–26.

Mannion, A.M. (1995c) Biodiversity, biotechnology and business. *Environmental Conservation* **22**, 201–27.

Mannion, A.M. (1995d) Agriculture, environment and biotechnology. *Agriculture, Ecosystems and Environment* **53**, 31–45.

Mannion, A.M. (1995e) Biotechnology and environmental quality. *Progress in Physical Geography* **19**, 192–215.

Mannion, A.M. (1997a) Biodiversity as an environmental and a heritage issue. Proceedings of the Third Conference on Agriculture in Higher Education, Montpellier, 1996 (in press).

Mannion, A.M. (1997b) Climate and vegetation. In R.D. Thompson and A. Perry (eds) *Applied Climatology: Principles and Practices*. Routledge, London pp. 123–40.

Mannion, A.M. and Bowlby, S.R. (1992) Introduction. In A.M. Mannion and S.R. Bowlby (eds), *Environmental Issues in the 1990s*. J. Wiley, Chichester, pp. 3–20.

Mannion, A.M. and Benton, L.T. (1997) *The early Holocene vegetation history of Moor Copse near Tidmarsh, Reading, Berkshire, UK*. Department of Geography, University of Reading, Geographical Papers B No 39, pp. 25.

Marcus, L.F. and Berger, R. (1984) The significance of radiocarbon dates for Rancho La Brea. In P.S. Martin and R.G. Klein (eds) *Quaternary Extinctions: A Prehistoric Revolution*. University of Arizona Press, Tucson, pp. 159–83.

Margulis, L. and Lovelock, J.E. (1974) Biological modulation of the Earth's atmosphere. *Icarus* **21**, 471–89.

Marin, M., Pedregosa, A., Rios, S., Ortiz, M.L. and Laborda, F. (1995) Biodegradation of diesel and heating oil by *Acinetobacter calcoaceticus* MMS – its possible applications for bioremediation. *International Biodeterioration and Biodegradation* **35**, 269–85.

Marion, J.L. and Cole, D.N. (1996) Spatial and temporal variation in soil and vegetation impacts. *Ecological Applications* **6**, 520–30.

Marion, J.L. and Rogers, C.S. (1994) The applicability of terrestrial visitor impact management strategies to the protection of coral reefs. *Ocean and Coastal Management* **22**, 153–63.

Markgraf, V. (1993) Younger Dryas in southernmost South America – an update. *Quaternary Science Reviews* **12**, 351–55.

Markgraf, V., McGlone, M. and Hope, G. (1995) Neogene palaeoenvironmental and paleoclimatic change in southern ecosystems – a southern perspective. *Trends in Ecology and Evolution* **10**, 143–47.

Marren, P. (1990) *Britain's Ancient Woodland: Woodland Heritage*. David and Charles, Newton Abbot, and the Nature Conservancy Council, London.

Marrone, P.G. (1994) Present and furture use of *Bacillus thuringiensis* in integrated pest management systems – an industrial perspective. *Biocontrol Science and Technology* **4**, 517–26.

Marschall, E.A. and Crowder, L.B. (1996) Assessing population responses to multiple anthropogenic effects – a case study with brook trout. *Ecological Applications* **6**, 152–67.

Marshall, L.G. (1984) Who killed cock robin? An investigation of the extinction controversy. In P.S. Martin and R.G. Klein (eds) *Quaternary Extinctions: A Prehistoric Revolution*. University of Arizona Press, Tucson AZ, pp. 785–806.

Martin, J.H. (1990) Glacial–interglacial CO_2 change: the iron hypothesis. *Paleoceanography* **5**, 1–13.

Martin, J.H. *et al.* (1994) Testing the iron hypothesis in ecosystems of the equatorial Pacific Ocean. *Nature* **371**, 123–29.

Martin, P.S. (1984) Prehistoric overkill: the global model. In P.S. Martin and R.G. Klein (eds) *Quaternary Extinctions: A Prehistoric Revolution*. University of Arizona Press, Tucson AZ, pp. 354–403.

Martin, P.S. and Klein, R.G. (eds) (1984) *Quaternary Extinctions: A Prehistoric Revolution*. University of Arizona Press, Tucson AZ.

Marx, J.L. (1989) Heredity, genes and DNA. In J.L. Marx (ed.) *A Revolution in Biotechnology*. Cambridge University Press, Cambridge, pp. 1–14.

Maslin, M. (1994) Rift–where humans began. *New Scientist* **142**(No. 1928), 24–27.

Mason, S.L.R., Hather, J.G. and Hillman, G.C. (1994) Preliminary investigation of the plant macro-remains from Dolni Věstonice II, and its implications for the role of plant foods in Palaeolithic and Mesolithic Europe. *Antiquity* **68**, 48–57.

Mather, A.S. (1990) *Global Forest Resources*. Belhaven Press, London.

Mather, A.S. (1993a) Introduction. In A.S. Mather (ed.) Afforestation. *Policies, Planning and Progress*. Belhaven Press, London, pp. 1–12.

Mather, A.S. (1993b) Afforestation in Britain. In A.S. Mather (ed.) *Afforestation. Policies, Planning and Progress*. Belhaven Press, London, pp. 13–33.

Mathewes, R.W. (1993) Evidence for Younger Dryas-age cooling on the north Pacific coast of America. *Quaternary Science Reviews* **12**, 321–31.

Mathewes, R.W., Heusser, L.E. and Patterson, R.T. (1993) Evidence for a Younger Dryas-like cooling event on the British Columbia coast. *Geology* **21**, 101–4.

Matsuura, S. (1995) China air pollution and Japan response to it. *International Environmental Affairs* **7**, 235–48.

Matthews, J.A. (1991) The Late Neoglacial ('Little Ice Age') glacier maximum in southern Norway: new ^{14}C-dating evidence and climatic implications. *The Holocene* **1**, 219–33.

Maunder, W.J. (1994) *Dictionary of Global Climate Change*, 2nd edn. UCL Press, London.

Mayewski, P.A., Meeker, L.D., Whitlow, S., Twickler, M.S., Morrison, M.C., Alley, R.B., Bloomfield, P. and Taylor, K. (1993) The atmosphere during the Younger Dryas. *Science* **261**, 195–97.

Mazurski, K.R. (1990) Industrial pollution: the threat to Polish forests. *Ambio* **19**, 70–74.

Meadows, M. (1996) Estimating landfill methane emissions. *Energy Conversion and Management* **37**, 1099–1104.

Meck, R.A. (1996) Complete decay of radionuclides – implications for low-level waste disposal in municipal landfills. *Health Physics* **70**, 706–11.

Melillo, J.M., McGuire, A.D., Kicklighter, D.W., Moore, B., Vorosmarty, C.J. and Schloss, A.L. (1993) Global climate change and terrestrial net primary productivity. *Nature* **363**, 234–40.

Melillo, J.M. *et al.* (1995) Vegetation ecosystem modeling and analysis project – community biogeography and biogeochemistry models in a continental-scale study of terrestrial ecosystem responses to climate-change CO_2 doubling. *Global Biogeochemical Cycles* **9**, 407–37.

Melillo, J.M., Prentice, I.C., Farquar, G.D., Schulze, E.-D. and Sala, O.E. (1996) Terrestrial biotic responses to environmental change and feedbacks to climate. In J.T. Houghton, L.G. Meira Filho, B.A. Callander, N. Harris, A. Kattenberg and K. Maskell (eds) *Climate Change 1995: The Science of Climate Change.* Cambridge University Press, Cambridge, pp. 445–81.

Mellaart, J. (1967) *Çatal Hüyük: A Neolithic Town in Anatolia.* Thames and Hudson, London.

Mellaart, J. (1994) Western Asia during the Neolithic and the Chalcolithic (about 12,000–5,000 years ago). In S.J. De Laet, A.H. Dani, J.L. Lorenzo and R.B. Nunoo (eds) *History of Humanity, Volume I: Prehistory and the Beginnings of Civilization.* Routledge, London, and UNESCO, Paris, pp. 425–40.

Mellanby, K. (1992) *The DDT Story.* British Crop Protection Council, Farnham

Mellars, P.A. (1987) *Excavations on Oronsay: Prehistoric Ecology on a Small Island.* Edinburgh University Press, Edinburgh.

Meltzer, D.J. (1993) Pleistocene peopling of the Americas. *Evolutionary Anthropology* **1**, 157–69.

Meltzer, D.J., Adovasio, J.M. and Dillenhay, T.D. (1994) On a Pleistocene human occupation at Pedra Furada, Brazil. *Antiquity* **68**, 695–714.

Mendoza, M. (1990) Global assessment of desertification: world atlas of thematic indicators of desertification, proposal document. In *Desertification Revisited: proceedings of an ad hoc consultative meeting on the assessment of desertification.* UNEP DC/PAC, Nairobi, pp. 289–94.

Mepham, T.B. (1993) Approaches to the ethical evaluation of animal biotechnologies. *Animal Production* **57**, 353–59.

Mercer, D. and Puttnam, D. (1988) *Rural England: Our Countryside at the Crossroads.* MacDonald Queen Anne Press, London.

Mercier, N., Valladas, H., Bar-Yosef, O. and Vandermeersch, B., Stringer, C. and Joron, J.-L. (1993) Thermoluminescence data for the Mousterian burial site of Es-Skhul, Mt. Carmel. *Journal of Archaeological Science* **20**, 169–74.

Mercier, N., Valladas, H. and Valladas, G. (1995) Flint thermoluminescence dates from the CFR laboratory at Gif: contributions to the study of the chronology of the Middle Palaeolithic. *Quaternary Science Reviews (Quaternary Geochronology)* **14**, 351–64.

Merrington, G. and Alloway, B.J. (1994) The transfer and fate of Cd, Cu, Pb and Zn from 2 historic metalliferous mine sites in the UK. *Applied Geochemistry* **9**, 677–87.

Mesnage, V. and Picot, B. (1995) The distribution of phosphate in sediments and its relationship with eutrophication of a Mediterranean coastal lagoon. *Hydrobiologia* **297**, 29–41.

Messer, J. (1987) The sociology and politics of land degeneration in Australia. In P. Blaikie and H. Brookfield (eds) *Land Degradation and Society.* Methuen, London, pp. 232–38.

Metcalfe, S. (1995) Holocene environmental change in the Zacapu Basin, Mexico: a diatom-based record. *The Holocene* **5**, 196–208.

Metcalfe, S.E., Street-Perrott, F.A., O'Hara, S.L., Hales, P.E. and Perrott, R.A. (1994) The palaeolimnological record of environmental change: examples from the arid frontier of Mesoamenca. In A.C. Millington and K. Pye (eds) *Environmental Change in Drylands: Biogeographical and Geomorphological Perspectives.* J. Wiley, Chichester, pp. 131–45.

Michael, P.W. (1994) Alien plants. In R.H. Groves (ed.) *Australian Vegetation*, 2nd edn. Cambridge University Press, Cambridge, pp. 57–83.

Michelena, R.O. and Irurtia, C.B. (1995) Susceptibility of soil to wind erosion in LaPampa Province, Argentina. *Arid Soil Research and Rehabilitation* **9**, 227–34.

Michels, K., Sivakumar, M.V.K. and Allison, B.E. (1995) Wind erosion control using crop residue 1. Effects on soil flux and soil properties. *Field Crops Research.* **40**, 101–10.

Micklin, P.P. (1994) The Aral Sea problem. *Proceedings of the Institution of Civil Engineers-Civil Engineering* **102**, 114–21.

Micklin, P. (1992) Water management in Soviet Central Asia: problems and prospects. In J. Stewart (ed.) *The Soviet Environment: Problems, Policies and Politics.* Cambridge University Press, Cambridge, pp. 88–114.

Micklin, P.P. (1988) Desiccation of the Aral Sea: a water management disaster in the Soviet Union. *Science* **241**, 1170–76.

Middleton, B.A. (1995) Seed banks and species richness potential of coal slurry ponds reclaimed as wetlands. *Restoration Ecology* **3**, 311–18.

Midgley, M. (1992) Towards a more humane view of the beasts? In D.E. Cooper and J.A. Palmer (eds) *The Environment in Question.* Routledge, London, pp. 28–36.

Midgley, M. (1994) The end of anthropocentrism. In R. Atfield and A. Belsey (eds) *Philosophy and the Natural Environment*. Royal Institute of Philosophy Supplement 36. Cambridge University Press, Cambridge, pp. 103–12.

Mildenhall, D.C. (1994) Early to mid Holocene pollen samples containing mangrove pollen from Sponge Bay East-Coast, North Island, New Zealand. *Journal of the Royal Society of New Zealand* **24**, 219–30.

Miller, D.E. and van der Merwe, N.J. (1994) Early metal working in sub-Saharan Africa: a review of recent research. *Journal of African History* **35**, 1–36.

Miller, J.D., Anderson, H.A., Roy, D. and Anderson, A.R. (1996) Impact of some initial forestry practices on the drainage waters from blanket peatlands. *Forestry* **69**, 193–203.

Miller, J.R., Rowland, J., Lechler, P.J., Desilets, M. and Hsu, L.C. (1996) Dispersal of mercury-contaminated sediments by geomorphic processes, Sixmile Canyon, Nevada, USA. Implications of site characterization and remediation of fluvial environments. *Water Air and Soil Pollution* **86**, 373–8.

Miller, N.F. (1991) The Near East. In W. van Zeist, K. Wasylikowa and K.-E. Behre (eds) *Progress in Old World Palaeoethnobotany*. A.A. Balkema, Rotterdam, pp. 133–60.

Miller, R.F. (1996) Allerød – Younger Dryas coleoptera from western Cape Breton Island, Nova Scotia, Canada. *Canadian Journal of Earth Sciences* **33**, 33–41.

Mingay, G.E. (1981) Introduction: rural England in the industrial age. In G.E. Mingay (ed.) *The Victorian Countryside* (2 vols). Routledge, London, Vol. 1, pp. 3–16.

Minhas, P.S. (1996) Saline water management for irrigation in India. *Agricultural Water Management* **30**, 1–24.

Mishra, J.K., Joshi, M.D. and Devi, R. (1994) Study of desertification process in Aravalli environment using remote-sensing techniques. *International Journal of Remote Sensing* **15**, 87–94.

Misra, V. and Bakre, P.P. (1994) Organochlorine contaminants and avifauna of Mahala water reservoir, Jaipur, India. *Science of the Total Environment* **144**, 145–51.

Misra, V.N. (1987) Middle Pleistocene adaptations in India. In O. Soffer (ed.) *The Pleistocene Old World: Regional Perspectives*. Plenum, New York, pp. 99–117.

Mitchell, G.E., Penny, L.F., Shotton, F.W. and West, R.G. (1973) *A Correlation of Quaternary Deposits in the British Isles*. Geological Society of London, Special Report 4.

Mitchell, J.F.B., Davis, R.A., Ingram, W.J. and Senior, C.A. (1995a) On surface temperature, greenhouse gases and aerosols: models and observations. *Journal of Climate* **8**, 2364–86.

Mitchell, J.F.B., Johns, T.C., Gregory, J.M. and Tett, S.F.B. (1995b) Transient climate response to increasing sulphate aerosols and greenhouse gases. *Nature* **376**, 501–4.

Mitchell, R.D. (1987) The colonial origins of Anglo-America. In R.D. Mitchell and P.A. Groves (eds) *North America: Historical Geography of a Changing Continent*. Hutchinson, London, pp. 73–120.

Mitsch, W.J. and Gosselink, J.G. (1993) *Wetlands*. Van Nostrand Reinhold, New York.

Mnatsakanian, R.A. (1992) *Environmental Legacy of the Former Soviet Republics*. Centre for Human Ecology, Edinburgh.

Mohnen, V.A. (1988) The challenge of acid rain. *Scientific American* **259**, 14–22.

Molina, M.J. (1996) Polar ozone depletion (Nobel lecture). *Angewandte Chemie* **35**, 1778–85 (international edition in English).

Molleson, T. (1994) The eloquent bones of Abu Hureyra. *Scientific American* **271**, 60–65.

Molloy, K. and O'Connell, M. (1987) The nature of the vegetational changes at about 5000 BP with particular reference to the elm decline: fresh evidence from Connemara, western Ireland. *New Phytologist* **106**, 203–10.

Molto, J.C., Viana, E., Pico, Y. and Font, G. (1995) The effect of urban pollution on lead levels in air of the city of Valenciá (Spain)–May 1989–October (1990) *Science of the Total Environment* **162**, 111–17.

Mommersteeg, H.J.P.M., Loutre, M.F., Young, R., Wijmstra, T.A. and Hooghiemstra, H. (1995) Orbital forced frequencies in the 975 000-year pollen record from Tenagi-Philippon (Greece) *Climate Dynamics* **11**, 4–24.

Monsalve, M.V., Derestrepo, H.G., Espinel, A., Correal, G. and Devine, C. (1994) Evidence of mitochondrial-DNA diversity in South American aboriginals. *Annals of Human Genetics* **58**, 265–73.

Mooney, H.A., Fuentes, E.R. and Kronberg, B.I. (eds) (1993) *Earth System Responses to Global Change*. Academic Press, San Diego CA.

Moore, A.M.T. (1992) The impact of accelerator dating at the early village of Abu Hureyra on the Euphrates. *Radiocarbon* **34**, 850–58.

Moore, A.M.T. and Hillman, G.C. (1992) The Pleistocene to Holocene transition and human economy in southwest Asia: the impact of the Younger Dryas. *American Antiquity* **57**, 482–94.

Moore, D.M. (1983) Human impact on island vegetation. In W. Holzner, M.J.A. Werger and I. Ikusima (eds) *Man's Impact on Vegetation*. Junk, The Hague, pp. 237–48.

Moore, P.A., Daniel, T.C., Sharpley, A.N. and Wood, C.W. (1995) Poultry manure management – environmentally sound options. *Journal of Soil and Water Conservation* **50**, 321–27.

Morey, D.F. (1992) Size, shape and development in the evolution of the domestic dog. *Journal of Archaeological Science* **19**, 181–204.

Morgan, L.L. (1996) Children and lead – a model of care for community health and primary care providers. *Family and Community Health* **19**, 42–48.

Morgan, R.P.C. (1986) *Soil Erosion and Conservation.* Longman, London.

Morgan, R.P.C. (1995) *Soil Erosion and Conservation*, 2nd edn. Longman, Harlow.

Morgan, W.B. and Solarz, J.A. (1994) Agricultural crisis in sub-Saharan Africa: development constraints and policy problems. *Geographical Journal* **106**, 57–73.

Moringa, H., Itota, C., Isezaki, N., Goto, H., Yaskawa, K., Kusakab, H., Liu, J., Gu, Z., Yuan, B. and Cong, S. (1993) O-18 and C-13 records for the last 14,000 years from lacustrine carbonates of Siling-co (lake) in the Qinghai-Tibetan plateau. *Geophysical Research Letters* **20**, 2909–12.

Mörner, M. (1987) Rural economy and society in Spanish South America. In L. Bethell (ed.) *Colonial Spanish America.* Cambridge University Press, Cambridge, pp. 286–314.

Morris, J. (1996) Recycling versus incineration – an energy conservation approach. *Journal of Hazardous Materials* **47**, 277–93.

Mosimann, T. (1985) Geo-ecological impacts of ski piste construction in the Swiss Alps. *Applied Geography* **5**, 29–37.

Moss, B. (1988) *Ecology of Freshwaters: Man and Medium*, 2nd edn. Blackwell, Oxford.

Moss, B. (1996) A land awash with nutrients – the problem of eutrophication. *Chemistry and Industry* **11**, 407–11.

Mott, R.J. (1994) Wisconsinan late-glacial environmental change in Nova Scotia: a regional synthesis. *Journal of Quaternary Science* **9**, 155–60.

Mott, R.J., Grant, D.R., Stea, R. and Occhietti, S. (1986) Late-glacial climatic oscillation in Atlantic Canada equivalent to the Allerød/Younger Dryas event. *Nature* **328**, 247–50.

Mott, R.T. and Stea, R.R. (1994) Late-glacial (Allerød Younger Dryas) buried organic deposits, Nova Scotia, Canada. *Quaternay Science Reviews* **12**, 645–57.

Moura-Costa, P. (1996) Tropical forestry practices for carbon sequestration: a review and case study from Southeast Asia. *Ambio* **25**, 279–83.

Mozaffari, M. and Sims, J.T. (1994) Phosphorus availability and sorption in an Atlantic coastal plain watershed dominated by animal-based agriculture. *Soil Science* **157**, 97–107.

Muhs, D.R., Kennedy, G.L. and Rockwell, T.K. (1994) Uranium-series ages of marine terrace corals from the Pacific coast of North America and implications for last-interglacial sea level history. *Quaternary Research* **42**, 72–87.

Muniz, I.P. (1991) Freshwater acidification: its effects on species and communities of freshwater microbes, plants and animals. *Proceedings of the Royal Society of Edinburgh* **B97**, 227–54.

Murdiyarso, D. and Wasrin, U.R. (1995) Estimating land-use change and carbon release from tropical forest conversion using remote sensing techniques. *Journal of Biogeography* **22**, 715–21.

Murphy, P. (1994) The Anglo-Saxon landscape and rural economy: some results from sites in East Anglia and Sussex. In J. Rackham (ed.) *Environment and Economy in Anglo-Saxon England.* Council for British Archaeology Research Report 89, pp. 23–39.

Murray, O. (1994) Life and Society in classical Greece. In J. Boardman, J. Griffin and O. Murray (eds) *The Oxford History of the Classical World.* Oxford University Press, Oxford, pp. 204–33.

Muscutt, A.D. and Withers, P.J.A. (1996) The phosphorus content of rivers in England and Wales. *Water Research* **30**, 1258–68.

Mwebazandawula, L. (1994) Changes in relative abundance of zooplankton in northern Lake Victoria, East Africa. *Hydrobiologia* **272**, 259–64.

Myers, N. (ed.) (1993a) *The Gaia Atlas of Planet Management*, 2nd edn. Gaia Books, London.

Myers, N. (1993b) Tropical forests: the main deforestation fronts. *Environmental Conservation* **20**, 9–16.

Myers, N. (1993c) Population, environment and development. *Environmental Conservation* **20**, 205–16.

Mytton, L.R. and Skøt, L. (1993) Breeding for improved symbiotic nitrogen fixation. In M.D. Hayward, N.O. Bosemark, I. Romagosa and M. Cerezo (eds) *Plant Breeding: Principles and Prospects.* Chapman and Hall, London, pp. 451–72.

Naeem, S., Thompson, L.J., Lawler, S.P., Lawton, J.H. and Woodfin, R.M. (1994) Declining biodiversity can alter the performance of ecosystems. *Nature* **368**, 734–36.

Naeem, S., Thompson, L.J., Lawler, S.P., Lawton, J.H. and Woodfin, R.M. (1995) Empirical evidence that declining species diversity may alter the performance of terrestrial ecosystems. *Philosphical Transactions of the Royal Society of London.* **B347**, 249–62.

Nagorsen, D.W., Keddie, G. and Hebda, R.J. (1995) Early Holocene black bears, *Ursus Americanus*, from Vancouver Island. *Canadian Field Naturalist* **109**, 11–18.

Nair, A., Mandapati, R., Dureja, P. and Pillai, M.K. (1995) DDT and HCH load in mothers and their infants in Delhi, India. *Bulletin of Environmental Contamination and Toxicology* **56**, 58–64.

Nair, P.K.R. (1993) *An Introduction to Agroforestry.* Kluwer Academic, Dordrecht, in co-operation with the International Centre for Research in Agroforestry (ICRAF), Nairobi.

Nair, S.M. and Balchand, A.N. (1993) Nearshore abundance of dissolved and sedimented forms of phosphorus. *Hydrobiologia* **252**, 71–81.

Nam, S.-I., Stein, R., Grobe, H. and Hubberten, H. (1995) Late Quaternary glacial–interglacial changes in sediment composition at the East Greenland continental margin and their palaeoceanographic implications. *Marine Geology* **122**, 243–62.

Nanson, G.C., Chen, X.Y. and Price, D.M. (1995) Aeolian and fluvial evidence of changing climate and wind patterns during the past 100 ka in the Western Simpson Desert, Australia. *Palaeogeography, Palaeoclimatology, Palaeoecology* **113**, 87–102.

Nebel, B.J. and Wright, R.T. (1996) *Environmental Science: The Way the World Works*, 5th edn. Prentice Hall, Upper Saddle River NJ.

Nelson, C.S., Hendy, C.H., Jarrett, G.R. and Guthbertson, A.M. (1985) Near-synchroneity of New Zealand alpine glaciations and northern hemisphere continental glaciations during the past 750 Ka. *Nature* **318**, 361–63.

Nelson, S.M. (ed.) (1995) *The Archaeology of Northeast China*. Routledge, London.

Nene, Y.L. (1996) Sustainable agriculture – future hope for developing countries. *Canadian Journal of Plant Pathology* **18**, 133–40.

Neumann, K. (1991) In search for the green sahara: palynology and botanical macro-remains. *Palaeoecology of Africa* **22**, 203–12.

Newnham, R.M., Delange, P.J. and Lowe, D.J. (1995) Holocene vegetation, climate and history of a raised bog complex, northern New Zealand based on palynology, plant macrofossils and tephrochronology. *The Holocene* **5**, 267–82.

Newton, I., Davis, P.E. and Moss, D. (1996) Distribution and breeding of red kites *Milvus milvus* in relation to afforestation and other land use in Wales. *Journal of Applied Ecology* **33**, 210–24.

Newton, R.M., Burns, D.A., Blette, V.L. and Driscoll, C.T. (1996) The effect of whole catchment liming on the episodic acidification of 2 Adirondack streams. *Biogeochemistry* **32**, 299–322.

Nicholls, N., Gruza, G.V., Jouzel, J., Karl, T.R., Ogallo, L.A. and Parker, D.E. (1996) Observed climate variability and change. In J.T. Houghton, L.G. Meira Filho, B.A. Callander, N. Harris, A. Kattenberg and K. Maskell (eds) *Climate Change 1995: The Science of Global Change*. Cambridge University Press, Cambridge, pp. 133–92.

Nickling, W.G. and Wolfe, S.A. (1994) The morphology and origin of nabkhas, region of Mopti, Mali, West Africa. *Journal of Arid Environments* **28**, 13–30.

Nilsson, A. (1992) *Greenhouse Earth*. J. Wiley, Chichester.

Nord, A.G. and Tronner, K. (1995) Effect of acid rain on sandstone – the Royal Palace and the Riddarholm Church, Stockholm. *Water Air and Soil Pollution* **85**, 2719–24.

Nixon, S.W. (1995) Coastal marine eutrophication – a definition, social causes, and future concerns. *Ophelia* **41**, 199–219.

Norby, R.J., Wullschleger, S.D. and Gunderson, C.A. (1996) Tree response to elevated CO_2 and implications for forests. In G.W. Koch and H.A. Mooney (eds) *Carbon Dioxide and Terrestrial Ecosystems*. Academic Press, San Diego CA, pp. 1–21.

Norton, D.A. and Palmer, J.G. (1992) Dendroclimatic evidence from Australasia. In R.S. Bradley and P.D.

Jones (eds) *Climate Since AD 1500*. Routledge, London. (Reprinted with revisions in 1995), pp. 463–82.

Nostrand, R.L. (1987) The Spanish borderlands. In R.D. Mitchell and P.A. Groves (eds) *North America: the Historical Geography of a Changing Continent*. Hutchinson, London, pp. 48–64.

Nour, S. (1996) Groundwater potential for irrigation in the East Oweinat, Western Desert, Egypt. *Environmental Geology* **27**, 143–54.

Nwoboshi, L.C. (1987) Regeneration success of natural management, enrichment planting, and plantations of native species in West Africa. In F. Mergen and J.R. Vincent (eds) *Natural Management of Tropical Moist Forests*. Yale University School of Forestry and Environmental Studies, New Haven CT, pp. 71–91.

O'Brien W.F. (1990) Prehistoric copper mining in southwest Ireland: the Mt. Gabriel-type mines. *Proceedings of the Prehistoric Society* **56**, 269–90.

O'Connor, D. (1993) Urbanism in bronze age Egypt and northeast Africa. In T. Shaw, P. Sinclair, B. Andah and A. Okpoko (eds) *The Archaeology of Africa: Food, Metals and Towns*. Routledge, London, pp.570–86.

Odum, E.P. (1975) *Ecology*. Holt, Rinehart and Winston, New York.

Odum, E.P. (1993) *Ecology and Our Endangered Life-Support Systems*, 2nd edn. Sinauer Associates, Sunderland MA, USA.

Oelschlaeger, M. (1991) *The Idea of Wilderness from Prehistory to the Age of Ecology*. Yale University Press, New Haven CT.

Ohta, S., Uchijima, Z. and Oshima, Y. (1995) Effect of $2 \times CO_2$ climatic warming on water temperature and agricultural potential in China. *Journal of Biogeography* **22**, 649–55.

Okafor, E.E. (1993) New evidence on early iron-smelting from southeastern Nigeria. In T. Shaw, P. Sinclair, B. Andah and A. Okopo (eds) *The Archaeology of Africa: Food, Metals, and Towns*. Routledge, London, pp. 432–48.

Okafor, N. (1994) Biotechnology and sustainable development in Sub-Saharan Africa. *World Journal of Microbiology and Biotechnology* **10**, 243–48.

Oldeman, L.R. (1994) The global extent of soil degradation. In D.J. Greenwood and I. Szabolcs (eds) *Soil Resilience and Sustainable Land Use*. CAB International, Wallingford, pp.99–118.

Onasanya, L.O., Ajewole, K. and Adeyeye, A. (1993) Lead content in roadside vegetation as indicators of atmospheric pollution. *Environment International* **19**, 615–18.

Onyeanusi, A.E. (1986) Measurements of impact of tourist off-road driving on grasslands in Masai Mara National Reserve, Kenya: a simulation approach. *Environmental Conservation* **13**, 325–29.

Opdyke, N.D., Glass, B., Hays, J.D. and Foster, J. (1966) Palaeomagnetic study of Antarctic deep-sea cores. *Science* **154**, 349–57.

O'Riordan, T. (1991) The new environmentalism and sustainable development. *The Science of the Total Environment* **108**, 5–15.

O'Riordan, T. (1995) Managing the global commons. In T. O'Riordan (ed.) *Environmental Science for Environmental Management*. Longman, Harlow, pp. 347–69.

Orlando, P., Perdelli, F., Christina, M.L., Oberto, C., Viglione, D., Palmieri, S., Vari, A. and Dibello, F. (1994) Blood lead levels in shopkeepers and car traffic pollution in Liguria, Italy. *European Journal of Epidemiology* **10**, 381–85.

Oswal, M.C. (1994) Water conservation and dryland crop production in arid and semiarid regions. *Annals of Arid Zone* **33**, 95–104.

Otieno, A. and Rowntree, K. (1987) A comparative study of land degradation in Machakos and Baringo Districts, Kenya. In A.C. Millington, A.C. Mutiso and J.A. Binns (eds) *African Resources* (2 vols). Department of Geography, University of Reading, Geographical Paper 97, Vol.2, *Management*, pp. 30–47.

Otte, M. (1994) Europe during the Upper Palaeolithic and Mesolithic. In S.J. De Laet, A.H. Dani, J.L. Lorenzo and R.B. Nunoo (eds) *History of Humanity, Volume I: Prehistory and the Beginnings of Civilization*. Routledge, London, and UNESCO, Paris, pp. 207–23.

Overpeck, J.T., Petersen, L.C., Kipp, N., Imbrie, J. and Rhind, D. (1989) Climate change in the circum-North Atlantic region during the last deglaciation. *Nature* **338**, 553–57.

Overton, M. (1991) The determinants of crop yields in early modern England. In B.M.S. Campbell and M. Overton (eds) *Land, Labour and Livestock: Historical Studies in European Agricultural Productivity*. Manchester University Press, Manchester, pp. 284–322.

Oyarzun, C.E. (1995) Land use, hydrological properties, and soil erodibilities in the Bio-Bio River basin, Central Chile. *Mountain Research and Development* **15**, 331–38.

Pagdin, C. (1995) Assessing tourism impacts in the third world – a Nepal case study. *Progress in Planning* **44**, 1193-98.

Paillard, D. and Labeyrie, L. (1994) Role of the thermohaline circulation in the abrupt warming after the Heinrich events. *Nature* **372**, 162–64.

Palmer, R. (1977) The agricultural history of Rhodesia. In R. Palmer and N. Parsons (eds) *The Roots of Rural Poverty in Central and Southern Africa*. Heinemann, London.

Pan, Y., McGuire, A.D., Kicklighter, D.W. and Melillo, J.M. (1996) The importance of climate and soils for estimates for net primary production - a sensitivity analysis with the terrestrial ecosystem model. *Global Change Biology* **2**, 5–23.

Pankhurst, C.F. (1994) Biological indicators of soil health and sustainable productivity. In D.J. Greenland and I. Szabolcs (eds) *Soil Resilience and Sustainable Land Use*. CAB International, Wallingford, pp. 331–51.

Panopoulos, N.J., Hatziloukas, E. and Affendra, A.S. (1996) Transgenic crop resistance to bacteria. *Field Crops Research* **45**, 85–97.

Pappas, M.G. (1996) The biotechnology of gene therapy. *Drug Development and Industrial Pharmacy* **22**, 791–803.

Parés, J.M. and Pérez-González, A. (1995) Palaeomagnetic age of hominid fossils at Atapuerca archaeological site, Spain. *Science* **269**, 830–32.

Parilla, G., Lavin, A., Bryden, H., Garcia, M. and Millard, R. (1994) Rising temperatures in the subtropical North Atlantic ocean over the past 35 years. *Nature* **369**, 48–51.

Parker, D.E., Jones, P.D., Folland, C.K. and Bevan, A. (1994) Interdecadal changes of surface temperature since the late nineteenth century. *Journal of Geophysical Research* **99**, 14373–99.

Parry, M. (1990) The potential effect of climate changes on agriculture and land use. In F.I. Woodward (ed.) *Global Climate Change: The Ecological Consequences*. Academic Press, London, pp. 63–91.

Parry, M.L. (1990) *Climate Change and World Agriculture*. Earthscan, London.

Parry, M.L., Hossell, J.E., Jones, P.J., Rehonan, T., Tranter, R.B., Marsh, J.S., Rosenzweig, C., Fischer, G., Carson, I.G. and Bunce, R.G.H. (1996) Integrating global and regional analyses of the effects of climate change: a case study of land use in England and Wales. *Climatic Change* **32**, 185–98.

Paszczynski, A. and Crawford, R.L. (1995) Potential for bioremediation of xenobiotic compounds by white-rot fungus *Phanerochaete chrysosporium*. *Biotechnology Progress* **11**, 368–79.

Patch, R.W. (1994) Imperial politics and local economy in colonial America. *Past and Present* **143**, 77–107.

Pauls, K.P. (1995) Plant biotechnology for crop improvement. *Biotechnology Advances* **13**, 673–93.

Paulson, D.D. (1994) Understanding tropical deforestation – the case of Western Samoa. *Environmental Conservation* **21**, 326–32.

Pearce, D. and Moran, M.D. (1994) *The Economic Value of Biodiversity*. International Union for Nature Conservation, Cambridge and Earthscan, London.

Pearce, F. (1992) *The Dammed*. Bodley Head, London.

Pearce, D.W. (1993) *Economic Values and the Natural World*. Earthscan, London.

Pearsall, D. and Piperno, D.R. (1990) Antiquity of maize cultivation in Ecuador: summary and reevaluation of the evidence. *American Antiquity* **55**, 324–37.

Peckol, P., Demeoanderson, B., Rivers, J., Valiela, I., Maldonado, M. and Yates, J. (1994) Growth, nutrient uptake capacities and tissue constituents of the macroalgae *Cladophora vagabunda* and *Gracilaria tikvahiae* related to site-specific nitrogen loading rates. *Marine Biology* **121**, 175–85.

Pedersen, K. (1996) Investigation of subterranean bacteria in deep crystalline bedrock and their importance for the disposal of nuclear waste. *Canadian Journal of Microbiology* **42**, 382–91.

Peel, C.J. (1996) The role of industry in international animal agriculture. *Journal of Animal Science* **74**, 1382–85.

Peglar, S.M. (1993a) The development of the cultural landscape around Diss Merc, Norfolk, UK, during the past 7000 years. *Review of Palaeobotany and Palynology* **76**, 1–47.

Peglar, S.M. (1993b) The Mid-Holocene *Ulmus* decline at Diss Mere, Norfolk, UK: a year-by-year pollen stratigraphy from annual laminations. *The Holocene* **3**, 1–13.

Pelletier, G. (1993) Somatic hybridisation. In M.D. Hayward, M.O. Bosemark, I. Romagosa and M. Cerezo (eds) *Plant Breeding: Principles and Prospects*. Chapman and Hall, London, pp. 93–106.

Penck, A. and Brückner, E. (1909) *Die Alpen im Eiszeitalter*. Tauchnitz, Leipzig.

Pennington, W. (1977) The Late Devensian flora and vegetation of Britain. *Philosophical Transactions of the Royal Society of London* **B280**, 247–71.

Penny, D., Steel, M., Waddell, P.J. and Hendy, M.D. (1995) Improved analyses of human mtDNA sequences support a recent African origin for *Homo sapiens*. *Molecular Biology and Evolution* **12**, 863–82.

Perkins, J. and Holochuk, N. (1993) Pesticides: historical changes demand ethical choices. In D. Pimentel and H. Lehman (eds) *The Pesticide Question: Environment, Economics and Ethics*. Chapman and Hall, New York, pp. 370–417.

Perry, I. and Moore, P.D. (1987) Dutch elm disease as an analogue of neolithic elm decline. *Nature* **326**, 72–73.

Peteet, D.M., Daniels, R.A., Heusser, L.S., Vogel, J.S., Southon, J.R. and Nelson, D.E. (1993) Late-glacial pollen macrofossils and fish remains in northeastern U.S.A. – The Younger Dryas oscillation. *Quaternary Science Reviews* **12**, 597–612.

Peteet, D.M., Daruels, R.A., Heusser, L.E., Vogel, J.S., Southon, J.R. and Nelson, D.E. (1994) Wisconsinan Late-glacial environmental change in southern New England: a regional synthesis. *Journals of Quaternary Science* **9**, 151–54.

Peters, D., Egger, J. and Entzian, G. (1995) Dynamical aspects of ozone mini-hole formation. *Meteorology and Atmospheric Physics* **55**, 205–14.

Petterson, R.B. (1996) Effect of forestry on the abundance and diversity of arboreal spiders in the boreal spruce forest. *Ecography* **19**, 221–28.

Pfister, C. (1992) Monthly temperature and precipitation in central Europe 1525–1979: quantifying documentary evidence on weather and its effects. In R.S. Bradley and P.D. Jones (eds) *Climate Since AD 1500*. Routledge, London. (Reprinted with revisions in 1995), pp. 118–42.

Phillips, W.D. Jr., and Phillips, C.R. (1992) *The Worlds of Christopher Columbus*. Cambridge University Press, Cambridge.

Philips (1992) *Philips Atlas of the World* (2nd Edition). Philips, London.

Piazza, A., Rendine, S., Minch. E., Menozzi, P., Mountain, J. and Cavalli-Sforza, L.L. (1995) Genetics and the origin of European Languages. *Proceedings of the National Academy of Sciences of the United States of America* **92**, 5836–40.

Pickard, J. (1994) Post-European changes in creeks of semi-arid rangelands, 'Polpah Station', New South Wales. In A.C. Millington and K. Pye (eds) *Environmental Change in Drylands: Biogeographical and Geomorphological Perspectives*, J. Wiley, Chichester, pp. 271–83.

Pickens, R.D. (1996) Add-on control techniques for nitrogen oxide emissions during municipal waste combustion. *Journal of Hazardous Materials* **47**, 195–204.

Pignatti, S. (1993) Impact of tourism on the mountain landscape of central Italy. *Landscape and Urban Planning* **24**, 49–53.

Pike, J.G. (1995) Some aspects of irrigation system management in India. *Agricultural Water Management* **27**, 95–104.

Pilbeam, D. (1968) The earliest hominids. *Nature* **219**, 1335.

Pilcher, J.R. (1991) Radiocarbon dating for the Quaternary scientist. *Quaternary Proceedings* **1**, 27–33.

Pillans, B. (1994) Direct marine-terrestrial correlations, Wanganui Basin, New Zealand: the last 1 million years. *Quaternary Science Reviews* **13**, 189–200.

Pimentel, D. (1993) Overview. In D. Pimentel (ed.) *World Soil Erosion and Conservation*. Cambridge University Press, Cambridge, pp. 1–5.

Piper, D.J.W., Mudie, P.J., Aksu, A.E. and Skene, K.I. (1994) A 1 Ma record of sediment flux south of the Grand Banks used to infer the development of glaciation in southeastern Canada. *Quaternary Science Reviews* **13**, 23–37.

Pirrone, N., Keeler, G.J. and Nriagu, J.O. (1996a) Regional differences in worldwide emissions of mercury to the atmosphere. *Atmospheric Environment* **30**, 2981–2987.

Pirrone, N., Keeler, G.J., Nriagu, J.O. and Warner, P.O. (1996b) Historical trends of airborne trace metals in Detroit from 1971 to 1992 *Water Air and Soil Pollution* **88**, 145–65.

Plochl, M. and Cramer, G. (1995) Coupling global models of vegetation structure and ecosystem processes: an example from Arctic and boreal ecosystems. *Tellus* **47b**, 240–50.

Plaut, G., Ghil, M. and Vautard, R. (1995) Interannual and interdecaded vulnerability in 335 years of central England temperatures. *Science* **268**, 710–13.

Plöchl, M. and Cramer, G. (1995) Coupling global models vegetation structure and ecosystem processes: an example from Artic and boreal ecosystems. *Tellus* **476**, 240–50

Poethke, H.J., Seitz, A. and Wissel, C. (1996) Species survival and metapopulations: conservation implications from ecological theory. In J. Settele, C. Margules, P. Poschlod and K. Henle (eds) *Species Survival in Fragmented Landscapes*. Kluwer Academic, Dordrecht, pp. 81–92.

Pollard, A.J. and Baker, A.J.M. (1996) Quantitative genetics of zinc hyperaccunulation in *Thlaspi caerulescens*. *New Phytologist* **132**, 113–18.

Poola, R.B., Nagalingam, B. and Gopalakrishnan, K.V. (1994) Performance studies with biomass-derived high octane fuel additives in a 2-stroke spark-ignition engine. *Biomass and Bioenergy* **6**, 369–79.

Popham, M. (1994) The collapse of Aegean civilization at the end of the Late Bronze Age. In B. Cunliffe (ed.) *The Oxford Illustrated Prehistory of Europe*. Oxford University, Oxford, pp. 277–303.

Porvari, P. (1995) Mercury levels of fish in Tucuri hydroelectric reservoir in River Moju in Amazonia, State of Para, Brazil. *Science of the Total Environment* **175**, 109–17.

Postel, S. (1993) Water and agriculture. in P.H. Gleick (ed.) *Water in Crisis: A Guide to the World's Fresh Water Resources*. Oxford University Press, Oxford, pp. 56–39.

Postel, S. (1994) Irrigation expansion slowing. In L.R. Brown, H. Kane and D.M. Roodman (eds) *Vital Signs 1994*. Norton, New York, pp. 44–45.

Postel, S. (1996) Forging a sustainable water strategy. In L.R. Brown (ed.) *State of the World 1996*. Earthscan, London, pp. 40–59.

Potter, T.W. (1987) *Roman Italy*. British Museum, London.

Potter, T.W. and Johns, C. (1992) *Roman Britain*. British Museum, London.

Potts, G.R. (1986) *The Partridge: Pesticides, Predation and Conservation*. Collins, London.

Pounds, N.J.G. (1990) *An Historical Geography of Europe*. Cambridge University Press, Cambridge.

Prather, M., Derwent, R., Ehhalt, D., Fraser, P., Sanhueza, E. and Zhou, X. (1995) Other trace gases and atmospheric chemistry. In J.T. Houghton, L.G. Meira Filho, J. Bruce, H. Lee, B.A. Callander, E. Haites, N. Harris and K. Maskell (eds) *Climate Change 1994: Radiative Forcing of Climate Change and an Evaluation of the IPCC IS92 Emission Scenarios*. Cambridge University Press, Cambridge, pp. 73–126.

Pretty, J.N., Thompson, J. and Kiara, J.K. (1995) Agricultural regeneration in Kenya – the catchment approach to soil and water conservation. *Ambio* **24**, 7–15.

Price, T.D. (1987) The mesolithic of Western Europe. *Journal of World Prehistory* **1**, 225–305.

Primavera, J.H. (1991) Intensive prawn farming in the Philippines: ecological, social, and economic implications. *Ambio* **20**, 28–33.

Primrose, S.B. (1991) *Molecular Biotechnology*. Blackwell, Oxford.

Principe, P. (1991) *Monetizing the Pharmacological Benefits of Plants*. US Environmental Protection Agency, Government Printing Office, Washington, DC.

Prove, B.G., Doogan, V.J. and Truong, P.N.V. (1995) Nature and magnitude of soil erosion in sugarcane land on the wet tropical coast of north-eastern Queensland. *Australian Journal of Experimental Agriculture* **35**, 641–49.

Pullin, R.S.V. (1996) Biodiversity and aquaculture. In F. di Castri and Y. Younés (eds) *Biodiversity, Science and Development: Towards a New Partnership*. CAB International, Wallingford, and the International Union of Biological Sciences, Paris, pp. 409–23.

Pupacko, A. (1993) Variations in Northern Sierra Nevada streamflow – implications of climate change. *Water Resources Bulletin* **29**, 283–90.

Purrington, C.B. and Bergelson, J. (1995) Assessing weediness of transgenic crops – industry plays plant ecologist. *Trends in Ecology and Evolution* **10**, 340–42.

P'yankova, L. (1994) Central Asia in the Bronze Age: sedentary and nomadic cultures. *Antiquity* 68, 355-372.

Pyne, S.J. (1991) *Burning Bush: A Fire History of Australia*. Henry Holt, New York.

Qiao, Y.L. (1995) An application of aerial remote sensing to monitor salinization at Xinding basin. *Advances in Space Research* **18**, 133–39.

Quine, T.A. and Walling, D.E. (1991) Rates of soil erosion on arable fields in Britain: quantitive data from caesium-137 measurements. *Soil Use and Management* **7**, 169–77.

Quine, T.A., Navas, A., Walling, D.E. and Machin, J. (1994) Soil erosion and redistribution on cultivated and uncultivated land near Las Bardenas in the central Ebro River basin, Spain. *Land Degradation and Rehabilitation* **5**, 41–55.

Quivira, M.P. and Dillehay, T.D. (1988) Monte Verde, south-central Chile: stratigraphy, climate change and human settlement. *Geoarchaeology* **3**, 177–91.

Rackham, J. (1994) Economy and Environment in Saxon London. In J. Rackham (ed.) *Environment and Economy in Anglo-Saxon England*. Council for British Archaeology Research Report 89, pp. 126–35.

Rackham, O. (1980) *Ancient Woodland: its History, Vegetation and Uses in England*. Edward Arnold, London.

Rackham, O. (1986) *The History of The Countryside*. Dent, London.

Rackham, O. (1990) *Trees and Woodland in the British Landscape: the Complete History of Britain's Trees, Woods and Hedgerows*. Weidenfeld and Nicolson (revised edition).

Rackham, O. (1994a) *The Illustrated History of The Countryside*. Weidenfeld and Nicolson, London.

Rackham, O. (1994b) Trees and Woodland in Anglo-Saxon England: the documentary evidence. In J. Rackham (ed.) *Environment and Economy in Anglo-Saxon England*. Council for British Archaeology Research Report 89, pp. 7–11.

Radford, B.J., Key, A.J., Robertson, L.N. and Thomas, G.A. (1995) Conservation tillage increases soil-water storage, soil animal populations, grain yield, and response to fertilizer in the semiarid subtropics. *Australian Journal of Experimental Agriculture* **35**, 223–32.

Radmer, R.J. and Parker, B.C. (1994) Commercial applications of algae – opportunities and constraints. *Journal of Applied Phycology* **6**, 93–98.

Raftery, B. (1994) *Pagan Celtic Ireland*. Thames and Hudson, London.

Raghunathan, K. and Gullett, B.K. (1996) Role of sulfur in reducing PCDD and PCDF formation. *Environmental Science and Technology* **30**, 1827–34.

Rahman, S.M., Khalil, M.I. and Ahmed, M.F. (1995) Yield, water relations and nitrogen utilization by wheat in salt-affected soils of Bangladesh. *Agricultural Water Management* **28**, 49–56.

Raina, P., Joshi, D.C. and Kolarkar, A.S. (1993) Mapping of soil degradation by using remote sensing on alluvial plain, Rajasthan, India. *Arid Soil Research and Rehabilitation* **7**, 145–61.

Rainina, E.I., Efremenco, E.N., Varfolomeyev, S.D., Simonian, A.L. and Wild, J.R. (1996) The development of a new biosensor based on recombinant *Escherichia coli* for the direct detection of organophosphorus neurotoxins. *Biosensors and Bioelectronics* **11**, 991–1000.

Raisbeck, G.M., Yiou, F. Bourles, D., Lorius, C., Jouzel, J. and Barkov, N.I. (1987) Evidence for two intervals of enhanced [10]Be deposition in Antarctic ice during the last glacial period. *Nature* **326**, 273–77.

Rajasuriya, A. and White, A.T. (1995) Coral reefs of Sri Lanka – review of their extent, condition and management status. *Coastal Management* **23**, 77–90.

Ramaswamy, V., Schwarzkopf, M.D. and Randel, W.J. (1996) Fingerprint of ozone depletion in the spatial and temporal pattern of recent lower stratosphere cooling. *Nature* **382**, 616–18.

Ramsay, G.D. (1982) *The English Woollen Industry 1500–1750*. Macmillan, London.

Rana, B.C. and Kumar, J.I.N. (1993) A composite rating of trophic status of certain ponds of Gujarat, India. *Journal of Environmental Biology* **14**, 113–20.

Randsborg, K. (1991) *The First Millennium AD in Europe and the Mediterranean*. Cambridge University Press, Cambridge.

Rappaport, R. (1992) *Controlling Crop Pests and Diseases*. Macmillan, New York.

Rapport, D.J., Regier, H.A. and Hutchinson, T.C. (1985) Ecosystem behaviour under stress. *American Naturalist* **125**, 617–40.

Rasmussen, T.L., Vanweering, T.C.E. and Labeyrie, L. (1996) High resolution stratigraphy of the Faeroe–Shetland channel and its relation to North Atlantic paleoceanography – the last 87 Kyr. *Marine Geology* **131**, 75–88.

Rast, W. and Thornton, J.A. (1996) Trends in eutrophication research and control. *Hydrological Processes* **10**, 295–313.

Ravindranath, N.H. and Somashekhar, B.S. (1995) Potential and economics of forestry options for carbon sequestration in India. *Biomass and Bioenergy*. **8**, 323–36.

Rawlings, D.E. and Silver, S. (1995) Mining with microbes. *Bio/technology* **13**, 773–78.

Rawson, E. (1994) The expansion of Rome. In J. Boardman, J. Griffin and O. Murray (eds) *The Oxford History of the Classical World*. Oxford University Press, Oxford, pp. 417–37.

Raybould, A.F. and Gray, A.J. (1994) Will hybrids of genetically modified crops invade natural communities? *Trends in Ecology and Evolution* **9**, 85–89.

Raymo, M.E. (1994) The initiation of northern hemisphere glaciation. *Annual Review of Earth and Planetary Sciences* **22**, 353–83.

Raymo, M.E. and Ruddiman, W.F. (1992) Tectonic forcing of late Cenozoic climate. *Nature* **359**, 117–22.

Raymo, M.E., Ruddiman, W.F. and Froelich, P.N. (1988) Influence of late Cenozoic mountain building on ocean geochemical cycles. *Geology* **16**, 649–53.

Raynaud, D., Chappellaz, J., Barnola, J.M., Korotkevich, Y.S. and Lorius, C. (1988) Climatic and CH_4 cycle implications of glacial–interglacial CH_4 change in the Vostok ice core. *Nature* **333**, 655–57.

Rea D.K. and Hovan, S.A. (1995) Grain size distribution and depositional processes of the mineral component of abyssal sediments: lessons from the North Pacific. *Paleoceanography* **10**, 251–58.

Reavie, E.D., Smol, J.P. and Carmichael, N.B. (1995) Postsettlement eutrophication histories of 6 British Columbian (Canada) Lakes. *Canadian Journal of Fisheries and Aquatic Sciences* **52**, 2388–2401.

Rees, J. (1987) Agriculture and horticulture. In J. Wacher (ed.) *The Roman World* (2 vols). Routledge, London, pp. 481–503.

Rees, R.M. and Ribbens, J.C.H. (1995) Relationships between afforestation, water chemistry and fish stocks in an upland catchment in south-west Scotland. *Water Air and Soil Pollution* **85**, 303–8.

Rehana, Z., Malik, A. and Ahmad, M. (1996) Genotoxicity of the Ganges water at Narora (UP), India. *Mutation Research – Genetic Toxicology* **367**, 187–93.

Reid, W.V. (1992) How many species will there be? In T.C. Whitmore and J.A. Sayer (eds) *Tropical Deforestation and Species Extinction*. Chapman and Hall, London, pp. 55–73.

Reid, W.V. (1996) Gene co-ops and the biotrade – translating genetic resource rights into sustainable development. *Journal of Ethnopharmacology* **51**, 75–92.

Reid, W.V. and Miller, K.R. (1989) *Keeping Options Alive: The Scientific Basis for Conserving Biodiversity*. World Resources Institute, Washington, DC.

Reid, W.V. et al. (1993) *Biodiversity Prospecting: Using Genetic Resources for Sustainable Development*. World Resources Institute, US, Instituto Nacional de

Biodiversidad (INbio), Costa Rica, Rainforest Alliance, US, and African Centre for Technology Studies, Kenya.

Reilly, J. (1994) Crops and climate change. *Nature* **367**, 118–19.

Reiners, W.A., Bouwman, A.F., Parsons, W.F.J. and Keller, M. (1994) Tropical rain forest conversion to pasture – changes in vegetation and soil properties. *Ecological Applications* **4**, 363–77.

Repetto, R. (1988) Subsidized timber sales from national forest lands in the United States. In R. Repetto and M. Gillis (eds) *Public Policies and the Misuse of Forest Resources*. Cambridge University Press, Cambridge, pp. 353–83.

Repo, T., Hanninen, H. and Kellomaki, S. (1996) The effects of long-term elevation of air temperature and CO_2 on the frost hardiness of Scots Pine. *Plant Cell and Environment* **19**, 209–16.

ReVelle, P. and ReVelle, C. (1992) *The Global Environment: Securing a Sustainable Future*. Jones & Bartlett, Boston.

Reyes, J.G.G., Jasso, A.M. and Lizatraga, C.V. (1996) Toxic effects of organochlorine pesticides on *Penaeus vannamei* – shrimps in Sinaloa, Mexico. *Chemosphere* **33**, 567–75

Reyes, M.D. and Martens, R. (1994) Geoecology of Laguna de Bay, Philippines I: techno-commercial impact on the trophic level structure of the Laguna de Bay aquatic ecosystem 1968–1980. *Ecological Modelling* **75**, 497–509.

Richard, P.J.H. (1994) Wisconsinan late-glacial environmental change in Québec: a regional synthesis. *Journal of Quaternary Science* **9**, 165–70.

Richards, D.A., Smart, P.L. and Edwards, R.L. (1994) Maximum sea levels for the last glacial period from U-series ages of submerged speleothems. *Nature* **367**, 357–60.

Richards, E.G. (ed.) (1987) *Forestry and the Forest Industries*. Martinus Nijhoff, Dordrecht.

Richards, J.F. (1990) Land transformation. In B.L. Turner II, W.C. Clark, R.W. Kates, J.F. Richards, J.T. Mathewes and W.B. Meyer (eds) *The Earth as Transformed by Human Action*. Cambridge University Press, Cambridge, pp. 163–78.

Richards, P. (1985) *Indigenous Agricultural Revolution*. Hutchinson, London.

Richardson, B.C. (1992) *The Caribbean in The Wider World, 1492–1992*. Cambridge University Press, Cambridge.

Richardson, C.W. and King, K.W. (1995) Erosion and nutrient losses from zero tillage on a clay soil. *Journal of Agricultural Engineering Research* **61**, 81–86.

Richardson, D.M., Vanwilgen, B.W., Higgens, S.I., Trindersmith, T.H., Cowling, R.M. and McKell, D.H. (1996) Current and future threats to plant biodiversity on the Cape Peninsula, South Africa. *Biodiversity and Conservation* **5**, 607–47.

Rickard, J. (1988) *Australia: A Cultural History*. Hutchinson, London.

Riegman, R. (1995) Nutrient-related selection mechanisms in marine phytoplankton communities and the impact of eutrophication on the planktonic food web. *Water Science and Technology* **32**, 63–75.

Ries, J.B. (1996) Landscape damage by skiing at the Schauinsland in the Black Forest, Germany. *Mountain Research and Development* **16**, 27–40.

Ries, R.D. and Perry, S.A. (1995) Potential effects of global climate warming on brook trout growth and prey consumption in Appalachian streams. *Climate Research* **5**, 197–206.

Riley, M. (1995) Tourism development under close control – the case of the Falkland Islands. *Tourism Management* **16**, 471–74.

Roberts, J.M. (1992) *History of the World*, 7th edn. Helicon, London.

Roberts, M.B., Stringer, C.B. and Parfitt, S.A. (1994) A hominid tibia from Middle Pleistocene sediments at Boxgrove. *Nature* **369**, 311–13.

Roberts, R.G., Jones, R., Spooner, N.A., Head, M.J., Murray, A.S. and Smith, M.A. (1994) The human colonisation of Australia: optical dates of 53,000 and 60,000 years bracket human arrival at Deaf Adder Gorge, Northern Territory. *Quaternary Geochronology (Quaternary Science Reviews)* **13**, 575–83.

Robertson, J.B. (1996) Why have earth scientists failed to find suitable nuclear waste disposal sites? *Geotimes* **41**, 16–19.

Robinson, D.A. (1978) *Soil Erosion and Soil Conservation in Zambia: A Geographical Appraisal*. Zambia Geographical Association, Occasional Study No. 9.

Robinson, J.J. and McEvoy, T.G. (1993) Biotechnology – the possibilities. *Animal Production* **57**, 335–52.

Rodhe, H. and Herrera, R. (eds) (1988) *Acidification in Tropical Countries*. J. Wiley, Chichester.

Rodman, P.S. and McHenry, H.M. (1980) Bioenergetics of hominid bipedalism. *American Journal of Physical Anthropology* **52**, 103–6.

Roebroeks, W. and van Kolfschoten, T. (1994) The earliest occupation of Europe: a short chronology. *Antiquity* **68**, 489–503.

Rolland, N. (1992) The Palaeolithic colonisation of Europe: an archaeological and biogeographic perspective. *Trabajos de Prehistoire* **49**, 69–111.

Rolston, H. III (1994) Value in nature and the nature of value. In R. Attfield and A. Belsey (eds) *Philosophy and the Natural Environment*. Royal Institute of Philosophy Supplement 36. Cambridge University Press, Cambridge, pp. 13–30.

Roosevelt, A.C. (1984) Population, health and the evolution of subsistence: conclusions from the conference. In M.N. Cohen and G.A. Armelagos (eds) *Paleopathology and the Origins of Agriculture*. Academic Press, Orlando FL, pp. 559–83.

Rosch, M. (1996) New approaches to prehistoric land-use

reconstruction in southwestern Germany. *Vegetation History and Archaeobotany* **5**, 65–97.

Rosenzweig, C. and Parry, M.L. (1994) Potential impact of climate change on world food supply. *Nature* **367**, 133–38.

Rosenzweig, C., Parry, M.L. and Fischer, G. (1995) World food supply. In K.M. Strzepak and J.B. Smith (eds) *As Climate Changes: International Impacts and Implications.* Cambridge University Press, Cambridge, pp. 27–54.

Ross, I.S. (1988) The use of micro-organisms for the removal and recovery of heavy metals from aqueous effluents. In R. Greenshields (ed.) *Resources and Applications of Biotechnology.* Macmillan, Basingstoke, pp. 100–109.

Roszak, T. (1993) *The Voice of the Earth.* Bantam Press, London.

Rotmans, J., Hulme, M. and Downing, T.E. (1994) Climate change implications for Europe: an application of the ESCAPE model. *Global Environmental Change* **4**, 97–124.

Rousseau D.-D., Parra, I., Cour, P. and Clet, M. (1995) Continental climatic changes in Normandy (France) between 3.3 and 2.3 M yr BP. *Palaeogeography, Palaeoclimatology, Palaeoecology* **111**, 373–83.

Rowan, J.S., Barnes, S.J.A., Hetherington, S.L., Lambers, B. and Parsons, F. (1995) Geomorphology and pollution – the environmental impacts of mining, Leadhills, Scotland. *Journal of Geochemical Exploration* **52**, 57–65.

Rowell,, D.P., Folland, C.K., Maskell, K. and Ward, M.N. (1995) Variability of summer rainfall over tropical north Africa (1906–92): observations and modelling. *Quarterly Journal of the Royal Meteorological Society.* **121**, 669–704.

Rowland, F.S. (1996) Stratospheric ozone depletion by chlorofluorocarbons (Nobel lecture). *Angewandte Chemie* **35**, 1786–98 (international edition in English).

Rowlands, I.H. (1995) *The Politics of Global Atmospheric Change.* Manchester University Press, Manchester.

Rowley-Conwy, P. (1993) Season and reason: the case for a regional interpretation of Mesolithic settlement patterns. In G.L. Peterkin, H.M. Bricker and P. Mellars (eds) *Hunting and Animal Exploitation in the Later Palaeolithic and Mesolithic of Eurasia.* American Anthropological Association, pp. 179–88.

Ruas, M.-P. (1992) The archaeobotanical record of cultivated and collected plants of economic importance from medieval sites in France. *Review of Palaeobotany and Palynology* **73**, 301–14.

Ruddiman, W.F. and Kutzbach, J.E. (1991) Plateau uplift and climatic change. *Scientific American* **264**, 42–50.

Ruddiman, W.F. and McIntyre, A. (1981) The North Atlantic during the last deglaciation. *Palaeogeography, Palaeoclimatology, Palaeoecology* **35**, 145–214.

Ruddiman, W.F. and Raymo, M.E. (1988) Northern hemisphere climate regimes during the past 3 Ma: possible tectonic connections. *Philosophical Transactions of the Royal Society of London* **B318**, 411–30.

Rudel, T. and Roper, J. (1996) Regional patterns and historical trends in tropical deforestation, 1976–1990: a quantitative comparative analysis. *Ambio* **25**, 160–66.

Rugh, C.L., Wilde, H.D., Stack, N.M., Thompson, D.M., Summers, A.O. and Meagher, R.B. (1996) Mercuric ion reductions and resistance in transgenic *Arabidopsis thaliana* plants expressing a modified bacterial mera gene. *Proceedings of the National Academy of Sciences of the United States of America* **93**, 3182–87.

Rukang, W. and Lanpo, J. (1994) China in the period of *Homo habilis* and *Homo erectus*. In S.J. De Laet, A.H. Dani, J.L. Lorenzo and R.B. Nunoo (eds) *History of Humanity, Volume I: Prehistory and the Beginnings of Civilization.* Routledge, London, and UNESCO, Paris, pp. 86–88.

Rukang, W. and Lin, S. (1983) Peking Man. *Scientific American* **248**, 78–86.

Rukang, W. and Xingren, D. (1985) *Homo erectus* in China. In W. Rukang and J.W. Olsen (eds) *Palaeoanthropology and Palaeolithic Archaeology in the People's Republic of China.* Academic Press, London, pp. 79–89.

Runnels, C.N. (1995) Environmental degradation in ancient Greece. *Scientific American* **272**, 72–75.

Russell-Wood, A.J.R. (1987) The gold cycle, *c.*1690–1750. In L. Bethell (ed.) *Colonial Brazil.* Cambridge University Press, Cambridge, pp. 190–243.

Rytomaa, T. (1996) 10 years after Chernobyl. *Annals of Medicine* **28**, 83–87.

Saeki, K., Fujimoto, M., Kolinjim, D. and Tatsukawa, R. (1995) Mercury concentrations in hair from populations in Wau-Bulolo area, Papua New Guinea. *Archives of Environmental Contamination and Toxicology* **30**, 412–17.

Sahin, V. and Hall, M.J. (1996) The effects of afforestation and deforestation on water yields. *Journal of Hydrology* **178**, 293–309.

Saigne, C. and Legrand, M. (1987) Measurements of methanesulphonic acid in Antarctic ice. *Nature* **330**, 240–42.

Saiko, T.A. (1995) Implications of the disintegration of the former Soviet Union for desertification control. *Environmental Monitoring and Assessment* **37**, 289–302.

Sakson, M. and Miller, U. (1993) Diatom assemblages in superficial sediments from the Gulf of Riga, Eastern Baltic Sea. *Hydrobiologia.* **269**, 243–49.

Salata, G.G., Wade, T.L., Sericano, J.L., Davis, J.W. and Brooks, J.M. (1995) Analysis of Gulf of Mexico bottle-nosed dolphins for organochlorine pesticides and PCBs. *Environmental Pollution* **88**, 167–75.

Sallares, R. (1991) *The Ecology of the Ancient Greek World.* Duckworth, London.

Salway, P. (1993) *The Oxford Illustrated History of Roman Britain.* Oxford University Press, Oxford.

Samanta, G., Chatterjee, A., Das, D., Samanta, G., Chowdhury, P.P., Chanda, C.R. and Chakraborti, D. (1995) Calcutta pollution 5: lead and other heavy metal contaminants in a residential area from a factory producing lead ingots and lead alloys. *Environmental Technology* **16**, 223–31.

Sanchez, J., Vaquero, M.C. and Legorburu, I. (1994) Metal pollution from old lead zinc mine works – biota and sediment from Oiartzun Valley. *Environmental Technology* **15**, 1069–76.

Sangodoyin, A.Y. (1993) Field evaluation of the possible impact of some pesticides in the soil and water environment in Nigeria. *Experimental Agriculture* **29**, 227–32.

Santee, M.L., Read, W.G., Waters, J.W., Froidevaux, L., Manney, G.L., Flower, D.A., Jarnot, R.E., Harwood, R.S., and Peckham, G.E. (1995) Interhemispheric differences in polar stratospheric HNO_3, ClO, and O_3. *Science* **267**, 849–52.

Santer, B.D., Taylor, K.E., Wigley, T.M.L., Penner. J.E., Jones, P.D. and Cubasch, U. (1995) Towards the detection and attribution of an anthropogenic effect on climate. *Climate Dynamics* **12**, 77–100.

Santer, B.D., Wigley, T.M.L., Barnett, T.P. and Anyamba, E. (1996) Detection of climate change and attribution of causes. In J.T. Houghton, L.G. Meira Filho, B.A. Callander, N. Harris, A. Kattenberg and K. Maskell (eds) *Climate Change 1995: The Science of Climate Change*. Cambridge University Press, Cambridge, pp. 407–43.

Sarich, V.M. and Wilson, A.C. (1967) Immunological time scale for human evolution. *Science* **158**, 1200–1204.

Sarnaik, S. and Kanekar, P. (1995) Bioremediation of color of methyl violet and phenol from industry waste effluent using *Pseudomonas* spp. isolated from factory soil. *Journal of Applied Bacteriology* **79**, 459–69.

Sasson, A. (1996) Biotechnologies and the use of plant genetic resources for industrial purposes: benefits and constraints for developing countries. In F. di Castri, and T. Younés (eds) *Biodiversity, Science and Development: Towards a New Partnership*. CAB International, Wallingford, and the International Union of Biological Sciences, Paris, pp. 469–87.

Satake, K., Tanaka, A. and Kimura, K. (1996) Accumulation of lead in tree trunk bark pockets as pollution time capsules. *Science of the Total Environment* **181**, 25–30.

Sauer, C.O. (1952) *Agricultural Origins and Dispersals*. American Geographical Society, Bowman Memorial Lecture 5, Series 2.

Saunders, D. and Hobbs, R. (1989) Corridors for conservation. *New Scientist* **121**(1649), 63–68.

Saunders, J.J. (1987) Britain's newest mammoths. *Nature* **330**, 419.

Schäbitz, F. (1994) Holocene climatic variations in northern Patagonia, Argentina. *Palaeogeography, Palaeoclimatology, Palaeoecology* **109**, 287–94.

Schama, S. (1995) *Landscape and Memory*. Harper Collins, London.

Schelske, C.L. and Hodell, D.A. (1995) Using carbon isotopes of bulk sedimentary organic matter to reconstruct the history of nutrient loading and eutrophication in Lake Erie. *Limnology and Oceanography* **40**, 918–29.

Schimel, D., Enting, I.G., Heimann, M., Wigley, T.M.L., Raynaud, D., Alves, D. and Siegenthaler, U. (1995) CO_2 and the carbon cycle. In J.T. Houghton, L.G. Meira Filho, J. Bruce, H. Lee, B.A. Callander, E. Haites, N. Harris and K. Maskell (eds) *Climate Change 1994: radiative forcing of climate change and an evaluation of the IPCC IS92 emission scenarios*. Cambridge University Press, Cambridge, pp. 35–71.

Schindler, D.W. (1988) Effects of acid rain on freshwater ecosystems. *Science* **239**, 149–57.

Schindler, D.W., Jefferson Curts, P., Parker, B.R. and Stainton, M.P. (1996) Consequences of climate warming and lake acidification for UV-B penetration in North American boreal lakes. *Nature* **379**, 705–8.

Schlesinger, W.H. and Gramenopoulos, N. (1996) Archival photographs show no climate induced changes in the Sudan, 1943–1994. *Global Change Biology* **2**, 137–41.

Schmitt, J. and Linder, C.R. (1994) Will escaped transgenes lead to ecological release? *Molecular Ecology* **3**, 71–74.

Schneider, R.R., Muller, P.J. and Ruhland, G. (1995) Late Quaternary surface circulation in the east equatorial south Atlantic: evidence from alkenone sea surface temperatures. *Paleoceanography* **10**, 197–219.

Schneider, S.H. and Boston, P.J. (eds) (1991) *Scientists on Gaia*. MIT Press, Cambridge MA.

Schofield, O., Kroon, B.M.A. and Prezelin, B.B. (1995) Impact of ultraviolet-B radiation on photosystem-11 activity and its relationship to the inhibition of carbon fixation rate for Antarctic ice algae communities. *Journal of Phycology* **31**, 703–15.

Schrenk, F., Bromage, T.G., Betzler, C.G., Ring, U. and Juwayeyi, Y.M. (1993) Oldest *Homo* and Pliocene biogeography of the Malawi Rift. *Nature* **365**, 833–36.

Schulmeister, J. and Lees, B.G. (1995) Pollen evidence from tropical Australia for the onset of ENSO-dominated climate at *c*. 4000 BP. *The Holocene* **5**, 10–18.

Schwartz, S.B. (1987) Plantation and peripheries, *c*.1580–1750. In L. Bethell (ed.) *Colonial Brazil*. Cambridge University Press, Cambridge, pp. 67–144.

Schwarzacher, W. (1993) *Cyclostratigraphy and the Milankovitch Theory*. Elsevier, Amsterdam.

Schweingruber, F.H. (1988) *Tree Rings: Basics and Applications of Dendrochronology*. Reidel, Dordrecht.

Scott, L. (1990) Palynological evidence for Late Quaternary environmental change in southern Africa. *Palaeoecology of Africa* **2**, 259–68.

Scott, L., Cooremans, B., de Wet, J.S. and Vogel, J.C. (1991) Holocene environmental changes in Namibia inferred from pollen analysis of swamp and lake deposits. *The Holocene* **1**, 8–13.

Scullion, J. and Malinovszky, K.M. (1995) Soil factors affecting tree growth on former opencast coal sites. *Land Degradation and Rehabilitation* **6**, 239–49.

Sedjo, R.A. (1993) The carbon cycle and global forest ecosystem. *Water Air and Soil Pollution* **70**, 295–307.

Senoo, K., Nishiyama, M. and Matsumoto, S. (1996) Bioremediation of gamma-HCH-polluted field soil by inoculation. *Soil Science and Plant Pollution* **42**, 11–19.

Senshui, Z. (1985) The early palaeolithic of China. In W. Rukang and J.W. Olsen (eds) *Palaeoanthropology and Palaeolithic Archaeology in the People's Republic of China*. Academic Press, London, pp. 147–86.

Sericano, J.L., Wade, T.L., Brooks, J.M., Atlas, E.L., Fay, R.R. and Wilkinson, D.L. (1993) National status and trends mussel watch program – chlordane-related compounds in Gulf of Mexico oysters. *Environmental Pollution* **82**, 23–32.

Sernander, R. (1908) On the evidence of postglacial changes of climate furnished by the peatmosses of northern Europe. *Geologiska Föreningens Stockholm Förhandlingar* **30**, 465–78.

Servant, M., Maley, J., Tureq, B., Absy, M.L., Brenac, P., Fournier, P. and Ledru, M.P. (1993) Tropical forest changes during the late Quaternary in African and South American lowlands. *Global and Planetary Change* **7**, 25–40.

Shackleton, N.J. (1987) Oxygen isotopes, ice volume and sea level. *Quaternary Science Reviews* **6**, 183–90.

Shackleton, N.J., Imbrie, J. and Hall, M.A. (1983) Oxygen and carbon isotope record of East Pacific core V19-30: implications for the formation of deep water in the Late Pleistocene North Atlantic. *Earth and Planetary Science Letters* **65**, 233–44.

Shackleton, N.J. *et al.* (1984) Oxygen isotope calibration of the onset of ice rafting and history of glaciation in the North Atlantic region. *Nature* **307**, 620–23.

Shackleton, N.J. and Opdyke, N.D. (1973) Oxygen-isotope and paleomagnetic stratigraphy of Pacific core V19-30: implications for the formation of deep water in the late Pleistocene North Atlantic. *Earth and Planetary Science Letters* **65**, 233-44.

Shailaja, M.S. and Singbal, S.Y.S. (1994) Organochlorine pesticide compounds from the Bay of Bengal. *Estuarine Coastal and Shelf Science* **39**, 219–26.

Shalev, S. (1994) The change in metal production from the Chalcolithic period to the Early Bronze Age in Israel and Jordan. *Antiquity* **68**, 630–37.

Shane, L.C.K. and Anderson, K.H. (1993) Intensity, gradients and reversals in late-glacial environmental change in east-central North America. *Quaternary Science Reviews* **12**, 307–20.

Sharma, A., Martin, M.J., Okabe, J.F., Trugho, R.A., Dhanjal, N.K., Logan, J.S. and Kumar, R. (1994) An isologous porcine promotor permits high level of expression of human haemoglobin in transgenic swine. *Bio/Technology* **12**, 55–59.

Sharma, D.P., Rao, K.V.G., Singh, K.N., Kumbhare, P.S. and Oosterbaan, R.J. (1994) Conjunctive use of saline and nonsaline irrigation waters in semiarid regions. *Irrigation Science.* **15**, 25–33.

Shemesh, A., Macko, S.A., Charles, C.D. and Rau, G.H. (1993) Isotopic evidence for reduced productivity in the glacial Southern Ocean. *Science* **262**, 407–10.

Shemesh, A., Burkle, L.H. and Hays, J.D. (1995) Late Pleistocene oxygen isotope records of biogenic silica from the Atlantic sector of the Southern Ocean. *Paleoceanography* **10**, 179–96.

Sherratt, A. (1994a) The transformation of early agrarian Europe: the later Neolithic and Copper Ages, 4500–2500 BC. In B. Cunliffe (ed.) *The Oxford Illustrated Prehistory of Europe*. Oxford University Press, Oxford, pp. 167–201.

Sherratt, A. (1994b) The emergence of élites. Earlier Bronze Age Europe, 2500–1300 BC. In B. Cunliffe (ed.) *The Oxford Illustrated Prehistory of Europe*. Oxford University Press, Oxford, pp. 244–76.

Shiva, V. (1993) *Monocultures of the Mind*. Zed Books, London, and Third World Network, Penang.

Shotton, F.W., Keen, D.H., Coope, G.R., Currant, A.P., Gibbard, P.L., Aalto, M., Peglar, S.M. and Robinson, J.E. (1993) The Middle Pleistocene deposits of Waverley Wood Pit, Warwickshire, England. *Journal of Quaternary Science* **8**, 293–325.

Shuler, M.L. (1995) *Baculovirus Expressing Systems and Biopesticides*. Wiley-Liss, New York.

Šibrava, V. (1986) Correlation of European glaciations and their relation to the deep sea record. *Quaternary Science Reviews* **5**, 433–41.

Šibrava, V., Bowen, D.Q. and Richmond, G.M. (eds) (1986) Quaternary glaciations in the northern hemisphere. *Quaternary Science Reviews* **5**.

Siegel, F.R., Slaboda, M.L. and Stanley, D.J. (1994) Metal pollution loading, Manzalah Lagoon, Nile Delta, Egypt: implications for aquaculture. *Environmental Geology* **23**, 89–98.

Siegenthaler, U. and Sarmiento, J.L. (1993) Atmospheric carbon dioxide and the ocean. *Nature* **365**, 119–25.

Silver, W.L., Brown, S. and Lugo, A.E. (1996) Effects of changes in biodiversity on ecosystem function in tropical forests. *Conservation Biology* **10**, 17–24.

Simmonds, P.D., Cunnold, D., Dollard, G., Davies, T., McCulloch, A. and Derwent, R. (1993) Evidence of the phase-out of CFC use in Europe over the period 1987–1990. *Atmospheric Environment* **27A**, 1397–1407.

Simmons, I.G. (1975) The ecological setting of mesolithic man in the highland zone. In J.G. Evans, S. Limbrey and H. Cleere (eds) *The Effect of Man on the Landscape: The Highland Zone*. Council for British Archaeology Research Reports, Oxford, No.11, pp. 57–63.

Simmons, I.G. (1993a) *Interpreting Nature: Cultural Constructions of the Environment*. Routledge, London.

Simmons, I.G. (1993b) *Environmental History: A Concise Introduction*. Blackwell, Oxford.

Simmons, I.G. (1993c) Vegetation change during the Mesolithic in the British Isles: some amplifications. In F.M. Chambers (ed.) *Climate Change and Human*

Impact on the Landscape: Studies in Palaeoecology and Environmental Archaeology. Chapman and Hall, London, pp. 109–18.

Simmons, I.G. (1995a) Nature, culture and history: 'all just supply, and all relation'. In D.F. Cooper and J.A. Palmer (eds), *Just Environments.* Routledge, London, pp. 59–71

Simmons, I.G. (1995b) Hunting on the English moorlands: then and now. In I.G. Simmons and A.M. Mannion (eds) *The Changing Nature of the People–Environment Relationship: Evidence from a Variety of Sources.* Proceedings of the International Geographical Union Commission on Historical Monitoring of Environmental Changes Meeting, Přihrazy, Czech Republic, published by Department of Social Geography and Regional Development, Charles University, Prague, Czech Republic, pp. 11–17.

Simmons, I.G. (1995c) The history of the early human environment. In B.Vyner (ed.) *Moorlands and Monuments: Studies in Honour of Don Spratt and Raymond Hayes.* Council for British Archaeology Research Report 101, pp. 5–15.

Simmons, I.G. and Innes, J.B. (1985) Late mesolithic land-use and its impacts in the English uplands. *Biogeographical Monographs* **2**, 7–17.

Simmons, I.G. and Innes, J.B. (1987) Mid-Holocene adaptations and later mesolithic forest disturbance in Northern England. *Journal of Archaeological Science* **14**, 383–403.

Simmons, I.G. and Innes, J.B. (1996a) Disturbance phases in the mid-Holocene vegetation at North Gill, North York Moors: form and process. *Journal of Archaeological Science* **23**, 183–91.

Simmons, I.G. and Innes, J.B. (1996b) Prehistoric charcoal in peat profiles at North Gill, North Yorkshire Moors, England. *Journal of Archaeological Science* **23**, 193–97.

Simmons, I.G. and Innes, J.B. (1996c) The ecology of an episode of prehistoric cereal cultivation on the North Yorkshire Moors, England. *Journal of Archaeological Science* **23**, 613–18.

Simons, E.L. (1964) The early relatives of man. *Scientific American* **211**, 51–65.

Simpson, R.W. and Xu, H.C. (1994) Atmospheric lead pollution in an urban area of Brisbane, Australia. *Atmospheric environment* **28**, 3073–82.

Singer, A.J. and Shemesh, A. (1995) Climatically linked carbon isotope variation during the past 430,000 years in Southern Ocean sediments. *Paleoceanography* **10**, 171–77.

Singh, C.B., Aujla, T.S., Sandhu, B.S. and Khera, K.L. (1996) Effect of transplanting date and irrigation regime on growth yield and water use in rice (*Oryza sativa*) in northern India. *Indian Journal of Agricultural Sciences* **66**, 137–41.

Singh, D.K. and Agarwral, H.C. (1995) Persistence of DDT and nature of bound residues in soil at higher altitude. *Environmental Science and Technology* **29**, 2301–4.

Singh, G., Singh, N.T. and Abrol, I.P. (1994) Agroforestry techniques for the rehabilitation of degraded salt-affected lands in India. *Land Degradation and Rehabilitation* **5**, 223–42.

Singh, G., Ram, B., Narain, P., Bhushan, L.S. and Abrol, I.P. (1992) Soil erosion rates in India. *Journal of Soil and Water Conservation* **47**, 97–99.

Singh, J. and Singh, J.P. (1995) Land degradation and economic stability. *Ecological Economics* **15**, 77–86.

Singh, S., Kar, A., Joshi, D.C., Kumar, S. and Sharma, K.D. (1994) Desertification problem in western Rajasthan. *Annals of Arid Zone* **33**, 191–202.

Sistani, K.R., Mays, D.A. and Taylor, R.W. (1995) Biogeochemical characteristics of wetlands developed after strip mining for coal. *Communications in Soil Science and Plant Analysis* **26**, 3221–29.

Sithole, S.D., Moyo, N. and Macheka, M. (1993) An assessment of lead pollution from vehicle emissions along selected roadways in Harare (Zimbabwe). *International Journal of Environmental Analytical Chemistry* **53**, 1–12.

Slaughter, R.A. (1994) Why we should care for future generations. *Futures* **26**, 1077–85.

Sluyter, A. (1995) Intensive wetland agriculture in Mesoamerica: space, time and form. *Annals of the Association of American Geographers* **84**, 557–84.

Smil, V. (1993a) *China's Environmental Crisis: An Inquiry into the Limits of National Development.* M.E. Sharpe, Armonk, New York.

Smil, V. (1993b) Afforestation in China. In A.S. Mather (ed.) *Afforestation. Policies, Planning and Progress.* Belhaven Press, London, pp. 105–17.

Smith, A.G. (1970) The influence of mesolithic and neolithic man on British vegetation: a discussion. In D. Walker and R.G. West (eds) *Studies in the Vegetational History of the British Isles: Essays in Honour of Harry Godwin.* Cambridge University Press, Cambridge, pp. 81–96.

Smith, A.G. (1981) The neolithic. In Simmons, I.G. and Tooley, M.J. (eds) *The Environment in British Prehistory.* Duckworth, London, pp.125-209.

Smith, A.K. and Pollard, D.A. (1996) The best available information – some case studies from New South Wales, Australia, of conservation related management responses which impact on recreational fishers. *Marine Policy* **20**, 261–67.

Smith, A.M. (1992) Holocene paleoclimate trends from palaeoflood analysis. *Global and Planetary Change* **97**, 235–40.

Smith, B.D. (1995) *The Emergence of Agriculture.* W.H. Freeman, New York.

Smith, C. (1992) *Late Stone Age Hunters of the British Isles.* Routledge, London.

Smith, D.M. (1986) *The Practice of Silviculture*, 8th edn. J. Wiley, New York.

Smith, F.D.M., May, R.M., Pellew, R., Johnson, T.H. and Walter, K.R. (1993) Estimating extinction rates. *Nature* **364**, 494–96.

Smith, T.M., Leemans, R. and Shugart, H.H. (1992) Sensitivity of terrestrial carbon storage to CO_2-induced climate change: comparison of four scenarios based on general circulation models. *Climate Change* **21**, 367–84.

Smith, T.M., Halpin, P.N., Stugart, H.H. and Secrett, C.M. (1995) Global forests. In K.M. Strzepek and J.B. Smith (eds) *As Climate Changes: International Impacts and Implications*. Cambridge University Press, Cambridge, pp. 146–79.

Smith, W.E. and Smith, A.M. (1975) *Minamata*. Holt, Rinehart and Winston, New York.

Snelder, D.J. and Bryan, R.B. (1995) The use of rainfall simulation tests to assess the influence of vegetation density on soil loss on degraded rangelands in Baringo District, Kenya. *Catena* **25**, 105–16.

Sokal, R.R., Oden, N.L. and Wilson, C. (1991) Genetic evidence for the spread of agriculture in Europe by demic diffusion. *Nature* **351**, 143–45.

Sommaruga, R., Conde, D. and Casal, J.A. (1995) The role of fertilizers and detergents for eutrophication in Uruguay. *Fresnius Environmental Bulletin* **4**, 111–16.

Southam, G. and Beveridge, T.J. (1994) The *in vitro* formation of placer gold by bacteria. *Geochimica et Cosmochimica* **58**, 4527–30.

Southward, A.J., Hawkins, S.J. and Burrows, M.T. (1995) 70 years' observations of changes in distribution and abundance of zooplankton and intertidal organisms in the western English Channel in relation to rising sea temperature. *Journal of Thermal Biology* **20**, 127–55.

Sowers, T. and Bender, M. (1995) Climate records covering the last deglaciation. *Science* **269**, 210–14.

Sowunmi, M.A. (1991) Late Quaternary environments in equatorial Africa: palynological evidence. *Palaeoecology of Africa* **22**, 213–38.

Spalding, R.F. and Exner, M.E. (1993) Occurrence of nitrate in groundwater – a review. *Journal of Environmental Quality* **22**, 392–402.

Spelman, C.A. (1994) *Non-Food Uses of Agricultural Raw Materials: Economics, Biotechnology and Politics*. CAB International, Wallingford.

Spence, J.R., Langor, D.W., Niemela, J., Carcamo, H.A. and Currie, C.R. (1996) Northern forestry and carabids – the case for concern about old-growth species. *Annales Zoologici Fennici* **33**, 173–84.

Spencer, H. (1864) *First Principles*. D. Appleton, New York.

Stanley, D.J. and Warne, A.G. (1993) Sea level and initiation of predynastic culture in the Nile delta. *Nature* **363**, 435–38.

Stefanovits, P. (1994) Soil degradation in Hungary. In D.J. Greenland and I. Szabolcs (eds) *Soil Resilience and Sustainable Land Use*. CAB International, Wallingford, pp. 119–127.

Stephenson, J.R. and Warnes, A. (1996) Release of genetically-modified microorganisms into the environment. *Journal of Chemical Technology and Biotechnology* **65**, 5–14.

Sterling, S. (1992) Rethinking resources. In D.E. Cooper and J.A. Palmer (eds) *The Environment in Question*. Routledge, London, pp. 226–37.

Stevens, S.F. (1993) Tourism, change, and continuity in Mount Everest region, Nepal. *Geographical Review* **83**, 410–27.

Stewart, P. (1993) Afforestation in Algeria. In A.S. Mather (ed.) *Afforestation: Policies, Planning and Progress*. Belhaven Press, London, pp. 92–104.

Stewart, P.J. (1987) *Growing Against the Grain: United Kingdom Forest Policy 1987*. Council for the Protection of Rural England, London.

Stockton, D. (1994) The founding of the empire. In J. Boardman, J. Griffin and O. Murray (eds) *The Oxford History of the Classical World*. Oxford University Press, Oxford, pp. 531–59.

Stoermer, E.F., Wolin, J.A. and Schelske, C.L. (1993) Paleolimnological comparison of the Laurentian Great Lakes based on diatrons. *Liminology and Oceanography* **38**, 1311–16.

Stohlgren, T.J. and Parsons, D.J. (1986) Vegetation and soil recovery in wilderness campsites closed to visitor use. *Journal of Environmental Management* **10**, 375–80.

Stork, N.E. (1993) How many species are there? *Biodiversity and Conservation* **2**, 215–32.

Stringer, C.B. (1992) Reconstructing recent human evolution. *Philosophical Transactions of the Royal Society of London* **337**, 217–24.

Stringer, C.B. (1994) Out of Africa – a personal history. In M.H. Nitecki and D.V. Nitecki (eds), *Origins of Anatomically Modern Humans*. Plenum, New York, pp. 149–74.

Stuart, A.J. (1982) *Pleistocene Vertebrates in the British Isles*. Longman, London.

Stuiver, M. and Braziunas, T.F. (1993) Sun, ocean, climate and atmospheric $^{14}CO_2$: an evaluation of causal and spectral relationships. *The Holocene* **3**, 289–305.

Subak, S. (1994) Methane from the house of Tudor and the Ming Dynasty: anthropogenic emissions in the sixteenth century. *Chemosphere* **29**, 843–54.

Sukumar, R., Ramesh, R., Pant, R.K. and Rajagopalan, G. (1993) A $\delta^{13}C$ record of late Quaternary climate change from tropical peats in southern India. *Nature* **364**, 703–6.

Sumida, S. (1996) OECD's biosafety work on large-scale releases of transgenic plants. *Field Crops Research* **45**, 187–94.

Susman, R.L. (1994) Fossil evidence for early hominid tool use. *Science* **265**, 1570–73.

Sussman, R.W. (1993) A current controversy in human evolution. *American Anthropologist* **95**, 9–13.

Sutcliffe, A.J. (1985) *On the Track of Ice Age Mammals*. British Museum, London.

Swannell, R.P.J., Lee, K. and McDonagh, M. (1996) Field evaluations of marine oil spill bioremediation. *Microbiological Reviews* **60**, 342–47.

Sykes, M.T. and Prentice, I.C. (1995) Boreal forest futures

– modeling the controls on tree species range limits and transient responses to climate change. *Water Air and Soil Pollution* **82**, 415–28.

Szeicz, J.M., MacDonald, G.M. and Duk-Rodkin, A. (1995) Late Quaternary vegetation history of the central Mackenzie Mountains, Northwest Territories, Canada. *Palaeogeography, Palaeoclimatology, Palaeoecology* **113**, 351–71.

Tallis, J.H. (1991) *Plant Community History: Long-term Changes in Plant Distribution and Diversity*. Chapman and Hall, London.

Tamisier, A. and Grillas, P. (1994) A review of habitat changes in the Camargue – an assessment of the effects of the loss of biological diversity on the wintering waterfowl community. *Biological Conservation* **70**, 39–47.

Tanabe, S., Subramanian, A., Ramesh, A., Kumaran, P.L. and Miyazaki, N. and Tatsukawa, R. (1993) Persistant organochlorine residues in dolphins from the Bay of Bengal, South India. *Marine Pollution Bulletin* **26**, 311–16.

Tanaka, K., Yamagata, T., Masangkay, J.S., Faruque, M.O., Vubinh, D., Salundik, F., Mansjoerr, S.S., Kawamoto, Y. and Namikawa, T. (1995) Nucleotide diversity of mitochondrial DNAs between the savanna and the river types of domestic water buffalos, *Bubalis bubalis*, based on restriction endonuclease cleavage patterns. *Biochemical Genetics* **33**, 137–48.

Tandeter, E. (1987) Forced and free labour in late-colonial Potosí. In E.P. Archetti, P. Cammack and P. Roberts (eds) *Sociology in 'Developing Societies', Latin America*. Macimillan, Basingstoke, pp. 26–33.

Tang, H.T. (1987) Problems and strategies for regenerating dipterocarp forests in Malaysia. In F. Mergen and J.R. Vincent (eds) *Natural Management of Tropical Moist Forests*. Yale University Press School of Forestry and Environmental Studies, New Haven CT, pp. 23–45.

Tansley, A.G. (1935) The use and abuse of vegetational concepts and terms. *Ecology* **16**, 284–307.

Tattersall, G.J. and Wright, P.A. (1996) The effects of ambient pH on nitrogen excretion in early life stages of the American toad (*Bufo americanus*). *Comparative Biochemistry and Physiology A* **113**, 369–74.

Taylor, D.M. (1993) Environmental change in montane southwest Uganda: a pollen record for the Holocene from Ahakagyezi swamp. *The Holocene* **3**, 324–32.

Taylor, K.C., Lamorey, G.W., Doyle, G.A., Alley, R.B., Grooks, P.M., Mayewski, P.A., White, J.W.C. and Barlow, L.K. (1993) The 'flickering switch' of late Pleistocene climate change. *Nature* **361**, 432–36.

Tchernov, E., Horwitz, L.K., Ronen, A. and Lister, A. (1994) The faunal remains from Evron Quarry in relation to other Lower Palaeolithic hominid sites in the Southern Levant. *Quaternary Research* **42**, 328–39.

Tebo, B.M. (1995) Metal precipitation by marine bacteria: potential for biotechnological applications. In J.K. Setlow (ed.) *Genetic Engineering Principles and Methods, Vol. 17*. Plenum, New York, pp. 231–63.

Tegart, W.J. McG., Sheldon, W. and Griffiths, D.C. (1990) *Climate Change: the IPCC impacts assessment*. Cambridge University Press, Cambridge.

Tengberg, A. (1995) Nebkha dunes as indicators of wind erosion and land degradation in the Sahel zone of Burkina Faso. *Journal of Arid Environments* **30**, 265–82.

Thacker, J.R.M. (1993/94) Transgenic crop plants and pest control. *Science Progress* **77**, 207–19.

Thirsk, J. (1987) *England's Agricultural Regions and Agrarian History, 1500–1750*. Macmillan, London.

Thomas, D.S.G. and Middleton, N.J. (1994) *Desertification: Exploding the Myth*. J. Wiley, Chichester.

Thomas, M.F. and Thorp, M.B. (1995) Geomorphic response to rapid climatic and hydrologic change during the late Pleistocene and early Holocene in the humid and sub-humid tropics. *Quaternary Science Reviews* **14**, 193–207.

Thompson, F.M.L. (1985) Towns, industry and the Victorian Landscape. In S.R.J. Woodell (ed.) *The English Landscape: Past, Present and Future*. Oxford University Press, Oxford, pp. 168–87.

Thompson, L.G., Mosley-Thompson, E., Dansgaard, W. and Grootes, P.M. (1986) The Little Ice Age as recorded in the stratigraphy of the tropical Quelccaya ice cap. *Science* **234**, 361–64.

Thompson, L.G., Mosley-Thompson, E.P., Davis, M.E., Bolzan, J.F., Dai, J., Yao, T., Gunderstrup, N., Wu, X., Klein, L. and Xie, Z. (1989) Holocene–Late Pleistocene climatic ice core records from Qinghai-Tibetan Plateau. *Science* **246**, 474–77.

Thompson, L.G., Mosley-Thompson, E., Davis, M.E., Lin, P.-N., Henderson, K.A., Cole-Dai, J., Bolzan, J.F. and Liu, K.-B. (1995) Late glacial stage, and Holocene tropical ice core records from Huascarán, Peru. *Science* **269**, 46–50.

Thompson, P.B. (1993) Genetically modified animals: ethical issues. *Journal of Animal Science* **71**(suppl. 3), 51–56.

Thouveny, J., de Beaulieu, J.L., Bonifay, E., Creer, K.M., Guiot, J., Icole, M., Johnsen, S., Jouzel, J., Reille, M., Williams, T. and Williamson, D. (1994) Climatic variations in Europe over the past 140 Kyr deduced from rock magnetism. *Nature* **371**, 503–6.

Thunell, R.C. and Miao, D.M. (1996) Sea-surface temperature of the western equatorial Pacific during the Younger Dryas. *Quaternary Research* **46**, 72–77.

Tiffen, M., Mortimore, M. and Gichuki, F. (1994) *More People, Less Erosion: Environmental Recovery in Kenya*. J. Wiley, Chichester.

Tilman, D. and Downing, J.A. (1994) Biodiversity and stability in grasslands. *Nature* **367**, 363–65.

Tilman, D., Wedin, D. and Knops, J. (1996) Productivity and sustainability influenced by biodiversity in grassland ecosystems. *Nature* **379**, 718–20.

Timberlake, S. (1988) Bronze Age mining at Cwmystwyth: the radiocarbon dates. *Archaeology in Wales* **28**, 50.

Times Books (1993) *Atlas of World History*, 4th edn. Times Books, London.

Tohme, J., Toro, O., Vargas, J. and Debouck, D.G. (1995) Variability in Andean nuna common beans (*Phaseolus vulgare* Fabaceae). *Economic Botany* **49**, 78–95.

Tolba, M.K., El-Kholy, O.A., El-Hinnawi, E., Holdgate, M.W., McMichael, D.F. and Munn, R.E. (eds) (1992) *The World Environment 1972–1992*. Chapman and Hall, London.

Toman, M.A. and Ashton, P.M.S. (1996) Sustainable forest ecosystems and management – a review. *Forest Science* **42**, 366–77.

Toy, J.J. and Hadley, R.F. (1987) *Geomorphology and Reclamation of Disturbed Lands*. Academic Press, New York.

Treumann, R.A. (1991) Global problems, globalization, and predictability. *World Futures* **31**, 47–53.

Trimble, S.W. (1992) The Alcovy River swamps: the result of culturally accelerated sedimentation. In L.M. Dilsaver and C.E. Colten (eds) *The American Environment: Interpretations of Past Geographies*. Rowman and Littlefield, Lanham, MD, pp. 21–32.

Trinkaus, E. (1983) *The Shanidar Neanderthals*. Academic Press, New York.

Tripathi, A. (1994) Airborne lead pollution in the city of Varanasi, India. *Atmospheric Environment* **28**, 2317–23.

Troels-Smith, J. (1960) Ivy, mistletoe and elm: climatic indicators – fodder plants: a contribution to the interpretation of the pollen zone border VII–VIII. *Danmarks Geologiske Undersøgelse Series 4*, **4**, 1–32.

Troels-Smith, J. (1956) Neolithic period in Denmark and Switzerland. *Science* **124**, 376–879.

Tsavalos, A.J. and Day, J.G. (1994) Development of media for the mixotrophic heterotrophic culture of *Brachiomonas submarina*. *Journal of Applied Phycology* **6**, 431–33.

Tsujimoto, Y., Masuda, J., Fukuyama, J. and Ito, H. (1994) N_2O emissions at solid waste disposal sites in Osaka City. *Journal of the Air and Waste Management Association* **44**, 1313–14.

Tucker, C.J., Dregne, H.E. and Newcomb, W.W. (1991) Expansion and contraction of the Sahara Desert from 1980 to 1990. *Science* **253**, 299–301.

Tucker, M. (1995) Carbon dioxide emissions and global GDP. *Ecological Economics* **15**, 215–23.

Tudge, C. (1995) Human origins: a family feud. *New Scientist* **146** (No.1978), 24–28.

Turner, J. (1979) The environment of northeast England during Roman times as shown by pollen analysis. *Journal of Archaeological Science* **6**, 285–90.

Turner, J., Innes, J.B. and Simmons, I.G. (1993) Spatial diversity in the mid-Flandrian vegetation history of North Gill, North Yorkshire. *New Phytologist* **123**, 599–647.

Turner, M. and Corlett, R.T. (1996) The conservation value of small isolated fragments of lowland tropical rain forest. *Trends in Ecology and Evolution* **11**, 330–33.

Turnock, D. (1993) Romania. In F.W. Carter and D. Turnock (eds) *Environmental Problems in Eastern Europe*. Routledge, London, pp. 135-163.

Tyagi, N.K., K.C., Pillai, N.N. and Willardson, I.S. (1993) Decision support for irrigation system improvement in saline environment. *Agricultural Water Management*. **23**, 285–301.

Tyson, P.D. and Lindesay, J.A. (1992) The climate of the last 2000 years in southern Africa. *The Holocene* **2**, 271–78.

Tzedakis, P.C. (1993) Long-term tree populations in northwest Europe through multiple Quaternary climatic cycles. *Nature* **364**, 437–440.

Udo, R.K., Areola, O.O., Anyoade, J.O. and Afolayan, A.A. (1990) Nigeria. In B.L.Turner II, W.C. Clark, R.W. Kates, J.F. Richards, J.T. Mathews and W.B. Meyer (eds) *The Earth as Transformed by Human Action*. Cambridge University Press, Cambridge, pp. 589–603.

Umali, D.L. (1993) *Irrigation Induced Salinity: A Growing Problem for Development and Environment*. World Bank, Washington, DC.

Unger, P.W. (1996) Common soil and water conservation practices. In M. Agassi (ed.) *Soil Erosion, Conservation and Rehabilitation*. Marcel Dekker, New York, pp. 239–66.

United Nations Conference on Environment and Development (1992) *The Global Partnership for Environment and Development: A Guide to Agenda 21*, UNCED, Geneva.

United Nations Environment Programme (UNEP) (1987) *Montreal Protocol on substances that deplete the ozone layer*. UNEP, Nairobi.

United Nations Environment Programme (1991a) *Status of Desertification and Implementation of the United Nations Plan of Action to Combat Desertification*. United Nations, Nairobi.

United Nations Environment Programme (1991b) *Environmental Impacts of Production and Use of Energy*. Tycooly International, Dublin.

United Nations Environment Programme (1991c) *Environmental Data Report 1990/91*. Blackwell, Oxford.

United Nations Programme (UNEP) (1992) *World Atlas of Desertification*. Edward Arnold, London.

United Nations Environment Programme (1993) *Environmental Data Report 1993/94*. Blackwell, Oxford

Unwin, P.T.H. (1990) Towns and trade 1066–1500. In R.A. Dodgshon and R.A. Butlin (eds). *An Historical Geography of England and Wales*, 2nd edn. Academic Press, London, pp. 123–49.

Utting, P. (1993) *Trees, People and Power: Social Dimensions of Deforestation and Forest Protection in Central America*. Earthscan, London.

Vail, L. (1983) The political economy of East-Central Africa. In D. Birmingham and P.M. Martin (eds) *History of Central Africa*. Longman, London, Vol.2. pp. 200–250.

Valladas, H., Reyss, J.L., Joron, J.L., Valladas, G., Bar-Josef, O. and Vandermeersch, B. (1988) Thermoluminescence dating of Mousterian 'Proto-Cro-Magnon' remains from Israel and the origin of modern man. *Nature* **331**, 614.

Valoch, K. (1994a) Europe (excluding the USSR) in the period of *Homo sapiens neanderthalensis* and contemporaries. In S.J. De Laet, A.H. Dani, J.L. Lorenzo and R.B. Nunoo (eds) *History of Humanity, Volume I: Prehistory and the Beginnings of Civilization*. Routledge, London, and UNESCO, Paris, pp. 136–44.

Valoch, K. (1994b) Archaeology of the Neanderthalers and their contemporaries: an overview. In S.J. De Laet, A.H. Dani, J.L. Lorenzo and R.B. Nunoo (eds) *History of Humanity, Volume I: Prehistory and the Beginnings of Civilization*. Routledge, London, and UNESCO, Paris, pp.107–15.

van Andel, T.H. and Runnels, C.N. (1995) The earliest farmers in Europe. *Antiquity* **69**, 481–500.

Van Blarcum, S.C., Miller, J.R. and Russell, G.L. (1995) High latitude river runoff in a doubled CO_2 climate. *Climatic Change* **30**, 7–26.

Vance, R.E., Beaudoin, A.B. and Luckman, B.H. (1995) The paleoecological record of 6 ka BP climate in the Canadian prairie provinces. *Geographie Physique et Quaternaire* **49**, 81–98.

Vanclay, J.K. (1993) Saving the tropical forest: needs and prognosis. *Ambio* **22**, 225–31.

Vandaele, K. and Poesen, J. (1995) Spatial and temporal patterns of soil erosion rates in an agricultural catchment, central Belgium. *Catena* **25**, 213–26.

van der Hammen, T. and Absy, M.L. (1994) Amazonia during the last glacial. *Palaeogeography, Palaeoclimatology, Palaeoecology* **109**, 247–61.

van der Hammen, T., Wijmstra, T.A. and Zagwijn, W.H. (1971) The floral record of the Late Cenozoic of Europe. In Turekian, K.K. (ed.) *The Late Cenozoic Glacial Ages*. Yale University Press, New Haven CT, pp. 391–424.

Van der Merwe, N.J. (1982) Carbon isotopes, photosynthesis and archaeology. *American Scientist* **70**, 596–606.

van Emden, H.F. and Peakall, D.B. (1996) *Beyond Silent Spring: Integrated Pest Management and Chemical Safety*. Chapman and Hall, London.

van Zeist, W. and de Roller, G.J. (1991/1992) The plant husbandry of aceramic Çayönü, S.E. Turkey. *Palaeohistoria* **33/34**, 65–96.

Varley, A. (1994) The exceptional and the everyday: vulnerability and analysis in the International Decade for Natural Disaster Reduction. In A.Varley (ed.) *Disasters, Development and Environment*. J. Wiley, Chichester, pp. 1–11.

Vavilov, N.I. (1992) *Origin and Geography of Cultivated Plants*. Cambridge University Press, Cambridge. (This is a collection of Vavilov's 1920–1940 papers which have been translated by D. Löve.)

Veldkamp, E. (1994) Organic carbon turnover in 3 tropical soils under pasture ofter deforestation. *Soil Science Society of America Journal* **58**, 175–80.

Venema, H.D. and Schiller, E.J. (1995) Water resources planning for the Senegal River Basin. *Water International* **20**, 61–71.

Venkatesh, S., Fournier, D.F., Waterland, L.R. and Carroll, G.J. (1996) Evaluation of mineral-based additives as sorbents for hazardous trace metal captive and immobilization in incineration processes. *Hazardous Waste and Hazardous Materials* **13**, 73–94.

Venosa, A.D., Suidan, M.T., Wrenn, B.A., Strohmeier, K.L., Haines, J.R., Eberhart, B.L., King, D. and Holder, E. (1996) Bioremediation of an experimental oil spill on the shoreline of Delaware Bay. *Environmental Science and Technology* **30**, 1764–75.

Vertes, C., Lakatosvarsanyi, M., Vertes, A., Meisel, W. and Horvath, A. (1995) Electrochemical and Mössbauer spectroscopy studies of corrosion of iron in a medium modelling acid rain. *ACH Models in Chemistry* **132**, 597–606.

Villa, P. (1994) Europe. Lower and Middle Pleistocene archaeology. In S.J. De Laet, A.H. Dani, J.L. Lorenzo and R.B. Nunoo (eds) *History of Humanity, Volume 1: Prehistory and the Beginnings of Civilization*. Routledge, London, and UNESCO, Paris, pp. 44–61.

Villalba, R. (1994) Climatic fluctuations in midlatitudes of South America during the last 1000 years: their relationships to the southern oscillation. *Revista Chilena de Historia Natural* **67**, 443–51.

von Post, L. (1916) Om skogsträdspollen i sydsvenka forfmosselagerföljder (föredragsreferat). *Geologiska Föreningens Stockholm Förhandlingar* **38**, 384–94.

von Richtofen, B.F. (1882) On the mode of origin of the loess. *Geological Magazine* **9**, 293–305.

Vrba, E.S. (1992) Mammals as a key to evolutionary theory. *Journal of Mammology* 73, 1-28.

Vrba, E.S. (1996) Climate, heterochrony, and human evolution. *Journal of Anthropological Research* **52**, 1–28.

Wace, N. (1988) Naturalised plants in the Australian landscape. In R.L. Heathcote (ed.) *The Australian Experience: Essays in Australian Land Settlement and Resource Management*. Longman, Melbourne, pp. 139–50.

Wacher, J. (1987) *The Roman Empire*. Dent, London.

Waddle, D.M. (1994) Matrix correlation tests support a single origin for modern humans. *Nature* **368**, 452–54.

Wadhams, P. (1995) Arctic sea-ice extent and thickness. *Philosophical Transactions of the Royal Society of London* **A352**, 301–9.

Wainwright, M. (1992) *An Introduction to Fungal Biotechnology*. J. Wiley, Chichester.

Walker, M.J.C. (1993) Holocene (Flandrian) vegetation change and human activity in the Carneddau area of upland mid-Wales. In F.M. Chambers (ed.) *Climate Change and Human Impact on the Landscape*. Chapman and Hall, London, pp.169–83.

Walker, M.J.C., Coope, G.R. and Lowe, J.J. (1993) The Devensian (Weichselian) late-glacial palaeoenvironmental record fron Gransmoor, East Yorkshire, England. *Quaternary Science Reviews* **12**, 659–80.

Walker, M.J.C., Bohncke, S.J.P., Coope, G.R., O'Connell, M., Usinger, H. and Verbruggen, C. (1994) The Devensian/Weichselian late-glacial in northwest Europe (Ireland, Britain, north Belgium, the Netherlands, northwest Germany). *Journal of Quaternary Science* **9**, 109–18.

Waller, M.P. (1993) Flandrian vegetational history of south-eastern England. Pollen data from Pannel Bridge, East Sussex. *New Phytologist* **124**, 345–69.

Walmsley, D.J. and Lewis, G.J. (1993) *People and Environment: Behavioural Approaches in Human Geography*, 2nd edn. Longman, Harlow.

Walton, J.R. (1990) Agriculture and rural Society. In R.A. Dodgshon and R.A. Butlin (eds) *An Historical Geography of England and Wales*, 2nd edn. Academic Press, London, pp. 323–50.

Wang, B.L. and Allard, M. (1995) Recent climatic trend and thermal response of permafrost at Salluit, Northern Quebec, Canada. *Permafrost and Periglacial Processes* **6**, 221–33.

Wang, W.X. and Wang, T. (1995) On the origin and trend of acid precipitation in China. *Water Air and Soil Pollution* **85**, 2295–2300.

Wangwacharakul, V. and Bowonwiwat, R. (1995) Economic evaluation of CO_2 response options in forestry sector: the case of Thailand. *Biomass and Bioenergy* **8**, 293–308.

Wanpo, H., Ciochon, R., Yumin, Gu, Larick, R., Qiren, F., Schwarcz, H., Yonge, C., de Vos, J. and Rink, W. (1995) Early *Homo* and associated artifacts from Asia. *Nature* **378**, 275–78.

Wansard, G. (1996) Quantification of paleotemperature changes during isotope stage 2 in the La Draga continental sequence (NE Spain) based on the Mg/Ca ratio of fresh-water ostracods. *Quaternary Science Reviews* **15**, 237–45.

Ward, D. (1987) Population growth, migration and urbanization, 1860–1920. In R.D. Mitchell and P.A. Groves (eds) *North America: The Historical Geography of a Changing Continent*. Hutchinson, London, pp. 299–320.

Ward, S.C., Koch, J.M. and Ainsworth, G.L. (1996) The effect of timing of rehabilitation procedures on the establishment of a jarrah forest after bauxite mining. *Restoration Ecology* **4**, 19–24.

Wardle, K.A. (1994) The palace civilizations of Minoan Crete and Mycenean Greece 2000–1200 BC. In B. Cunliffe (ed.) *The Oxford Illustrated Prehistory of Europe*. Oxford University Press, Oxford, pp. 202–43.

Wasylikowa, K., Harlan, J.R., Evans, J., Wendorf, F., Schild, R., Close, A.E., Krolik, M. and Housley, R.A. (1993) Examination of botanical remains from early Neolithic houses at Nabta Playa, Western Desert, Egypt, with special reference to sorghum grains. In T. Shaw, P. Sinclair, B. Andah and A. Opoko (eds) *The Archaeology of Africa: Food, Metals and Towns*. Routledge, London, pp. 154–64.

Watson, A.J., Law, C.S., Van Scoy, K.A., Millero, F.J., Yao, W., Friederich, G.E., Liddicoat, M.I., Wanninkhof, R.H., Barber, R.T. and Coale, K.H. (1994) Minimal effect of iron fertilization on sea-surface carbon dioxide concentrations. *Nature* **371**, 143–45.

Watson, R.T., Meira Filho, L.G., Sanhueza, E. and Janetos, A. (1992) Sources and sinks. In J.T. Houghton, B.A. Callander and S.K. Varney (eds) *Climate Change 1992: The Supplementary Report to the IPCC Scientific Assessment*. Cambridge University Press, Cambridge, pp. 25–46.

Watts, W.A. (1967) Interglacial deposits in Kildromin Townland, near Herbertstown, Co. Limerick. *Proceedings of the Royal Irish Academy* **65B**, 339–48.

Watts, W.A. (1971) Postglacial and interglacial vegetation history of southern Georgia and central Florida. *Ecology* **52**, 676–90.

Watts, W.A. (1977) The Late Devensian vegetation of Ireland. *Philosophical Transactions of the Royal Society of London* **B280**, 273–93.

Watts, W.A. (1980a) The Late Quaternary vegetation history of the southeastern United States. *Annual Review of Ecology and Systematics* **11**, 387–409.

Watts, W.A. (1980b) Regional variation in the response of vegetation to late glacial climatic events in Europe. In J.J. Lowe, J.M. Gray and J.E. Robinson (eds) *Studies in the Late Glacial of North-west Europe*. Pergamon, Oxford, pp. 1–21.

Watts, W.A. (1985) Quaternary vegetation cycles. In K.J. Edwards and W.P. Warren (eds) *The Quaternary History of Ireland*. Academic Press, London, pp. 155–85.

Watts, W.A. and Hansen, B.S.C. (1994) Pre-Holocene and Holocene pollen records of vegetation history from the Florida peninsula and their climatic implications. *Palaeogeography, Palaeoclimatology, Palaeoecology* **109**, 163–76.

Watts, W.A., Allen, J.R.M., Huntley, B. and Fritz, S.C. (1996) Vegetation history and climate of the last 15,000 years at Laghi di Monticchio, southern Italy. *Quaternary Science Reviews* **15**, 113–32.

Wearing, C.H. and Hokkanen, H.M.T. (1994) Pest resistance to *Bacillus thuringiensis* – case studies of ecological crop assessment for Bt gene incorporation and strategies of management. *Biocontrol Science and Technology* **4**, 573–90.

Weaver, P.P.E. and Pujol, C. (1988) History of the last deglaciation in the Alboran Sea (western Mediterranean) and adjacent North Atlantic as revealed by coccolith floras. *Palaeogeography, Palaeoclimatology, Palaeoecology* **64**, 35–42.

Webb, N.R. and Thomas, J.R. (1994) Conserving insect

habitats in heathland biotopes: a questions of scale. In P.J. Edwards, R.M. May and N.R. Webb (eds) *Large-scale Ecology and Conservation Biology*. Blackwell, Oxford, pp. 129–51.

Weber, A. (1995) The Neolithic and Early Bronze Age of the Lake Baikal region: a review of recent research. *Journal of World Prehistory* **9**, 99–165.

Weiss, H., Courty, M.-A., Wetterstrom, W., Guichard, F., Senior, L., Meadow, R. and Curnow, A. (1994) The genesis and collapse of third millennium North Mesopotamian civilisation. *Science* **261**, 995–1004.

Wellburn, A. (1994) *Air Pollution and Climate Change: The Biological Impact*, 2nd edn. Longman, Harlow.

Wells, M.P. (1993) Neglect of biological riches – the economics of nature tourism. *Biodiversity and Conservation* **2**, 445–64.

Welsh, D. and Maughan, O.E. (1994) Concentrations of selenium in biota, sediments, and water in the Cibola National Wildlife Refuge. *Archives of Environmental Contamination and Toxicology* **26**, 452–58.

Wendorf, F., Close, A.E., Schild, R., Wasylikowa, K., Housley, R.A., Harlan, J.R. and Królik, H. (1992) Saharan exploitation of wild plants 8,000 years ago. *Nature* **359**, 721–24.

Wenming, Y. (1991) China's earliest rice agriculture remains. *Indo-Pacific Prehistory Association Bulletin* **10**, 118–26.

West, B. and Zhou, B.-X. (1988) Did chickens go north? New evidence for domestication. *Journal of Archaeological Science* **15**, 515–33.

Westbroek, P. (1991) *Life as a Geological Force: Dynamics of the Earth*. Norton, New York.

Westing, A.H. (1994) Population, desertification, and migration. *Environmental Conservation* **21**, 109–14.

Wetterstrom, W. (1993) Foraging and farming in Egypt: the transition from hunting and gathering to horticulture in the Nile valley. In T. Shaw, P. Sinclair, B. Andah and A. Okpoko (eds) *The Archaeology of Africa: Food, Metals and Towns*. Routledge, London, pp. 165–226.

Wheeler, D. (1995) Early instrumental weather data from Cadiz: a study of late eighteenth century and early nineteenth century records. *International Journal of Climatology* **15**, 801–10.

Wheeler, J.C. (1995) Evolution and present situation of the South American camelids. *Biological Journal of the Linnean Society* **54**, 271–95.

Wheeler, J.C., Russel, A.J.F. and Redden, H. (1995) Llamas and alpacas: preconquest breeds and post-conquest hybrids. *Journal of Archaeological Science* **22**, 833–40.

Whinam, J. and Comfort, M. (1996) The impact of commercial horse riding on sub-alpine environments at Cradle Mountain, Tasmania, Australia. *Journal of Environmental Management* **47**, 61–70.

White, J.W.C. (1993) Climate change: don't touch the dial. *Nature* **364**, 186.

White, J.W.C., Ciais, P., Figge, R.A., Kenny, R. and Markgraf, V. (1994) A high-resolution record of atmospheric CO_2 content from carbon isotopes in peat. *Nature* **367**, 153–56.

White, T.D., Suwa, G. and Asfaw, B. (1995) *Australopithecus ramidus*, a new species of early hominid from Aramis, Ethiopia (corrigendum). *Nature* **375**, 88.

Whiteman, C.A. and Rose, J. (1992) Thames River sediments of the British Early and Middle Pleistocene. *Quaternary Science Reviews* **11**, 363–76.

Whitlow, R. (1988) Potential versus actual erosion in Zimbabwe. *Applied Geography* **8**, 87–100.

Whitmore, T.C. and Sayer, J.A. (eds) (1992) *Tropical Deforestation and species Extinction*. Chapman and Hall, London.

Whitmore, T.J., Brenner, M., Engstrom, D.R. and Song, X.L. (1994) Accelerated soil erosion in watersheds of Yunnan Province, China. *Journal of Soil and Water Conservation* **49**, 67–72.

Whitmore, T.J., Brenner, M., Curtis, J.H., Dahlin, B.H. and Leyden, B.W. (1996) Holocene climatic and human influences on lakes of the Yucatan Peninsula, Mexico: an interdisciplinary palaeolimnological approach. *The Holocene* **6**, 273–87.

Whitmore, T.M. and Turner, B.L. II (1992) Landscapes of cultivation in Mesoamerica on the eve of the conquest. *Annals of the Association of American Geographers* **82**, 402–25.

Whitney, G.G. (1987) An ecological history of the Great Lakes forest of Michigan. *Journal of Ecology* **75**, 667–84.

Whitney, G.G. (1994) *From Coastal Widerness to Fruited Plain*. Cambridge University Press, Cambridge.

Wigley, T.M.L. (1991) Could reducing fossil-fuel emissions cause global warming? *Nature* **349**, 503–06.

Wigley, T.M.L. and Kelly, P.M. (1990) Holocene climatic change, ^{14}C wiggles and variations in solar irradiance. *Philosophical Transactions of the Royal Society of London* **330**, 547–60.

Wigley, T.M.L. and Raper, S.C.B. (1992) Implications for climate and sea level of revised IPCC emissions scenarios. *Nature* **357**, 293–300.

Wilkins, D.A. (1984) The Flandrian woods of Lewis (Scotland). *Journal of Ecology* **72**, 251–58

Willcox, G. (1996) Evidence for plant exploitation and vegetation history from 3 early Neolithic pre-pottery sites on the Euphrates (Syria). *Vegetation History and Archaeobotany* **5**, 143–52.

Williams, M. (1988) The clearing of the woods. In R.L. Heathcote (ed.) *The Australian Experience: Essays in Australian Settlement and Resource Management*. Longman, Melbourne, pp. 115–26.

Williams, M. (1989) *Americans and their Forests*. Cambridge University Press, Cambridge.

Williams, M. (1990) The clearing of the forests. In M.P. Conzen (ed.) *The Making of the American Landscape*. Unwin Hyman, Boston, pp. 146–68.

Williams, M., Fordyce, F., Paijitprapapon, A. and

Charoenchaisri, P. (1996) Arsenic contamination in surface drainage and groundwater in part of the southeast Asian tin belt, Nakhon Si Thammarat Province, southern Thailand. *Environmental Geology* **27**, 16–33.

Williams, P.W. (1996) A 230 KA record of glacial and interglacial events from Aurora Cave, Fjordland, New Zealand. *New Zealand Journal of Geology and Geophysics* **39**, 225–41.

Wilson, E.O. (ed.) (1988) *Biodiversity*. National Academy Press, Washington, DC.

Wilson, G.A. (1993) The pace of indigenous forest clearance on farms in the Catlins District, South Island, New Zealand, 1861–1991. *New Zealand Geographer* **49**, 15–25.

Winograd, I.J., Coplen, T.B., Landwehr, J.M., Riggs, A.C., Ludwig, K.R., Szabo, B.J., Kolescar, P.T. and Revesz, K.M. (1992) Continuous 500,000-year climate record from vein calcite in Devils Hole, Nevada. *Science* **258**, 255–60.

Wirrmann, D. and Mourguiart, P. (1995) Late Quaternary spatio-temporal limnological variations in the Altiplano of Bolivia and Peru. *Quaternary Research* **43**, 344–54.

WoldeGabriel, G., White, T.D., Suwa, G., Renne, P., de Heinzelin, J., Hart, W.K. and Heiken, G. (1994) Ecological and temporal placement of early Pliocene hominids at Aramis, Ethiopia. *Nature* **311**, 330–33.

Wolfe, D.W. and Erickson, J.D. (1993) Carbon dioxide effects on plants: uncertainties and implications for modelling crop response to climate change. In H.M. Kaiser and T.E. Drennan (eds) *Agricultural Dimensions of Global Climate Change*. St Lucie Press, Delray Beach, FL, pp. 153–78.

Wolff, W.J. (1993) Netherlands wetlands. *Hydrobiologia* **265**, 1–14.

Wolffe, E.W. and Suttie, E.D. (1994) Antarctic snow record of southern hemisphere lead pollution. *Geophysical Research Letters* **21**, 781–84.

Wolpoff, M.H., Thorne, A.G., Smith, F.H., Frayer, D.W. and Pope, G.G. (1994) Multiregional evolution: a world-wide source for modern human populations. In M.H. Nitecki and D.V. Nitecki (eds) *Origins of Anatomically Modern Humans*. Plenum, New York, pp. 175–99.

Wood, B. (1992) Origin and evolution of the genus *Homo*. *Nature* **355**, 783–90.

Wood, B. (1994) The oldest hominid yet. *Nature* **371**, 280–81.

Wood, B. and Turner, A. (1995) Out of Africa. *Nature* **378**, 239–40.

Wood Mackenzie. (1994) Agrochemical products part 1: the key agrochemical product groups. *Agrochemical Service*. Wood Mackenzie, London, products section update, p. 1.

Wood Mackenzie. (1995) Agricultural biotechnology. *Agrochemical Monitor* **120**, 1–20.

Wood MacKenzie (1996) Agrochemical products. *Agrochemical Service,* May 1996, products section update.

Wood, W.B. (1990) Tropical deforestation: balancing regional development demands and global environmental concerns. *Global Environmental Change* **1**, 23–41.

Woodman, P.C. (1977) Mount Sandel. *Current Archaeology* **59**, 372–76.

Woods, D. and Rawlings, D.E. (1989) Bacterial leaching and biomining. In J.L. Marx (ed.) *A Revolution in Biotechnology*. Cambridge University Press, Cambridge, pp. 82–93.

Woodward, C.L. (1996) Soil compaction and top soil removal effects on soil properties and seedling growth in Amazonian Ecuador. *Forest Ecology and Management* **82**, 197–209.

Woodward, F.I. (1992) Predicting plant response to global environmental change. *New Phytologist* **122**, 239–51.

Woodward, F.I. and Lee, S.E. (1995) Global scale forest function and distribution. *Forestry* **68**, 317–25.

Woodward, F.I., Smith, T.M. and Emanuel, W.R. (1995) Global land primary productivity and phytogeography model. *Palaeogeography, Palaeoclimatology, Palaeocology* **75**, 259–82.

World Commission on Environment and Development (1987) *Our Common Future*. Oxford University Press, Oxford.

World Conservation and Monitoring Centre (1992) *Global Biodiversity: Status of the Earth's Living Resources*. Chapman and Hall, London.

World Resources Institute (1989) *World Resources 1988-89: A Guide to the Global Environment*. Oxford University Press, Oxford.

World Resources Institute (1996) *World Resources 1996-97: A Guide to the Global Environment*. Oxford University Press, Oxford.

World Tourism Organisation. (1995) *Yearbook of Tourism Statistics*. World Tourism Organisation, Madrid.

Worona, M.A. and Whitlock, C. (1995) Late Quaternary vegetation and climate history near Little Lake, central Coast Range, Oregon. *Geological Society of America Bulletin* **107**, 867–76.

Worster, D. (1979) *Dust Bowl*. Oxford University Press, New York.

Wright, K.I. (1994) Ground-stone tools and hunter-gatherer subsistence in southwest Asia: implications for the transition of farming. *American Antiquity* **59**, 238–63.

Wright, R.F. and Hautis, M. (1991) Reversibility of acidification: soils and surface waters. *Proceedings of the Royal Society of Edinburgh* **B97**, 169–91.

Wright, R.F. and Schindler, D.W. (1995) Interaction of acid rain and global changes – effects on terrestrial and aquatic ecosystems. *Water Air and Soil Pollution* **85**, 89–99.

Wrigley, E.A. and Schofield, R.S. (1988) *The Population History of England 1541–1871: A Reconstruction*, 2nd edn. Cambridge University Press, Cambridge.

Wunsch, P., Greilinger, C., Bieniek, D. and Kettrup, A. (1996) Investigation of the binding of heavy metals in thermally treated residues from waste incineration. *Chemosphere* **32**, 2211–18.

Wymer, J.J. (1982) *The Palaeolithic Age*. Croom Helm, London.

Wynn, G. (1987) Forging a Canadian nation. In R.D. Mitchell and P.A. Groves (eds) *North America: The Historical Geography of a Changing Continent*. Hutchinson, London, pp. 373–409.

Wynn-Williams, D.D. (1996) Response of pioneer soil microalgal colonists to environmental change in Antarctica. *Microbial Ecology* **31**, 177–88.

Xinzhi, W. and Linghong, J.W. (1985) Chronology in Chinese palaeoanthropology. In W. Rukang and J.W. Olsen (eds) *Palaeoanthropology and Palaeolithic Archaeology in the People's Republic of China*. Academic Press, London, pp. 29–51.

Xiong, S.Y., Xiong, Z.X. and Wang, P.W. (1996) Soil salinity in the irrigated area of the Yellow River in Ningxia, China. *Arid Soil Research and Rehabilitation* **10**, 98–101.

Xu, D. (1995) The potential for reducing atmospheric carbon by large-scale afforestation in China and related cost/benefit analysis. *Biomass and Bioenergy* **8**, 337–44.

Yakowitz, H. (1993) Waste management: what now, what next? An overview of policies and practices in the OECD area. *Resources, Conservation and Recycling* **8**, 131–78.

Yan, N.D., Keller, W., Scully, N.M., Lean, D.R.S. and Dillon, P.J. (1996) Increased UV-B penetration in a lake owing to draught-induced acidification. *Nature* **381**, 141–43.

Yang, S. and Furedy, C. (1993) Recovery of wastes for recycling in Beijing. *Environmental Conservation* **20**, 79–82.

Yassaglou, N.J. (1990) Desertification in Greece. In J.L. Rubio and R.J. Rickson (eds) *Strategies to Combat Desertification in Mediterranean Europe*. Commission of the European Communities, Luxembourg, pp. 148–62.

Yellen, J.E., Brooks, A.S., Cornelissen, E., Mehlman, M.J. and Stewart, K. (1995) A middle stone age worked bone industry from Katanda, Upper Semliki Valley, Zaire. *Science* **268**, 553–56.

Yelling, J. (1990) Agriculture 1500–1730. In R.A. Dodgshon and R.A. Butlin (eds) *An Historical Geography of England and Wales*, 2nd edn. Academic Press, London, pp. 181–98.

Yiou, F., Raisbeck, G.M., Bourles, D., Lorius, C. and Barkov, N.I. (1985) [10]Be in ice at Vostok Antarctica during the last climatic cycle. *Nature* **316**, 616–17.

Yohe, G., Neumann, J., Marshall, P. and Ameden, H. (1996) The economic cost of greenhouse-induced sea-level rise for developed property in the United States. *Climatic Change* **32**, 387–410.

Young, A.R.M. (1996) *Environmental Change in Australia Since 1788*. Oxford University Press, Melbourne.

Young, D.A. and Bettinger, R.L. (1995) Simulating the global human expansion in the Late Pleistocene. *Journal of Archaeological Science* **22**, 89–92.

Young, K.R. (1996) Threats to biological diversity caused by *coca*/cocaine deforestation in Peru. *Environmental Conservation* **23**, 7–15.

Young, W.J., Marston, F.M. and Davis, J.R. (1996) Nutrient exports and land use in Australian catchments. *Journal of Environmental Management* **47**, 165–83.

Younger, P.L. (1995) Hydrogeochemistry of minewaters flowing from abandoned mine workings in County Durham. *Quarterly Journal of Engineering Geology* **28**, **52**, 5101–13.

Zagwijn, W.H. (1975) Variations in climate as shown by pollen-analysis especially in the Lower Pleistocene of Europe. In A.F. Wright and F. Moseley (eds) *Ice Ages Ancient and Modern*. Seel House, Liverpool, pp. 137–52.

Zagwijn, W.H. (1985) An outline of the Quaternary stratigraphy of the Netherlands. *Geologie en Mijnbouw* **64**, 17–24.

Zagwijn, W.H. (1992) The beginning of the ice age in Europe and its major subdivisions. *Quaternary Science Reviews* **11**, 583–91.

Zapataarias, F.J., Torrizo, L.B. and Ande, B. (1995) Current developments in plant biotechnology for genetic improvement – the case of rice (*Oryza sativa L*). *World Journal of Microbiology and Biotechnology* **11**, 393–99.

Zepp, R.G., Callaghan, T. and Erikson, D. (1995) Effects of increased solar ultraviolet radiation on biogeochemical cycles. *Ambio* **24**, 181–87.

Zhang, X., Quine, T.A., Walling, D.E. and Zhau, L. (1994) Application of the caesium-137 technique in a study of soil erosion on gully slopes in a yuan area of the Loess Plateau near Xifeng, Gansu Province, China. *Geografiska Annaler* **76A**, 103–20.

Zhao, M., Beveridge, N.A.S., Shackleton, N.J., Sarnthein, M. and Eglinton, G. (1995) Molecular stratigraphy of cores off northwest Africa – sea-surface temperature history over the last 80 ka. *Paleoceanography* **10**, 661–75.

Zhao, S. (1994) *Geography of China*. J. Wiley, New York.

Zhou, G.Y. (1995) Influences of tropical forest changes on environmental quality in Hainan Province, PR of China. *Ecological Engineering* **4**, 223–29.

Zhou, S.Z., Chen, F.H., Pan, P.T., Cao, J.X., Li, J.J. and Derbyhire, E. (1991) Environmental change during the Holocene in western China on a millennial timescale. *The Holocene* **1**, 151–56.

Zhou, W.J., An, Z.S. and Head, M.J. (1994) Stratigraphic division of Holocene loess in China. *Radiocarbon* **36**, 37–45.

Zhu, R.X., Laj, C. and Mazaud, A. (1994) The Matuyama-Brunhes and upper Jaramillo transitions recorded in a loess section at Weinan, north-central China. *Earth and Planetary Science Letters* **125**, 143–58.

Zielinski, G.A., Mayewski, P.A., Meeker, L.D., Whitlow, S., Twickler, M.S., Morrison, M., Meese, D.A., Gow, A.J. and Alley, R.B. (1994) Record of volcanism since

7,000 BC from the GISP2 Greenland ice core and implications for the volcano-climate systems. *Science* **264,** 948-952.

Zielinski, G.A., Germani, M.S., Larsen, G., Baillie, M.G.L., Whitlow, S., Twickler, M.S. and Taylor, K. (1995) Evidence of the Eldgjá (Iceland) eruption in the GISP2 Greenland ice core: relationship to eruption processes and climatic conditions in the tenth century. *The Holocene* **5,** 129–40.

Zielinski, G.A., Mayewski, P.A., Meeker, L.D., Whitlow, S. and Twickler, M.S. (1996) A 110,000 yr record of explosive volcanism from the GISP2 (Greenland) ice core. *Quaternary Research* **45,** 109–18.

Zohary, D. and Hopf, M. (1993) *Domestication of Plants in the Old World*, 2nd edn. Oxford University Press, Oxford.

Zonn, I.S. (1995) Desertification in Russia – problems and solutions (an example in the Republic of Kalmykia – Khalmg Tangch). *Environmental Monitoring and Assessment* **37,** 347–63.

Zvelbil, M. (1986) Mesolithic prelude and neolithic revolution. In M. Zvelbil (ed.) *Hunters in Transition.* Cambridge University Press, Cambridge, pp. 5–15.

Zvelbil, M. (1994) Plant use in the Mesolithic and its role in the transition to farming. *Proceedings of the Prehistoric Society* **60,** 35–74.

Index

Aboriginal cultures, 102, 157
Abu Hureyra site, 103, 111
Acheulian (tool-making tradition), 83, 85, 90, 95
Acid pollution from mining, 168, 169
Acid rain, 156, 186–93
Acidification, 23-4, 186–93, 274, 308
Adhémar, 3
Aerosols and global climate, 176
Afforestation in developed world, 266–75
Afforestation in developing world, 275–9
Africa, 36, 62, 289
    colonialism, 147–51
    desertification, 251–4
    environmental change, 27, 75–6
    lake level change, 68, 72
    site for domestication, 107–8
    soil erosion, 246, 250–1
Agassiz, 3
Agenda 21 of the Earth Summit, 312, 314
Agribusiness, 205, 296
Agricultural biotechnology, 289–96
Agricultural intensification in 1750–1850, 145-7
Agricultural Revolution, 141–2
Agriculture
    and Elm decline, 74–5
    and eutrophication, 220–5
    and global warming, 179, 184
    and industry, 143, 145
    and water quality, 261–4
    as energy source, 8
    developed world, 205–38
    developing world, 239–65
    development, 102–12
    during Neolithic, 113–15
    in future, 315–6
    in perspective, 308–9
    sustainable, 116, 296
Agriculture Act of 1947, 147

Agroecosystems, 6, 78
Agroforestry, 246, 250, 259, 280
Akkad civilisation, 115, 116
Alar® plant growth regulator, 233, 237
Alaska, 69
Alcovy River, United States, 156
Algal blooms, 222, 223
Algeria, 278
Alkalinisation, 255–61
Alkenone analysis, 17, 18, 66
Allerød period, 51–2
Aluminium, 23, 170–1, 190, 192
Amazon River, 167
Amazonia, 38
Anaconda mine, New Mexico, 163
Andes as site for early domestication, 108–9
Anglian (glaciation), 33–4
Angling, impact of, 283
Anglo-Saxon period, 137
Animal health products, 231, 296
Animal productivity, 296
Animal rights, 10
Animals
    affected by global warming, 182
    breeding, 146
    domestication, 109–12
    in historic Britain, 137
    introduced into Americas, 152
Antarctic, 21–2, 27, 30–32, 40, 48, 195
Apes as human ancestors, 79
Appalachian Region coal mines, 169
Aquatic ecosystems and acid pollution, 190
Aquifers, 260
Aral Sea, 229–31
Arctic, 30–32, 48, 64, 195
Ardipithecus ramidus, 81
Argentina, 247
Aristotle, 2

Arsenic, 167, 301
Art in prehistory, 96, 100, 124
Ash (from burning) as a resource, 141, 201
Asia, soil erosion in (*see also individual countries*), 246, 248–50
Asian republics, desertification in, 219
Astronomical forcing, 3, 29, 40
Athens as city-state, 129
Atmosphere, chemistry of, 5, 21
Attica, 130
Australia, 62, 63, 167, 229
    colonialism, 157–60
    desertification, 219
    environmental history, 27, 36, 38, 60, 70, 76, 87
    eutrophication, 224
    future climate, 178
    reclamation of mine-damaged land, 170–1
    soil erosion, 158, 214–15
    tourism, 285–6
Australopithecines, 81
Ayres Rock, Australia, tourism, 285–6
Azilian (tool making tradition), 95, 97

Bacillus thuringiensis (Bt), 292, 294
Bacteria, 169, 199
Bakhta (period of ice advance), 31
Baltic Sea, eutrophication of, 223
Bangladesh, 177
Barbarism, 96
Baringo district of Kenya, 149–50, 250
Barium, 20
Barley, domestication of, 106
Bauxite mining, 170–1
Beans, domestication of, 106, 108, 109
Beryllium analyses, 22, 28
Biodegradable insecticides, 236
Biodegradable waste, 199, 201
Biodiversity, 70, 105, 322
    loss of, 205–10, 233, 239–45, 272, 275
Biodiversity Treaty, 243, 312, 314
Biogas fuel, 298
Biogeochemical cycles, 187, 221
    carbon, 5, 22, 171, 307
    lead, 196–7
Biological control, 235
Biological magnification, 235, 236, 264
Biomass fuels, 6, 8, 295, 302
Biomining, 298
Biomolecular evidence for human evolution, 78–9
Biopesticides, 231, 292
Bioremediation, 315
Biosensors, 299

Biota influencing climate, 5
Biotechnology, 289–302
Bipedalism, 81
Bird populations, 208, 210, 227, 235, 237, 274
Black Death, 139, 140
Blytt, 3
Bølling oscillation, 48, 52–3
BOD (Biochemical oxygen demand), 299
Bone industry, 96
Bow and arrow, Mesolithic, 99
Brain size in genus *Homo*, 83
Brazil, 153, 240, 245
Brimpton (interstadial), 18
Britain, 199, 222, 282
    climate change, 32–4
    during Bronze Age, 121–2
    during Flandrian, 67–8
    during historic period, 133–47
    during Iron Age, 124, 126
    during late-glacial period, 49–51, 53–5
    during late Holocene, 73–4
    faunal changes, 60–2
    forestry, 272–4
    industrialisation, 143–5
    Mesolithic sites, 98–9
    soil erosion, 212–14
British colonialism in America, 151
Bromine, 194, 195
Bronze Age, 118–22
Brørup (interstadial), 14–15, 32
Brundtland report, 9, 312
Buckland, William, 3
Burkina Faso, 252, 253

Cadmium, 20, 166, 167, 263
Caesium, 203, 212
Calcium carbonate, 16, 17, 23, 30, 39, 191
Calowanie site, Poland, 101
Campsites damaged by visitors, 280
Canada
    during Holocene, 66, 69
    during late-glacial period, 51, 58–9
    forestry, 182, 267, 268, 272
    global warming, 178, 182
    soil erosion, 216
Canals, 256, 257, 263
Cancer, 195, 201, 203
Cane swamp, 158
CAP (Common Agricultural Policy), 147, 205, 288
Cape Verde Islands, 148
Capital investment, 141, 143
Carbon, biogeochemical cycle of, 5, 22, 171, 307

Carbon budget, 243, 274, 279
Carbon dioxide
   and chemical weathering, 30
   and forests, 207, 239, 272, 279
   atmospheric, 22, 46, 171, 172, 315
   crop response to, 174, 184
Carbon isotope analyses, 28
Caribbean settlement by Europeans, 151
Carrying capacity, 128
Cash crops, 148
Cassava, 152, 290
Cattle ranching, 154, 157, 244
Cattle, domestication of, 110
Cave paintings, 2
CCVM (Canadian Climate–Vegetation Model), 182
Celtic culture, 124
Cement manufacture, 171
Central America during late-glacial period, 59–60
Centres of domestication, 104–6
Centuriation, 132
Cereal production, 142, 146, 184–6
CFCs (Chlorofluorocarbons), 9, 171–3, 175, 194–6
Chalcolithic Age, 118
Chaos and forecasting, 10
Characas, 153
Charcoal, 60, 72, 133, 140, 155
   as record of occupation, 70, 89, 99–100
Chelfordian (interstadial), 18
Chemolithotrophs, 298
Chernobyl nuclear power plant, 203
Cherry, date of flowering, 24, 26
Chesapeake Bay, United States, 224
Chicken, domestication of, 111
Chile, 24, 60, 63, 71–2, 88, 248
China, 69, 187, 259
   afforestation, 278–9
   biogas production, 298
   climate history, 24–6, 76
   desertification, 253, 254
   fossil sites, 91–2
   future rice production, 184
   Neolithic communities, 116–17
   pesticide use, 263–4
   site for domestication, 107, 111
   soil erosion, 248–9
Chlorine, 194, 195, 201
Chlorofluorocarbons (CFCs), 9, 171–3, 175, 194–6
Cholera, 145
Cistercian monks as farmers, 138
Civilisation, emergence of, 113, 120
Classical World, 123
Clear-felling for woodchip industries, 159

Clearance for agriculture, 101, 154, 157, 159, 208, 239–45
Clements, Frederick, 4
Climate
   as agent of change, 6, 43, 304
   deterioration during Iron Age, 124
   influence on emergence of agriculture, 103
   influence on evolution, 82, 87
   prediction, 27, 177–186, 312–15
   role in ecosystems, 5
Climate Treaty, Rio 1992, 314, 320
Climatic change
   causes, 3, 28–30, 40, 176
   during Holocene, 63–7, 71-3
   during Quaternary period, 12–41
   evidence from ice cores, 21–4
Climatic optimum, 43, 64, 71
Climax formation, 4
Cloning, 290, 291
Closed systems, 5
Cloth industry, 140
Cloud cover, 23
Clouds, acidity of, 186, 187
Clover as rotational crop, 142
Clovis culture, 63, 87–8
Coal, 141, 143, 144–5, 163, 169
Coccolithophores, 18, 32, 66
Codeine, 243
Coinage, 127
Cold and warm stages of glaciation, 3, 12, 95
Coleopteran data, 48–9, 75
Colliery spoil, environmental impact of, 169
Colombia, 36–7, 60
Colonialism, 147–60, 310
Colorado River, 226, 228
Columbus, Christopher, 151
Compaction of soil, 210, 246
Congo region, 68, 75
Conservation and tourism, 285
Conservation in colonial period, 150
Contour stripping, 163
Cooling, global, 22, 29, 32, 175, 176, 178
Copper Age, 78, 118–9
Copper mining, 122, 133, 156, 163, 299
Coppicing, 133, 139, 140
Coral, 39, 166, 286
Coral reef hypothesis, 23
Cornwall, England, 143
Coryndon Definition, 151
Costa Rica, 59, 243, 248
Cotton, 106, 108–9, 292, 294
Crete, palace civilisations of, 120

Critical load of acid deposition, 188, 193
Croll, James, 3
Cromerian (interglacial), 14–15, 30–1, 33–4
Crop-fallow system in Mediterranean agriculture, 132
Crop diversity in historic Britain, 136, 137
Crop improvement using biotechnology, 289–96
Crop plants, major, 105–9
Crop productivity and global warming, 174, 184–5
Crop protection chemicals, 146, 231–8
Crop rotation, 142
Crops, salt-tolerant, 257, 259
Cryocratic phase of interglacial, 43–5
Cultural eutrophication see Eutrophication
Cultural explosion model, 86
Cyclodiene insecticides, 232, 235, 264
Cytheropteron testudo, 12

Dams, 256, 263
Danebury hill fort, 127
Dark Ages, 134
Darwin, Charles, 2
Dating techniques, 36, 39
Davis, William Morris, 2
DDT, 232, 233, 235, 263, 264
de Charpentier, Jean, 3
Deer farming, 111
Deer hunting, 98
Deforestation, 171
    developing world, 275-9
    during Bronze Age, 122
    during historic period, 130, 135, 152
    for agriculture, 206, 207, 239–45
Degradation of the environment, 210, 211, 245
Demic diffusion, 113
Dendrochronology, 24–8, 29
Denmark, 43, 44, 55, 57, 68, 74
Derelict land, reclamation, 168
Desalinisation, 227
Desertification, 116, 217–20, 251–5
Desulphurisation of fuels, 187, 193, 299
Detergents, 201
Determinism, environmental, 2, 5, 8, 95, 102, 134, 315
Deuterium analyses, 22
Development, sustainable - defined (*see also* sustainability), 9, 10
Devensian (stadial), 13–15, 61–2
Diamond mining, 166
Diatoms, 20, 187, 193, 223
Dieldrin insecticide, 232, 235, 264
Digitalin, 243

Diluvial theory, 3
Dioxins, 201
Direct drilling, 215, 237
Disease resistance of plants, 290, 295
Diseases imported by European colonists, 151, 152
Dispersion of plants, 182
Diss Mere, Norfolk, 75, 135
Dmanisi fossil, 83
DMS (Dimethylsulphide), 23
DNA (Deoxyribonucleic acid), 79, 85, 110, 290, 293
Dog, domestication of, 98, 109
Dome ice cores, Antarctica, 23
Domesday book, 137, 138
Domestication of plants and animals, 102-12
Donegal, Ireland, 55–6
Drainage during Bronze Age, 121
Drainage networks in the Fens, 139, 142
Drought, 251
Drugs from natural products, 242
Dry deposition (of acidity), 186, 187
Dry farming, 132
Dualism in geography, 2
Dune systems of United States, 66
Dust affecting climate, 23, 28, 29
Dust bowl conditions, 158, 217
Dutch Elm disease, 74
Dye 3 ice core, Greenland, 28

Earth Summit 1992, 9, 243, 295, 312, 320
Earth's orbit, 3, 29
Ecology, 4, 9, 44, 103, 276
Economics, 9, 101–2
Ecosystems, 4, 6–7, 11, 179, 209
Ecotourism, 285
Ecotron, 209
Ecuador, 248
EEC (European Economic Community), 205
Eemian (interglacial), 21, 31
Egypt, 115, 255, 260
Einkorn wheat, 106, 107, 112
El Abra (stadial), 60
El Niño, 76
Eldgjá volcano, Iceland, 29
Electricity, 201, 202–4
Elephant, 82
Elk, extinction of, 62, 63
Elm, 67, 68, 74–5
Elsterian (glacial stage), 14–15, 30–1
Emilani huxleyi, 18
Emmer wheat, 106, 107, 112
Enclosure of land, 142
Encomienda system, 153

Energy
    alternative sources, 295, 298, 302, 315
    for fertilisers, 221
    global consumption, 173, 174, 310, 312
    in ecosystems, 5, 6–7, 8
Engheno, 153
England, climate history (*see also* Britain), 27
Environment(s)
    as provider of resources, 306
    defined, 8
    human-dominated, 206
    unjust, 310, 320
Environmental change
    due to climate, 6, 43, 304
    due to fossil-fuel burning, 171–86, 186–93
    due to mineral extraction, 162–8
    due to waste disposal, 199–204
    during Holocene, 73–7
    social factors, 160–1, 251, 309–12, 316–21
Environmental determinism, 2, 5, 8, 95, 102, 134, 315
EPA (Environmental Protection Agency), 233
Ephedra pollen, 36
Equilibrium in systems, 5
Eratosthenes, 2
Erosion of coasts (*see also* Soil erosion), 177
Estancias, 153
Ethanol as biomass fuel, 302
Ethics, 10
Ethiopia, 250, 252
EU (European Union) agricultural policy, 147, 205, 288
Eucalyptus, 157, 170, 276
Europe (*see also individual countries*)
    climate history, 25–7, 43, 53–8, 65
    desertification, 220
    during historic period, 135
    during Iron Age, 123–5
    forestry, 272–4
    Mesolithic sites, 97–102
    phosphate pollution, 222–3
    soil erosion, 212–4
    spread of agriculture, 113–15
Eustatic changes in sea level, 38–40
Eustresses, 5
Eutrophication, 155, 261, 298
    defined, 222
    impact of agriculture, 220–5
    minimising effects, 224–5
Everest region, 288
Evolution, 2, 78–89
Extinction of species, 61–3, 87, 88, 159, 184

Extinction rates, 10, 209, 210, 240
Extractive industries *see* Mineral extraction

Fallow period, 148, 251
Famine, 141, 264
Farm abandonment, 141
Farming *see* Agriculture *and* Forestry
Faunal changes during late-glacial period, 60–3, 68
Feedback in systems, 5, 6, 307
Fenlands, 139, 142, 208
Fertilisers, 142, 146
    and water quality, 261–2
    artificial, 143, 211, 221–2, 272
    reducing need for, 293, 296
Finland, 223
Fire, 5, 154, 157, 159, 220, 282
    use by early humans, 85, 91, 93, 99–100, 101
Fish, 167, 184, 190, 196, 231, 236
Fish farming, 261
Fishing by early humans, 96
Fishing, environmental impact of, 283, 285
Fitzroya cupressoides, 24
Flandrian (warm stage), 13–15, 61, 67–71
Floods, 72, 248
Flora of Flandrian Britain, 67–8
Florida, United States, 38, 44–5
Flow Country, Britain, 274
Food procurement strategies, 80, 81, 96–7
Food production and global warming, 179–86
Food security, 116, 305, 316
Food trade, 116, 146, 205, 317
Food webs, 6, 235
Foods, new, 301
Foraminifera, 12–13, 17, 31, 39, 46
Forecasting, 10, 27, 176–9, 312–15
Forest/prairie boundary, 73
Forest of Dean, 140
Forestry in developed world, 266–75
Forestry in developing world, 275–9
Forests
    affected by global warming, 182–3
    African colonial, 148–9
    Australian, 158–9
    British, 67
    clearance, 154, 155, 206, 207, 239–45
    conservation, 150
    damaged by acid deposition, 191
    medieval, 138
    Mesolithic, 98
    regeneration, 170
    sink for carbon dioxide, 174
    succession, 32, 70, 156

Fossils, 3, 81, 83, 86, 91
Fossil fuel, 8, 310, 315
    leading to acidification, 186–93
    leading to global warming, 171–86
Fragmentation of habitats, 208
Frains Lake, Michigan, 155
France, 2, 32, 94
French colonialism in America, 151
Fuel-powered systems, 6, 143, 199
Fuelwood, 140, 154, 275, 288
Fungi used in bioremediation, 301
Fungicides, 232, 237
Funza boreholes, Colombia, 36–7
Fynbos, South Africa, 208

Gaia hypothesis, 5–6, 23, 176, 306
Gaian evolution, 83
Galapágos Islands, 287
Game-viewing, Kenya, 289
Garlic as indicator of trampling, 74
Gazelle, 110
GCM (General circulation model), 176, 177–9, 181–2, 314
General Revision Act of 1891, United States, 269
Genes, 79
Genetic data, 80, 113, 115
Genetic distances, 113
Genetic engineering, 289, 291, 293
Geneva Convention, 193
Geographical cycle, 3
GFDL model, 182-3
Giekie, Archibald and James, 3
Gilbert, 3
Girdling of trees, 155
GISP2 (ice core), 19, 21, 28–9, 46, 51, 64
GISS model, 182–3
Glacial theory, 40
Glacials, 3, 12–16, 30
Glaciation in Britain, 54
Glacio-eustatic changes in sea level, 38–40
GLASOD (Global Assessment of Soil Degradation), 246
Glass recycling, 200
Global cooling, 45–8, 195, 313
Global warming, 173, 175, 177–86, 190, 279, 314
Globalisation, 9, 285, 320, 321
Glyphosate herbicide, 232, 237
Goat, 110, 111, 148
Gold mining, 154, 167, 299
Grain yields in Britain, 142
Gransmoor, England, 49
Grasslands, 36, 68, 209

Great Green Wall, China, 278
Great Lakes, Canada, 156, 188, 202, 224
Great Plains, United States, 217
Greece, 32, 120, 129–30, 220
Green Revolution, 225, 255, 293
Greenhouse effect, enhanced, 22, 171–86, 306
Greenhouse gases, 30, 171–7, 243
Greenland ice cores, 19, 21, 28, 31, 46, 51, 64, 131
GRIP (ice core), 18, 19, 21, 46–8, 63
Ground-water, 199, 260, 261
Guantiva (interstadial), 60
Guatemala, 38
Guinea pig, domestication of, 112
GWP (Global warming potential), 173, 175

Haber–Bosch process, 221
Habitat conservation, 271, 316
Habitat destruction, 205–9, 239–45
Haciendas, 153
Hadrian's Wall, Scotland 133
Hallstatt culture, 123–7
HCFC (Hydrofluorocarbon), 171, 196
Heathlands, 208
Heavy metals
    pollution, 166, 167, 196–7, 202, 263
    remediation of pollution, 300
Hedgerows, 142, 147
Heinrich events, 18, 19, 38, 46
Hemdu culture, 117
Herbicides, 232, 237, 272, 295
Herodotus, 2
Hillforts, 124, 127
Hippopotamus, 61
Historic period, 129–61
Historical data, 12, 24–8
Hockham Mere site, England, 67
Holocene (warm stage), 13–15, 21, 62, 63–77, 100-1
Holstein (interglacial), 31
Hominid Corridor Reseach Project, 82
Homo sapiens sapiens as agent of change, 1, 6–7, 71, 73, 304
Homo spp. evolution, 78–89
Hongshang culture, 117
Hormone herbicides, 237
Horse, domestication of, 112
Horse riding, 281
Hoxnian (interglacial), 33–4, 61
Huascarán ice cores, Peru, 36, 48, 71
Huaynaputina volcano, Peru, 24
Human health, factors affecting, 192, 198, 202, 203, 204
Human race, fate of, 304

Humans as agents of change, 1, 6–7, 11, 71, 73, 304
Humus, loss of, 211
Hunter-gatherers, 2, 7, 96, 101
Hunting, 284
Hutton, James, 3
Hybridisation of DNA, 79
Hydrofluorocarbons (HCFCs), 171, 196
Hydrological cycle, 6, 178
Hydropolitics, 260
Hydroxyl radicals, 175
Hypsithermal period, 64

Ice, global amount, 18
Ice ages, 3, 22, 42
Ice cores, 12, 19, 21–4, 64, 71, 197
Ice sheets, 38, 42, 89, 177
Imperialism, 310
Incineration, 200–1
India, 66, 261, 264
    land degradation, 254, 257–9
    soil erosion, 249–50
Indonesia, 245
Indus Water Treaty, 256
Industrial Revolution, 140, 207
Industrialisation, 162–204, 221, 308, 315
    environmental impact, 143–6
Innovation and sustainability, 305
Insecticides, 232, 233, 263
Insects, 48, 75, 272, 292, 294
Insulin resistance, 85
Interglacial cycle, 42–3, 63
Interglacials, 3, 12
International Geological Correlation Programme,
    13–16
Interstadials, 12, 29
Intervention in agriculture, 205
Introduced species, 152, 157, 159, 160, 287
IPCC (Intergovernmental Panel on Climate Change),
    176
IPM (Integrated Pest Management), 264, 292
Ipswichian (interglacial), 13–15, 33–4, 61
Iraq, 96, 260
Ireland, 44–5, 53–6, 62, 122, 124, 126, 188
Iron as nutrient, 23
Iron Age, 122–8
Iron production, 132, 139, 140, 155
Irrigation
    global extent, 255–6
    leading to desertification, 219
    leading to salinisation, 225, 226, 229
    Neolithic, 113, 115
Islands at risk from rising sea levels, 178

Islands of the developing world, 244
Isostatic changes in sea level, 38–40
Israel, 32
ITTO (International Tropical Timber Organisation),
    277

Jamieson, 3
Japan, 76, 197, 199, 202
Jarrah forest, 170
Jebel Qafzeh fossils, 86
Jericho, 112

Kanapoi fossils, 81
Kant, Immanuel, 2, 10
Kazantsevo (interglacial), 31
Kennecott mine, 163
Kenya, 149–50, 250, 289
Killarney oscillation, 51
Kimberley diamond mine, 166
Kitoi culture, 117
Kopet dag area, 118–19

La Téne culture, 124-7
Laetoli fossils, 81, 84
Lagged response in ecosystems, 5
Lake Baikal, Russia, 117
Lake Eyre, Australia, 70, 76
Lake Okeechobee, Florida, 224
Lake Titicaca, Bolivia, 70
Lake Victoria, Africa, 261
Lake Windermere, England, 48–9, 52
Lakes
    change in level of, 36, 58, 68–9, 72
    pollution in, 155, 187–9, 192, 193
Land tenure, 148, 245
Landfill, 199
Landnam phase, 74
Landscape change, 4, 154, 156, 205–9, 239–45, 280
Landscape of medieval England, 138
Langelandselv (interglacial), 32
Language, origins of, 84, 88, 115
Lashford Lane Fen site, England, 99, 100
Late-glacial period, 43–52
Latifundia, 132
Lead mining, 133, 140, 166, 198
Lead pollution, 130–1, 168, 196–8
Legumes, 293
Leisure, 280
Lembus Forest, Kenya, 150
Lentil, domestication of, 106
Levallois tool-making technique, 94, 95
Lichens, 272

Life as a geological force, 5
Limestone, 192
Liming, to combat acidification, 192
Lithic industry, 93–7
Litter decomposition, 196
Little Ice Age, 29, 71, 73, 76, 139, 141, 314
Livestock grazing, 157
Loch Lang site, Scotland, 67
Loch Lomond stadial, 51
Loess, 3, 31
Loess plateau, China, 16, 76, 248–9
Loess stratigraphy, 16–17
Logging, 243
London, 75, 134, 199
Longgupo Cave fossils, 85
Los Millares site, Spain, 121
Lumber industry, 154
Lyell, Charles, 3

Machakos district of Kenya, 149–50, 250
Madagascar, 250
MAFF (Ministry of Agriculture, Fisheries and Food), 233
Magdalenian (tool-making tradition), 95, 97
Magnesium in soils, 191
Magnetic data, 64
Maize, domestication of, 106, 108
Makarovo site, 97
Malawi, 250
Malawi Rift fossils, 82
Malaysia, 276-7
Mammals, extinction of, 61–2, 88
Mammoth, 62, 88
Mangroves, 245
Manuring, 130
Maps as evidence for environmental change, 155, 156
Marine environment, 179, 190, 301
Marine resources, 99
Marsworth, England, 13–14, 61
Mauna Loa site, 171
Maya, 76
Mechanisation, 6, 146, 246, 251
Medicines, traditional, 240
Medieval optimum (warm period), 71
Mediterranean region, 2, 44–5, 69, 119–21, 123
Mediterranean Sea, 235, 288
Mediterranean zone in historic period, 131–2
Megalithic monuments, 116
Mercator, 2
Mercury accumulation gene, 300
Mercury pollution, 167, 201, 202

Mesocratic phase of interglacial, 43–5, 64, 68, 94, 98
Mesolithic period, 78, 97–102
Metal-using cultures, 118–28
Metals, pollution by, 166, 167, 196–7, 202, 263, 300
Meteorological data, 12, 24–8
Methane
    atmospheric, 22, 143, 172, 173, 175, 243
    from waste material, 199, 298
Methyl mercury, 202
Mexico, 76, 108–9, 111, 152, 226–7, 247
Microalgae, 302
Middle Ages, 134–9
Migration, 83–4, 85, 113, 135, 137
Milankovitch cycles, 22, 29, 40, 307
Milankovitch, Milutin, 3
Millet, domestication of, 106, 107
Mimosa shrub, 159
Mineral extraction, 162–8, 243, 298
Mining, 131, 154, 156, 162–71, 298, 299
Minoan civilisation, 120
Mitochondria, 85, 110
Models for climate, 176, 177–9, 314
Molecular clock, 79
Monasteries, power of, 138
Mongolia, 21
Monsoons, 66
Monte Verde, Chile, 63, 88
Montreal Protocol on CFCs, 9, 196
Mousterian tool-making technique, 90, 95
MSA (Methane sulphonic acid), 23
Mu Us area, China, 253, 254
Multiregion model for human evolution, 85
Murray River, 229
Mycenaean civilisation, 120
Mycopesticides, 292
Mycorrhizal networks, 169

NAA (Nitrate Advisory Area), 225
NADW (North Atlantic Deep Water), 20, 30, 46
National Parks of United States, 157
Natufian period, 110, 112
Neanderthal species, 86, 96
Near East site for domestications, 103, 107, 109, 111
Near East, Neolithic communities in, 112–13
Neodeterminism, 5
Neolithic cultures, 78, 102, 112–18
Nepal, 288
Netherlands, 32, 55, 178, 222
New England, 154
New Forest, England, 138
New Guinea, 38

New Zealand, 21, 34–5, 60, 214–15
   colonialism, 159–60
Newton, Isaac, 2
Niger, 250, 252
Nigeria, 148, 275
Nile Valley, 2, 115, 263
Nitrates, 199, 221, 223, 224
Nitrogen acids, 186, 187, 188, 189, 261
Nitrogen deficiency, 169
Nitrogen fixation, 142, 221, 290, 291
Nitrous oxide as greenhouse gas, 172–3, 175, 199, 243
No-tillage agriculture, 215, 216, 237
Noah's Ark hypothesis, 85
Norfolk Broads, England, 139, 222, 282
North America (*see also individual countries*), 109, 187, 214, 215
   colonialism, 151–7
   early humans, 87–8
   environmental history, 58–9, 62, 72–3
   eutrophication, 223–4
   forestry, 267–72
North Sea, 223
Norway, 31, 55
Nuclear power, 202–4, 290
Nutrients in soil, 191, 220, 246

Ocean cores, 12, 16–20, 40, 46
Oceans as sinks for carbon dioxide, 173–4
Odderade (interstadial), 14–15, 32
Odours, as environmental problem, 199
Oil-spills, 202, 301
Oil crops, 290, 295, 302
Oil extraction, 163
Older Dryas (cold period), 45
Oldowan (tool-making tradition), 83, 89–90
Oligocratic phase of interglacial, 43–5, 71
Olive, 129
Open systems, 5
Opencast mining, 163
Oppida (enclosed sites), 127
Orbit of the Earth, 3
Organochlorine insecticides, 232, 233, 264
Organophosphate insecticides, 232, 235
Oryx, 284
Otiran (glacial stage), 34
Overgrazing, 219, 254
Oxygen isotope analyses, 13, 16, 17, 21, 39, 47, 63
Ozone, 172, 175, 188, 193–6

Pakistan, 254, 255–7
Palace civilisations, 120

Palaeoanthropology, 78–89
Palaeobotanical data, *see* Pollen analysis
Palaeoecology, 3, 155
Palaeoenvironmental data, 18, 24, 59, 307, 313
Palaeolithic cultures, 78, 89–97
Pannage, 139
Paper production, 200, 267, 280
Paranthropus aethiopicus, 82
Paysannat schemes, 148
PCB (Polychlorinated biphenyl), 202, 300
Peatland, 74, 139, 274
Penck and Brückner, 3
People/environment relationships, 1, 8–11, 102
Permethrin insecticide, 232, 236
Perraudin, Jean-Pierre, 3
Peru, 21, 59–60, 70, 71, 153, 246, 248
Pesticides, 211, 231–8, 263–4, 290, 292
Petrol, lead in, 197, 198
PFE (Permanent forest estate) in Malaysia, 276
Pharmaceuticals from natural products, 240
Philippines, 200
Philosophy of nature, 2–3, 10
Phoenician metallurgy, 123
Phosphate fertilisers, 163, 169, 221, 222, 249
Phosphorus in aquatic systems, 20, 156, 199, 261
Photosynthesis, 305
Phytoplankton, 23, 196
Phytoremediation, 300
Pig, domestication of, 110, 111
PIRLA project, 187–8
Plant breeding, 146, 289–303
Plant domestication, 104–6
Plant use by early cultures, 97, 100
Plantation agriculture, 148, 152, 153
Plantation forestry, 276, 278–9
Plants and Gaia hypothesis, 6
Plants, medicines from, 242
Plasmid, 293
Playfair, John, 3
Pliocene–Pleistocene boundary, 12–13
Ploughing and afforestation, 274
Poland, 58, 101, 191
Polar regions, 30–32
Political factors in environmental change, 9, 205, 233, 260, 309–12, 316–21
Pollarding, 139
Pollen analysis
   British, 34, 48–50, 52, 53, 67, 99, 133–4, 135
   European, 20–1, 32, 56, 74, 132
   rest of world, 36–7, 38, 91, 155
Pollution and biotechnology, 298, 299
Pollution by acid emissions, 186–93

Population change in ecosystems, 5
Population growth in historic period, 143, 156, 244
Population growth, future, 310, 311, 316, 318-19, 322
Population pressure in prehistory, 105, 113
Portugal, 47, 288
Portuguese colonialism, 147, 153
Possibilism, 4
Possum, 160
Potash fertilisers, 141, 221
Potassium-Argon dating, 36
Potato, 106, 109, 145, 292
Pottery, 102, 112, 117, 127
Power relations of high-tech agriculture, 316
PPNA/PPNB periods, 112
Praetiglian glaciation, 13–15
Pre-Colombian cultures, 102, 152
Pre-Cromer (glacial stage), 14–15, 30
Precautionary principle, 10, 321, 322
PRIA (Pre-Roman Iron Age), 126
Productivity
    decline of, in soils, 246
    global future, 182
    increased through biotechnology, 294, 296
    of agriculture, 142, 174, 255–7, 316
    of oceans, 20, 23
Protocratic phase of interglacial, 43–5, 98
Pruteen®, 302
Ptolemy, 2
Public Health Act of 1875, 145
Pyrethroid insecticides, 232, 236

QBO (Quasi-biennial oscillation), 28
Quaternary period, 3, 12–41, 307
Quinoa, domestication of, 109
Quorn®, 302

Rabbit, 159
Radioactive wastes, 163, 168, 202–4
Radiocarbon dating, 97, 113–14
Railways in Britain, 144, 146
Rain forests, 70–1, 158–9, 240–5, 276
Rainfall as agent of erosion, 211
Rainfall, historic levels of, 72
Ranching, 154, 157, 244, 245
Rancho La Brea site, 62
Rape crop and biotechnology, 295
Reclamation
    desertified land, 254
    mine-damaged land, 168–71
    wetlands, 139, 142
Recombinant DNA, 293

Recreation, 280–4
Recycling of resources, 200, 298, 315
Reforestation, 156
Refrigerants: use of CFCs, 194
Refugia, 32, 70–1
Registration of crop protection chemicals, 234, 235
Rehabilitation of soils, 214, 216
Reservoirs, 262
Resource recovery, 298
Resource use per capita, 310, 320
Rhizobium species, 292–3
Rice, 106, 107, 184, 294
Rinderpest, 148
Riss glacial period, 13–15, 95
River flow modelled, 178
Road construction, 271, 272, 277
Rome, 131–4
Rotation of crops, 124, 142
Royal forests, 138
Rubber industry, 153
Rubbish disposal, 199–204, 288
Russia, 203, 214, 215, 219–20, 229–31, 288
Russian Plain during the Holocene, 68
Rwanda, 250

Saalian (glacial stage), 14–15, 30–1
Safari holidays, 289
Sahara desert, 38, 75
Sahel zone, 27, 38, 250, 251–2
Salinisation, 225–31, 246, 255–61
Salt-tolerant crops, 257, 259
Saltation, 247
Samos, Greece, 288
Sandstone, 192
Sanitation during Industrial Revolution, 145
Savagery, 96
Scandinavia, 30, 39, 187, 190
SCARP (Salinity Control Reclamation Project), 256
Scords Wood site, England, 75
Scotland (*see also* Britain), 54, 68, 133, 168, 188, 189
SCP (Single-cell proteins), 301
Sea-surface temperature, 46, 65–6
Sea levels, 3
    future, 177
    past, 34, 38–40, 87, 89, 99
Sediments, ocean, 16–20
Selenium, 227, 300
Sellafield site, 203
Semiarid regions, 259
Senegal river, 254
Sernander, 3

Serovo people, 117
Set-aside (agricultural policy), 225, 316
Settlement, permanent, 101, 112
Sewage, 199–204, 222, 261, 298
Shaitan (period of ice advance), 31
Shanidar burial site, Iraq, 96
Sheep, 110, 111, 157, 220
Sheetwash erosion see soil erosion, 158
Shell-midden sites, 99
Shelterbelts, 216, 254
Shifted cultivators, 245
Shifting cultivation causing deforestation, 243
Shrimp farming, 245, 264
Siberia, 24, 31
Silver mining, 131, 154, 167
Silviculture, 269–71, 276
Single-region model of evolution, 85
Sivapithecus fossils, 79
Skiing, 282–3
Slash and burn cultivation, 152
Slavery, 130, 147, 152, 153
Smelting, emission of heavy metals, 198
Social factors in environmental change, 160–1, 251, 309–12, 316–21
Sodium in ice cores, 23
Soil conservation, 209–17, 237, 245–51
Soil erosion
    developed world, 209–17
    developing world, 245–51
    during Bronze Age, 122
    during colonial period, 150, 155
    during historic period, 130
    factors influencing, 158, 211–12, 280, 282–3
Soil nutrient levels, 71, 124, 191, 220
Soils, remediation of metal-contaminated, 300
Solar-powered systems, 6
Somaclonal variation, 291
Soot, pollution by, 145
Sorghum, domestication of, 106, 108, 115
South Africa, 72, 75, 208, 219
South America, *see also individual countries*, 59–60, 71–2, 73, 88
    colonialism, 151–7
    site for domestication, 108–9, 111
Soviet Union, 203, 214, 215, 219–20, 229–31, 288
Soya bean cultivation, 293
Spain, 27
Spanish colonialism in South America, 151–2
Species diversity see Biodiversity
Speleotherm, 39
Spencer, Herbert, 2
Sport, 280, 282, 286

Squash, domestication of, 106, 109
SST (Sea Surface Temperatures), 27
Stadials, 12, 29
Stanton Harcourt (interglacial), 16, 33–4, 61
Star Carr site, Britain, 98
Sterkfontein fossils, 81
Stone-tool making, 78
Strabo, 2
Stratospheric ozone layer, 193–6
Straw for erosion control, 216
Streamflow to detect climatic change, 27
Strip mining, 163
Strobilurin fungicides, 237
Subsidence due to mining, 166
Sudan, 252
Sugar-cane, 152, 153
Sulphur acids, 186, 187, 188, 189, 191, 193
Sumerian civilisation, 115, 116
Sunspot cycles, 26, 28
Surtsey Island, 5
Susquehanna River, 224
Sustainability, 9, 10, 116, 305, 312
    improved using biotechnology, 296
    tourism, 285, 289
    wood production, 182, 267, 277
Swanscombe (interglacial), 33–4
SWAP (Surface Waters Acidification Programme), 187
Sweden, 31, 57, 68, 69
Switzerland, 55, 57
Systems, environmental, 4–6, 307

Tansley, Sir Arthur, 4
Tasmania, 34, 43
Taxes, environmental, 320
Taxol, 243
Technology
    agent of change, 6–7, 80
    history and pre-history, 113, 123, 133, 140, 153
    transfer of, 295, 310
Tefluthrin insecticide, 232, 236
Telocratic phase of interglacial, 43
Temperature reconstruction, 20, 25–6, 36, 48–50, 72
Temperature see Warming and Cooling
Terracing, 152, 216
Thailand, 167, 280
Thames river, 34
Thirty per cent club, 193
Thorium analyses, 40
Three Gorges Dam project, 249
Thresholds in ecosystems, 6

Tidmarsh Wood site, England, 99
Tierra del Fuego, 60
Timber production, 133, 150, 266–80
Tin, 122, 143
Tissue culture, 290, 291
Toad, 190
Tobacco, 154
Tobol (interglacial), 31
Tomato, 294
Tool use in prehistory, 83, 85, 89–93, 98, 118
Tortoise, 68
Totipotency, of plant cells, 290, 291
Tourism, 9, 284–9
Trade winds, 36
Traffic flow and lead levels, 198
Trampling of vegetation, 74, 280, 281
Transgenic species, 296, 301
Tree-ring data, 24–8, 29, 71–2
Tree species, decline in populations of, 74–5, 182–3
Trees, pioneer species, 67
Troposphere, 186
Troy, city of, 119
TSS (Tropical Shelterwood System), 276
Tudor influences on British landscape, 139–40
Tundra, 62, 67, 182
Typhus, 145

Ubeidiya Formation, Jordan, 94
Ukraine, 112
UNCED (Earth Summit 1992), 9, 243, 295, 312, 320
Ungulate animals, 158
United States, 44–5, 156, 214, 217, 280
    acid pollution, 188, 189
    climate, 25–7, 177
    during the Holocene, 58–9, 66, 72
    eutrophication, 223–4
    forestry, 267, 268–71
    heavy metal pollution, 167, 197
    salinisation, 225–9
    water resources, 227–9
Upton Warren (interstadial), 18
Uralic languages, 115
Uranium-series dating method, 16, 39
Uranium mining, 163, 168, 299
Urban-industrial systems, 6, 143, 162
Urban civilisations, 120
Urban pollution, 192, 199
Urbanisation, 123, 143
Uruk culture, 113
UV–B radiation, 190, 193–6

Valdai (stadial), 13–15
Varenius, 2
Varvarina Gora site, 97
Vegetation of Quaternary period, 36, 70
Vegetation succession, 43–5
Vegetation zones, global, 180–1
Vehicles, 197, 282, 289
Venetz, Ignace, 3
Vikings, 138, 151
Village, as focus of settlement, 138
Villas, Roman, 132
Vines, 26
Virus-free crops, 291
Viruses, 294
Volcanism, 24, 28
von Humboldt, 2
von Post, 3
von Richtofen, 3
Vostok ice core, 21–2, 23, 28, 48
Vrica, Italy, 12

Wales (see also Britain), 168
Wanganui Bay, New Zealand, 34–5
Warm stages, 12
Warming, global, 22, 24, 27, 40, 48, 171–86
Waste materials, 199–204, 298
Water
    agent of erosion, 211, 246
    global usage, 184–5, 225, 227, 259
    pollution, 167, 201
    quality, 199–200, 220–31, 261–4
Water hyacinth, 263
Waterlogging, 255–61
Waterpower, 140, 144
WCED (World Commission on Environment and Development), 9, 312
Weald, England, 140
Weathering, chemical, 30
Weichselian (stadial), 13–15, 30–1, 57
Western Samoa, 244
Wet deposition (of acidity), 186, 187
Wet farming, 132
Wetlands, 23, 139, 142, 169–70, 182, 207, 221
Wheat, 106, 142, 146
Wilderness areas of United States, 269, 281
Wildlife and tourism, 284, 287
Wind as agent of erosion, 158, 211, 217, 246
Wisconsin (stadial), 13–15
Wolstonian (glaciation), 13–15
Wood production, 133, 150, 266–80
Woodland, 135, 139, 140, 141, 303
Woodland clearance, 124, 133–4, 137

Woodworking in the Mesolithic, 99, 100
Wool industry in Britain, 140, 146
Würm (stadial), 13–15, 57

Xerothermic climate, 66
Xihondu site, China, 91, 92

Yangshao culture, 117
Yangtze River, 107, 116, 249
Yellow River, 116, 248–9, 260
Yeoman class, 140

York, England, 138
Younger Dryas (cold period), 18, 34, 45, 55, 62, 63, 73
Youngest Dryas (cold period), 105
Yulara resort, Australia, 285–6

Zhoukoudian fossils, 84, 85, 91, 92–3
Zimbabwe, 148, 250
Zinc, 166, 167
Zircon, fission track dating in, 36
Zyryanka (period of ice advance), 31